Cities of the World

Cities of the World

Regional Patterns and Urban Environments

SIXTH EDITION

EDITED BY
STANLEY D. BRUNN, JESSICA K. GRAYBILL, MAUREEN HAYS-MITCHELL, AND DONALD J. ZEIGLER

ROWMAN & LITTLEFIELD
Lanham • Boulder • New York • London

Executive Editor: Susan McEachern
Assistant Editor: Audra Figgins
Senior Marketing Manager: Karin Cholak
Marketing Manager: Kim Lyons
Production Editor: Alden Perkins

Credits and acknowledgments of sources for material or information used with permission appear on the appropriate page within the text.

Published by Rowman & Littlefield
A wholly owned subsidiary of The Rowman & Littlefield Publishing Group, Inc.
4501 Forbes Boulevard, Suite 200, Lanham, Maryland 20706
www.rowman.com

Unit A, Whitacre Mews, 26-34 Stannary Street, London SE11 4AB, United Kingdom

British Library Cataloguing in Publication Information Available

Library of Congress Cataloging-in-Publication Data

Cities of the world : regional patterns and urban environments / edited by Stanley D. Brunn, Jessica K. Graybill, Maureen Hays-Mitchell, and Donald J. Zeigler.—Sixth Edition.
pages cm
Revised edition of Cities of the world, 2012.
Includes bibliographical references and index.
ISBN 978-1-4422-4916-5 (cloth : alk. paper)—ISBN 978-1-4422-4917-2 (electronic) 1. Cities and towns. 2. City planning. 3. Urbanization. 4. Urban policy. I. Brunn, Stanley D., editor. II. Graybill, Jessica K., 1973– editor. III. Hays-Mitchell, Maureen, editor.
HT151.C569 2016
307.76—dc23

2015036537

∞™ The paper used in this publication meets the minimum requirements of American National Standard for Information Sciences—Permanence of Paper for Printed Library Materials, ANSI/NISO Z39.48-1992.

Printed in the United States of America

To a greener and more just future for planet Earth,
its cities and residents.

Contents

List of Illustrations

Figures

Boxes

Tables

Preface

In 1982, *Cities of the World* debuted. It presented an innovative approach to the study of urban geography. Renowned urban geographers, who were regional specialists, shared their knowledge of and insight into the history, patterns, challenges, and prospects for cities in eleven world regions. *Cities of the World* was an immediate success. Subsequent editions built on this model—revising, updating, modifying, and enhancing the approach. With each edition, the popularity of the book swelled. It is commonly found in courses on urban geography, urban and regional planning, as well as courses in global affairs, anthropology, history, and economics, at both the undergraduate and graduate levels.

Thirty-four years later, we present the sixth edition of *Cities of the World*—and we present it in color! Color photographs, regional maps, and graphics provide a more appealing and accurate depiction of many dimensions of the urban regions under study. Just as each subsequent edition of *Cities of the World* has embraced the changes encountered in the global and regional urban systems, so too does this sixth edition. In this, we deepen our focus on urban environmental issues, social and economic injustice, security and conflict, and daily life. Building on 2015 as the Year of Water, we have introduced urban water issues and concerns as a common undercurrent running through all chapters. Author teams explore how "water" affects cities and how cities affect water in their respective regions—from glacier loss to increasing aridity, sea-level rise, increased flooding, potable water scarcity, and beyond. We hope our new subtitle "Regional Patterns and Urban Environments" captures these innovations.

All thirteen chapters in this sixth edition have been substantially revised, and some introduce new author teams, whom we welcome warmly. They bring fresh perspectives and expertise to the project. Most authors have done extensive fieldwork in their region and also traveled extensively in both rural and urban areas. The organization of this edition is similar to the previous five. The "book end" chapters explore contemporary world urbanization (chapter 1) and the future of cities (chapter 13). The remaining eleven chapters are devoted to urbanization and cities in major world regions. Each chapter begins with two facing pages; on the left side, a regional map that shows the major cities and, on the right, a table of basic statistical information about cities and urbanization in each region and a list of ten salient points about that region's urban experience are provided. The regional chapters conclude with a list of references that can be used by the student and instructor for additional information about cities in that region or specific cities.

We owe a debt of gratitude to many individuals who played major roles in helping this sixth edition see the light of day. We thank all chapter authors for providing timely, insightful, and well-written chapters and Alexis Ellis for her valuable cartographic contribution, and Donna Gilbreath for her assistance in preparing the index. Susan McEachern of Rowman and Littlefield has provided longstanding support for this volume and previous ones. Her eye for detail, continuity, and change is unmatched. Susan's team at Rowman and Littlefield worked to ensure the high quality of this edition, and we thank them for their commitment, timely support, and attention to detail throughout the process. Finally, we thank our families whose enthusiastic and selfless support made this project enjoyable and possible.

As always, we welcome feedback from students and teachers on ways to ensure that subsequent editions will make learning about the world's cities and global urbanization more useful, appealing, challenging, and rewarding. We hope you enjoy this latest edition.

Stanley D. Brunn
Jessica K. Graybill
Maureen Hays-Mitchell
Donald J. Zeigler

Cities of the World

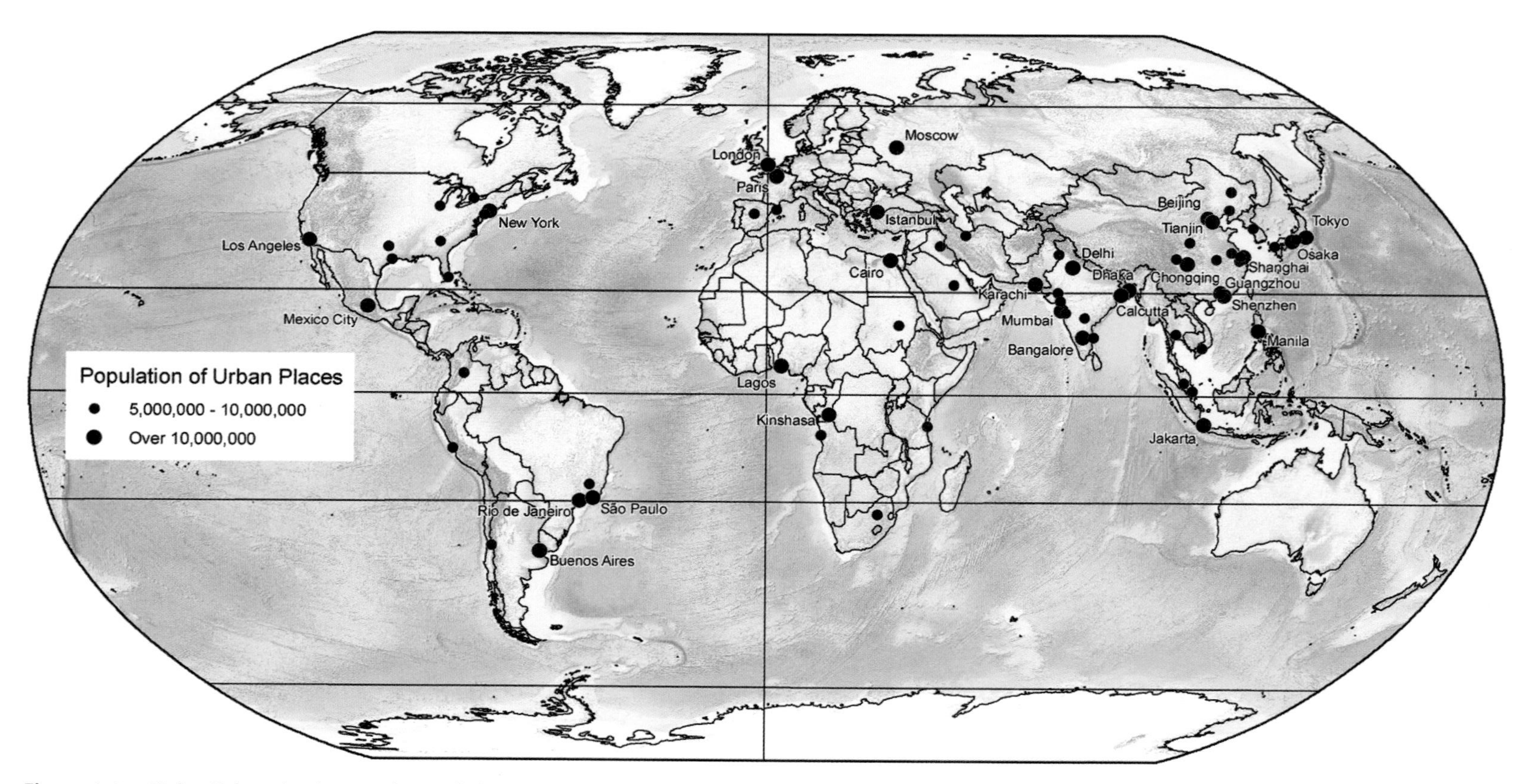

Figure 1.1 Major Urban Agglomerations of the World. *Source:* United Nations, World Urbanization Prospects: 2014 Revision.

1

World Urban Development

JESSICA K. GRAYBILL, MAUREEN HAYS-MITCHELL, DONALD J. ZEIGLER, AND STANLEY D. BRUNN

KEY URBAN FACTS

Total World Population (2015)	7.3 billion
Percent Urban	54%
Total Urban Population	3.9 billion
Most Urbanized Counties	Microstates such as Monaco and Nauru (100%) Singapore (100%) Belgium (98%)
Least Urbanized Countries	Burundi (12%) Papua New Guinea (13%) Uganda (16%)
Annual Urban Growth Rate (2010–2015)	0.9%
Number of Megacities (>10 million)	28
Agglomerations with 500,000 + Population	1009 (53% of world population)
Countries with Most Urban Agglomerations	China (278), United States (275), India (112), United Kingdom (140), Russia (59)
Cities with Highest Densities	Dhaka, Bangladesh (112,700/sq mi, 43,500/sq km) Hyderabad, Pakistan (104,300/sq mi, 40,300/sq km) Mumbai, India (83,900/sq mi, 32,400/sq km)
Largest Megacities (2014)	Tokyo (37.8 million) Delhi (24.9 million) Shanghai (23.0 million) Ciudad de México (20.8 million) São Paulo (20.8 million)
World Cities	88
Global Cities	London, New York, Tokyo

KEY CHAPTER THEMES

1. The world's population is growing rapidly, but the world's urban population is growing four times as fast.
2. In 2007, Earth became a majority-urban planet, yet the proportion of people living in cities varies widely from 12 percent in Burundi to 100 percent in Singapore.
3. The scale of urbanization is increasing as evidenced by the emergence of megacities, conurbations, and megalopolises around the world.
4. Some countries' patterns of urbanization follow the rank-size rule, while other countries are characterized by urban primacy or dual primacy.
5. The evolution of cities is best understood as a three-stage process: preindustrial cities, industrial cities, and postindustrial cities.
6. Cities are usually classified by function as market centers, transportation centers, or specialized service centers.
7. Four classic models have been proposed to explain the spatial organization of land uses within cities: concentric zone model, sector model, multiple nuclei model, and inverse concentric zone model.
8. Urban management issues revolve around the environment, population, urban services, race and ethnicity, housing, employment, privacy, governance, and globalization, among others.
9. As the world continues to urbanize, sustainable development challenges will be increasingly concentrated in cities.
10. Urban water issues, ranging from water quality and quantity to the challenges of sea-level rise, occur at the intersection of nature and society.

Comparing maps of the world from 2015 and 1900 would show two features that have become strikingly different over time and space: proliferation of independent nations and mushrooming numbers and sizes of cities. A century ago, about a dozen major empires divided the world; today, there are 195 independent countries, most carved out of previous empires. Continued disintegration of imperial realms gave birth to the world's most recent newly independent state, South Sudan, and its capital city of Juba. Likewise, a century ago, the number of the world's major cities was small and concentrated in the industrialized countries of Europe, North America, and Japan. Today, the greatest numbers of cities, and the largest cities, are found in former colonial regions of the developing world and in China (Figure 1.1). Around the year 1800, perhaps 3 percent of global population lived in urban places of 5,000 people or more. By 1900, more than 13 percent did and by 2000, this percent skyrocketed to more than 47 percent. In 2007, for the first time in human history, over half of Earth's human population made the ***city*** their home. The rapid pace of ***urbanization*** is accompanied by globalization and the creation of ***world cities***, the outcome of technological advances in transportation and communication (Box 1.1). Our global

Box 1.1 Globalization and World Cities

Peter Taylor, Northumbria University, England

Although there is a large literature on "world" or "global" cities, little evidence has been gathered on what actually makes such cities so important: their connections with other cities across the world. Thus, if world cities are indeed the crossroads of globalization, then we need to consider seriously how we measure intercity relations. It was just such thinking that led to the setting up of the Globalization and World Cities (GaWC) Network as a virtual center for world cities research. GaWC currently carries out three strands of research:

- Comparative City Connectivity Studies: These focus on relations between chosen cities as they respond to particular events. In one study, Singapore, New York, and London were compared in the way in which their service sectors responded to the 1997 Asian financial crisis. In another study, relations between London and Frankfurt were studied in the wake of the launch of the Euro currency. The generic finding of this work is that city competitive processes are generally much less important than cooperative processes carried out through office networks within the private sector.
- Elite Labor Migrations between Cities: Moving skilled labor around to different world cities is found to be a key globalization strategy for financial firms wanting to embed their businesses into the world-city network. For instance, London firms regularly send staff to Paris, Amsterdam, and Frankfurt to provide "seamless" service across European cities. The prime finding of this research is that a transnational space of flows is produced as a necessary prerequisite for firms accumulating financial knowledge.
- Global Network Connectivity of Cities: A world-city network is an amalgam of the office networks of financial and business service firms. This network has three levels: a network level of cities in the world economy, a nodal level of cities as global service centers, and a subnodal level of global service firms that are the prime creators of the world-city network. This specification directs a data collection that enables the global network connectivities of world cities to be calculated.

GaWC research goes beyond world-city formation to study world-city network formation. The focus has been on this complex process within economic globalization. For a full global urban analysis, further research is required within other important strands of globalization. The Study Group's website is: http://www.lboro.ac.uk/gawc/.

urban habitat is increasingly connected by the flows of people, goods, services, and capital that unite cities, people, and environments across time and space. Urban institutions drive globalization, the hallmark of twenty-first-century economic geography.

Yet, our urban planet also pulsates with problems at the human-nature interface (Figure 1.2). They range from local concerns about air quality, to regional problems of water quality and quantity, to global environmental change, including rising average global temperature and sea levels and changes to the global hydrosphere-atmosphere circulation patterns. Due to the interconnectedness of people, cities, and regions, urban problems must be approached on multiple scales. Climate change is transforming environments worldwide (albeit differentially) and the consequences of climate change are already being felt at the local level, such as in coastal cities confronted by rising sea level and in cities where temperature increases are expected over the next century. Chicago, for instance, is already planting more shade trees, choosing species that will thrive in a warmer climate. In 2014, in anticipation of warmer school days, the city installed air-conditioning in schools for the first time.

Worldwide urbanization has dramatic, revolutionary implications for the history

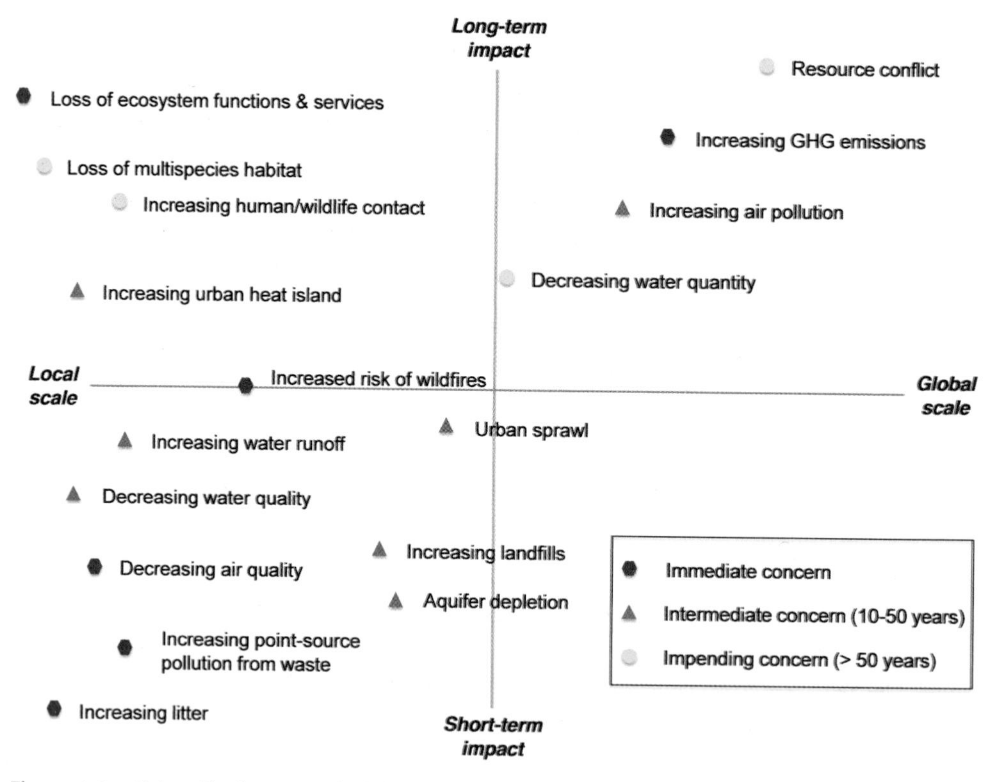

Figure 1.2 Urban Environmental Risks. This conceptual diagram indicates the generalized, possible risks and concerns for the environment of urban and urbanizing places at (i) local, regional and global scales and (ii) across short- and long-term time horizons. Because individual places will experience different suites of environmental concerns, this diagram is intended to pique discussion of possible urban environmental changes. *Source:* Jessica Graybill.

of civilization—as dramatic as were earlier agricultural and industrial revolutions. In the industrial countries of Europe, North America, Australia, and some of Asia—the more developed countries (MDCs)—urbanization accompanied and was the consequence of industrialization. Although far from being utopian, cities in those regions brought previously unimagined prosperity and longevity to millions. Industrial and economic growth combined with rapid urbanization to produce a demographic transformation that decreased population growth and enabled cities to expand apace with economic development. In the developing countries of Latin America, Africa, and most of Asia—the less developed countries (LDCs)—urbanization has occurred only partially due to industrial and economic growth. In many of these countries, it is primarily a result of rising expectations by rural people who migrate to cities seeking escape from misery. This rush to the cities, unaccompanied until very recently by significant declines in natural population growth, has resulted in the explosion of urban places in LDCs.

Although there are exceptions (most notably in South America where urbanization levels are high), most highly urbanized nations experience high standards of living. However, even in the MDCs, where life for most urban residents is incomparably better than for those living in the cities of LDCs, there are serious concerns about the future of the city. What, for example, is the optimal city size? What should be the role of the ***capital city***? Are cities getting too large to provide effective administration and humane and uplifting urban environments? Is the growth of ***megacities*** and ***metacities*** unmanageable? Will the ***megalopolis*** or ***conurbation*** become the norm for the twenty-first century? What will be the impact of ever-expanding ***urban agglomerations*** on human society; life-sustaining environmental systems that provide water, food, and multispecies habitats; resource development; and governments facing increased social disparities, cultural pluralism, and diversity of political expression? Can the ***sustainable city*** movement improve health and well-being for all of Earth's inhabitants? How does the rise of ***urbanism*** create new understandings and uses of, and desires for, nature by humanity?

THE WORLD URBAN SYSTEM: PROSPECTS UNTIL 2050

In 1800, the world stood on the brink of 1 billion. In only 130 years, humanity added a second billion, and in only 11 years, the last

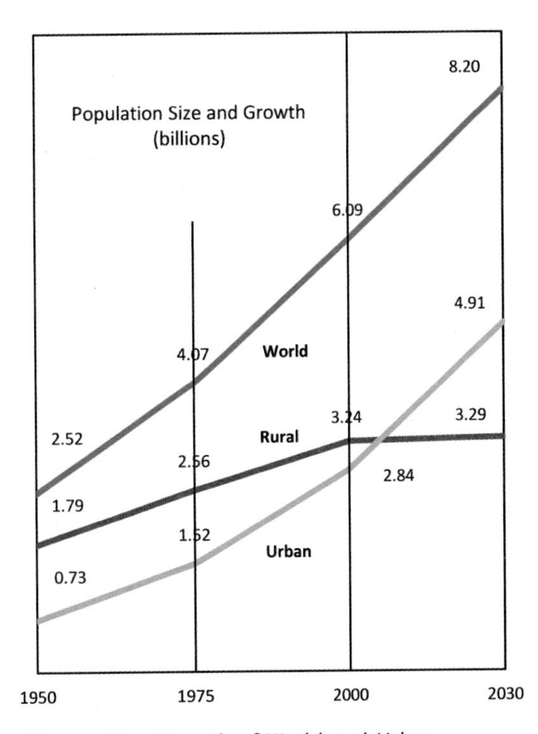

Figure 1.3 Growth of World and Urban Population, 1950–2030. *Source:* UN, World Urbanization Prospects: 2005.

billion. Between 1950 and 2008, the world's population increased more than 2.5 times, but the world's urban population increased almost 4.5 times (Figure 1.3). Fifteen years into the twenty-first century, world human population is over 7 billion and over half live in urban settings. This exponential growth of human population and the rise of cities as the dominant habitat for humans changes both urban and rural places, but at different rates in different places worldwide.

Economics, trade, culture, religion, and environment are just a few of the linkages that connect cities worldwide. ***Urban landscapes*** are also connected to nonurban people, places, and phenomena that impact the city, its growth (or decline), and the health and well-being of urban residents. The world urban system, then, describes people and their economic, social, cultural, and environmental activities, which connects urban people and phenomena from local to global scales. Some of the most important trends in our current world urban system are related to continued human population growth and spatial distribution.

The United Nation's *World Urbanization Prospects* (2014) provides a valuable overview of urbanization trends for the next several decades. Overall, world urban population is expected to increase by 62 percent by 2050, from 3.9 billion in 2014 to 6.3 billion in 2050. Virtually all of the expected growth in the world population will be concentrated in the ***urban areas*** of the lesser developed regions. Slightly counterintuitive to the knowledge that urban regions are growing is the fact that the rate of growth of the world urban population is slowing down. From 2025 to 2050, the urban growth rate is expected to decline to 1.3 percent per year. While urban population will increase, world rural population is expected to reach a maximum of 3.4 billion in 2020 and slowly decline thereafter, decreasing to 3.2 billion in 2050. These global trends are driven mostly by the dynamics of rural population growth in lesser developed regions. But, the sustained increase of the urban population combined with the pronounced deceleration of rural population growth will result in an increasing proportion of the population living in urban areas. Globally, the level of urbanization is expected to rise from 54 percent in 2015 to 66 percent in 2050.

Among cities of different sizes, the world urban population is not distributed evenly. Over half of the world's urban dwellers live in cities or towns with fewer than half a million inhabitants. The greatest numbers of people live in cities with less than 1 million people, but the phenomenon of the ***megacity*** is increasing worldwide. There are 28 megacities worldwide, each with at least 10 million inhabitants, accounting for 12 percent of the world urban population. The number of megacities is projected to increase to 41 by 2030, accounting for approximately 12 percent of the world urban population.

Until 1975, just three megacities existed worldwide: New York, Tokyo, and Mexico City. Today, Asia has 11 megacities, Latin America has 4, and Africa, Europe, and North America have 2 each. Tokyo, the capital of Japan, is the world's most populous urban agglomeration, comprised of Tokyo and 87 surrounding cities and towns. If it were a country, it would rank 35th in population size, at 37.2 million inhabitants. Tokyo now qualifies as a ***metacity*** according to the United Nation's description for urban agglomerations with more than 20 million inhabitants. Following Tokyo, the next largest urban agglomerations are Delhi, Shanghai, Mexico City, Mumbai, and São Paulo, each with 21 million inhabitants; and New York-Newark and Los Angeles in the

United States, with approximately 19 million inhabitants. In 2025, the world's most populous urban agglomeration will remain Tokyo, with 38 million inhabitants, although its population will scarcely increase. It will be followed by two major megacities in India: Delhi with 29 million inhabitants and Mumbai with 26 million. Megacities are experiencing very different rates of population change than other kinds of cities. Generally, their rate of growth is slow, at less than 1 percent per year. Megacities exhibiting these slow rates of growth include all those located in developed countries and the four megacities in Latin America.

Outdone only by the metacity, megacities represent the extreme of the distribution of cities by population size. They are followed by large cities with populations from 5 million to just under 10 million, which in 2015 numbered 43 and are expected to number 63 in 2030. Three-quarters of these "megacities in waiting," which house 8 percent of the world population, are located in developing countries. Cities with more than a million inhabitants but fewer than 5 million are numerous, and every fifth person, statistically, lives in this medium-sized city. Smaller cities, with populations from 500,000 to 1 million inhabitants, are even more numerous and account for about 10 percent of the overall urban population. The number of these cities is expected to decrease in the next couple of decades, to house just under half of the world's urban population.

Distribution of the urban population by city size class varies among world regions. Europe is exceptional in that 67 percent of its urban dwellers live in urban centers with fewer than 500,000 inhabitants and only 8 percent live in cities with 5 million inhabitants or more. Distribution of the urban population in Africa by size of urban settlement resembles that of Europe. In Asia, North America, and Latin America and the Caribbean, the concentration of people in large cities is marked: one in every five urban dwellers in those major areas lives in a large urban agglomeration.

Historically, the process of rapid urbanization started first in today's more developed regions. In 1920, just less than 30 percent of their population was urban and by 1950, more than half of their population was living in urban areas. In 2015, high levels of urbanization, surpassing 80 percent, characterized Australia, New Zealand, and Northern America. Europe, with 73 percent of its population living in urban areas, is the least urbanized in the developed world.

More developed regions of the world had a higher percentage of their populations living in urban areas in both 1950 and 2010. In absolute numbers, however, there were more urban dwellers in the MDCs in 1950; by the end of the century, this had changed (Table 1.1). Unfortunately, urban development has not kept up with urban growth throughout Middle and South America, Africa, and the Middle East, and much of Asia. Latin America and the Caribbean, for instance, have caught up to the MDCs in degree of urbanization, but economic development, health care, and education lag. Poor housing quality is a striking characteristic in these regions, standing in stark contrast to cities in the MDCs. Sub-Saharan Africa remains the least urbanized and the least developed region in the world. Latin America's urban explosion may be over, but in Africa, India, and China, the urban population explosion continues. Only in 2010 did China reach urban majority status and it is expected to gain momentum as the century proceeds.

While increases in urban population have been felt worldwide, the pace of urban change is most dramatic in the world's developing

Table 1.1 Urban Patterns in More Developed Regions and Less Developed Regions (in thousands)

	World pop 2014	*MDR Urban*	*LDR Urban*
Urban	3,880,128	980,403	289,9725
Rural	3,363,656	275,828	3,087,828
TOTAL	7,243,784	1,256,231	5,987,553
	World pop 2050 projected		
Urban	6,338,611	1,113,500	5,225,111
Rural	3,212,333	189,610	3,022,723
TOTAL	9,550,944	1,303,110	8,247,834

Source: United Nations, World Urbanization Prospects: 2014 Revision.

regions (Figure 1.4). Among the less developed regions, Latin America and the Caribbean have an exceptionally high level of urbanization, higher than that of Europe. Africa and Asia, in contrast, remain mostly rural, with 40 percent and 42 percent, respectively, of their populations living in urban areas. By mid-century, Africa and Asia are expected still to have lower levels of urbanization than other world regions. This is quickly changing in some countries; for example, the largest urban growth is expected in India, China, and Nigeria until at least 2050.

The greatest number of cities and the greatest number of large cities (3 million or more

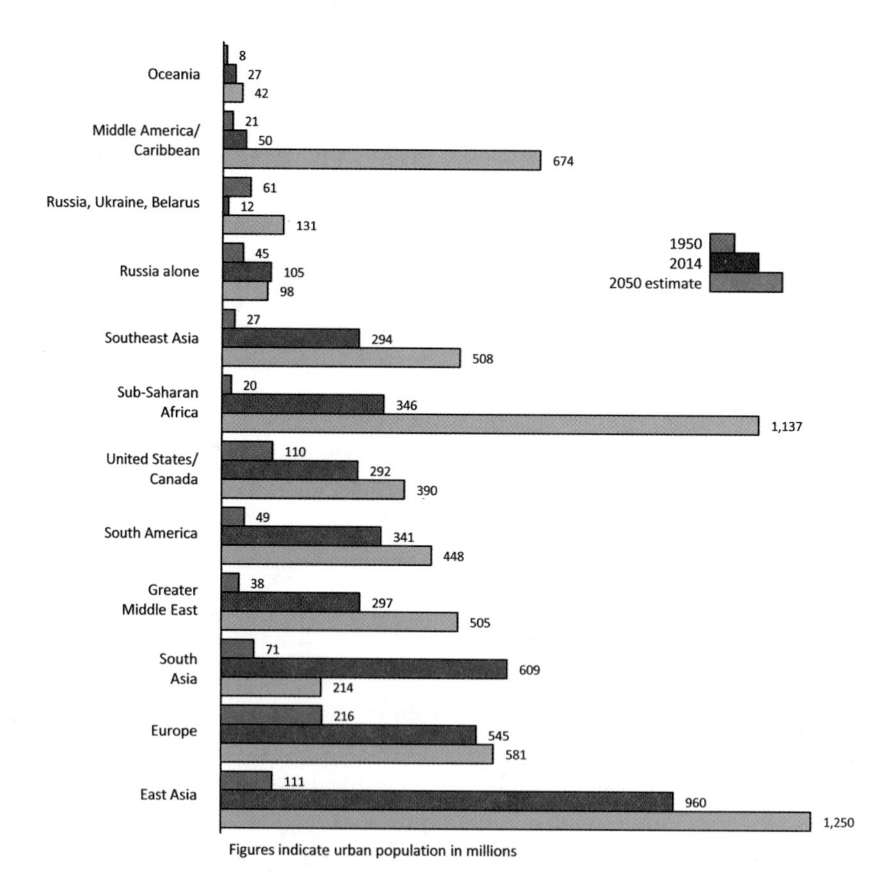

Figure 1.4 Urban Population of World Regions, 1950, 2014, 2050. *Source:* UN, World Urbanization Prospects: 2005 and 2014.

inhabitants) are now found in LDCs. This can also be seen in a list of the world's largest urban areas, where those in the LDCs now outnumber those in the MDCs. Of the 20 largest urban agglomerations in 1950, 13 were in the MDCs and 7 were in the LDCs. The 20 largest urban agglomerations in 2000 included only five in the MDCs, located in only three countries: Japan (Tokyo and Osaka), the United States (New York and Los Angeles), and France (Paris). Today, Mexico City is larger than three-quarters of the world's independent states. Since 1975, less developed regions are urbanizing at much more rapid rates compared to more developed regions, across all city sizes (Figure 1.5).

Amid the continuing trend of urban growth worldwide, some cities experience population decline over time, especially in regions where overall human population is not increasing or where urban economic viability is stagnant. Many cities currently experiencing decline are located in parts of Asia, Europe, and North America. Taking the long view of the overall process of urbanization, it is noteworthy that the movement of people out of rural into urban areas occurs at different rates and for different reasons in different countries, causing continuous evolution of the world urban system over time and space.

WORLD URBANIZATION: PAST TRENDS

Early Urbanization: Antiquity to Fifth Century CE

The first cities in human history were located in Mesopotamia, along the Tigris and Euphrates Rivers, probably about 4000 BCE. Cities were founded in the Nile Valley about 3000 BCE in the Indus Valley (present-day Pakistan), by 2500 BCE in the Yellow River Valley of China by 2000 BCE, and in Mexico and Peru by 500 CE (Figure 1.6). These early cities are thought to have been relatively small. Ur in lower Mesopotamia, for instance, was the largest city in

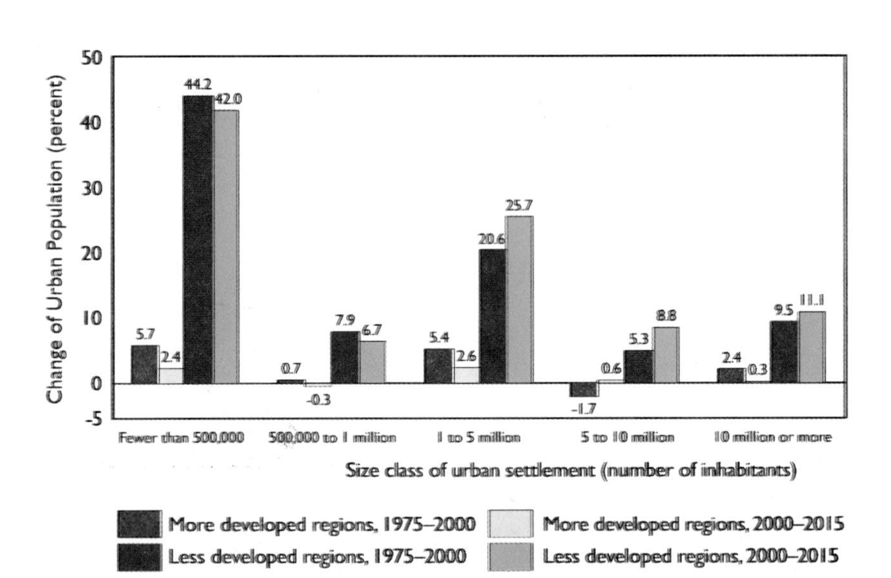

Figure 1.5 Urban Population in MDCs versus LDCs by Size Class of Urban Settlement, 1975–2015. *Source:* UN, World Urbanization Prospects: 2001 and 2014 Revisions.

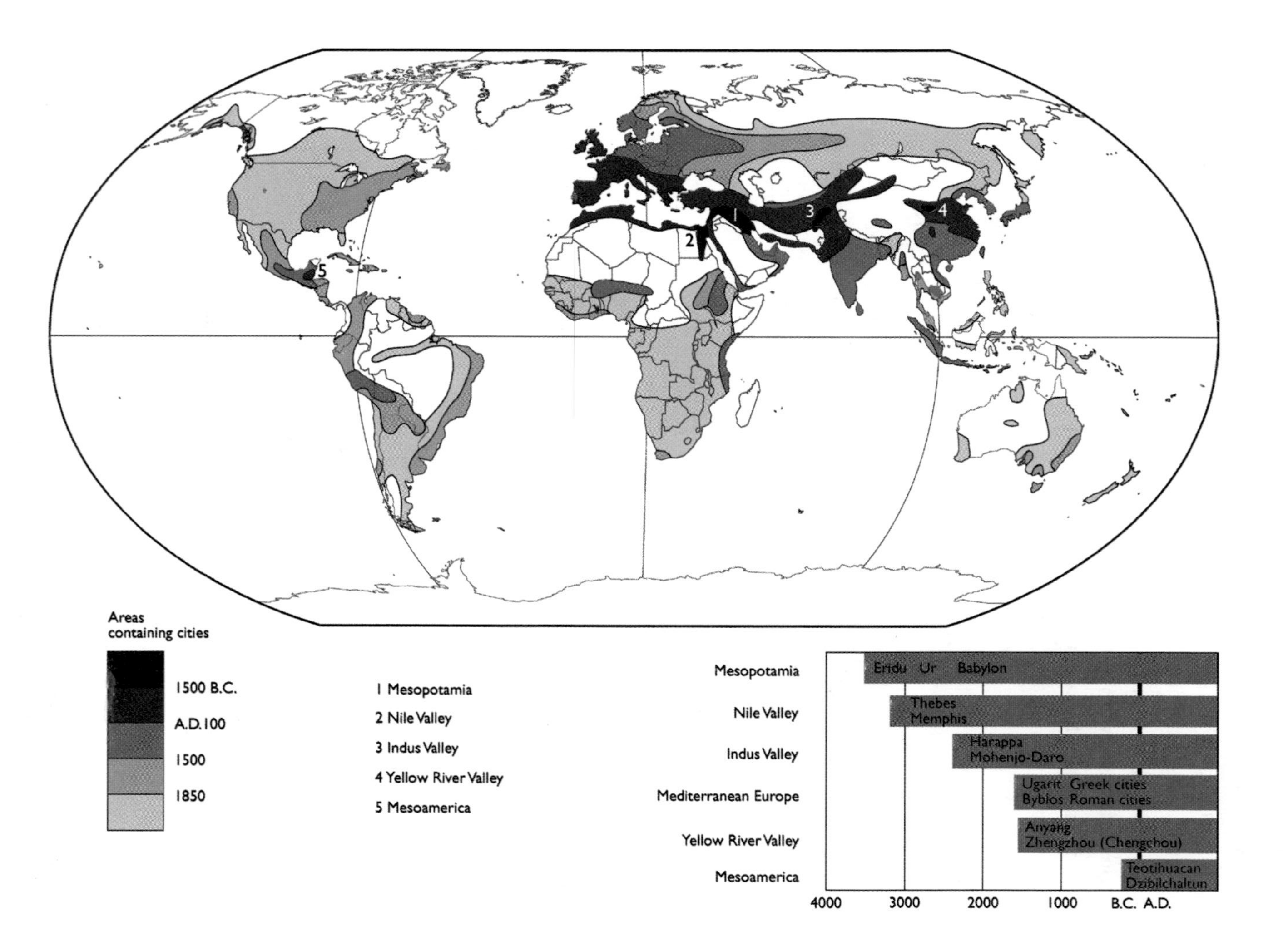

Figure 1.6 Spread of Urbanization, Antiquity to Modern Times. *Source:* Adapted from A. J. Rose, *Patterns of Cities* (Sydney: Thomas Nelson, 1967). 21. Used by permission.

the world 6,000 years ago with a population of about 60,000. In fact, most cities of antiquity held only 2,000 to 20,000 inhabitants without significant increase in the number of cities overall. The largest ancient city was Rome, which Peter Hall has called "the first great city in world history." In the second century CE, Rome may have had 1 million inhabitants, making it the world's first city of that size. Between the second and ninth centuries, however, Rome's population declined to less than 200,000. In fact, the world's largest cities in 100 CE were completely different from the largest cities in 1000 CE (Table 1.2).

Ancient cities appeared where nature and the state of technology enabled cultivators to produce more food and other essential goods than necessary for survival for themselves and their families. That surplus established a division of labor among specialized occupations and the beginning of commercial exchanges. Cities were the settlement form adopted by those members of society whose direct presence in places of agricultural production was not necessary. These cities were religious, administrative, and political centers and represented a new social order, but one that remained dynamically linked to rural society. In these ancient cities were specialists, such as priests and service workers, as well as a population that appreciated the arts and the use of symbols for counting and writing.

Table 1.2 The Largest Cities in History

	Largest Cities in the Year 100			*Largest Cities in the Year 1000*	
1	Rome	450,000	1	Cordova, Spain	450,000
2	Luoyang, China	420,000	2	Kaifeng, China	400,000
3	Seleucia (on the Tigris), Iraq	250,000	3	Constantinople (Istanbul), Turkey	300,000
4	Alexandria, Egypt	250,000	4	Angkor, Cambodia	200,000
5	Antioch, Turkey	150,000	5	Kyoto, Japan	175,000
6	Anuradhapura, Sri Lanka	130,000	6	Cairo, Egypt	125,000
7	Peshawar, Pakistan	120,000	7	Baghdad, Iraq	125,000
8	Carthage, Tunisia	100,000	8	Nishapur (Neyshabur), Iran	125,000
9	Suzhou, China	n/a	9	Al-Hasa, Saudi Arabia	110,000
10	Smyrna, Turkey	90,000	10	Pata (Anhilwara), India	100,000

	Largest Cities in the Year 1500			*Largest Cities in the Year 2000*	
1	Beijing, China	672,000	1	Tokyo, Japan	34,450,000
2	Vijayanagar, India	500,000	2	Ciudad de México (Mexico City), Mexico	18,066,000
3	Cairo, Egypt	400,000	3	New York–Newark, USA	17,846,000
4	Hangzhou, China	250,000	4	São Paulo, Brazil	17,099,000
5	Tabriz, Iran	250,000	5	Mumbai (Bombay), India	16,086,000
6	Constantinople (Istanbul), Turkey	200,000	6	Shanghai, China	13,243,000
7	Guar, India	200,000	7	Kolkata (Calcutta), India	13,058,000
8	Paris, France	185,000	8	Delhi, India	12,441,000
9	Guangzhou, China	150,000	9	Buenos Aires, Argentina	11,847,000
10	Nanjing, China	147,000	10	Los Angeles–Long Beach–Santa Ana, USA	11,814,000

Sources: Historical cities: Tertius Chandler, *Four Thousand Years of Urban Growth: An Historical Census* (St. David's University Press, 1987); http://geography.about.com/library/weekly/aa01201a.htm; Year 2000 data: UN, *World Urbanization Prospects: 2005 Revision*, http://www.un.org/esa/population/publications/WUP2005/2005wup.htm

Other attributes of early cities included taxation, external trade, social classes, and gender differences in the assignment of work. Farms, villages, and smaller towns surrounded each city, where exchange of goods, ideas, and people, and the complexity of technology and the division of labor was limited. Trade, then, was a basic function of ancient cities, which were linked to the surrounding rural areas and to other cities by a relatively complex system of production and distribution, as well as by religious, military, and economic institutions (Figure 1.7).

The Middle Period: Fifth to Seventeenth Century CE

From the fall of the Roman Empire to the seventeenth century, cities in Europe grew only slowly or not at all. The few large cities declined in size and function. Thus, the Western Roman Empire's fall in the fifth century CE marked the effective end of urbanization in Western Europe for over six hundred years.

The major reason for the decline of European cities was a decrease in spatial interaction. After the collapse of Rome and the dissolution of the empire that it commanded, urban localities became isolated, turning to self-sufficiency to survive. From the very beginning, cities have survived and increased in size because of trade with their rural hinterlands and with other cities, near and far. The disruption of the Roman transportation system, the spread of Islam in the seventh and eighth centuries, and the pillaging raids of the Norse in the ninth century almost completely eliminated trade between cities. These

Figure 1.7 The original adobe wall around Bukhara, Uzbekistan is several meters thick, a reminder of the ancient culture and history associated with this city along the Silk Route. *Source:* Photo by Jessica Graybill.

events, plus periodic attacks by Germanic and other northern groups, resulted in an almost complete disruption of urban and rural interaction. Both rural and urban populations declined, transportation networks deteriorated, entire regions became isolated, and people became preoccupied with defense and survival.

Although urban revival did occur six hundred years after the fall of the Roman Empire via fortified settlements and ecclesiastical centers, growth in population and production remained quite small. The reason is simply that exchange was limited—conducted largely with people of the immediate surrounding region. Most urban residents spent their lives within the city walls. Thus, urban communities developed very close-knit social structures. Power was shared between feudal lords and religious leaders. The economically active population was organized into guilds—for craftspersons, artisans, merchants, and others. Social status was determined by one's position in the guild, family, church, and feudal administration. Gender roles were also well defined. Despite the rigorous societal structure, merchants and the guilds saw innovative possibilities in "free cities," where a person could reach his or her full potential within a community setting.

Over time, commerce expanded and linked cities to expanding state power, resulting in a system called mercantilism. The purpose of mercantilism was to use the power of the state to help the nation develop its economic potential and population. Mercantile policies protected merchant interests by controlling trade subsidies, creating trade monopolies, and maintaining strong, armed forces to defend commercial interests. Cities were mercantilism's growth centers, and specialization and trade kept the system alive.

Mercantilism, though based on new economic practices, had one important element in common with the system of the previous period. It restrained and controlled individual merchants in favor of the needs of society. However, the rising middle class of merchants and traders opposed any restrictions on their profits. They opposed economic regulation and used their growing power to demand freedom from state control. They desired an end to mercantilism. As the power of the capitalists increased, the goal of the economy became expansion, and economic profit became the function of city growth. While the new market economy provided the means for social recognition, the social costs were high. The greatest hardship fell on those receiving the fewest benefits—women, poor farmers, and members of the rising industrial working class. The new force of capitalism pushed aside the last vestiges of feudal life and created a new central function for the city—industrialization. It was capitalism that ushered in the Industrial Revolution and led to the emergence of the industrial city.

While Europe experienced a process of decline and rebirth of the ***preindustrial city***, areas of the non-Western world experienced quite different patterns. In East Asia, for example, the city did not decline as in medieval Europe. In China, numerous cities founded in antiquity have remained continuously occupied and economically viable for centuries. Moreover, long before any city in Europe again grew to a size to rival ancient Rome, very large cities were thriving in East Asia. Changan (present-day Xi'an), for example, reputedly had a population of more than 1 million people when it was the capital of Tang China in the seventh century. Kyoto, the capital of Japan for over a thousand years (and modeled after ancient Changan), had a population exceeding 1 million by the middle of the eighteenth century. Although most of the ancient cities of Asia had populations of less than 1 million, they were still far larger than cities in Europe until the commercial-industrial revolutions there. The principal explanation for this historical pattern of urban growth lies in the very different cultures and geographical environments, plus the ***sites and situations*** of the great cities that anchored Asian civilizations.

Although empires waxed and waned in Asia, just as in Europe, premodern cities there continued to serve as vital centers of political administration, cultural and religious authority, and markets for agricultural surplus. Only with the arrival of Western colonialism did those societies and their cities begin to be threatened. Several centuries of Western colonialism in Asia added a new kind of city to the region, a Western commercial city sometimes grafted onto a traditional city, sometimes created anew. In either case, the new ***colonial city*** came eventually to dominate eastern Asia's ***urban landscape***. That dominance has continued into the contemporary period.

In the Greater Middle East, the preindustrial city also existed and thrived through the centuries, long before Europeans began colonizing the region. But once colonialism was fully asserted in the region, the same process of grafting and creating new Western commercial cities occurred, with consequences similar to those in eastern Asia.

In Sub-Saharan Africa and Latin America, the urban experience varied somewhat from that of much of Asia and the Middle East. In Latin America, the indigenous city—and the societies that created that city, such as the Mayan, Incan, and Aztec—was obliterated by Spanish conquest and colonization. The Spanish and the Portuguese thus created new cities in the vast realm of Latin America that reflected European cultures. In Sub-Saharan Africa, the indigenous cities of various African kingdoms, such as Mali, Songhay, Axum, and Zimbabwe, had existed for centuries but were also impacted by European colonialism. By the nineteenth century, they were largely destroyed, and Europeans created new commercial cities, usually coastal, that quickly dominated the region.

Hence, the European-created city became the model for urban growth and development worldwide with the materialization of this vision of the city in Europe and through its export to colonial empires after 1500 CE. In some regions, it was imposed on indigenous societies that were exterminated or displaced (as in North and South America, Australia, and the Pacific). In regions with long histories of indigenous cultures and urban life, it existed alongside and transformed indigenous cities (as in most of Asia, the Middle East, and Africa).

Industrial and Postindustrial Urbanization: Eighteenth Century to the Present

Only after the Industrial Revolution, which began around 1750, did significant worldwide urbanization occur. Industrial cities, drawing first on water power and then on steam generated from coal, saw an increase in the scale of manufacturing. The factory system was born, the demand for labor increased, and rural-to-urban migration swelled the size of cities, first in Great Britain, then in Europe and North America. By the nineteenth century, cities emerged as important places of population concentration. In 1900, only one nation, Great Britain, could be regarded as an urbanized society in the sense that more than half of its inhabitants resided in ***urban places***. During the twentieth century, however, the number of urbanized nations increased dramatically. In the United States, the 1920 census revealed that a majority of Americans lived in cities for the first time. The great industrial cities of Manchester and Birmingham blossomed in England during this period. Scotland saw the rise of Glasgow. In the United States, economies based on manufacturing built Chicago, Pittsburgh, Cincinnati, and St. Louis. In the early and mid-twentieth century, the automobile industry transformed another set of cities: Detroit in the United States, Turin in Italy, Tolliati in Russia, and Adelaide in Australia.

CITY FUNCTIONS AND URBAN ECONOMIES

City Functions

Some cities are born because a strategic location must be defended; others serve the demands of trade and commerce; others serve governmental administration or religious pilgrimage, and still others thrive from turning primary commodities into manufactured goods. Geographers have traditionally classified cities into three categories based on their dominant functions: (1) market centers (trade and commerce); (2) transportation centers (transport services); and (3) specialized service centers (such as government, recreation, or religious pilgrimage). Some cities serve a single function—the "textile cities" of the southeastern United States, for instance—but functional diversity is more common.

Cities categorized as market centers are also known as central places because they perform a variety of retail functions for the surrounding area. Central places offer multiple goods and services (from grocery stores and gas stations to schools and corporate headquarters). Small central places, or market centers, depend less on the characteristics of a particular site and more on being centrally located with respect to their market areas. These centers tend to be located within the trade areas of larger cities: people living in small cities must go to larger cities to make certain purchases for which there is not a sufficient market locally. There is thus a spatial order to settlements and their functional organization. Central place theory, as developed by geographer Walter Christaller in the 1930s, explained the regular size, spacing, and functions of urban settlements as they might be distributed across a fertile agricultural region, for instance. In central place theory, the largest cities, or highest-order centers, are surrounded by medium-sized cities that are in turn surrounded by small cities, all forming an integrated part of a spatially organized, nested hierarchy. The locational orientation of market centers is quite different from the locational orientation of transportation and specialized function cities.

Transportation cities perform break-of-bulk or break-in-transport functions along waterways, railroads, or highways. Where raw materials or semi-finished products are transferred from one mode of transport to another—for example, from water to rail or rail to highway—cities emerge as processing centers or as trans-shipment centers. Unlike central places, whose regularity in location is accounted for by marketing principles, transportation cities are located in linear patterns along rail lines, coastlines, or major rivers. Frequently, major transport cities are the focus of two or more modes of transportation, for example, the coastal city that is the hub of railways, highways, and shipping networks.

Today, of course, almost all cities have multiple transportation linkages. Exceptions tend to be isolated towns, such as mining centers in Siberia that may have only air or rail connections or primitive seasonal roads to the outside. Cities performing single functions, such as recreation, mining, administration, or manufacturing, are called specialized function cities. A very high percentage of the population participating in one or two related activities is evidence of specialization. Oxford, England, is a university town; Rochester, Minnesota, is a health-care town; Norfolk, Virginia, a military town; Canberra, Australia, a government town; and Cancún, Mexico, a tourist town. Specialization is also evident in cities where the extraction or processing of a resource is the major activity. Cities labeled as mining and manufacturing cities have much more specialization than those with diversified economic bases.

Sectors of the Urban Economy

The economic functions of a city are reflected in the composition of its labor force. Preindustrial societies are associated with rural economies. These economies have the largest percentages of their labor force engaged in the primary sector: agriculture, fishing, forestry, mining. Preindustrial cities have historically been commercial islands in seas of rural populations. The Industrial Revolution triggered the emergence of cities oriented to manufacturing, the secondary sector of the economy. It created a demand for labor in factories, and a larger percentage of the population began living in urban areas. As factory workers added their buying power to the city's economy, the service sector, or tertiary economic activities, grew as well. The quaternary sector, a more advanced stage of the service sector, consists of information- and intellect-intensive services, which play an increasingly important role in the world economy. Identifying the mix of primary, secondary, tertiary, and quaternary activities within urban regions as they have changed over time helps identify specific stages of humanity's economic evolution (Figure 1.8).

The association between urbanization and industrialization has been characteristic of Europe, North America, Japan, Australia, and New Zealand. That is, cities and industries grew synchronously. Across Africa, Asia, and Latin America, many urbanizing countries have not experienced a corresponding increase in the manufacturing sector of the economy. Instead, their service sectors—small retailers, government servants, teachers, professionals, and bankers—provide jobs for growing urban populations. Also included are many service workers in the informal economy, including people willing to perform odd jobs (watching parked cars or cleaning houses) and working in unskilled service occupations, such as street vending, scavenging, and laboring at construction sites. In the informal sector, barter and the exchange of services often take the place of monetary exchanges, thus bypassing government accounting and taxation.

Basic and Nonbasic Economic Activities

The economic base concept states that two types of activities or functions exist: those that are necessary for urban growth and those that exist primarily to supplement those necessary functions. The former are called basic economic activities. They involve the manufacturing, processing, or trading of goods or the providing of services for markets located outside the city's boundaries. Examples include automobile assembly and insurance underwriting. They are the key to economic growth. Economic functions of a city-servicing nature are called nonbasic functions. Grocery stores,

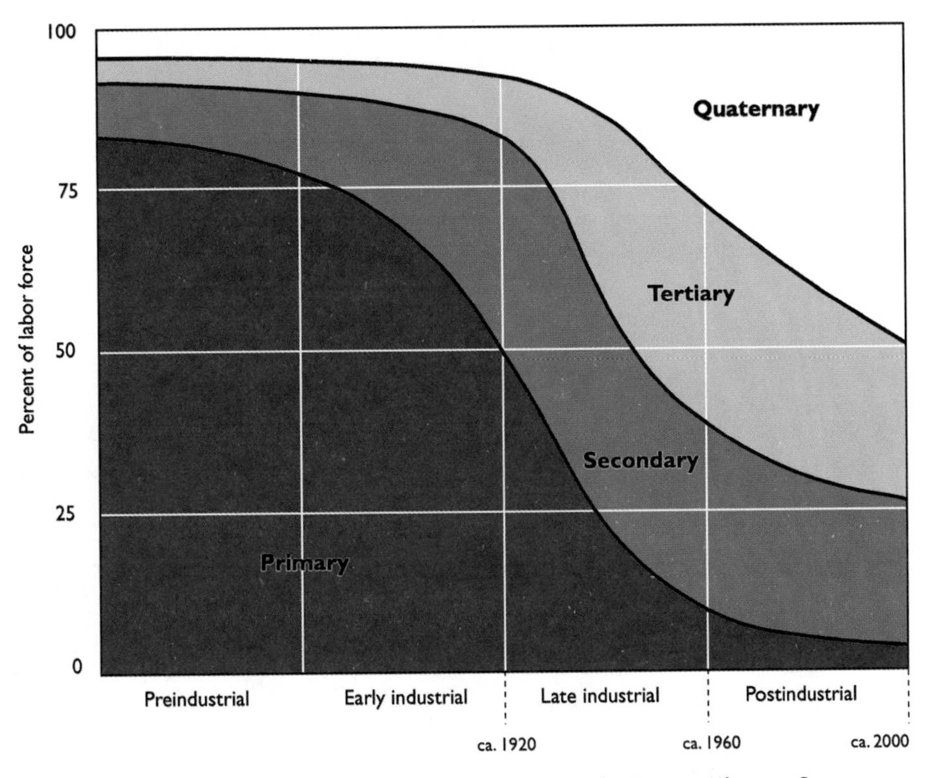

Figure 1.8 Labor Force Composition at Various Stages in Human History. *Source:* Adapted from Ronald Abler et al., *Human Geography in a Shrinking World* (Belmont, CA: Wadsworth, 1975), 49.

restaurants, beauty salons, and so forth are nonbasic economic activities because they cater primarily to residents within the city itself (Figure 1.9). Income generated by a city's economic base is channeled back into the city's nonbasic sector, where employees in those industries purchase groceries, gasoline, insurance, entertainment, and other everyday needs and wants.

The economic base of some cities is grounded in manufacturing industries, the secondary sector of the economy. Manchester, England, and Pittsburgh, Pennsylvania, are prime examples of older industrial cities whose growth and prosperity depended upon world markets for cotton textiles and steel, respectively. Since World War II, both cities have lost their manufacturing base and have sought to create service industries for which there is a larger market (Box 1.2). The economic base of the ***postindustrial city***, in fact, is to be found in the tertiary and quaternary sectors of the economy. Silicon valleys developed in the late twentieth century to service the needs of the computer industry. In the early twenty-first century, biotech valleys are becoming the economic base of choice. Cities such as Geneva, Singapore, San Francisco, and Boston are competing to have biotechnology firms move into their regions. Money from biomedical and pharmaceutical research provides an economic base tied to high-level applications of technology and brainpower.

Figure 1.9 Street peddlers in Shakhrisabz, Uzbekistan, the birthplace of Tamerlane (Timur) sell goods from China and Turkey to local Uzbek customers in this ancient Silk Route city. *Source:* Photo by Jessica Graybill.

As a city's economic base increases, it has a multiplier effect throughout the community. Growth (and conversely, decline) becomes a cumulative process in which growth begets growth (and vice versa). This is known as the principle of circular and cumulative causation. For instance, one of the major ways cities grew in the past was by attracting more manufacturing enterprises. Each new factory stimulated general economic development and population growth. Business output increased due to a greater demand for products. Rising profits increased savings, causing investments to rise. Increased productivity resulted in greater wealth. The growing population then reached a new level, or threshold, resulting in a new round of demands. Larger cities are able to offer a greater number and variety of services than smaller cities. Conversely, conditions can create negative circular and cumulative causation—a downward spiral. Thus, it is easy to understand why city mayors work so hard to promote their respective cities as favorable sites for investment and new business locations.

THEORIES ON THE SPATIAL STRUCTURE OF CITIES

Geographers have long been intrigued by the internal spatial structure of cities. Components of that structure include industrial zones, commercial districts, warehouse rows, residential areas, parks and open space, and transportation routes, among others. Multiple

Box 1.2 Jellied Eels for the Urban Palate

Timothy Kidd, Old Dominion University

The gastronomical scene of a city is often a significant cultural marker showcasing changes in ethnic composition, shifting taste preferences, and availability of ingredients. Cities often have their own unique dishes, restaurants, drinks, and street food. Street food ranges from the doner kebab in Istanbul to hot dogs in New York to balut in Manila, typically sold from stalls or vans in parts of the city with high foot traffic. The U.N. estimates that 2.5 billion people eat such street food everyday because it is faster and cheaper than restaurant meals. As fast food franchises have gone global, however, the nature of the take-away meal has been altered. While there are often local variations on foreign-based fast food menus (e.g., McDonald's *McAloo Tikki* potato burger in India), there is still a homogenized feel to such establishments. In some cases, including London, global franchises coupled with changing demographics have been deleterious for traditional street food purveyors.

In London's historically poor East End, the remnants of uniquely English fast food still exist. Perhaps most famous of all traditional dishes are jellied eels. The Thames River estuary provided an excellent habitat for eels, and fish traps known as eel bucks lined the waterway to reap the bounty. The plentiful eels became something of a delicacy among the poor, immigrant East Enders. Fishmongers sold them in great markets like Spitalfields to restaurants and to street hawkers known as "piemen." By the first half of the 1800s, however, London's industrial expansion brought severe pollution that had a decidedly negative effect on the Thames fishery.

The decline in fish stocks heralded the beginning of a new East End dining tradition: the eel, pie, and mash shop. These eateries served meat pies made from beef or mutton with mashed potatoes ("mash") doused with parsley-infused, green-eel gravy. Often the restaurants were similar in their interior décor, sporting tiled floors and walls, marble counter tops, and wooden benches. From the Victorian era until the end of World War II, eel, pie, and mash shops boomed. Today, there are probably fewer than 100 pie and mash shops left in the East End, many thriving on nontraditional items such as vegetarian pies.

In the late twentieth century, considerable numbers of Bangladeshis moved into the East End. Soon, curries and kedgeree became more popular than the traditional jellied eels. However, in the 1990s and 2000s, the traditional English fare had a bit of a renaissance when celebrity chefs like Gordon Ramsey began serving upscale versions. With more sit-down restaurants and pubs selling eels, the once-common street stalls saw business falter. Various media including the BBC lamented the 2013 closure of the famous Tubby Isaac's after 94 years of serving jellied eels from a food cart.

Chopped into bite size pieces with bones left in and boiled with herbs, the eels are left to cool in their own gelatinous stock and then served with chili vinegar. It is said to be an acquired taste. Perhaps preferences for pizza or chicken tikka are proving more popular for younger generations than this most traditional English dish. Eel aficionados, however, appear to have a new source for their delicacy: supermarket chains. Since the recession of 2008, large stores such as Morrison's, Sainsbury's, and Tesco have started selling jellied eels, with sales rapidly growing as customers seek cheap protein just as they did in the 1800s. What was once the quintessential London street food can now be found as far away as Scotland and Northern Ireland. In London, the East End's most popular street food lives on in toponyms like the culturally famous Eel Pie Island in the River Thames, and in the Eel Pie Recording Studios founded by guitarist Pete Townsend of The Who.

theories have been developed to describe and explain the pattern of land use and the distribution of population groups within cities. The four most widely accepted theories, or models, of city structure are the concentric zone model, the sector model, the multiple nuclei model, and the inverse concentric zone model (Figure 1.10). All evolved from observations of urban landscapes that suggested that different land uses were predictably, not randomly, distributed across the city.

The Concentric Zone Model

A concentric zone theory was first conceptualized by Friedrich Engels (co-author of the Communist Manifesto) in the mid-nineteenth century. In 1844, Engels observed that the population of the ***industrial city*** of Manchester, England, was residentially segregated based on class. He noted that the commercial district (offices plus retail and wholesale trade) was located in the center of Manchester and extended about half a mile radially. Besides the commercial district, Manchester consisted of unmixed working people's quarters, which extended a mile and a half (2.3 km) around the commercial district. Next, extending outward from the city, were the comfortable country homes of the upper bourgeoisie. Engels believed this general pattern to be more or less common to all industrial cities.

Engels may have described the pattern first, but most social scientists consider E. W. Burgess, a University of Chicago sociologist, to be the father of the concentric zone model. According to Burgess, the growth of any city occurs through radial expansion from the city core, forming a series of concentric rings, or a set of nested circles that represents successive zones of specialized urban land uses. The five zones that Burgess described during the 1920s, before the automobile transformed Chicago, were: (1) the central business district (CBD) and its retail and wholesale areas; (2) the zone of transition, characterized by stagnation and social deterioration; (3) the zone of factory workers' homes; (4) the zone of better residential units, including single-family dwellings and apartments; and (5) the commuter zone, extending beyond the city limits and consisting of suburbs and satellite communities.

The process that Burgess used to explain these concentric rings was called invasion and

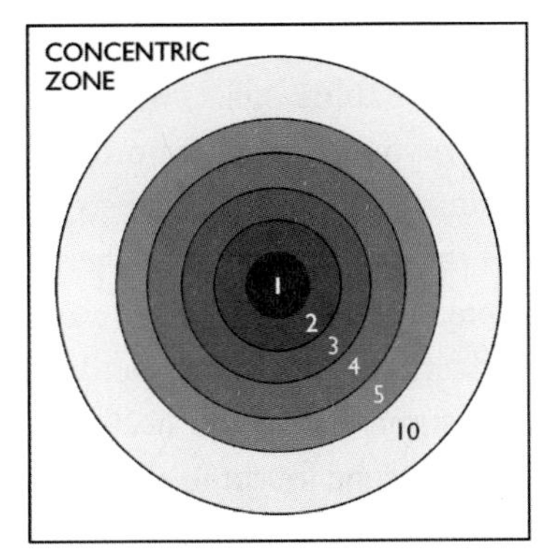

Figure 1.10 Generalized Patterns of Internal Urban Structure. *Source:* Adapted from various sources.

succession. Each type of land use and each socioeconomic group in the inner zone tends to extend its zone by invasion of the next outer zone. As the city expands, population groups are spatially redistributed by residence and occupation. Burgess further demonstrated that many social characteristics—the percentage of foreign-born groups, poverty, and delinquency rates—are spatially distributed in a series of gradients away from the central business district. Each tends to decrease outward from the city center.

The Sector Model

Homer Hoyt, an economist, developed the sector model in the 1930s. Hoyt examined spatial variations in household rent in 142 American cities. He concluded that general patterns of housing values applied to all cities and that those patterns tended to appear as sectors, not concentric rings. According to Hoyt, residential land use arranges itself along selected highways leading into the CBD, thus giving land-use patterns a directional bias. High-rent residences were the most important group in explaining city growth because they tended to pull the entire city in the same physical direction. New residential areas did not encircle the city at its outer limits, but extended farther and farther outward along select transportation axes, giving the land-use map the appearance of a pie cut into many pieces. The sectoral pattern of city growth is partially explained by a filtering process. When new housing is constructed, it is located primarily on the outer edges of the high-rent sector. The homes of community leaders, new offices, and stores are attracted to the same areas. As inner, middle-class areas are abandoned, lower-income groups filter into them. By this process, the city grows over time in the direction of the expanding high-rent residential sector.

The Multiple Nuclei Model

In 1945, two geographers, Chauncy Harris and Edward Ullman, developed a third model to explain urban land-use patterns, the multiple nuclei model. According to their theory, cities tend to grow around several distinct nodes, thus forming a polynuclear (many-centered) pattern, which is explained by four factors. (1) Certain activities are limited to particular sites because they have highly specialized needs. For example, a retail district must be widely accessible, which is best found in a central location, while a manufacturing district needs transportation facilities. (2) Related activities or economic functions tend to cluster in the same district because their activities are more efficient as a cohesive unit. An example is spatial clustering of automobile dealers, tire shops, and auto repair and glass shops. (3) Some related activities, by their very nature, repel each other. A high-class residential district, for example, will normally locate away from a heavy manufacturing district. (4) Certain activities, unable to generate enough income to pay the high rents of certain sites, may be relegated to more inaccessible locations. Examples may include some specialty shops.

The number of distinct nuclei occurring within a city is likely to be a function of city size and age of development. Auto-oriented cities, which often have a distinctly horizontal rather than vertical form, include industrial parks, regional shopping centers, and suburbs layered by age of residents, income, and housing value. Rampant urban sprawl is likely to be reflected in a mixed pattern of industrial, commercial, and residential areas in peripheral locations. Geographer Peirce Lewis describes this sprawling urban landscape as the ***galactic metropolis*** because the nucleations resemble a galaxy of stars and planets. Some of those nucleations become cities in the suburbs, what some call edge cities. Edge cities are, in effect, the CBDs of newly emerging urban centers scattered through the newer suburban ring surrounding older central cities. This pattern reinforces what has been described as the typical urban spatial model for most U.S. urban areas, the doughnut model. In this model, the hole in the doughnut is the central city (with poor, mostly nonwhite, blue-collar, working-class residents, large numbers of whom are on welfare, and a declining tax base and economy). The ring of the doughnut is the suburbs (rich, mostly white, middle- and upper-class, white-collar employment, and an expanding tax base and economy). Old, often declining manufacturing tends to be found in the hole; new, often high-tech manufacturing tends to be located in the suburban ring, in the edge cities. In the twenty-first century, however, city centers are reinventing themselves by building more residential quarters, upgrading infrastructure, invigorating the shopping experience, showcasing the arts (Box 1.3), and emphasizing mixed land-use development.

The Inverse Concentric Zone Model

The preceding three theories of urban spatial structure apply primarily to cities of the MDCs and to American cities in particular. Many cities in the LDCs follow somewhat different growth patterns. A frequent one is the inverse concentric zone pattern, which is a reversal of the concentric zone model. Cities where this pattern exists have been called ***preindustrial***; that is, they are primarily administrative and/or religious centers (or were at the time of their founding). In such cities, the central area is the place of residence of the elite class. The poor live on the periphery. Unlike most cities in the MDCs, social class in these places is inversely related to distance from the center of the city. The reasons for this pattern are twofold: (1) the lack of adequate and dependable transportation systems, which thus restricts the elites to the center of the city so they can be close to their places of work, and (2) the functions of the city, which are primarily administrative and religious/cultural, are controlled

Box 1.3 Performance Art and Psychogeography

Katrinka Somdahl-Sands, Rowan University

The performance of *Go! Taste the City* took place along seven blocks of Nicollet Avenue in Minneapolis. The BodyCartography Project, and Olive Bieringa as its representative, transgressed the norms of city streets by adding a new practice: dance. Bieringa actively engages with city streets to expose those edges within urban environments that lie below the surface of social consciousness. For her Minneapolis audience, she made it possible to experience the city in an entirely new way.

The most common way of being aware of our bodies moving through city spaces is as a pedestrian. The habitual practice of walking unintentionally conveys societal conventions regarding the "appropriateness" of certain bodily actions. The way one walks can reinforce (or disrupt) cultural practices of racial, ethnic, class, and gender differentiation. The act of walking, usually an unconscious act, becomes a kind of performance. Dance, on the other hand, actively and consciously manipulates the aspects that make walking a powerful point of study. It uses the physicality of the body to articulate complex thoughts and feelings that cannot be easily put into words—represented. Dancing in city streets questions the rationalist view that dominates much of modern life.

Down Nicollet Avenue, Olive Bieringa skipped, slid, and swirled; raced, did headstands, and rolled around on the ground; swayed in the wind produced by passing automobiles; darted in and out of the dappled sunlight produced by street-side trees; shifted effortlessly between moments of quiet and bustle; laid in the street until an approaching bus forced her back to the sidewalk; and danced on mounds of dirt produced by construction projects. All of her self-propelled movement created a particular sense of place.

The Situationists International's (SI) notion of psychogeography and use of the derive, or drift, help us situate Bieringa's performance. Psychogeography explores the hidden, nonphysical connections between spaces, acknowledging how the built environment conditions our access and feelings on city streets. Drift is intended as a nonverbal discourse on urbanism's terrain and the encounters found there. Bieringa's dance was responding to the psychogeography of Nicollet Avenue. She actively interacted with elements and individuals of the city that most take for granted or try to avoid. Bieringa's race, gender, and occupational status were explicitly juxtaposed against the expectations of comportment. When she behaved "normally" she was invisible, yet when she spun through intersections or crawled in a park she was *seen*. Women are expected to be workers or consumers in public space. Participating in city life in other ways becomes a fundamentally transgressive and transformative act. Bieringa was not the stereotypical woman in the city. She used her female body to disrupt (transgress) what women "do" on city streets. Consequently, Bieringa was able to reveal the power relations behind why some bodies are more noticeable than others.

Nicollet Avenue took on a whole new meaning for those who saw it under the careful guidance of Olive Bieringa. By engaging audience members, both at the level of the mind and the body, she asked her audience to consciously consider how meaning was entwined in their urban environment. One does not need to "dance" down a street to become conscious of the psychogeography that is already there; one only needs to slow down enough to see them.

by the elite and concentrated in the center of the city (with its government buildings, cultural institutions, places of worship, etc.).

As many developing countries industrialize, newer growth industries tend to locate on city peripheries, often in industrial parks or enterprise zones established by the government for the purpose of attracting domestic and foreign investors. City centers tend to be too congested for industrial plants of any considerable size and urban elites often do not want large industrial plants near their places of work and residence. Hence, emerging gradually in many of the larger cities of the LDCs is the pattern of the multiple nuclei model, with new industrial parks serving as the nuclei. In other words, the inverse concentric zone pattern, while still valid in many LDCs, is merging with the multiple nuclei pattern.

As useful as these four models are, they must be viewed as generalizations about an extremely complex mix of factors that influence urban land use (Figure 1.11). Elements of more than one model are present in any city. Moreover, each of the models must be viewed as dynamic because ongoing changes in economic functions, social and administrative services, transportation, and population groups continually alter the size and shape of specific sectors or zones (Figure 1.12). Furthermore, the complexities of applying these theories multiply when working with non-Western cultures, economic systems, and urban places. Nowhere is this more apparent than in China and countries of the former Soviet bloc, where various forms of the ***socialist city*** were created and where internal spatial structures were quite unlike those described by any of these four theories. The legacy of those socialist patterns lingers, even as free-market forces transform those cities.

URBAN CHALLENGES

Managing the Environment

Air and water pollution, excessive noise levels, water quality, visual blight, and land clearing for urban expansion are among the many serious urban environmental problems worldwide (Figure 1.2). Cities in the MDCs have the means to address these problems, but cities in the LDCs often regard such concerns as less important than immediate economic and well-being concerns. For instance, cities with the greatest air pollution are no longer affluent London (formerly nicknamed "the Big Smoke") and Los Angeles (once dubbed "Smog Central"), but developing cities like Shanghai, Mexico City, and São Paulo. An additional environmental problem arises from the expansion of urban areas into nearby agriculturally productive land. Urban expansion swallows up about 2 million acres (809,600 hectares) in China and 1 million acres (404,600 hectares) in the United States of farmland annually. Moreover, global climate change adds new dimensions to urban environmental problems experienced as water

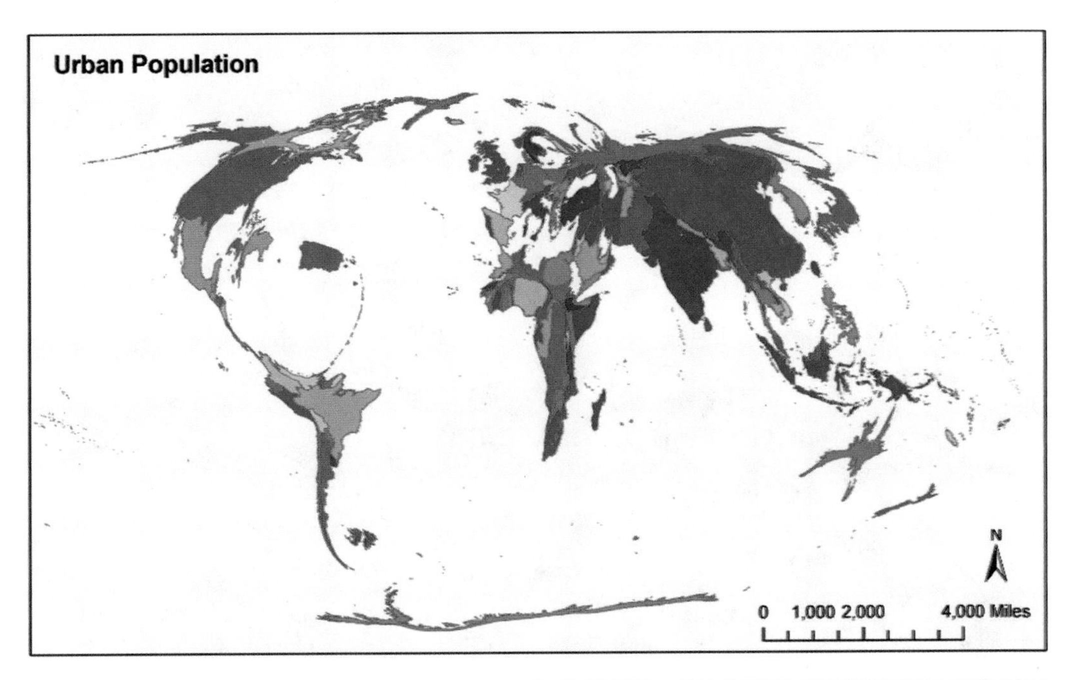

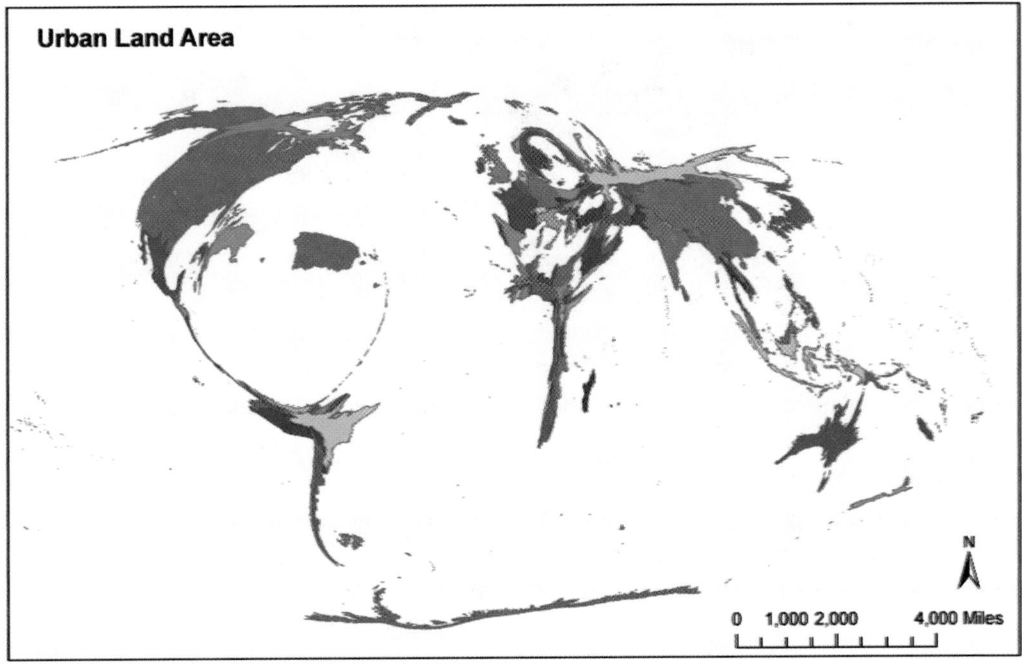

Figure 1.11 These cartograms indicate the amount of territory classified as urban in countries worldwide (not all countries are included). *Source:* Demographia World Urban Areas 11th Annual Edition: 2015:01, January 2015. http://www.demographia.com/db-worldua.pdf

Figure 1.12 The Frontenac Hotel was built by the Canadian Pacific Railway. As the most well-known signature architecture of Quebec City, it still functions as tourist magnet even though most no longer come by train. *Source:* Photo by D. J. Zeigler.

becomes scarcer; heat waves more frequent, prolonged, and severe; and sea levels rise. Global climate change will affect cities differently depending on their physical locations (e.g., coastal or inland, desert or tropical), but scientific consensus by the world's top climate scientists, as reported in the Intergovernmental Panel on Climate Change Fifth Assessment Report in 2014, indicates changing environmental conditions for most cities within the next century (Box 1.4).

Managing Population Size and Growth

Excessive size of urban regions, in population and in geographical area, might best be described as a cause of problems than itself a problem. Size presents a particularly severe challenge in less developed regions, where the economic base of cities is inadequate to cope with stresses created by increasing human habitation of urban places. A concomitant factor of excessive size is overcrowding, meaning too many people occupy too little space, but it does not always equate to population density. Some cultures—and some people—are more adapted to high-density living than others. It is sometimes difficult to comprehend the magnitude and effects of the mass of humanity that lives in—and is moving to—larger cities such as Manila, Shanghai, or Cairo, especially for citizens of MDCs today,

Box 1.4 Cities and Stormwater Runoff

Michael H. Finewood, Pace University

With the goal of restoring the biological, chemical, and physical integrity of the country's waterways, the Clean Water Act (CWA) is the United States' seminal water pollution law. It has been successful in returning many rivers and streams to a swimmable, fishable, and drinkable status. However, the design of the CWA differentiates between two types of pollution. *Point sources* flow from a specific source (e.g., a pipe) and are federally regulated; they are permitted and enforced. The CWA does a relatively good job at managing these sources of pollution. On the other hand, *nonpoint sources* flow diffusely across the landscape during and after rain or snowmelt. They are regulated in a more haphazard manner because the CWA must rely on a combination of best management practices, federal encouragement, lawsuits, grants, and technical assistance.

Metropolitan regions are among the largest sources of nonpoint source pollution in the United States, making it a critical issue for metropolitan environmental planning. For eastern, postindustrial cities (e.g., Pittsburgh, Cleveland, and New York City), nonpoint source pollution from *stormwater runoff* embodies the wicked nature of urban water governance. When it rains enough in these cities, impervious surfaces such as roads and rooftops prevent water from percolating into the soil. Instead, stormwater tends to pool and run off hardened surfaces to various low points in the city, such as drains or nearby waterways.

There are two fundamental challenges here. First, stormwater runoff overwhelms sewage treatment plants. In many cities, the volume of runoff has increased as our cities have expanded outward, hardening the landscape with roads and rooftops. Existing urban water conveyance systems and treatment plants were not designed to handle growing volumes of stormwater; they can't keep up with the growth. This is particularly deleterious when sewer and stormwater systems are combined. In this case, when enough water overwhelms the system, it introduces sewage and other pollutants into local rivers and streams. And, this leads to the second point.

When stormwater runs across hardened surfaces, it picks up pollutants from cars, power plant emissions, and other sources, which then make their way into rivers and streams. As cities experience what can often be just a fraction of an inch of rain, the systems overflow, and sewage *and* polluted stormwater back up or run off into local waterways. Add to this scenario climate change and increasingly fragmented communities, and it becomes clear as to why this is such a wicked problem. Nonetheless, federal, state, and local regulations compel metropolitan regions to address the issue.

But here is the good thing. There are some innovative ways that communities are trying to address stormwater challenges. From technical approaches, such as *sustainable drainage systems*, to community-driven approaches, such as *ecodistrict planning*, municipalities are realizing that multidisciplinary, collaborative, and community-based projects have created the most overall benefits, while meeting water quality regulations.

In other words, urban water governance strategies that are locally negotiated, with a wide range of stakeholders participating in the process, can produce outcomes that create multiple values for communities. Think here of a river walk that serves as a recreational space, riparian restoration, and expanded floodplain to mitigate flooding; or green infrastructure that captures stormwater before it is introduced into the system, while simultaneously enhancing biodiversity and increasing community greenspace.

who are accustomed to less hectic paces of urbanization and are unfamiliar with extreme overcrowding in their cities. Government-directed land-use planning is often used as an antidote to the uncontrolled size and growth of cities; and, in some countries, planning has included ***new towns*** as instruments of population redistribution.

The rate of population growth—or decline—may challenge cities as well. Some cities, especially in LDCs, are growing so fast that economic development cannot keep up. Hyperurbanization describes what is happening in the world's most rapidly growing cities. Indeed, introduction of the term "metacity" by the United Nations indicates that growth of the very largest urban places, with over 20 million inhabitants, will only continue as urbanization continues. Conversely, some cities in the developed world may be stagnating or declining. Whether in Japan, Russia, Germany, or the United States, the problems of no-growth cities are very similar. They are often home to industries with outdated technology, high production costs, expensive labor, an aging workforce, and products with declining demand. In the developing world, these cities are generally victims of deindustrialization. They have not contended with transition from manufacturing to service-based economies.

In growing urban places, concern exists about the future of ecosystem functions and services for all species. For example, as cities in the arid and semi-arid western United States continue to increase in population and spatial extent, many regional and city leaders have two major concerns: (1) providing adequate water supplies, both in quantity and quality, for rising demand and (2) managing increased human-wildlife contact as suburbanization encroaches on natural habitats and water supplies of, for example, bobcats, mountain lions, and coyotes. A drastic example of water supply concerns is that of Lake Powell, on the Colorado Rover, the second largest reservoir in the United States. After years of diminishing snowfall (and thus river flow and lake fill) in the Rocky Mountains, this reservoir is now filled to only 45 percent of its capacity, which was full only 15 years ago. The Colorado River and Lake Powell serve ranches and agricultural zones across the western states, including those in California, which provides most of the nation with fresh produce year round. These water features are also a key component in the water delivery system for the ever-growing desert cities in the Southwestern United States, such as Las Vegas in Nevada and Phoenix in Arizona.

Managing Urban Services

With so many people in cities, local governments are challenged to provide all of the necessary human services for residents—education,

Figure 1.13 Even in rich cities such as Macao, one of China's Special Administrative Regions, scavengers find a niche in the urban ecosystem by collecting cardboard and other items that have value as recyclables. *Source:* Photo by D. J. Zeigler.

health care, pharmacies, clean water, sewage disposal, garbage pick-up, police and fire protection, disaster relief, public parks, mass transit, among numerous other essential services (Figure 1.13). How can a city in the developing world that is doubling in population approximately every ten years maintain the economic growth needed to provide for so many new inhabitants, particularly when those new residents are poor? Even in the world's most developed countries, providing social and environmental services to sprawling, energy-inefficient suburban and exurban regions strains municipal budgets. For example, many urban school districts in the United States struggle to fund students every school year, creating social inequalities within cities and deepening urban-suburban divides. Environmentally, cities may struggle to provide sufficiently clean water supplies for an ever-growing population, as demand rises but water supplies do not.

Managing Slums and Squatter Settlements

Most cities in the developing world have slums or squatter settlements, poorer communities that are not fully integrated, socially or economically, into the development process. Squatter settlements have various names in different countries: *barriadas* or *asentimientos humanos* in Peru, *favelas* in Brazil, *geçekondu* in Turkey, *bustees* in India, and *bidonvilles* in former French colonies. Slums tend to be found in old, run-down areas of inner cities (sometimes, paradoxically, on very valuable land). Squatter settlements, usually located on the outskirts of cities in the developing world, are typically newer and comprised of makeshift dwellings erected without official permission on land not owned by squatters. They may be constructed of cardboard, tin, adobe bricks, mats, sacks, or any other available materials. They tend to lack essential services, sometimes even electricity. Because of their rudimentary construction and lack of a plan for long-term development, slums are often sites for rampant environmental and health problems, often related to the presence of open sewage and other waste that, left untreated, infest water sources and create unsanitary living conditions. Concern especially arises due to population mobility: as people move through a city to work, buy necessary items and enjoy themselves, diseases can spread quickly throughout an urban community. While slums are not, of course, the only source of diseases, the closeness of large human populations in cities with slums brings into question how these regions will plan for overall well-being and health amid poverty and overcrowding.

Managing Society

Perhaps one of the most insidious effects of hyperurbanization throughout the world is a reduced sense of social responsibility and connection to the environment. As people compete for space and services, antisocial or even sociopathic attitudes may emerge. City life can bring out the worst in human behavior. People exhibit social pathologies when they must queue for services, think nothing of despoiling public property, disregard traffic regulations, or show a disregard for fellow citizens' rights. When large cities provide neither a sense of community nor a respected police presence, crime soars. Social norms that constrain unscrupulous behavior in rural areas may be absent in cities.

On the other hand, city life can also bring out the best in human behavior, especially after a crisis. Entire communities can come together after tragic events, whether caused by humans or nature. For example, high-profile shooting incidents in the United States, Norway, and France have increased discussion of how and why such horrific events occur, and how society might act to reduce them. Urban environmental devastation and crisis, such as Hurricane Sandy that struck the East Coast of the United States in 2012, or the 2015 earthquakes in the Kathmandu valley in Nepal show that urban human populations can, and do, come together in times of need and often for the good of society. As cities continue to grow and continue to confront human-made and environmental disasters, urban leaders and planners will be challenged to develop systems of people and places (Box 1.5) that can help develop a sense of responsibility and ethical citizenship.

Managing Unemployment

Virtually everything connected with the city is related to the economic health of its population, and economic well-being is dependent on people having jobs. In capitalist economies, however, employment is not guaranteed. The result is often unemployment or underemployment. In LDCs, too many people may be competing for too few jobs, driving down labor costs. In MDCs, large segments of urban populations may lack the skills to find jobs in high-end service sectors. Sometimes, the result is underemployment: people take jobs that are not commensurate with their skills. These jobs pay less than a living wage, meaning that some must take more than one job to survive and others must supplement their income with employment in the informal sector of the economy with long hours and no fringe benefits. In cities of the developing world, underemployment rates of 30–40 percent are not uncommon. Women and children, recent migrants, and the elderly are often the most victimized by employment concerns.

Box 1.5 Planning for Blue Space

Don Zeigler, Old Dominion University

Thanks to water, "blue space" makes its appearance in almost every city. Many of the world's largest urban areas are located near the "ocean blue," and most others are located near "rivers blue." And, even if a city is not situated on the shore of a sea or river, there is likely to be a spring in town or at least a little "fountain blue." In the Industrial Era, blue space was often sacrificed to economic development. Thanks to the research of Wallace Nichols, recently published in the book *Blue Mind* (Little Brown, 2014), we now know that was a big mistake: Water makes us healthy. We always knew that life was impossible without water, but now we know that good mental health derives from water as well. Experiments with human subjects have shown that urban landscape images bring on feelings of suffocation, and rural landscape images bring on feelings of peacefulness. How do you trick the human mind into feeling peaceful in a city? By using water to its best advantage, by building parks along riverfronts, by encouraging urban water sports, by infusing the language of fountains and gurgling rivulets into a city's soundscape, and by planning for viewsheds that take in water vistas. Neuroscientists tell us that the sights and sounds of water trigger the release of dopamine and oxytocin in the human body. The result is that people are happier.

Figure 1.14 Neuroscientists now tell us that the presence of water sharpens the intellect and enhances feelings of well-being. Selecting a place along the Charles River in Boston might be the best thing a student could do to maximize study time. *Source:* Photo by D. J. Zeigler.

So, don't underestimate the role of water in making cities better places to live (Figure 1.14). Water resources have more than economic benefits; the social, emotional, and cognitive benefits are equally as important in a society that depends on the "creative class" for new ideas. If cities have bodies of natural water in their midst, they should be treasured, opened up to the public, and developed as necessities of urban life. If cities have polluted their near-shore waters, industrialized their flood plains, or buried their streams in culverts, they should be brought back to life as natural resources that are essential to urban mental health. These are big ideas, new ideas, and transformative ideas. They are currently being promoted by the Blue Mind Movement, an extensive network of scientists with specialties in human emotions and cognition. In 2015, their annual conference in Washington, DC, operated under the theme of "Urban Blue."

Managing Racial and Ethnic Issues

Unemployment, underemployment, and other factors breed a variety of subsidiary problems related to race, ethnicity, and class status. For example, increasing ethnic diversity in some parts of Europe and racial diversity in the United States sometimes creates tension among parts of the populations in these countries who perceive that cultural diversity may threaten their culture, way of life, or ability to prosper economically. Such tensions are high in the United Kingdom related to long-term migrant (or guest) workers from other cultures, such as Indians, Pakistanis, and Bangladeshis. The United States is also seeing a resurgence of high emotions about the role of race and the place of diverse peoples in the so-called American melting pot (Figure 1.15).

Additionally, the number of people considered to be refugees worldwide reached the highest number ever in 2015, over 20 million people. According to the United Nations, developing countries host over 85 percent of the world's refugees, a burden that only adds complexity to the developing world's urban places, including refugee camps that act like cities. For more developed places with relative economic prosperity, such as the United States, illegal immigration continues as migrants, primarily from Mexico and other Latin American countries, come seeking a better life. These people, along with legal immigrants and refugees, commonly settle in cities, as do Cuban refugees in Miami, Florida. Compounded with preexisting underlying racial tensions, refugees and new migrants may come into conflict with (1) community elites who find their power diluted, (2) groups with whom they compete for jobs, often other minorities, and (3) majorities who have completely different languages, religions, and worldviews. Throughout the world, many cities must manage severe centrifugal forces generated by cultural diversity.

Managing Privacy

The tentacles of wireless communication penetrate deeply into our private lives, especially as increasing amounts of information about us are stored in "The Cloud," reservoirs of information stored on computer servers in cyberspace. We voluntarily provide so much information to both public and private sectors and assume that it is secure and appropriately

Figure 1.15 The March on Washington for Jobs and Freedom took place under the leadership of Dr. Martin Luther King, Jr., and others in 1963. The 50th anniversary of the march and the "I Have a Dream" speech took place in 2013 to keep the dream alive. *Source:* Photo by D. J. Zeigler.

used. Yet, once "out there," this data is beyond our control. Likewise, there are also involuntary means of collecting information about us. We may be forced to yield biometric data, making it possible, for instance, for cities to track residents. With e-tracking devices such as cell phones, information is created about our whereabouts 24 hours a day, and geographic information systems (GIS) can be used—by private or governmental entities—to map it out and store it permanently. Moreover, surveillance in cities is becoming increasingly common. We have accepted Closed-Circuit TV cameras on our streets or airport security checks because we feel they make us safe. But, they also deprive us of privacy (Figure 1.16). An example of this conundrum is the proposed use of body cameras for on-duty police officers as a way to monitor appropriate police responses when dealing with situations of potential conflict, especially regarding race, ethnicity, or gender.

Managing Modernization and Globalization

One phenomenon sweeping cities, especially larger ones, is the dilemma of Westernization versus modernization. The problem facing the LDCs is how to raise living standards without abandoning traditional cultural values and ways of life. Some might argue that tradition and modernization are incompatible, that modernization automatically entails change, and that change will likely take the form of Westernization. Certainly, there are ample signs of Westernization (some call it homogenization or globalization) of the world's major cities in the forms of skyscrapers, modern

Figure 1.16 Banksy is a well-known graffiti artist whose works materialize on the urban landscape while no one is watching. In London, his unauthorized critique of CCTV appeared overnight on Royal Mail Service property. *Source:* Photo by D. J. Zeigler.

architecture, the automobile society, advertising, mass consumption, and so forth. Nonetheless, as anyone who has lived in cities of the LDCs can attest, traditional cultural values and lifestyles persist even in the most modern metropolis. Almost the entire world—rural and urban—is adjusting to changing global economies. Globalization means the movement of products, money, information, and human talent around the world in ever-larger quantities at ever-lower costs and in ever less time.

Mayors and governing councils must now think globally and locally. Companies produce items for global, not local or regional, markets. Money is transferred electronically from major financial centers in Europe to Asia and from North to South America. Trade barriers are being reduced between countries. Transnational corporations and nongovernmental organizations promote a "world without borders." The net result is an easier flow of people, money, credit, and products across boundaries that once separated people with different ideologies and economies.

There are potential problems because change in the flow of goods and people also means change in the ways humans use their environments. As globalization continues, local to global environments also transform as we demand more from them and create new production, consumption, and waste patterns.

For example, when Starbucks first opened in Seattle, Washington, in 1971, the company purchased coffee beans directly from growers and sold them locally in Seattle. Today, Starbucks is a global company that purchases, roasts, and sells coffee beans, coffee drinks, and many other consumer items worldwide. Largely because of Starbucks' branding campaigns, the globalization of "coffee-house culture" has increased the ecological footprint of coffee production (more beans produced in ecologically marginal areas), consumption (more coffee—and water to make it—consumed worldwide), and waste (coffee grounds, disposable cups). This example shows that while globalization means we can travel anywhere in the world and feel somewhat at home, there are also environmental concerns that are often hidden from consideration in our increasingly global lifestyle.

Managing Traffic

Another obvious effect of urbanization, produced largely by increasing numbers of motor vehicles, is traffic congestion. Superficially, this might be understood as a mere nuisance, an aggravation of less consequence than survival-level problems such as employment, housing, and social services. Nonetheless, traffic congestion is a serious dilemma that chokes off the movement of people and goods in many cities to the point of standstill. Consequences include economic inefficiency (time loss and resource waste), social stress, and pollution, all of which diminish a city's development potential and bespeak concerns about climate change at the largest scales. Health and well-being are also affected by traffic. As numbers of vehicles increase in cities, incidents of asthma, other respiratory concerns, and carbon monoxide levels also increase. Less critical, but important for leading a satisfying urban life, is consideration of noise pollution from automobile traffic. One glance at traffic in the ***primate city*** of Moscow, Russia, indicates how trying it can be to live in places suffering the consequences of car culture.

Managing Urban Governance

Urban governments worldwide face challenges related to the balance between revenues and expenditures. Few governments have the funds to meet every need, so priorities must be established. Those priorities may be set by higher levels of government (sometimes authoritarian) or by the democratic process locally. In any case, the task of governing or administering services to a growing city is daunting, whether the city is New York or Mumbai. In some countries, like the United States, problems arise because urban areas are fragmented among so many jurisdictions, some overlapping. In other countries, many in the developing world, government bureaucracies are bloated with excess employees, are suspected of serving elites, and generally have a hard time combating pressing urban problems.

Increasingly, urban governance includes decision making about the environment in and around urban areas. As human populations in cities increase and as consumption of ecological goods and services—land, water, and local and regional biomes—increases in and around urban areas, it is recognized that environmental governance is closely related to urban governance. Every year, cities increase their attempts to address environmental concerns for and with urban residents, worldwide. While cities in the MDCs are currently further ahead in addressing urban environmental governance, especially in Europe, some cities in the LDCs, notably Curitiba in Brazil, are taking steps that will make them forerunners for addressing environmental concerns and some aspects of sustainability.

CONCEPTS, TERMS, AND DEFINITIONS

Geographers approach the study of cities through the discipline's major subfields: economic, political, cultural, and environmental geography (Figure 1.17). Geography's intra- and interdisciplinary approach to the study of the drivers and outcomes of urban processes and patterns requires an understanding of concepts and terms used in urban geography, a broad subfield of geography devoted to the study of urban people, places, and phenomena. The concepts and definitions in the glossary below have been italicized and bolded throughout this chapter.

Capital City

"Capital" comes from the Latin word for "head," caput. Capital cities are literally "head cities," the headquarters of government functions. Every country has a national capital and a few (e.g., South Africa, Bolivia) have more than one. Capital cities are seats of political power, decision-making centers, and loci of national sovereignty. Their landscapes are

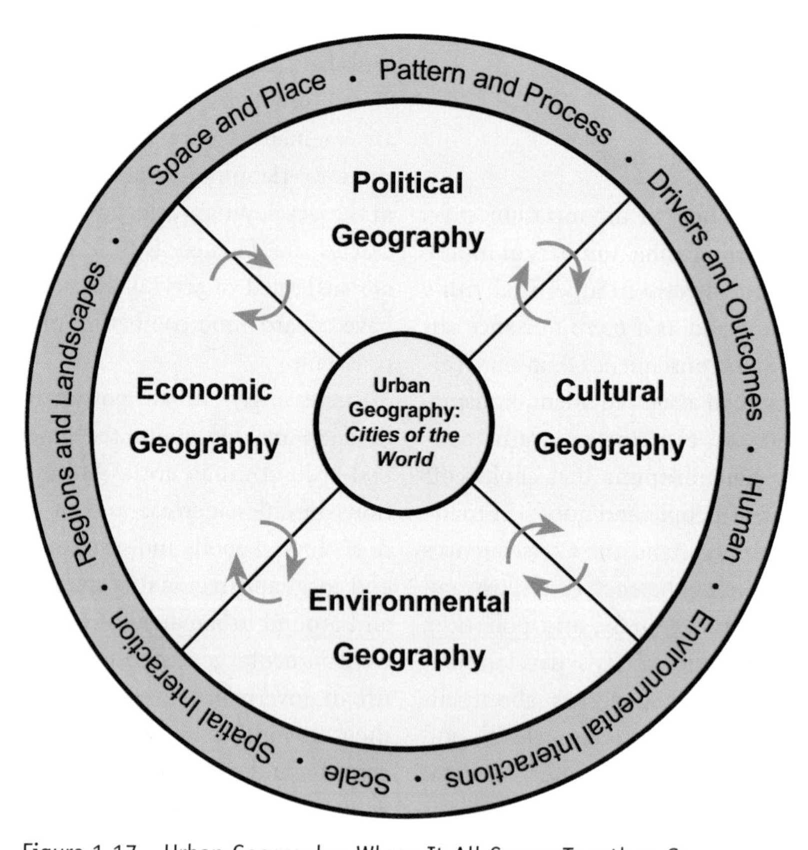

Figure 1.17 Urban Geography: Where It All Comes Together. *Source:* Authors.

charged with the symbols of solidarity, real or imagined; their museums are attics of the nation; and their locations symbolize the central role they play in national urban systems. In some countries, national capitals share power with provincial capitals. As a class of cities, they are among the best known worldwide.

City

The term "city" is essentially a political designation referring to a large, densely populated place that is legally incorporated as a municipality. However, a settlement of any size may call itself a city, whether it is large or small. Towns are generally smaller than cities. Sometimes, geographers refer to "the city" as not just a place, but a concept denoting multiple phenomena associated with urban settings.

Colonial City

Although virtually gone today, the colonial city has profoundly impacted urban patterns throughout much of the world, especially in places that were dominated by European imperial powers, beginning around 1500 CE and culminating in the nineteenth and early twentieth centuries. The colonial city was unique because of its focus on commercial functions, its peculiar situation requirements, and blend of Western urban forms with traditional indigenous cultures. Two different types of colonial cities existed, depending on the ages of the colonial enclave and the native or indigenous settlement. In one type, the European city was created where no significant urban place had existed previously. This led to in-migration of local peoples, drawn by economic opportunities created under colonial rule. Examples include Mumbai, Hong Kong, and Nairobi. In the other type, the European city was grafted onto an existing indigenous urban place, becoming the dominant growth pole for that city and typically overwhelming the original indigenous center in size and importance. Examples include Shanghai, Delhi, and Tunis and Mexico City, built atop Tenochtitlan, the Aztec capital. Either type of colonial city would eventually give rise to a dual city: one part modern and Western and another part traditional and indigenous.

Conurbation

As urban areas expand, they engulf smaller cities in the urban expansion zone, turn nearby towns into full-fledged cities, sometimes stimulate the development of new cities, and bump into other expanding urban areas. Of twentieth-century European origin, this term is commonly used in Europe, for example, to speak of the Randstad conurbation in the Netherlands or the Rhine-Ruhr conurbation in Germany. In the United States, the Dallas-Fort Worth Urban Area is a good example of a conurbation.

Galactic Metropolis

Geographer Peirce Lewis coined the term "galactic metropolis" to describe how economic and spatial structure—especially transportation and communications technologies in the United States—reinforced connections among seemingly disparate spatial elements that created a geometry that favored urban centers. For example, urban places are linked closely with surrounding orbits of suburban communities; small towns are drawn to urban phenomena by the gravitational forces of nearby cities; and rural areas are part of ubiquitous political discourse, television entertainment, and news coverage that ignores the boundaries between types and sizes of places.

Industrial City

An industrial city has an economy based on the production of manufactured goods, sometimes light industrial products (e.g., food, textiles, footwear) and sometimes heavy industrial items (e.g., motor vehicles, appliances, ships, machinery). Factories and foundries anchor these urban landscapes. Although small-scale manufacturing characterized even preindustrial cities, the invention of the steam engine begat ever-larger factories and cities.

Megacity

Megacity is used colloquially to designate the very largest urban places, usually conceptualized as an urban core and its peripheral expansion zone. A city with more than 10 million inhabitants may be called a megacity. In 1950, only New York City exceeded 10 million. Today, 28 cities worldwide fit that category, and one out of every 8 people worldwide lives in a megacity.

Megalopolis

Originating in twentieth-century North America, geographer Jean Gottmann first applied the term "megalopolis" in 1961 to the urbanized northeastern seaboard of the United States from Boston to Washington. Its coinage focused attention on a new scale of urbanization to describe the coalescence of metropolitan areas at the regional scale. That coalescence is channeled along transportation corridors that connect cities. It is evident in the magnitude of vehicle traffic, telephone calls, electronic exchanges, and air transport among cities strung along a megalopolitan corridor. Megalopolis, like metropolis, is derived from the ancient Greek word for city, *polis*.

Metacity

The United Nations describes the metacity, or hypercity, as a spatially sprawling conurbation of more than 20 million people. When Tokyo grew larger than 20 million in the 1960s, it became the first metacity in the world. By 2020, eight additional metacities are anticipated: Mumbai, Delhi, Mexico City, Sao Paulo, New York, Dhaka, Jakarta, and Lagos. Metacities are described as polycentric and with diffuse governance, which allows for continued growth to occur throughout. Often, the central city stagnates due to the shift of the economic base to city edges and suburban places, where space is less regulated; and due to the daily commutes from densely populated outlying villages or suburbs to the multiple nuclei for economic activity located across the metacity. As humans continue to choose urban habitats, the biological and physical worlds in and near urban places also change. Urban ecologists—interdisciplinary scientists who are interested in understanding the forms and functions of urban environments—see metacities as a new form of urbanization with distinct ecological and social attributes that we must research in new ways if we are to understand how urban places may become more sustainable for humans and other species in the long term.

Metropolis and Metropolitan Area

The term "metropolis" originally meant the "mother city" of a country, state, or empire. Today, it is used loosely to refer to any large city. A metropolitan area includes a central city (or cities) plus all surrounding territory—urban, suburban, or rural—that is integrated with the urban core (usually measured by commuting patterns). In the United States, the term Metropolitan Area (MA) is officially used, of which there are three types: Metropolitan Statistical Areas (MSAs), Primary Metropolitan Statistical Areas (PMSAs), and Consolidated Metropolitan Statistical Areas (CMSAs). The U.S. Bureau of the Census has been designating metropolitan areas since 1950. While terminology and criteria have changed since then, the core definition has remained the same, in that metropolitan areas (1) have an urban core of at least 50,000 people, (2) include surrounding urban and rural territory that is socially and economically integrated with the core, and (3) are built from county (or county-equivalent) units. In Canada, their counterparts are officially termed Census Metropolitan Areas (CMAs), which have an urban core of at least 100,000 people.

New Town

The new town, narrowly interpreted, is a twentieth-century phenomenon and refers to a comprehensively planned urban community built from scratch with the intent of becoming as self-contained as possible by encouraging the development of an economic base and a full range of urban services and facilities. New towns emerge for multiple reasons: relieving urban overcrowding; controlling urban sprawl; providing optimum living environments for residents; serving as growth poles for the development of peripheral regions; creating or relocating a national or provincial capital. The new-town movement began in Britain and later diffused to other European countries, the United States, the Soviet Union, and to newly independent countries in the post–World War II era. The idealized form of the new town, in the West at least, tends to follow the British Garden City concept, with its emphasis on manageable population size, pod-like housing tracts, neighborhood service centers, mixed land uses, much green space, pedestrian walkways, and a self-contained employment base (akin to premodern villages). Three types of new towns have been developed: (1) suburban-ring cities such as Reston, Virginia, in the United States; (2) new capitals such as Brasilia, Brazil; and (3) economic growth poles such as Ciudad Guyana, Venezuela.

Preindustrial City

The preindustrial city—sometimes called the traditional city—identifies a city that was founded and grew before the nineteenth and twentieth centuries and thus typically had quite different characteristics from industrial cities. Elements of the traditional city are still part of urban landscapes, particularly in the developing world, even though there are no longer any pure preindustrial cities in existence. Remnants of the traditional city include central markets (many survive in Europe, Asia, and Latin America and are slowly returning in the United States), pedestrian quarters where streets are too narrow for cars, walls and gates now serving as visual reminders of the past, and intimidating architecture (palaces and cathedrals) that preceded industrialization.

Postindustrial City

A relatively new type of city, the postindustrial city, has emerged, particularly in the world's wealthiest countries. Its economy is not tied to manufacturing but instead to high service sector employment. Cities that are mainly the headquarters for corporations or for governmental and intergovernmental organizations are examples, as are those specializing in research and development (R&D), health and medicine, and tourism/recreation. With increasing numbers of people employed in tertiary and quaternary occupations, especially in fields such as finance, health, leisure, R&D, education, and telecommunications and in various levels of government, cities with concentrations of these activities have economic bases that contrast sharply to cities with industrial economic origins.

Primate City

A type of city defined solely by size and function is the primate city. Coined by geographer Mark Jefferson in the 1930s, this term refers to the tendency of some countries to have one exceptionally large city that is economically dominant and culturally expressive of national identity. A true primate city is at least twice as large as the second-largest city, but the gap is often larger. Paris, for instance, is seven times larger than France's second-largest city Lyon and Moscow is four times larger than St. Petersburg in Russia. In general, primacy is more typical of the developing world where primate cities exceed the size of the next largest city by many times and typically are former colonial capitals. In a few instances, countries may be characterized by dual primacy, where two large cities share a dominant role, such as Rio de Janeiro and São Paulo in Brazil. The presence of a primate city in a country usually suggests an imbalance in development: a progressive core, defined by the primate city and its environs, and a lagging periphery on which the primate city may depend for resources and migrant labor. Some understand the relationship between core and periphery as a parasitic one.

Rank-Size Rule

The rank-size rule represents an alternative to primacy. This concept evolved out of empirical research on the relationships among cities of different population sizes in a country. The rule states that the population of a particular city should be equal to the population of the country's largest city divided by its rank. In other words, the fifth-largest city in a country should be one-fifth the size of the largest city. A deviation from this ranking may mean that the urban system is unbalanced and possibly characterized by urban primacy.

Site and Situation

Why are cities located where they are? What are the drivers, patterns, processes, and outcomes of urban growth? Concepts used by urban geographers to answer these and related questions are site and situation. Site refers to the physical characteristics of the place where a city originated and evolved, such as surface landforms, underlying geology, elevation, water features, coastline configuration, and other aspects of physical geography. Montreal's site, for instance, is defined by the Lachine Rapids, historically the upstream limit of oceangoing commerce. Paris' site is defined by an island in the Seine River, known as the Île de la Cité, which gave the city defensive advantages and offered an easy bridging point across the Seine. New York City's site is defined by a deepwater harbor. A city's origin is often wrapped up in site characteristics.

Situation refers to the relative location of a city. It connotes a city's connectedness with other places and surrounding regions. Some cities are centrally located at trade route junctions while others are isolated. A city's growth and decline depend more on situation than on-site characteristics. In fact, a good relative situation can compensate for a poor site. Venice, for instance, triumphed as a center of the Renaissance, not because of its site (so water saturated it is sinking), but because of its relative location at the head of the Adriatic Sea with access to good passes through the Alps. New York City emerged as the United States' most populous city in the early nineteenth century, not because of a superior site, but because of a superior situation. After the opening of the Erie Canal in 1825, New York had access to the resource-rich interior of the United States via the Great Lakes.

Figure 1.18 These heroic statues in front of the opera house in Novosibirsk, Russia, are typical of former socialist cities. Statues, paintings, posters were all designed to inspire the populace to sacrifices lives of personal comfort for the sake of national welfare. *Source:* Photo by D. J. Zeigler.

Socialist and Post-socialist City

Cities that evolved under communist regimes in the former Soviet Union, Eastern Europe, China, North Korea, Southeast Asia, and Cuba have given us the concept of the socialist city. Communism was characterized by massive government involvement in the economy, coupled with the absence of private land ownership and free markets. Communism produced cities that were distinct in form, function, and internal spatial structure (Figure 1.18). Although most communist regimes collapsed in the late twentieth century, central planning and the command economy have left a lasting, visible impression on these urban landscapes.

However, most socialist cities of the world are now experiencing rapid change and a post-socialist city is emerging. Cities evolving under post-socialist regimes are breaking away from the urban plans that were so strictly enforced by communist or socialist governments. Socialism's compact, comprehensively planned cities were structured internally and regionally to be self-sufficient, but this is changing today as individuals and businesses make their own decisions about where to locate residences and businesses in freer market economies. New growth trends are reforming socioeconomic and political processes in addition to the built environment. Although China is still under Communist Party rule, for example, competitive enterprise transforms its urban landscapes. Only North Korea, and to some extent Cuba, continue to maintain cities under the principles of communism.

Suburbia

Suburbia is a product of twentieth-century, automobile-centric development. While there is no single definition, the United States Bureau of the Census provides some key defining characteristics of suburban areas: an intermediate population density, landscapes dominated by trees, grassy yards and single-family owner-occupied houses; separation of residential, commercial, and industrial land uses; political jurisdictions that are independent of the urban area's incorporated central cities. Planned suburban spaces in North America began with the rise of industrialism at the end of the nineteenth century and the societal preference for personalized greenspace around private homes, especially for the growing class of wealthy industrialists. Today, suburban development occurs worldwide as cities continue to expand outwards and as access to automobile ownership rises.

Sustainable City

With increasing knowledge of our human impact on global environments—terrestrial, marine, and atmospheric—there is increasing interest in finding gentler ways of living on our planet that consider its actual carrying capacity. A sustainable city should meet the needs of the present without sacrificing the ability of future generations to meet their own needs. Ambiguity in the phrasing of this idea, however, leads to variable interpretations and actions toward sustainability in urban settings worldwide. In an ideal scenario, a sustainable city attempts to create well-being for multiple species in four domains: ecology, economics, politics, and culture.

No city is yet a sustainable city, but thousands of cities around the world are lessening their environmental impacts by reducing reliance on fossil fuels and thus greenhouse gases emissions, reducing consumption, recycling urban wastes, utilizing water conservation technologies, erecting energy-efficient buildings, reinvigorating mass transportation systems, encouraging walking and cycling, increasing population densities to reduce sprawl, expanding open space, planting trees and flowers, and cultivating reliance on local food webs. Curitiba in Brazil and Seattle and Portland in the U.S. Pacific Northwest were early leaders of the sustainable city movement. For these exemplary cities, one of the key elements of urban sustainability has been the involvement of urban residents in decision-making processes about the future of humans and the environment. Indeed, without inclusion of citizens in knowledge building and governance, our cities may never approach becoming sustainable.

Urbanism

Urbanism is a broad concept that generally refers to all aspects—political, economic, and social—of the urban way of life. Urbanism is not the process of urban growth, but rather the end result of urbanization. It suggests that the urban way of life is dramatically different than the rural way of life: as people leave the country and move to the city, their lifestyles and livelihoods change.

Urbanization

Urbanization has two main phases. The important variables in the first phase are population density and economic functions. A place does not become urban until its workforce is detached from the soil; trade, manufacturing, and service provision dominate the economies of urban places. The important variables in the second phase are social, psychological, and behavioral. As a population becomes increasingly urban, for instance, family sizes become smaller because the value placed upon children changes.

Urban Agglomeration

As an urban place increases in size and population, existing and newly created metropolitan areas are commonly adjoined as a result of urban and suburban sprawl. This results in a spatially extensive clustering of urban areas that border one or more central cities. While the definition of an urban agglomeration varies worldwide, the physical contiguity created by continued urban and suburban expansion around a central urban place, or places, provides a general understanding of this term.

Urban Area

As cities expand, boundaries between urban, suburban, and rural areas become increasingly blurred, especially in industrialized countries where automobile transportation fosters urban sprawl. Thus, the urban(ized) area is defined as the built-up area where buildings, roads, and essentially urban land uses predominate, even beyond the political boundaries of cities and towns. The urban area is commonly considered to be a city and its suburbs.

Urban Place

As a place increases in population, it eventually becomes large enough that its economy is no longer tied to agriculture or other primary activities. At that point, a rural place becomes an urban place. The minimum number of inhabitants for a place to be considered urban varies significantly from country to country. In Denmark and Sweden, only 200 people are required for a place to be classified as urban; in New Zealand the figure is 1,000; in Argentina 2,000; in Ghana 5,000; and in Greece 10,000. In the United States, the Bureau of the Census defines places as urban if they have at least 2,500 people.

Urban Landscapes

Urban landscapes, visible and invisible, are the manifestations of the thoughts, deeds, and actions of human beings. They are charged with clues to the economic, cultural, political, and environmental values of the people who created them (Figure 1.19). At the macroscale, geographers may look at the vertical and horizontal dimensions of the landscape—at city

Figure 1.19 Matsu's followers in Taipei love parades. With their big ears, these maidens remind everyone to listen to the voices of enlightened beings. Matsu is the goddess honored over and above all others on the island of Taiwan. *Source:* Photo by D. J. Zeigler.

skylines and urban sprawl. At the microscale, they may look at architectural styles; signage; activity patterns near busy intersections; urban foodways; or resource use (for instance, water savings). Interpreting, analyzing, and critiquing the landscape is a traditional theme of urban geography.

World City

World cities function as command-and-control centers of the world economy (Box 1.1). They offer advanced, knowledge-based producer services (businesses serving businesses), particularly in the fields of accounting, insurance, advertising, law, technical expertise, and the creative arts. The top tier cities, defined by their financial centrality, are called global cities, of which there are three: New York, London, and Tokyo. One rung lower are second-tier world cities: Paris, Frankfurt, Los Angeles, Chicago, Hong Kong, and Singapore, among others. Beyond that are several dozen more cities—Amsterdam, Moscow, Sydney, Toronto, San Francisco, and others—which draw strength from particular mega-regions or from particular cultural and economic niches. Even cities such as Mecca and Jerusalem may be termed world cities because their influence is felt worldwide within particular religious communities.

SUGGESTED READINGS

Amin, A., and N. Thrift. 2002. *Cities: Reimagining the Urban.* Cambridge, England: Polity, 2002. Challenges the notion that the contemporary city is separate from the country and offers a model of the New Urbanism.

Beatley, T. 2010. *Biophilic Cities: Integrating Nature into Urban Design and Planning.* Washington, DC: Island Press. Argues that urban greening is not only about nature itself but also about humans interacting more with natural landscapes.

Davis, M. 2007. *Planet of Slums.* New York: Verso. A documentation of poverty in cities of the developing world.

Forman, R. T. 2014. *Urban Ecology: Science of Cities.* New York: Cambridge University Press. Presents models, patterns, and examples of human-dominated ecosystems in cities and presents new urban ecology principles.

Hall, P. 1998. *Cities in Civilization.* New York: Pantheon. Looks at the world's great cities during their golden ages, with an emphasis on culture, innovation, and the arts.

Knox, P., ed. 2014. *Atlas of Cities.* Princeton, NJ: Princeton University Press. A taxonomy of cities that looks at different aspects of their physical, economic, social, and political structures.

LeGates, R. T., and F. Stout, eds. 2011. *The City Reader.* New York: Routledge. An anthology of classic and contemporary titles about urban history, design planning, social, and environmental problems.

Parnell, S., and S. Oldfield, eds. 2014. *The Routledge Handbook on Cities of the Global South.* New York: Routledge. A discussion on the meaning of the city in, or of, the global south.

Sassen, S. 2001. *The Global City: New York, London, Tokyo*, 2nd ed. Princeton, NJ: Princeton University Press. Explores the world's three leading centers for international transactions and their impact on the global urban hierarchy.

Soderstrom, M. 2006. *Green City: People, Nature and Urban Places.* Montreal: Véhicule. An examination of 11 cities and their interactions with the natural environment.

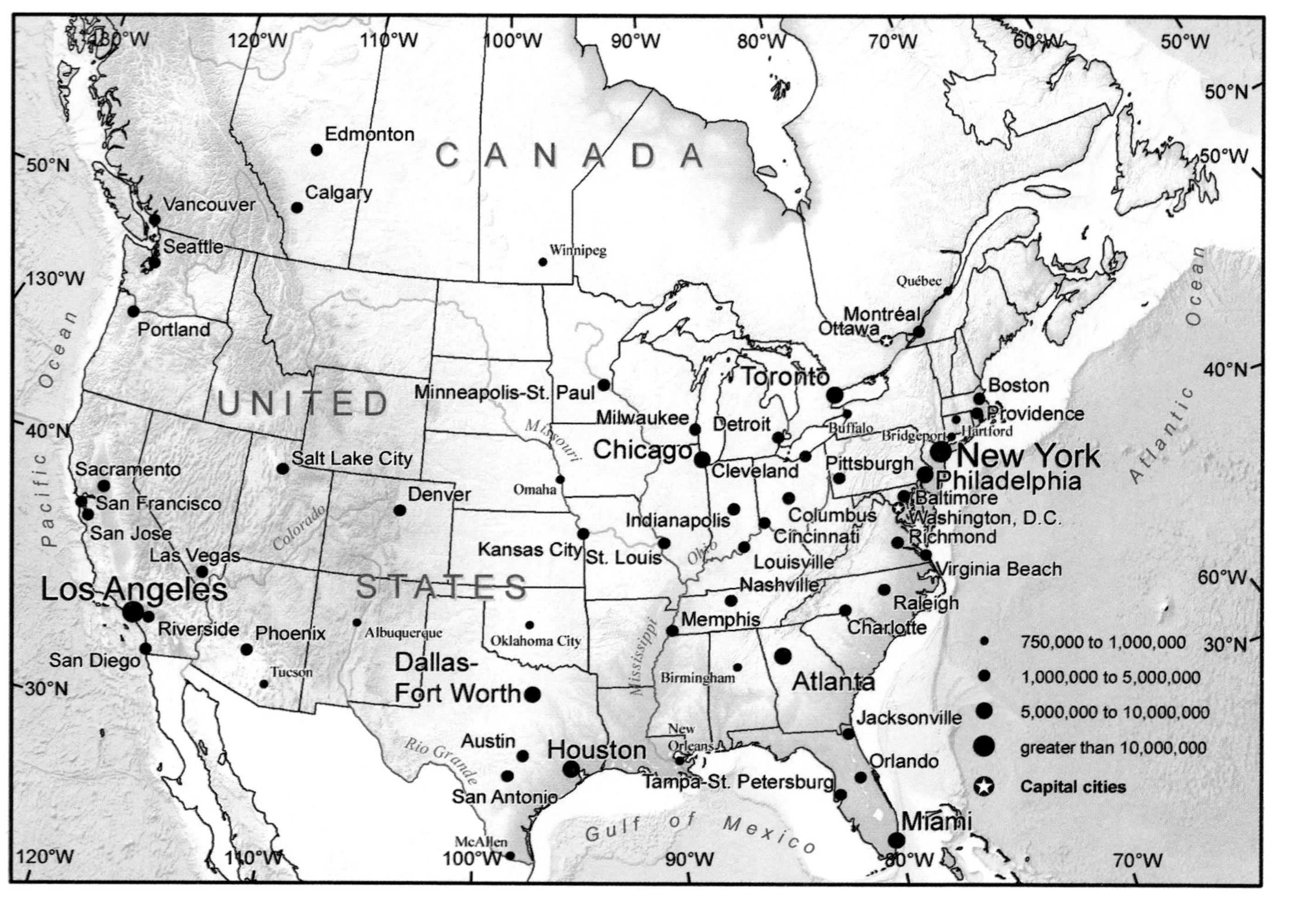

Figure 2.1 Major Urban Agglomerations of the United States and Canada. *Source:* United Nations, Department of Economic and Social Affairs, Population Division (2014), World Urbanization Prospects: 2014 Revision, http://esa.un.org/unpd/wup/.

2

Cities of the United States and Canada

LISA BENTON-SHORT AND NATHANIEL M. LEWIS

KEY URBAN FACTS

Total Population	361 million
Percent Urban Population	82%
Total Urban Population	292 million
Most Urbanized Country	Canada (82%)
Least Urbanized Country	United States (81%)
Number of Megacities	2
Number of Cities of More Than 1 Million	50 cities
Largest Cities (*Metacities*)	*New York* (19 m), *Los Angeles* (12 m), Chicago (9 m), Toronto (6 m)
Number of World Cities (Highest Ranking)	14 (New York, Chicago, Toronto, Los Angeles, San Francisco, Washington, Miami, Boston, Atlanta)
Global City	New York

KEY CHAPTER THEMES

1. After 1950, the United States and Canada became metropolitan societies, as the majority of their populations came to live in metropolitan areas.
2. The United States and Canada comprise one of the most urbanized regions of the world, with an urban hierarchy composed of small, medium, large, and multimillion cities.
3. The most pronounced changes in urban land-use patterns include a declining core and expanding suburbs; however, in some cities, the core has seen a resurgence.
4. Cities in the United States and Canada have been recently shaped by the intensification of economic globalization and greater competition in the global urban hierarchy.
5. Reliance on the automobile and weak investment in public transportation has resulted in cities with low population densities and growing sprawl.

6. A long history of industrialization has left cities dealing with numerous environmental issues, including air, land, and water pollution, all of which threaten to erode the quality of life for many urban residents.
7. Immigration, now from a more diverse set of countries than ever, is transforming numerous cities in North America as immigrants gravitate toward long-established immigrant magnets and newly emerging gateway cities.
8. In a post-9/11 world, heightened concerns about security have been transforming urban space through the installation of security cameras and the fortressing of selected spaces.
9. The United States and Canada are recovering from the 2008 economic recession, but many are still learning to cope with deindustrialization and the rise of the service economy.
10. Cities in drier regions of the continent face water scarcity challenges; cities along the coast face rising sea-level challenges; and cities everywhere face water quality challenges.

Many urban scholars suggest that the world is in the midst of the Third Urban Revolution, a complex phenomenon that began in the middle of the twentieth century. It is defined by a massive increase in urban populations, the development of megacities and giant metropolitan regions, and the global redistribution of economic activities. As former manufacturing cities decline, new industrial cities, service-sector hubs, and tech-poles emerge elsewhere. Cities of the United States and Canada (hereafter, "North America" in this chapter) embody the dynamic trends of this Third Urban Revolution (Figure 2.1). In both countries, a rising percentage of the population resides in cities. In 2015, over 85 percent of the U.S. population lived in urban areas. In Canada, this figure was slightly less, though almost half of its urban population lived in just six cities: Toronto, Montreal, Vancouver, Calgary, Ottawa, and Edmonton (Figure 2.2). Experts predict that by 2050 the proportion of North America's population that is urban could be greater than 90 percent. Without a doubt, urbanism is the norm in the United States and Canada. But urban norms continue to change: Central cities have become the loci of the new urban spectacle; inner cities are peppered with sites of gentrified renaissance as well as rampant poverty and crime; inner suburbs are showing signs of decline; and exurban development continues apace as gated communities and mixed-use developments sprawl into the former countryside. The new lexicon that has emerged to describe many North American cities—"postmodern," "global," "networked," "hybrid," "splintered"—offers some hint as to the rich complexity and deep contradictions of the Third Urban Revolution. Yet, much remains to be said and done before we can make sense of the new forms of urbanism that characterize the twenty-first-century North American city.

Large city-regions are emerging as the new building blocks of national and global economies. The largest one in the United States is the urbanized Northeastern seaboard, a region referred to as Megalopolis by Jean Gottmann in the 1950s. Megalopolis stretches from the southern suburbs of Washington, DC, north through Baltimore, Philadelphia, and New York to Boston and its northern suburbs. It is responsible for 20 percent of the nation's gross domestic product. In 1950, the area that came to be known as Megalopolis had a population of almost 32 million people. Today, the population is over 50 million.

Figure 2.2 Toronto's unique City Hall was built in the 1960s to brand the city. Canada's largest city is also a hub for international travelers, such as these young men from India. *Source:* Photo by D. J. Zeigler.

North American megalopolitan regions, modeled conceptually after Gottmann's Megalopolis, are defined as clustered networks of metropolitan regions that have populations of more than 10 million. In North America, there are 11 megalopolitan regions (Table 2.1). Collectively, they constitute only 20 percent of the nation's land surface but comprise 67 percent of the population. From 2010 to 2040, they will account for approximately three-quarters of all predicted growth in population and construction.

Yet another term used to describe areas transformed by global investment, sophisticated communications, and widespread corporate and individual mobility is the *global city-region*. In an era of globalization, cities increasingly function as the nodes of the global economy; and it is more appropriate to see cities or networks of cities in their regional (rather than local) contexts. As a result of globalization, city-regions such as Los Angeles, San Diego, Seattle, and New York are transitioning from national or regional economic capitals to more integrated cities of the world. Toronto, which has traditionally been the focus of an East-West Canadian economy, has increasingly turned its attention to North-South economic opportunities created by the North American Free Trade Agreement. Distinguishing features of global city-regions are global businesses and an elite work force, contrasting sharply with disadvantaged, insular residential cities.

Numerous transformations that are occurring in contemporary North American cities contrast with long-held myths. One myth is that cities in the United States and Canada lack historic character. Yet many cities, notably Savannah, Charleston, Boston, Montreal, and Quebec City, have active preservation and restoration programs. Another myth is that the

Table 2.1 Megalopolitan Areas of the United States and Canada

Area	*Anchor Cities*
Cascadia	Vancouver, Seattle, Portland, Eugene
NorCal	San Francisco, San Jose, Oakland, Sacramento
Southland	Los Angeles, San Diego, Las Vegas
Valley of the Sun	Phoenix, Tucson
I-35 Corridor	Kansas City, Oklahoma City, Dallas, San Antonio
Gulf Coast	Houston, New Orleans, Mobile
Piedmont	Birmingham, Atlanta, Charlotte, Raleigh
Peninsula	Tampa, Orlando, Fort Lauderdale, Miami
Midwest	Chicago, Madison, Detroit, Indianapolis, Cincinnati
Northeast	Richmond, Washington, Philadelphia, New York, Boston
Golden Horseshoe	Toronto, Hamilton, Buffalo

Source: Adapted from the Metropolitan Institute at Virginia Tech.

ubiquitous skyscrapers, freeways, shopping malls, office parks, and bland suburban boxes described as "distinct" to North American cities (Figure 2.3) are actually found in cities throughout the world.

U.S. and Canadian cities vary tremendously in size, form, and fortune. In the United States, recent urban growth has been robust in western and southwestern cities, while many cities in the East and Midwest such as Detroit and Buffalo have seen economic and demographic decline. In Canada, the fastest growing metropolitan areas are the Prairie cities of Calgary and Edmonton; but within urban regions, there has been a deconcentration of population in central cities and growth in suburbs and exurban zones (i.e., former rural areas).

HISTORICAL OVERVIEW

North America's cities are almost all less than 300 years old. Their origin and evolution are best understood by examining them in the historical context.

Colonial Mercantilism: 1700–1840

Beginning in the late sixteenth century, the Spanish, Dutch, French, and British established colonies in eastern North America in order to gain access to raw materials. Each European power exercised control over commercial trade networks to maintain its own advantage. These regulations resulted in an export-based market where American commodities such as sugar, timber, furs, and tobacco were developed to satisfy Europe's changing patterns of consumption. During the colonial era, North American cities were very small in both population and physical size. They served primarily as export centers for raw materials destined for Europe. The largest cities during this era were found along the Atlantic coast (e.g., Boston and Philadelphia) and along rivers (e.g., Quebec City and Montreal). Along the St. Lawrence River, Montreal controlled the northern route into the center of North America. Near the mouth of the Mississippi River, New Orleans controlled the southern route into the continent.

Figure 2.3 Skyscrapers, such as the Wrigley Building in Chicago, became the cathedrals of urban commerce as steel-frame construction and the elevator enabled the design of ever taller buildings. *Source:* Photo by D. J. Zeigler.

The growth of cities during the colonial era was greatly influenced by the popularity of various North American exports. Quebec City was founded by Samuel de Champlain in 1608. The French settlement served mainly fur traders and missionaries. Settlers established relations with the indigenous Algonquins, who traded beaver pelts in return for metal knives, axes, cloth, and other goods. Since beaver pelts were highly prized and expensive in European markets, trade flourished between the settlers, Algonquins, and French merchants. The physical layout of Quebec is typified by narrow, winding streets and a city wall—complete with watchtowers—built by the conquering British.

The city of New Orleans owes its development to the Mississippi River. French traders and Jesuit priests traveled along the Mississippi in search of pelts, converts, and allies. The city of New Orleans was founded by a French merchant company in 1718. The first settlers in New Orleans encountered the watery geography of a giant delta: half marsh, half mud, a floating, spongy raft of shifting vegetation. The city was located 120 miles (73 km) from where the river flows into the Gulf of Mexico. Sited at a bend in the river close to Lake Pontchartrain, the settlement's advantageous location enabled the portage of goods from the lake to the city. It was easier to ship goods to the lakeshore and then overland to the city than to sail up the ever-shifting Mississippi River. As an outpost of the French empire, New Orleans became part of a global network of colonial possessions that stretched across the Americas, Africa, and Asia. Even so, the city grew slowly. A 1764 map shows that one-third of the planned-city blocks were empty. The city plan was marked by the French tradition of "long lots" fronting on the river, rather than irregular or square parcels of land.

In contrast to New Orleans and Quebec City, Philadelphia was designed using a grid system that incorporated four large market squares. English Quaker William Penn founded the city in 1681 and sited it at the confluence of the Delaware and Schuylkill Rivers. Penn's grid was a reaction to the disorder stemming from the narrow alleyway and curved roads that he had observed in his hometown of London. His symmetrical, orderly plan became the template for Philadelphia's later growth. The city became a busy shipping port for external goods (feed, food, and tobacco destined for England) as well as a market town for frontier products (rifles and Conestoga wagons) used in the westward expansion. Philadelphia also gained status as a major banking center and became home to the first U.S. stock exchange in 1790.

For much of the colonial era, cities were "walking cities." They rarely covered more than a few square miles or had more than 100,000 people. The outward expansion of many North American cities was limited by geographic features such as rivers and hills. Planners often avoided the high ground because it was difficult to pump water or to get horse-drawn fire services uphill. Early forms of public transport, primarily carts and carriages, were also ill-equipped to handle steep slopes. Economic growth, however, would later bring new forms of transportation, and cities would begin to grow outwards and upwards.

Industrial Capitalism: 1840–1970

During the era of industrial capitalism, the U.S. and Canadian economies shifted away from exporting raw materials to manufacturing finished goods. An industrial economy is one dominated by mechanized factory production, which began in the United States in 1793 (Figure 2.4). In 1800, approximately 7 percent of the U.S. population was urban. By 1900, it was more than 40 percent, and by 1920, the country had become a majority-urban nation. Urban growth went hand-in-hand with industrialization. The economic foundations of the industrial city were the coal mine, the iron furnace, and the reliable power of the steam engine. All of this industrialization was accompanied by unprecedented growth in cities, in terms of both population size and geographic area. By 1830, the eastern seaboard cities of New York, Philadelphia, and Baltimore were the main industrial cities in the United States. Industrial expansion also fueled growth in North America's interior. Cities located along rivers and lakes took advantage of advances in transportation, such as canals and railroads, to become important hubs in the distribution of goods (Figure 2.5). By the 1860s, Buffalo, Pittsburgh, St. Louis, Chicago, Toronto, and Cincinnati emerged as key gateways alongside already established Montreal and New Orleans. In Canada, Winnipeg became the hub of rail service for the west, and Edmonton and Calgary emerged as major regional service centers by the 1870s.

Between1885 and 1935, the U.S. economy completed its transformation from an agricultural and mercantile base to an industrial-capitalist one. In the early twentieth century, the

Figure 2.4 Slater's Mill is today an historical landmark in Pawtucket, Rhode Island; it marked the beginning of the factory system in the United States.
Source: Photo by D. J. Zeigler.

Figure 2.5 The Erie Canal, running through downtown Syracuse, New York, was critical in pushing New York City to the top of the U.S. urban hierarchy.

emergence of powerful national corporations and large-scale assembly-line manufacturing prompted robust economic growth. While many of the biggest cities were still located in the Northeast, midwestern cities such as Chicago, Detroit, and Cleveland had grown into vital industrial centers by the 1920s. During this time, Canada also experienced significant growth in its western cities, spurred by the discovery and development of oil and natural gas fields near the urban centers of Calgary and Edmonton.

Industrialization in North American cities involved more than just the proliferation of factories. Many key inventions had changed the look of the cities and transformed their spatial patterns. The use of iron, and then steel, in building construction launched the era of skyscrapers. In the 1880s, the electric street-trolley helped make mass transit possible and laid the foundations for twentieth-century suburbanization by allowing people to live farther away from city centers. Consequently, most industrial cities grew outwards at the edges as well as upwards in the center.

The end of World War II marked a significant turning point for cities in North America. Many U.S. and Canadian corporations merged or expanded into large multinational firms and achieved dominance in the North American market. The United States became the world's largest and richest economy. The late 1940s also began an era of widespread automobile ownership and suburbanization. Reliance on private vehicles and weak investment in public transportation resulted in low population densities and increased sprawl, particularly for cities in the West. Cities such as Los Angeles, San Diego, Houston, Phoenix, Dallas, Denver, and Vancouver grew horizontally as much as they did vertically. From 1950 onward, the United States became simultaneously more urban and more suburban. Urban regions continued to grow overall, but there was also an exodus of people moving out from city cores to newly built suburbs. To lure suburbanites back to the city, municipalities undertook immense urban renewal and infrastructure projects including highways, bridges, and civic centers. By the late 1960s, however, three significant changes materialized that would have profound changes on cities throughout North America: globalization, deindustrialization, and decentralization (Figure 2.6).

Postindustrial Capitalism: 1975–present

By the 1970s, the era of global capitalism had begun and many corporations were moving manufacturing operations out of North America to developing countries where lower labor costs and tax breaks promised higher profit margins and larger market shares. In cities such as Pittsburgh, Syracuse, Buffalo, Akron, Cleveland, and Detroit, companies fired or relocated workers, closed factories, and moved out of the region or country. These industrial cities became "Rustbelt" cities, vibrant manufacturing centers that had degenerated into ghost towns. Even high-growth cities such as Los Angeles and San Francisco struggled to cope with the social and economic consequences of a decline in manufacturing-based employment. This decline marked a critical shift in the North American economy. Michael Moore's 1989 documentary *Roger and Me* chronicled massive job losses and factory closings in Flint, Michigan, which was then home to General Motors. GM laid off 40,000 people in Flint in the 1980s, a figure that represented half of Flint's GM workforce and one of the largest layoffs in American history. For many industrial cities, high unemployment rates continue to impact local economies. In

Figure 2.6 Signs of deindustrialization, such as this abandoned steel mill, marked the landscapes of industrial-era cities such as Pittsburgh during the 1980s. *Source:* Photo by D. J. Zeigler.

response, political and economic leaders have begun creating strategies to bolster employment and investment, especially in sectors such as services and tourism.

At the same time that many cities in the United States saw economic decline, others experienced rapid growth. Cities such as Seattle, Orlando, Miami, Phoenix, and San Diego successfully blended an existing industrial base with an expanding service sector. Newer city-regions, such as Atlanta, Charlotte, Dallas-Fort Worth, and Silicon Valley (an urban techno-pole between San Francisco and San Jose), came into their own at this time. Silicon Valley is home to Apple and Hewlett-Packard, leaders in the high-technology sector that emerged after the 1980s.

The economies and populations of many cities became increasingly decentralized during this era. Decentralization occurs when city centers lose either population or jobs. A decline in the tax base, which limits municipal funding for social services, results in disinvestment in education and infrastructure. Residents and jobs leave the center of the city for the suburbs or other metropolitan and even nonmetropolitan areas. 1970 marked the first time there was an actual decline in central-city populations, and by 1980 the trend was intensifying. Most people were relocating to the suburbs.

The rise of the service sector, also known as the tertiary sector, has also played a critical role in urban development. Finance, insurance, real estate, education, medical services, wholesale and retail trade, and information and communication technologies, among others, have replaced manufacturing as key components of North American urban economies. The service sector is highly diverse. Some jobs are low-level and pay minimum wage or slightly more; they include data entry, cleaning services, and retail (e.g., servers at a restaurant). Other jobs, referred to as quaternary activities, generate far higher wages. These

Box 2.1 Neoliberal-Parasitic Economies in Chicago

Geographers David Wilson and Matthew Anderson have written about the rise of neoliberal-parasitic economies. This has coincided with an increase in Latino immigration to U.S. cities. Chicago's experience is notable as its Latino population rose from essentially nonexistent in 1970 to 754,000 in 2005. Latino's are now 30 percent of Chicago's total population. With Mexican-origin residents alone occupying roughly 30 percent of the jobs in Chicago's formal and informal service sector, this pool of immigrant labor has provided local businesses with an expanding source of dependable low-wage labor over the past ten years. The neighborhood of Little Village has become both the "Mexico of the Midwest" and the epicenter for parasitic economy formation in Chicago.

A "parasitic economy" is a set of businesses that target spatially immobile people and operates by plundering them despite their limited base of resources and wealth. In the domains of retail provision, housing provision, and work, people are exploited in similar ways as the sub-prime mortgage market has targeted other low-income populations (i.e., African-Americans).

Embedded in this landscape, along with day-labor sites, temporary agencies, and convenience stores, is a host of parasitic institutions, such as payday lenders, pawn shops, and check cashers (Figure 2.7). Payday lenders service populations in need of credit through short-term cash loans typically at high interest rates, often as high as 20 percent. With bills often due before payday, these loans are in high demand as few other options are afforded to this spatially confined population. However, due to high interest rates, many are unable to pay and are forced to refinance, deepening their indebtedness. Pawn shops also provide fixed-term

Figure 2.7 Pawn shops are examples of the parasitic economies that mark the poorer sections of many American cities and suburbs.
Source: Photo by Lisa Benton-Short.

loans to customers who use assets (rather than credit checks) as collateral to back the loans. These loans are also often renewed at high interest rates to avoid appropriation of the collateral by pawnshop owners.

Check cashers, perhaps the most parasitic of these institutional forms, serve those paid by check but who lack bank accounts (due to bad credit, indebtedness, etc.). These agencies cash checks for fees that may range from 3 to 10 percent of the check's value. These businesses also write checks for customers (for $10–15 fees) as a means of paying bills. With no bank access, this has become a particularly popular service. The largest of these, in assets and stores, are Western Union, Segue, and Money Express, together numbering 19 stores in Little Village and representing the penetration of global financial capital into this local formation's operation.

Residents who repeatedly use these services (i.e., turning checks into cash or other checks), due to few other options, accumulate additional costs in fees that often amount to thousands of dollars annually. Yet, with employment typically at $8.00/hour, many have no realistic way to pay off their loans, which keeps them perpetually indebted. Arriving in America with dreams of a better life, and returning money to households in Mexico, they often become disillusioned with their new circumstances.

Source: David Wilson and Matthew Anderson. 2014. "Urban Economic Restructuring," in L. Benton-Short (ed.), *Cities of North America: Contemporary Challenges in US and Canadian Cities*. Rowman and Littlefield.

include research and development, brokerage services, banking, medicine, law, advertising, computer engineering and software development. Richard Florida has referred to those employed in these types of service-sector jobs as the "creative class." Cities have expended tremendous efforts to attract the "creative class" to reignite their economies.

MODELS OF URBAN STRUCTURE

There are two general features that characterize most North American cities. The first is the grid system. The second is sprawl—the low-density horizontal spread of development from the central city.

The average city in the United States and Canada has the following land-use distribution: 30 percent residential; 10 percent industrial/manufacturing; 4 percent commercial; 20 percent roads and highways; 15 percent public land, government buildings, and parks; and 20 percent vacant or undeveloped land (Figure 2.8). The patterning of these land-use categories differs according to the age of the city. For example, cities established prior to 1840 tend to have very dense, compact cores. In cities that developed in the early twentieth century, industrial activities might be located outside the core area to take advantage of advances in transportation facilities such as railroads. Still other cities that saw development occur after 1950 have deconcentrated cores with expansive residential zones due to the automobile and the emergence of suburbs. The majority of North American cities, however, have both high-rent and low-rent

Figure 2.8 Roads and highways take up an enormous one-fifth of urban land in the United States, exemplified by this iconic photo of the Los Angeles freeway system. *Source:* Photo by Lisa Benton-Short.

residential areas in the inner core, and a gradually increasing gradient of housing prices as one moves from the inner suburbs to the outer suburbs. Patterns of expansion and land-use sometimes follow the concentric zones model developed by Burgess in the 1920s, the sector model developed by Hoyt in the 1930s, or the multiple nuclei model developed by Harris and Ullman in the 1940s.

The grid plan, which imposes a rectangular street grid on urban space, dates from antiquity. The grid provides a simple, rational format for allocating land by setting streets at right angles to one another. Despite a multitude of geographies and topographies, many cities laid out on a grid share common design features: a lack of sensitivity to the physical environment, the imposition of the grid regardless of topography, a focus on straight lines (geometry over geography), and an underlying sense of control over urban space. The 1734 map of Savannah shows the rigid adoption of the grid (Figure 2.9). Similarly, San Francisco imposed a grid on one of the most varied topographies of any North American city.

The grid plan became nearly universal in the construction of new towns and cities in the United States and Canada. It allowed for the rapid subdivision of large parcels of land. As U.S. cities grew outward, however, particularly after the mid-twentieth century, the grid became less prevalent. Suburbs offer a free-form pattern of growth, complete with cul-de-sacs and winding lanes that contrast sharply with the grid-plan city. When cities and suburbs merge, the grid merges into a series of loops, curves, and open spaces; and its rigidity begins to disappear.

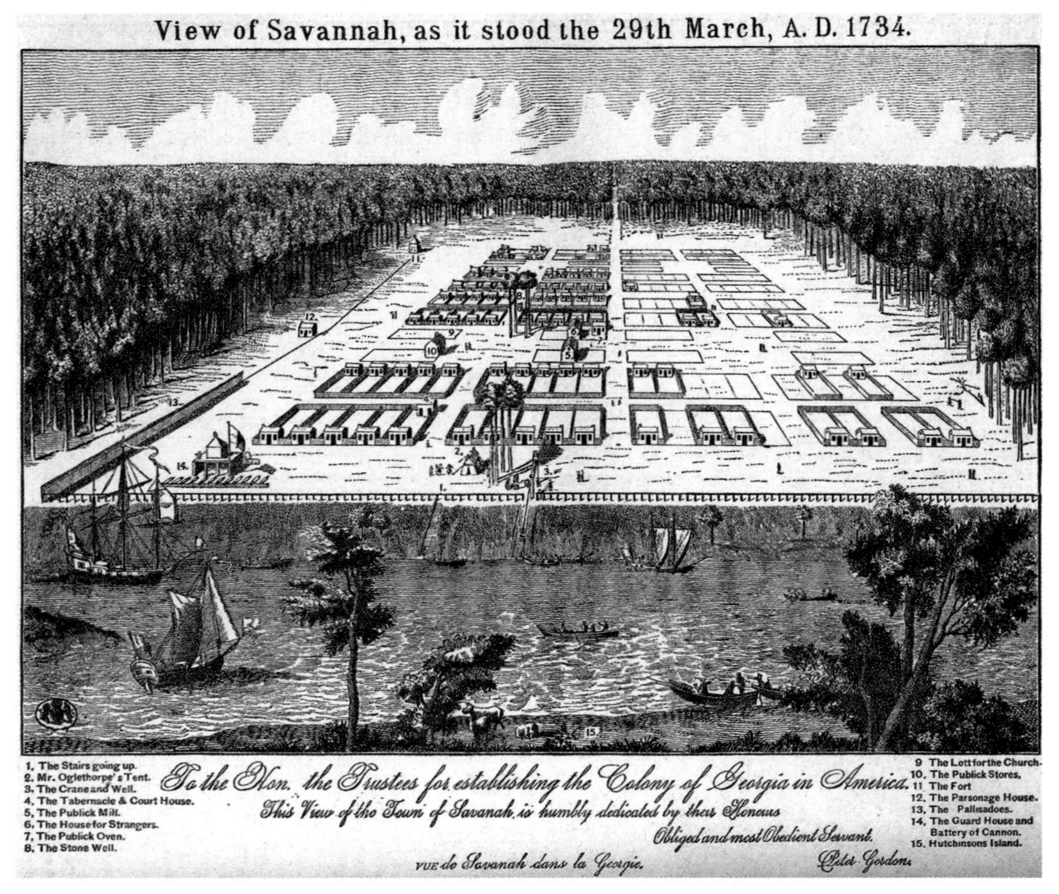

Figure 2.9 "View of Savannah, as it stood the 29th March, A.D. 1734." *Source:* Report on the Social Statistics of Cities, Compiled by George E. Waring, Jr., U.S. Census Office, Part II, 1886. Courtesy of the University of Texas Libraries, University of Texas at Austin (http://www.lib.utexas.edu/maps/historical/savannah_1734.jpg).

In contrast, under the influence of New Urbanism, some recent suburbs have returned to the grid system. New Urbanism longs nostalgically for lost community and the high-density cities of the past. New Urbanist design principles emphasize walkability, mixed-use "town centers" and higher-density residential areas. Ironically, New Urbanism, which is a response to suburban sprawl, has had less impact on redesigning cities than it has on the redesign of suburbs. The most cited example of New Urbanism is the town of Celebration, Florida, a community initially planned and built by the Disney Corporation. Celebration's design elements include low-rise, high-density residential areas where garages are at the back of the residences; walkways and porches allow for pedestrian movement; and a vibrant, car-free, mixed-use downtown facilitates more human interaction in public spaces.

The *edge city* offers yet another model of urban structure. Edge cities consist of predominately large-square-footage office space located beyond the central city. The journalist Joel Garreau coined the term "edge cities" and defined them as having:

- more than 5 million square feet of office space, enough to house up to 50,000 office workers (as many as some traditional downtowns).
- more than 600,000 square feet of retail space, the size of a medium shopping mall.
- more jobs than bedrooms.
- no resemblance to cities developed before 1960.

These edge cities attract large numbers of service-sector workers during the day, but empty each night as residential areas are scarce. Garreau identified 123 places as being true edge cities, including two dozen such areas in greater Los Angeles, 23 in metro Washington, DC, and 21 in greater New York City. Tyson's Corner, Virginia, west of Washington, is an example of an edge city.

Today, edge cities are being rethought. Since edge cities were built in and around major highway intersections, traffic congestion in these areas has become a problem. Pedestrian access is poor and public transportation is nearly absent. Additionally, as development continues in and around edge cities, they may "merge" into megalopolitan areas that comprise multiple residential and commercial/business nodes well outside the urban core. The recent demand for urban living has had an impact on edge cities. Many "commercial strips" and edge cities that were once obsolete are now increasing in density. One of the most interesting transformations is occurring in Tyson's Corner, which planners propose to transform into a walkable, sustainable urban center. By 2050, Tyson's will be home to 100,000 residents and 200,000 jobs. Plans call for new attention to urban design and the pedestrian realm; growth will be focused at Metro stops; and new parks and public facilities will be constructed. The superblocks of streets that are six lanes wide will be broken up as the city transitions from an auto-oriented suburban place into a more pedestrian-oriented urban destination. It is possible the edge cities of the twentieth century will be transformed into new spatial forms during the next decades (Box 2.2).

DISTINCTIVE CITIES

New York City: A Global Metropolis

New York City is the largest city in the United States and the one most frequently identified as the financial and cultural capital of the country. With 8.4 million people in its five boroughs, and over 19 million in the metropolitan area, New York is the only metropolis in North America to rival the megacities of Asia and Latin America in terms of population size. As a true "global city," New York is often the starting point for global economic and cultural trends. Recent events, however, have shown that New York—like any other city—is far from invincible. The World Trade Center attacks of 2001, the blackout of 2003, and the real estate crisis of 2008 and 2009 have shown that New York City's global and regional connectedness can be a source of vulnerability as well as status.

Despite these setbacks, New York continues to lead the world in commerce and industry. The Port of New York and New Jersey, traditionally the main connection between the Atlantic Ocean and the interior of North America (via the Erie Canal), now handles the third-highest tonnage of ship cargo in the United States and is counted among the world's top-20 busiest ports. New York—along with Los Angeles—continues to dominate the nation's manufacturing base with over 20,000 establishments, but it has also gained

Box 2.2 The Death of the Shopping Mall?

Figure 2.10 As architecture critic Michael Sorkin has observed, "Like the suburban house that rejects the sociability of front porches and sidewalks for private back yards, malls look inward, turning their backs on the public street." *Source:* Photo by Lisa Benton-Short.

A trend has been underway for some time—the fall and overhaul of the American shopping mall. From 1970 to 2000, enclosed shopping malls were constructed all over North America (Figure 2.10). Long trips to downtown department stores had fallen out of favor and the suburbs were the destination for new, clean, and air-conditioned shopping. During these three decades, about 1,500 malls were built in America's suburbs.

But, since 2000, the mall has been on the decline. With shoppers and investment returning to the downtown, suburban malls are dying. In the next decade, as many as half of the malls in America could be torn down or reconfigured. In many suburbs, in fact, dead malls are already highly visible. Signs fade, facades crumble, and weeds sprout in parking lots. There is even a website that keeps track of such sorry decay: Deadmalls.com defines a "dead mall" as one with a high vacancy rate, low consumer traffic, and that may appear dated or deteriorating. But, does all of this mean the death of the shopping mall?

Developers in the United States have begun building a new type of shopping center known as the "lifestyle center." Lifestyle centers were developed, in part, to counter the problems associated with the spatial segregation of suburban sprawl. Ironically, developers of lifestyle centers looked to traditional downtowns as an inspiration. Buildings are often made to look like multiple storefronts that have evolved over time. Shops open directly to the sidewalk. Lifestyle centers feature mixed uses (retail, office, and residential space), pedestrian walkability, and are designed to open toward the street, thereby creating outside public spaces that include wide sidewalks, plazas, parks and public squares. Some might include libraries, movie theatres, or civic amenities. Unlike traditional shopping malls with anchor department stores like Nordstrom or Macy's, lifestyle centers focus more on specialty retailers such as Ann Taylor, Williams Sonoma, Talbots, Banana Republic, Nicole Miller, Eddie Bauer, Pottery Barn, Liz Claiborne, The Gap, and Restoration Hardware

Lifestyle centers have become common in both affluent suburbs and revitalized downtowns. They are now among the most popular retail formats in America. In 2002, there were only 30 lifestyle centers in the United States; by 2004 there were 120; today about 19 lifestyle centers are constructed each year. Lifestyle centers continue to proliferate, even as conventional shopping malls are declining.

traction and thrived in the increasingly valuable "creative industries" (Figure 2.11). These sectors are concentrated in different portions of the city, with music and theater in Times Square, fashion in the Garment District, interior design and architecture in Chelsea and SoHo, and advertising on Madison Avenue. These industries not only benefit from city tax incentives for television and film, but also from the talent available at nearby educational institutions, such as Julliard (music), the Pratt Institute (design), and the Fashion Institute of Technology.

New York sits in a strategically important location where mainland New York State meets the Atlantic Ocean, Long Island, southwestern Connecticut, and northeastern New Jersey. This region is commonly called the "tri-state area." The hard metamorphic rock making up Manhattan Island has allowed for intense vertical development of skyscrapers. Central Park, a product of the urban parks movement of the Progressive Era, is the only extensive open space on the island of Manhattan. First inhabited by the Algonquin Indians, the island was purchased and settled by the Dutch in 1624 and named New Amsterdam. By 1800, New York City, with its 60,000 residents, had become the largest city in the country. It has maintained this position since then, now forming the center of the Boston-to-Washington Megalopolis. Along with Tokyo and London, New York is one of the three traditional "global cities" that anchor world trade, commerce, banking, and stock transactions.

Even as New York becomes ever more globally interconnected, the city is in many ways socially, economically, and spatially fragmented. Despite its world-city status, New York has always struggled with its reputation for poor sanitation and high crime. Many geographers have argued that these problems, rather than being truly corrected, are temporarily suppressed, confined to particular areas of the city, or relocated to the outer boroughs and suburbs. New York has frequently located environmentally hazardous projects, such as expressways and incinerators, in poor and racialized neighborhoods (e.g., South Bronx, Sunset Park) where public purview is reduced

Figure 2.11 Peter Woytuk sculptures, playing off of New York's nickname, the Big Apple, became a public art exhibit that extended all along Broadway, this one on the Upper West Side. *Source:* Photo by D. J. Zeigler.

and resistance is less likely. The city has also tried to reconfigure itself as a safe, livable city and tourist destination. In the 1990s, New York implemented "quality of life" laws that criminalized panhandling and homelessness while policing and securitizing (e.g., through closed-circuit cameras) highly trafficked and visible public spaces. Even places that previously served as important sites of public protest, such as the steps of City Hall, are now frequently fenced off or guarded.

Gentrification has undoubtedly driven the fragmentation of New York City. Although gentrification is commonly understood as the product of gradual, localized, ground-up improvements to neighborhood properties, much of the gentrification in New York since 2000 has been driven by corporate developers and municipal interests. The New York Urban Development Corporation, for example, has forcibly purchased many Times Square properties for redevelopment. Privately managed business improvement districts (BIDs) are now responsible for many of the public space improvements (e.g., beautification, signage). As many economic geographers have observed, the competitive advantage of the creative industries located in New York is contingent on the successful "branding" of the neighborhoods in which they are located. By the same token, previously decaying neighborhoods such as the Meatpacking District (Manhattan) and Williamsburg (Brooklyn) strive to attract creative firms and workers by cultivating their status as the city's next arts destination or nightlife center. Although this intra-urban neighborhood competition has expanded the range of living space available for the upper class and the upwardly mobile, it has also priced out middle-class residents of neighborhoods such as Park Slope (Brooklyn), Harlem, and the Lower East side of Manhattan. Critics of gentrification have claimed that rapid redevelopment, securitization, and branding of urban space creates "Disneyfied" fantasy cities and "entertainment machines" that put tourists and business tenants ahead of residents.

Immigration has been central to both the identity of New York City and the individual identities of its neighborhoods. Over 12 million immigrants from Ireland, Germany, Italy, Poland, Greece, and elsewhere arrived in New York City during the late 1800s and early 1900s. During the past three decades, however, newcomers from Latin America and Asia have comprised the bulk of the immigrant population. Despite being one of most ethnically diverse areas in the United States, New York City is still highly segregated. Real estate agents, who serve as gatekeepers to the city's properties, often sort new immigrants toward neighborhoods dominated by their respective ethno-racial groups. Such practices, however, reinforce self-segregating tendencies, ethnoracial ghettos (e.g., Puerto Ricans in the Bronx, Dominicans in Washington Heights), and mutual antipathies between groups. New York thus remains a city of extremes. Within a relatively small area, extreme wealth meets with extreme poverty, global integration encounters local fragmentation, and individualistic economic gain sits side by side with the increasing management and regulation of public space. These dichotomies, however, are likely to guarantee New York's place as a fascinating site of geographic study for years to come.

Los Angeles: Outward Glitz, Inner Turmoil

Los Angeles contrasts sharply with New York City. The large, sprawling city covers 498 square miles (1290 sq km), has multiple business districts (e.g., Hollywood, Beverly Hills), and depends on complex, often congested, networks of highways to connect the various "nuclei" of the city. The sprawl of the city, typified by concrete structures, superhighways, and low-lying residential and commercial developments (e.g., strip malls) extends north through the San Fernando Valley ("The Valley") and eastward into the "Inland Empire" of San Bernardino and Riverside Counties. With a metropolitan-area population of about 13 million, the city of Los Angeles itself is the nation's second largest, a position that it took from Chicago during the 1980s. Yet, like New York, Los Angeles is also a city of extreme contrasts. Both the immense wealth of Beverly Hills and the endemic poverty and disorder of South Central Los Angeles have been fixtures in the U.S. media and the American cultural imagination. A young, politically liberal population in the arts and entertainment sector stands in stark contrast to the Republican families (and the conspicuous consumption) featured in the *Real Housewives* of Orange County and of Beverly Hills. And in a city perhaps less beholden to dominant notions of social class and pedigree than New York City, Boston, or Chicago, gated communities—intended to protect and contain wealth—abound in many areas.

Since the mid-twentieth century, Los Angeles has been perhaps best known for its role as a premier "entertainment machine." Coined by Richard Lloyd and Terry Nichols Clark, the entertainment machine comprises the various industries that cities use to respond to postindustrial elites that want to experience their own cities as if they were tourists. These practices impact considerations about the proper nature of amenities to provide in contemporary cities. While Lloyd and Clark used

Chicago as their case study, Los Angeles differs in that it is an entertainment machine not only for its own residents, but for a global film and television industry. Following the arrival of the Southern Pacific and Santa Fe railroad lines in 1876 and 1885, and the construction of an artificial port in 1914, Los Angeles became a West Coast metropolis that would truly rival New York and Philadelphia. The port, now referred to as Los Angeles-Long Beach, provided an export hub for the West's growing oil industry as well as an entrepôt for cruise ships and imported goods from Asia. Today, Los Angeles is known for being a national leader in clothing and luxury goods manufacturing, but especially film and television. Three major film studios (Paramount, 20th Century Fox, and Universal) are located within the city limits, while NBC television studios are located in Burbank, and CBS in Culver City. Many prominent cable television networks (e.g., Bravo, MTV, and VH1) are also located in Los Angeles, and the reality shows that now dominate North American cable television often depict mansion-and-driver lifestyles associated with the city's wealthiest jurisdictions. The demographics of Hollywood in particular reflect the legions of starving actors aiming to land a job with the film studios and television networks. The median household income hovers around $35,000, low for Los Angeles, and the average household has about two dwellers. Renters occupy well over 90 percent of the housing units. The percentages of never-married men (55 percent) and never-married women (40 percent) are among the country's highest.

Although Los Angeles has a shorter history of international immigration than New York—most of the city's migrants were from within the United States before the 1960s—the city is one of the most diverse in the United States. More than a third of the city's population is foreign born, with the most recent waves of immigrants coming from Latin America—particularly Mexico—and Asia. The newness of immigration also poses challenges for Los Angeles. Most immigrants are not English speaking, and both Los Angeles and California have debated whether to continue integrating foreign languages (especially Spanish) into school curricula and signage, or to make English the sole official language. New immigrants in the Los Angeles area also tend to be poor and experience discrimination in housing, employment, and education. While these immigrants help support the quintessentially "southern California lifestyle" by working in service industries ranging from gardening to dry cleaning, they often live in spatially and economically marginalized neighborhoods such as East Los Angeles and Compton. The economic disadvantage of the city's most vulnerable groups has created levels of organized and gang-related crime that are among the highest in the country. According to the Los Angeles Police Department, the city is home to 45,000 gang members, organized into about 450 gangs. Among them are the Crips and Bloods (African American), the Sureños, (Mexican), and Mara Salvatrucha (mainly of Salvadoran descent). The preponderance of organized crime groups has led to the city being referred to as the "Gang Capital of America."

Los Angeles is also marked by a number of environmental problems, such as traffic congestion and pollution. In this sprawling, automobile-dependent city, most residents commute and more than 7 out of 10 workers drive to work alone. All of the streetcar lines in Los Angeles were closed in 1963 in favor of freeway development. Although the city reinstituted a commuter rail system in 1990, its five lines—designed mostly to connect downtown Los Angeles and the surrounding suburbs—are insufficient to provide an adequate alternative to car transport. The traffic congestion, coupled with an unfavorable basin location and dry climate, has made Los Angeles the smoggiest city in the country. Although smog alerts have decreased since the 1970s—when there were almost 100 per year—the National Lung Association has consistently ranked Los Angeles as the first or second most polluted city in the United States. Los Angeles also has insufficient water resources to support its population. Transfers of water from northern California and the Colorado River have reached their limits, and when combined with the frequent droughts, the only alternatives seem to be conservation and a turn toward the sea. In 1990, the first desalinization plant opened along the California coast. Finally, the ever-present threat of earthquakes is announced by several low-grade tremors each year. The potential for devastation is real, and emergency planning is a priority item for schools, businesses, and police.

Detroit and Cleveland: Shrinking Cities

Overall, the U.S. national economy has seen growth and prosperity; however, some cities confront a declining or stagnant economy and a shrinking population. In 1950, the population of municipal Cleveland, Ohio, was 900,000, but by 2014 it had fallen to 390,000. Detroit, Michigan, was once home to both GM and Ford and was dubbed "Motor City" or "Motown." It was home to 1.8 million residents in 1950. Today, its population is well less than half of that number. These two former urban industrial giants have experienced a severe reversal of fortune.

In Detroit, as high-paying manufacturing jobs became scarce and both corporations and property owners began to leave, many residents lost their homes due to foreclosure and eviction. Unemployment, the crack cocaine epidemic of the 1980s and 1990s, and resulting drug-related violence and property crimes gave Detroit unwelcome notoriety as one of the most crime-ridden cities in North America. Detroit's woes have continued. In 2010, it was estimated that about one-third of the city, some 40 square miles (104 sq km), was vacant. The current joke is that the only expanding business in Detroit is demolition.

Redevelopment has become a buzzword in Detroit since, but redevelopment strategies have garnered mixed results. In the mid-1990s, three casinos opened in Detroit's downtown. In 2000, Comerica Park replaced historic Tiger Stadium as the home of the Detroit Tigers, and in 2002 the NFL Detroit Lions returned to a new downtown stadium, Ford Field. The 2004 opening of "The Compuware" gave downtown Detroit its first significant new office building in a decade. The city hosted the 2005 Major League Baseball All-Star Game and Super Bowl XL in 2006, both of which prompted more improvements to the downtown area. Currently, Detroit is constructing a riverfront promenade park as part of the Detroit International Riverfront Project and is working with Canada on a second bridge that will connect Detroit and Windsor, Ontario.

Even so, the new infrastructure has not necessarily improved economic growth. Detroit remains one of the nation's poorest cities. Currently, almost 40 percent of residents live below the poverty level, and the population (83 percent African American, 9 percent white, and 7 percent Hispanic) remains highly segregated. Abandoned housing ranks as one of the city's most persistent problems. In 2010, a total of 78,000 housing units were vacant or abandoned and 55,000 of those were in foreclosure. Detroit, already suffering from deindustrialization in the twentieth century, was also one of the U.S. cities hit hardest by the 2008 foreclosure crisis.

Cleveland has also struggled against the legacy of deindustrialization in order to reinvent itself in a more competitive global economy. Initiatives to rebuild Cleveland have replicated the formulae that many Northeastern and Midwestern cities have employed: new museums, sports stadiums, convention centers, renovated industrial warehouse districts for housing and retail, and waterfront development. Pundits dubbed these efforts "the Cleveland Comeback." One of the city's most successful projects has been the Rock and Roll Hall of Fame and Museum, which opened to the public in 1995. The building, located on the shore of Lake Erie, was crucial in regenerating Cleveland's waterfront area. New downtown stadiums for the city's professional sports teams have also aided revitalization. The Gateway Sports Complex cost $360 million and included an open-air stadium for baseball and an indoor arena for basketball. Currently, the city is redeveloping the waterfront along both Lake Erie and the Cuyahoga River as a destination for tourists and locals alike. The city has also become a regional and national player in health services and biomedical technologies by capitalizing on the wealth of educational and medical facilities in the region. Both the Cleveland Clinic and University Hospitals have announced billions of dollars of investment in new facilities. Despite these efforts, some experts claim that the Cleveland comeback has stalled. Between 2000 and 2007, Cleveland suffered one of the largest proportional population losses in the country, shrinking by 8 percent. In addition, many of Cleveland's inner suburbs continue to decline and overall urban growth remains negligible. The case studies of Detroit and Cleveland show that both cities continue to experience mixed results in efforts to realign and reinvigorate their economies.

Montreal: Moving Uphill from Upheaval

While most North American cities' economic and demographic destinies have been shaped by global economic trends, Montreal's have been shaped by cultural identity and the resultant political clashes. In French-speaking Quebec, beginning in the 1960s, many residents abandoned the Roman Catholic Church and traditional family structures and now frequently opt for common-law relationships instead. Consequently, the provincial fertility rate—the number of children per woman of childbearing age—has stayed below 1.5. Yet Montreal was perhaps less affected by the modernization imperatives of its 1960s' government than it was by the Quebec separatist movement that emerged at the same time. For most of Canada's history, Montreal had served as the unofficial corporate and cultural capital of the country, hosting the World Expo in 1967 and the summer Olympics in 1976. But even as these events were held, Montreal was in a state of unrest. Between 1963 and 1970, the paramilitary Quebec Liberation Front became responsible for over 160 violent incidents, which killed eight people and injured many more, including the bombing of the Montreal Stock Exchange in 1969 and the October Crisis of 1970 where a British Trade Commissioner was kidnapped and an anti-separatist Quebec labor minister was murdered. This extremism, however, belied more mainstream efforts to ensure that the province of Quebec, and especially Montreal, remained French. In 1976, Law 101 made French the official language of government, commerce, and educational instruction in Montreal, and took extra efforts to ensure that newcomers and their children would have to learn French. Unfortunately, many corporations and their workers, who were used to operating in an English business milieu, decided to leave. Major firm headquarters, such as those for TD Canada Trust and Canadian Pacific Rail, left the city. Between 1990 and 2011, Montreal lost 21 of its Canadian top-500 companies (a loss of 25 percent), with most moving to Toronto.

Referenda for independence in 1980 and in 1995 cemented Montreal's fate as a corporate second-fiddle to Toronto. While both referenda failed (albeit only barely), they created an atmosphere of hostility between many French Quebeckers and the remaining Anglo minority as well as immigrants—who had been blamed in French media for the failures of the dual referenda. Despite the post-1980 provincial immigration policies designed to prioritize French language ability and connection to the province rather than solely education and skills (as in the rest of Canada), immigration to Montreal has grown during the past two decades—even if not to the same extent as in Toronto. In his book *Creating Diversity Capital*, Blair Ruble argues that the growth of immigration in ethnically and racially bifurcated cities such as Montreal (French-Canadian/Anglo-Canadian) and Washington, DC (black/white) has allayed some of the tensions between opposing groups and allowed for forms of cultural hybridity; for example, the crowning of a bilingual, West Indian queen in Montreal's St. Patrick's Day parade.

Yet immigrants have not necessarily been the key player in Montreal's slow rebound from the decline of the late twentieth century. The new global order that prioritizes cultural cachet over corporate rankings and specialized merchants over mass manufacturing means that Montreal's unique history and culture—coupled with cheap rents from housing subsidies and the after-effects of decline—have made it the home base for creative and design-based industries in Canada. Given its eclectic architecture and broad availability of film services and crew members (it is home to the National Film Board of Canada), Montreal is now a popular filming location for U.S. studios. The city is also home to festivals such as Just for Laughs and the Montreal Jazz Festival, as well as global cultural enterprises like Cirque du Soleil. In recognition of these arts-based industries and a budding fashion district in the Plateau neighborhood, Montreal was named a UNESCO City of Design in 2006. The growth of the arts, however, is also having spillovers into hi-tech industries. With the help of government subsidies supporting video game designers, the city has attracted world-leading game developers and publisher studios such as Ubisoft, Eidos Interactive, Artificial Mind and Movement, Bioware, and Strategy First. The case of Montreal shows that cultural differentiation has evolved from a pitfall to a strength in the new global economy.

Ottawa: A Capital of Compromise

More a tool of political compromise than an economic boomtown, Canada's federal capital, Ottawa, has long acted as a nexus of compromise between the many facets of Canadian identity: French and English, urban and rural, Upper Canada (Ontario) and Lower Canada (Quebec and the Atlantic provinces), and the interests of federal bureaucrats and a burgeoning technology sector. Ottawa and its Quebec neighbor, Gatineau, comprise the fourth-largest census metropolitan area (CMA) in Canada with1.24 million residents.

Ottawa was designated the capital of Canada by Britain's Queen Victoria in 1857, even before confederation and independence in 1867. Already an important lumber town at the northern terminus of the Rideau Canal linking the Ottawa River and Lake Ontario, Ottawa was chosen for its defendable strategic position (i.e., away from the U.S. border) and its location on the Ontario-Quebec border, midway between Toronto and Quebec City. The location of compromise is evident in the linguistic diversity of Ottawa-Gatineau today. About 40 percent of the population declares itself French-English bilingual, while another 45 percent claim to speak English only and 15 percent claim to speak French only. But, living in an urban area straddling two provinces also creates challenges. Many residents live in one province and work in another, and frequently negotiate the traffic laws, rental practices, and tax structures of two different jurisdictions. In addition to legislative and linguistic differences in the area, there are political differences between downtown Ottawa residents and the rural and suburban residents that joined the city following amalgamation in 2001. Ottawa politicians frequently avoid projects that may irritate more politically conservative residents in West Carleton, Osgoode, Greely, and other towns that pay Ottawa taxes but do not always receive Ottawa municipal services.

An additional challenge is the dominance of the federal government in the development of the area. The Ottawa-Gatineau metropolitan area is also part of the National Capital Region, a federal jurisdiction managed by a corporation called the National Capital Commission (NCC). Some see the NCC, which manages the federal government's vast properties in the area, as a hindrance to developing Ottawa into a world-class city. Buildings cannot exceed the elevation of the National Parliament and most development has been geared toward creating a "green capital" dotted with parks and fringed with a belt of woodlands and farmlands. Ottawa, however, has refashioned its historic and national emblems as focal points of development. In 2006, Confederation Boulevard was built to help pedestrians access attractions on both sides of the Ottawa River, including the Parliament, the Rideau Canal locks, and the Museum of Civilization in Gatineau. The historic Byward market in the center of downtown Ottawa—originally a supply stop for the lumber camps around Ottawa—is now the center of a revitalized downtown nightlife and shopping district.

In Ottawa, the cooperation of government and industry has led to the city-region becoming a tech-pole now known as "Silicon Valley North." The Ottawa-Gatineau hi-tech sector, which is focusing on information technology, telecommunications, and nanotechnology, has employed anywhere between 50,000 and 85,000 workers per year during the past decade and includes well-known firms such as Corel, Nortel, and Adobe. Although the "government town" is usually seen as the antithesis of an open-market environment fostering technology and innovation, the Canadian federal government has actually been central in the development of the region's tech-pole. The government is not only the largest user of information technology in the area, it has also funded hi-tech research and innovation in the laboratories of the National Research Council. Other public-private bodies, such as the Ottawa-Carleton Research Institute and the Ottawa Capital Network, provide the funding and programming to ensure that local innovations can be leveraged into the creation of local startup firms. By capitalizing on these networks and Ottawa's highly educated talent pool (30 percent have university degrees),

Ottawa has developed its own version of Silicon Valley—one committed to local research, development, and reinvestment.

Washington, DC: A New Immigrant Gateway

No longer are the world's migrants moving to the older, established destinations of the nineteenth and early twentieth centuries. In the twenty-first century, many choose to settle in cities where new economic growth is providing opportunities for both high-skilled and low-skilled workers. Washington (coextensive with the federal District of Columbia) and its suburbs in Maryland and Virginia have become home to hundreds of thousands of new immigrants in the past 15 years. Despite the economic recession of 2008, Washington, DC, saw modest economic growth in federal government jobs, contract work, and information technology. Military-funded aerospace firms such as Northrop-Grumman and Lockheed Martin have established East Coast headquarters in DC to be close to the Department of Defense. More recently, the Dulles "High Tech Corridor," which stretches westward from Arlington, Virginia, toward Dulles International Airport, has attracted high-skilled software engineers and other high-technology workers. The firm AOL employs 5,000 in its headquarters there. To address a shortage of high-skilled workers in the 1990s, the U.S. Department of Homeland Security created an H1 Visa specifically to target foreigners with college degrees in computer-related fields. As a result, many skilled immigrants from India, South Korea, Hong Kong, and mainland China moved to the Washington, DC, region during this time. The continued growth of a prosperous, global "creative class" has led to a simultaneous demand for domestic and manual labor. Many immigrants from El Salvador, Bolivia, Peru, Brazil, Mexico, and Guatemala have also come to DC to work as nannies,

Figure 2.12 Migrants make their presence felt in numerous ways. In this case, there are sufficient Brazilian immigrants for a Brazilian service at this Baptist Church outside Washington, DC. *Source:* Photo by Lisa Benton-Short.

landscapers, construction workers, and hotel and restaurant staff (Figure 2.12).

Washington has become a magnet for immigrants from diverse countries. In 1970, only 4.5 percent of the population of Washington, DC, was foreign born. By 2009, Washington was home to 1 million foreign-born residents, which accounted for 20 percent of the urban population. Cities such as Las Vegas, Orlando, Atlanta, and Charlotte, along with DC, have experienced the bulk of their immigration in the past two decades, making them new immigrant gateways.

Immigrants arriving in these new gateways, however, do not follow the same settlement patterns and processes seen in more established gateway cities such as New York and Chicago. The conventional model of spatial assimilation assumes that immigrants first cluster with their own ethnic groups in center-city enclaves such as Chinatowns and Little Italys and—as they gain higher levels of education and income—leave these enclaves to reside in suburban areas with higher social status and larger homes. In the Washington area, many immigrants move directly to the suburbs rather than the central city. In addition, many immigrants live in moderate- to high-income neighborhoods, not in the poorest ones. In fact, many are settling in places that only 30 years ago were mostly white and had very few foreign-born residents. Now, the historic image of a city polarized into "Black and White" no longer holds true, and city leaders and residents are grappling with how to include and support increasingly diverse communities.

New Orleans: Vulnerable City

Disasters remind us of just how vulnerable cities are to environmental forces. To New Orleans, hurricanes and floods are not new. The city was originally located close to where the Mississippi River flows into the Gulf of Mexico, along a river bend south of Lake Pontchartrain. It was sited on a relatively high piece of land where French traders had already been encamped. This area became known as the French Quarter. The high water table created construction difficulties and the volatile river regularly flooded. In the early twentieth century, improvements in pumping technology encouraged more development in the lower lying areas, which tended to be settled by working-class, poor, and minority populations. Still, the river that acts as the city's economic lifeblood also threatens to destroy it.

In 2005, the city of New Orleans was devastated by Hurricane Katrina. Hurricanes produce more than an inch of rain per hour, high winds that can tear buildings and other structures away from their moorings, and damage to infrastructure ranging from washed out or flooded roads to power lines that are torn apart by wind and waves. Storm surges associated with the high winds can reach over 20 feet (6 m) above ordinary sea level. For most coastal cities, flooding is a serious problem. For a city below sea level, like much of New Orleans, it is potentially catastrophic. It was not the ferocious winds of Katrina that damaged New Orleans, but the storm surge that breached the levees. The city center flooded when portions of the levees at 17th Street and Industrial Canal collapsed. Almost 80 percent of the city was under water, sometimes over 20 feet (6 m) in depth. An estimated 1,000 people were killed, most of them drowned by rapidly rising floodwaters.

At first glance, Hurricane Katrina seemed like a natural disaster since a hurricane is a force of nature. But Katrina's impacts and effects were intensified by socioeconomic conditions. The flooding of the city was caused by poorly designed levees that could not withstand a predictable storm surge. Levee pilings that should have been 15 feet (4.5 m) high had settled in the unstable soil to only 12–13 feet (3.7–4 m) above sea level. It was not Katrina alone that caused the disaster but the shoddy engineering and poor design of inadequately funded public works.

As Katrina approached New Orleans, evacuation orders were finally given. But there was little provision made for those without cars or for the city's most vulnerable residents. Many disabled, poor, black, and elderly residents were trapped. Thousands made their way to the Superdome and the Convention Center, where they remained for days, a stunning indictment of social and racial inequality in New Orleans. The same inequalities were reflected in the damage caused by Hurricane Katrina. Flooding disproportionately affected poor neighborhoods such as the Lower Ninth Ward. The more affluent, predominantly white sections, such as the French Quarter, Audubon Park, and the Garden District had been built at higher elevations and escaped flood damage. Most of the high-poverty tracts were flooded. In closer detail, the "natural" disaster appears to have been a social disaster as well. Environmental disasters become social in the way they are handled and how the distribution of their effects reflects social differences.

While most of the rest of the country has long moved on to other headlines, the rebuilding of New Orleans is not over (Figure 2.13). Five years after Katrina, the damage of the hurricane lived on in abandoned houses, in empty storefronts, and in the ongoing debates about how to rebuild the city. It is estimated that

Figure 2.13 In areas that were flooded during Katrina, houses have been raised above flood level in anticipation of future threats. *Source:* Photo by Lisa Benton-Short.

about 236,000 people left New Orleans due to Hurricane Katrina; many did not return. The lack of jobs continues to be one reason why the city has not fully recovered its pre-Katrina population, and block after block in the Lower Ninth Ward remains empty. While government-sponsored rebuilding efforts have been slow, nonprofit organizations initiated many grassroots programs. The Make It Right NOLA Foundation, established by actor Brad Pitt, has had tremendous success building affordable homes for working families. By 2015, the Foundation had rebuilt more than 100 eco-friendly homes, allowing more than 350 people to return to the Lower Ninth Ward.

URBAN PROBLEMS AND PROSPECTS

Globalization and the Urban Hierarchy

A major driver of urban change today is the growing linkages between cities and global trends. During the past two decades, many cities have become more competitive at the global scale and processes of economic globalization have restructured those cities spatially. New financial districts, luxurious residential areas, and unprecedented property booms indicate success in the global marketplace and a competitive position in the urban hierarchy. Almost all cities are affected by globalization and try to secure international investment, but not all cities become world cities. International banks, global department store chains, and other high-end retail establishments provide visual reminders of globalization's effects on the North American city.

Consider the changing locations of *Fortune 500* company headquarters. In the 1950s and 1960s, large Northeastern and Midwestern cities such as Chicago, Boston, Philadelphia, Pittsburgh, and Montreal, were home to the world's largest industrial companies. In 1960, New York had six of the top ten Fortune 500 headquarters, including Standard Oil, General Electric, U.S. Steel, Mobil Oil, Texaco, and Western Electric. Today, the number of *Fortune 500* corporate headquarters in New York has fallen by half. Firms such as General Electric and Xerox have moved to the suburbs or the Sunbelt. Recently, Hertz car rentals moved its headquarters from New Jersey to Fort Myers, Florida, following a merger with Thrifty, while Volkswagen moved its North American headquarters from Detroit to suburban Washington, DC. The biggest growth in corporate headquarters is occurring in Orlando and West Palm Beach (Florida), Greensboro (North Carolina), Atlanta, Dallas, and Houston. This changing geographic distribution is the result of many factors: company relocations, rises and falls of local firms, and the merging of companies. The lower costs for office space and housing in medium-sized cities such as West Palm Beach and Greensboro provide another reason for a company to relocate. Most firms, however, have chosen metropolitan areas of at least 1 million.

Because key drivers of globalization such as finance and hi-tech have been concentrated in Europe, North America, and East Asia, cities in these regions tend to dominate the global urban hierarchy. Some cities have benefited from globalization and have eclipsed their rivals. For example, many corporate headquarters have moved from Montreal to Toronto following the rise of Quebec's separatist movement. While Toronto has become the major conduit between Canada and the international capital markets, Calgary and Edmonton serve as crucial links to the more specialized international petroleum industry. Quebec City has also transitioned from specializing in agricultural processing and basic manufacturing to a more diverse, postindustrial economy. The North American Free Trade Agreement of 1992 has resulted in an increase of exports from Quebec City to the United States. In addition to a fairly robust aerospace industry, tourism, information technology, and biotechnology comprise some of the growth sectors for Quebec City. Other cities have fallen down the hierarchy. Detroit, Cleveland, Buffalo, and Pittsburgh have experienced a decline, as automobile factories and steel mills have closed. These cities struggle to compete for coveted global linkages and networks that promise to reinvigorate their economies. In Canada, Thunder Bay, St. John's, and Halifax contend with the challenges of urban economies based on natural resources.

There are many different articulations of the urban hierarchy (Table 2.2). Some cities compete for financial command functions—stock markets, banks, multinational corporate headquarters, and other forms of capital exchange. New York is the most important city in this regard, followed by Chicago and Toronto. Other articulations include multicultural command centers such as Vancouver, Toronto, Miami, and Los Angeles, which forge important linkages to other regions through their immigrant populations. Some cities have found a niche in the hierarchy by establishing transportation connections through global airline networks. These cities include Toronto, Los Angeles, Seattle, Memphis (home to Federal Express), Atlanta (home to UPS), and Anchorage, a stop along military and cross-Arctic air routes. Some cities are resurrecting their position in the urban hierarchy. Following massive factory closures and job losses associated with deindustrialization, cities such as Pittsburgh and Cleveland have successfully realigned their economies with global markets. While some of the old industrial companies remain, these cities now capitalize on health-care facilities that have gained national reputations. In contrast to Detroit's continued decline since the collapse of the U.S. auto industry, Cleveland and Pittsburgh have moved back up the hierarchy—albeit in more specialized roles.

Globalization and Localization

As North American cities become part of global circuits of people and capital, they also seek out ways to maintain control over their local character and identity. To attract ever more global investment, they market local attributes and amenities. Globalization and localization therefore act as competing forces in most North American cities. The increasing density of connections between migrants, their countries of origin, and North American gateway cities has led to the emergence of cross-border practices and identities typically referred to as "transnational." Cities, eager to capitalize on the cachet and dynamism of an international population have also branded themselves as transnational or "global." Transnationalism is celebrated in places like the San Jose-Santa Clara metropolitan area of California, commonly known as Silicon Valley. Here, Indian, Chinese, and Israeli professionals who were educated in the engineering and computer science programs of local institutions like Stanford University and UC Berkeley—and who stayed in the area to work—were central in establishing Silicon Valley as the chief U.S. hi-tech cluster and forming local auxiliary organizations such as the Silicon Valley Indian Professionals Association.

Many cities have tried to parlay the diversity of their populations into identities as global or cosmopolitan cities." In

Table 2.2 The World's Most Globally Engaged, Competitive, and Connected Cities

A. K. Kearney's Global Engagement	*The Economist's Global City Competitiveness*	*P. Taylor's Global Network Connectivity*
New York	**New York**	London
London	London	**New York**
Paris	Singapore	Hong Kong
Tokyo	Paris	Paris
Hong Kong	Hong Kong	Tokyo
Los Angeles	Tokyo	Singapore
Chicago	Zurich	**Chicago**
Seoul	**Washington, DC**	Milan
Brussels	**Chicago**	**Los Angeles**
Washington DC	**Boston**	**Toronto**
Singapore	Frankfurt	Madrid
Sydney	**Toronto**	Amsterdam
Vienna	**San Francisco**	Sydney
Beijing	Geneva	Frankfurt
Boston	Sydney	Brussels
Toronto	Amsterdam	Sao Paulo
San Francisco	**Vancouver**	**San Francisco**
Madrid	**Los Angeles**	Mexico City
Moscow	Stockholm	Zurich
Berlin	Seoul	Taipei
Shanghai	**Montreal**	Mumbai
Buenos Aires	**Houston**	Jakarta
Frankfurt	Copenhagen	Buenos Aires
Barcelona	Vienna	Melbourne
Zurich	**Dallas**	**Miami**
Amsterdam	Dublin	Kuala Lumpur
Stockholm	Madrid	Stockholm
Rome	**Seattle**	Bangkok
Dubai	**Philadelphia**	Prague
Montreal	Berlin	Dublin
	Atlanta	

North American cities in bold.

1998, Toronto—known for its "hyperdiverse" population adopted the motto "Our Diversity, Our Strength." The mere presence of transnational immigrants, however, does not imply that their skills or cultural contributions are valued equally to those of the native-born Canadians. Yet, not every city responds to transnationalism. In Vancouver, the upsurge of immigrants from Hong Kong in the 1980s and 1990s created a "moral panic" over both the city's identity and Canadian identity at large. Ironically, some claimed that the once-colonized citizens of Hong Kong were becoming the colonizers of British Columbia. The 1990s became a flashpoint for anti-Asian discourse in Vancouver: Media claimed that the Asian domination of suburbs like Richmond was creating "white flight" and that Hong

Box 2.3 Suburbs Still in Crisis

While it is true that the U.S. and Canadian society is highly suburban, most people do not appreciate the diversity of the suburban experience. For many years, scholarly and popular discussions of poverty focused on the inner-city. Suburbs were idealized as sites of success, opportunity, and prosperity. In recent decades, suburbs of North America have become sites of immense change. Some have flourished, but some have declined. One element of change that has received a lot of media and scholarly attention is the rise in suburban poverty. Brookings Institution scholars Alan Berube and Elizabeth Kneebone reported that by 2005 the suburban poor outnumbered the city poor by about one million. With the advent of the 2008 recession, suburban poverty has risen dramatically. By 2010, there were an estimated 15 million Americans living under the poverty line in the suburbs. In her book *Once the American Dream*, urban scholar Bernadette Hanlon examines the downward trajectory for many suburbs, especially in the inner ring, and offers reasons for their decline: an aging housing stock, foreclosures, severe fiscal problems, slowed population growth, and increasing poverty.

An example of a declining suburb is Essex outside Baltimore, Maryland. Located along the waterfront, this suburb has a long history of aerospace and aircraft production. The aircraft and aerospace manufacturer Glenn L. Martin Company located there in 1929, drawing thousands of workers from around Maryland and other parts of the United States. Gaining government contracts to build airplanes, the company employed some 53,000 workers during its heyday right around World War II. After the war, it had to downsize, but residents found other jobs with a nearby steel manufacturer. The suburb of Essex once had a strong industrial base that supported the local economy.

Beginning about the 1970s, decline in Essex began as industry slowly began to disappear. About 30 percent of the workforce of Essex was employed in manufacturing in 1980; this declined to 14 percent by 2000 and to about 8 percent by 2009. The loss of manufacturing jobs; the construction of shopping malls in suburbs a little further out; and contamination from local sewage plants each negatively impacted the economy, society, and image of Essex. Income decline and increased poverty occurred. In 1970, the median household income of Essex was about $54,000, but dropped to an estimated $49,700 by 2009. In addition, poverty rates increased dramatically from six percent in 1970 to almost 16 percent by 2009.

Source: Bernadette Hanlon. 2014. "Suburban Forms and Their Challenges," in L. Benton-Short (ed.), *Cities of North America: Contemporary Challenges in US and Canadian Cities*. Rowman and Littlefield.

Kong immigrants were building contemporary-style "monster homes" in neighborhoods that had previously been dominated by "Anglo" Tudor-style homes. Asian immigrants were even blamed for driving up the cost of rent and the competitiveness of schools.

Many local jurisdictions have tried to take control of the immigrant streams that bring them newcomers. In Canada, new policies have been set largely by the provinces (though often led by city-based employers). Since the early 2000s, federally sanctioned programs have given the provinces control over immigrant attraction, selection, settlement, and integration. Provincial laws often relaxed language requirements for new immigrants and allowed individual employers to make nominations intended to incentivize immigration to economically and demographically declining cities (e.g., Winnipeg and Halifax) rather than the "MTV" cities (Montreal, Toronto, Vancouver) that account for 80 percent of immigration to Canada. In the United States, county-based and municipal ordinances range from "sanctuary" policies (e.g., accepting Mexican consular cards as ID) to those aimed at apprehending undocumented immigrants (e.g., checking the immigration status of anyone arrested). Pro-immigration ordinances are clustered in the West and in central cities (areas with traditionally high levels of immigration), while anti-immigration measures are clustered in the South, rural areas, and suburbs—areas generally experiencing accelerating immigration.

North American cities also employ local urban development strategies to manage global flows of capital. Two relatively new strategies are urban villages and business improvement districts. While these strategies seek to differentiate certain neighborhoods amid a blandly upscale, cosmopolitan landscape, they also brand these places as worthwhile places to visit, live in, and invest in. A global trend in and of itself, the urban village concept was popularized in the United Kingdom in the late 1990s and typically delineates an area for medium-density development, pedestrian-friendly measures, and mixed-use zoning rather than the single-use zoning that has given rise to industrial parks and suburbs. Self-contained villages, or "urban campuses," are often criticized, however, because they displace groups such as artists who may be unable to afford redeveloped studio space and may fail to align with new neighborhood visions. The BID, developed in both Canada and the United States in the early 1970s, is a quasi-governmental association in which businesses and property owners within a delineated area pay a surtax for collective, privately provided services such as street cleaning, trash pickup, beautification, business recruitment, and security. In cities such as Washington, DC, large, corporatized BIDs created during a period of dysfunctional urban governance in the early 2000s have taken it upon themselves to fully redesign many postindustrial areas for the upwardly mobile buyer. Some have even taken over city-provided services, even hiring off-duty police officers as private security guards to patrol areas that are still "in transition." On the positive side, BIDs are seen to enhance neighborhood amenities and to provide services that city budgets cannot fund. On the negative side, it is unclear whether BIDs accelerate gentrification and how they protect social equity while promoting economic development.

Immigration and Increasing Diversity

The millions of economic migrants to North American cities provide a window onto the reconfiguration of urban networks. Because most immigrants initially go to cities, many of North America's "immigrant gateways" are hyperdiverse and globally linked through transnational networks. Hyperdiverse immigrant cities are those places where the percentage of foreign-born residents exceeds the national level and where there is no single dominant country of origin among the immigrant population. Immigrant gateways are growing in number because of globalization and the acceleration of migration driven by income differentials, social networks, and various state policies designed to recruit skilled and unskilled laborers.

As large numbers of foreign-born residents mix with more established populations, North American cities become the places where global differences are both celebrated and contested. Immigrants add to a city's global competitiveness by enhancing cross-border business connections, linguistic capabilities, and attractiveness to tourists. With birth rates declining in both Canada and the United States, migration is now an even more important determinant of urban growth or decline. Although immigration is a global phenomenon, some regions of the world receive significantly more immigrants than others. In North America, long an established region of immigrant settlement, the rates of immigration are still among the highest in the world. Although the three largest immigrant destinations are New York City, Toronto, and Los Angeles, there are 60 other metropolitan regions with more than 100,000 foreign-born residents. In Canada, immigrants go primarily to Vancouver, Montreal, and Toronto. But even in smaller Canadian cities such as Ottawa and Calgary, over 20 percent of the population is foreign born. In the United States, immigrants are targeting newer gateways such as Washington, DC, Phoenix, Charlotte, and Atlanta.

At the dawn of the twentieth century, New York City was the premier immigrant gateway in the United States, and nearly all immigrants were European. The city was linguistically and ethnically diverse, but not racially diverse. In the first years of the twenty-first century, New York has become one of the most racially and ethnically diverse places on the planet. Of the top ten sending countries, which represent half of New York City's foreign-born population, only one—Italy—is European. The other countries, which are mostly Latin American, include the Dominican Republic, Jamaica, Mexico, Guyana, Ecuador, Haiti, Colombia, and China (Figure 2.14).

A similar pattern holds true for Toronto. In 2011, 49 percent of the city's population was foreign born, one of the highest percentages for any major metropolitan area. Nine countries account for half of the foreign-born population, led by China, then India, the United Kingdom, Italy, the Philippines, Jamaica, Portugal, Poland, and Sri Lanka. In New York, Toronto, and many other metropolitan areas, there is a growing tendency for immigrants to come from a broader range of sending countries and for North American cities to become ever more diverse.

Not all immigrant gateways are hyperdiverse, however. Mexican immigrants account for about half the foreign-born population in Los Angeles, Chicago, Houston, and Dallas. In the Los Angeles market, four of the top ten television shows are in the Spanish language. Similarly, foreign-born Cubans dominate in Miami. Immigrants from mainland China and Hong Kong account for over a quarter of the foreign-born population in Vancouver. It is fair to say that North American cities will continue to be home to many of the world's immigrants well into the twenty-first century.

Women in the City

North American cities are being reconfigured to ensure safe harbor for the world's elite workers and their wealth. Many cities aspire to

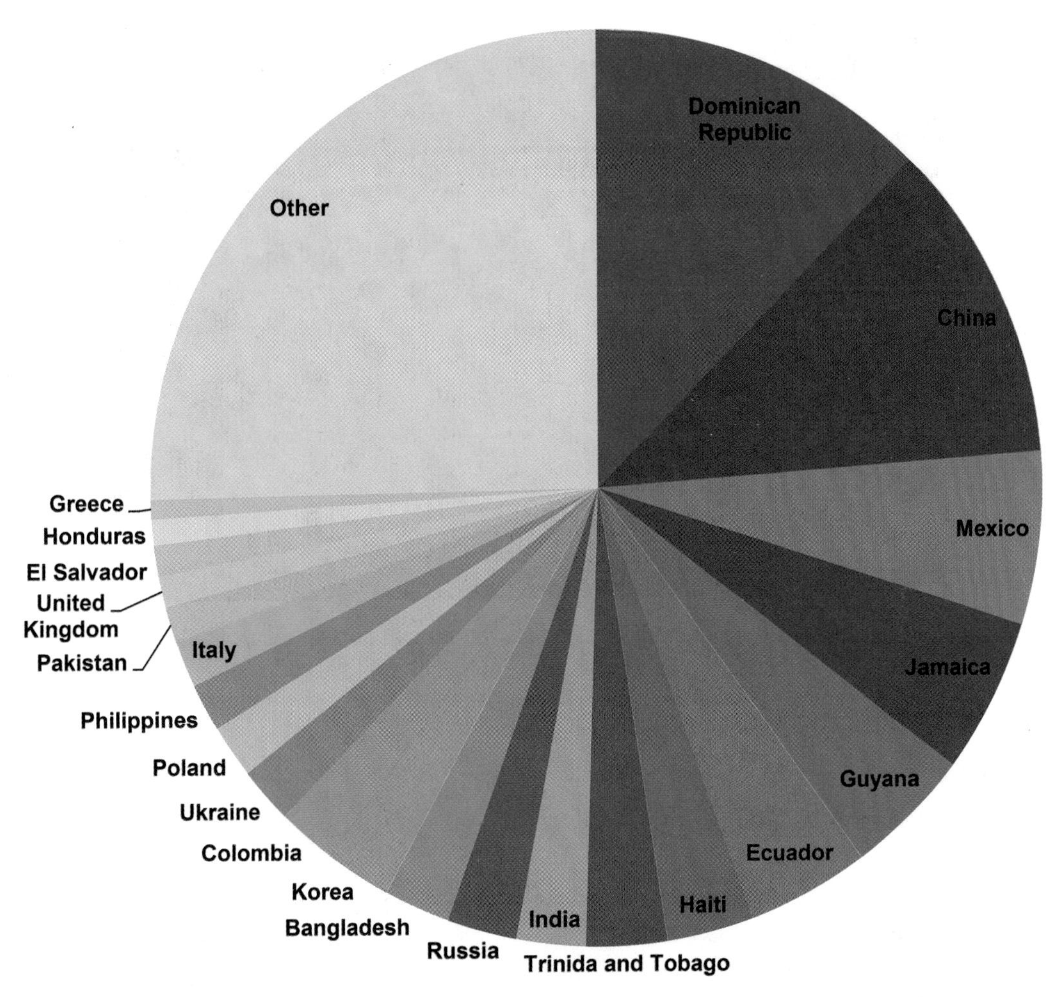

Figure 2.14 New York's Foreign-Born Population. *Source:* M. Price and Lisa Benton-Short. 2007. Globalization, Urbanization and Migration dataset (http://gstudynet.org/gum/).

a cosmopolitan landscape, marked by loft living, ever more arts and entertainment venues, and the expansion of downtown residential zones (Figure 2.15). At the same time, many cities have placed more emphasis on groups such as women that have not always felt safe in the city. Increasingly, women are both the buyers of in-town housing and the targets of real estate marketing campaigns. In Toronto, female one-person households are the largest group of condominium owners, and women are featured in the vast majority of advertisements in publications such as *Condo Weekly*. The trend toward women purchasing condos in Toronto has been celebrated as a new phase of liberation, in which women have joined the institution of home ownership. Typically incentivized through lending practices that favor young buyers, home ownership has long been cast as a step toward becoming a responsible Canadian citizen. Many women, especially in single-income households, see the purchase of property as a means to financial security or capital accumulation in a

Figure 2.15 In 2014, DC hosted its first international pop-up picnic, called Diner en Blanc, for 1500 people. The concept, which originated in Paris, requires that guests wear all-white clothing and bring their own food and chairs. *Source:* Photo by Lisa Benton-Short.

volatile economy. A condominium may also be more affordable than a house for a single woman, eschewing some of the pressure to be in a dual-income partnership before purchasing property. The ads that market condos to women echo these ideas, employing images of the liberated, creative, independent woman enjoying nightlife with her female friends, working on an artistic endeavor, or holding a meeting at a coffee house.

Efforts to include women may have a financial imperative as well as a humanistic one. First, real estate agents often market certain buildings to women on the basis of security: 24-hour concierge, key-card entry, security cameras, and gated underground parking. In this way, developers exploit preexisting, gendered expectations about fear in the city. Second, the extent to which condo ownership actually facilitates women's inclusion in public life may be quite narrow. The securitized, multiple-amenity condo may facilitate community *within* the building rather than interaction with the surrounding city. Gyms, patios, on-site coffee shops, and other quasi-public areas of a condo building may become women's central social spaces while they carefully pick and choose where and when to engage with the city around them. Moreover, these types of buildings are geared toward a female consumer who can both afford to purchase the condo and use the revitalized downtown core as she sees fit, while less privileged women (e.g., poor, single mothers) may be displaced into the decaying inner suburbs. Third and finally, women are less frequently involved in the organization of development processes. Real estate companies, property owners, and the executive boards of the business improvement areas managing the neighborhoods where these condos are located tend to be mostly male. These men are therefore articulating the supposed needs and demands of the female market.

Urban LGBTQ Communities

A similar tension exists for lesbian, gay, bisexual, transgender, and queer (LGBTQ) people. The centrality of the city in queer cultures is historically rooted in the rural-to-urban migrations of the industrial era, during which single men and women were freed from nuclear family structures. Given that homosexuality was illegal in Canada and the United States for most of the twentieth century, early gay and lesbian life in cities centered on unmarked bars and lounges, private parties, and informal public meeting places. More recently, however, the expression of nonnormative sexual identities in cities has come to be associated with "the gay village." Early gay villages reflected a territorialization strategy in which LGBT people—usually gay men—bought up businesses and residential property in peripheral neighborhoods in an effort to demonstrate the visibility and upward mobility that they felt were integral to making claims for equal rights. Some claim that this territorialization stage, which occurred in many landmark villages (e.g., the Castro in San Francisco, Church Street in Toronto) during the 1970s and 1980s, was just a preliminary phase in the "evolution" of gay villages. The evolutionary model suggested that gay villages would become more "mainstream" as their commercial districts became attractive to heterosexual residents and the growing acceptance of queer people rendered the village less necessary. Since the 1969 Stonewall Rebellion in New York City (Figure 2.16), where LGBT people protested a police raid on a local gay bar, gay and lesbian rights have advanced significantly in North America. In 2005, Canada legalized gay marriage at the federal level. In 2015, the United States followed suit when the Supreme Court upheld the rights of same-sex couples to legally marry nationwide.

Yet, the implications of these changes for LGBTQ identities, and for gay villages specifically, seem far from uniform. In cities like Montreal and Chicago, gay villages have been heavily branded and marketed by business associations and municipal governments. They have not only evolved into established commercial districts, they have also been criticized as "boys towns" that exclude lesbians and the broader array of queer identities. Chicago's village is actually named Boystown. Another criticism is that they crassly commodify sexual identity for the consumption of cosmopolitan heterosexual consumers while offering few venues or outlets for queer people who lack high incomes or whose cultural identities fall outside of the Euro-American mainstream. In a few cases, gay villages may be stripped of their "gay" identity by municipal governments trying to re-market the area to the upwardly mobile heterosexual mainstream. Urban planner Petra Doan finds that Atlanta has purposefully "de-gayed" the neighborhood of Midtown through rent increases, property speculation, closures of bars, and denials of applications for LGBTQ events.

In Canada, where gay rights are more established (homosexuality decriminalized in 1968, as opposed to 2003 in the United States), "the village" has become a hot topic in the gay media of many cities. In Toronto, many suggest that because safe space is no longer needed in a new era of full gay rights, the idea of a gay village has become irrelevant altogether. Two hundred miles away, in the Canadian capital of Ottawa, a very different story emerged. In a city more known for its Cold-War era expulsions of gay men from the

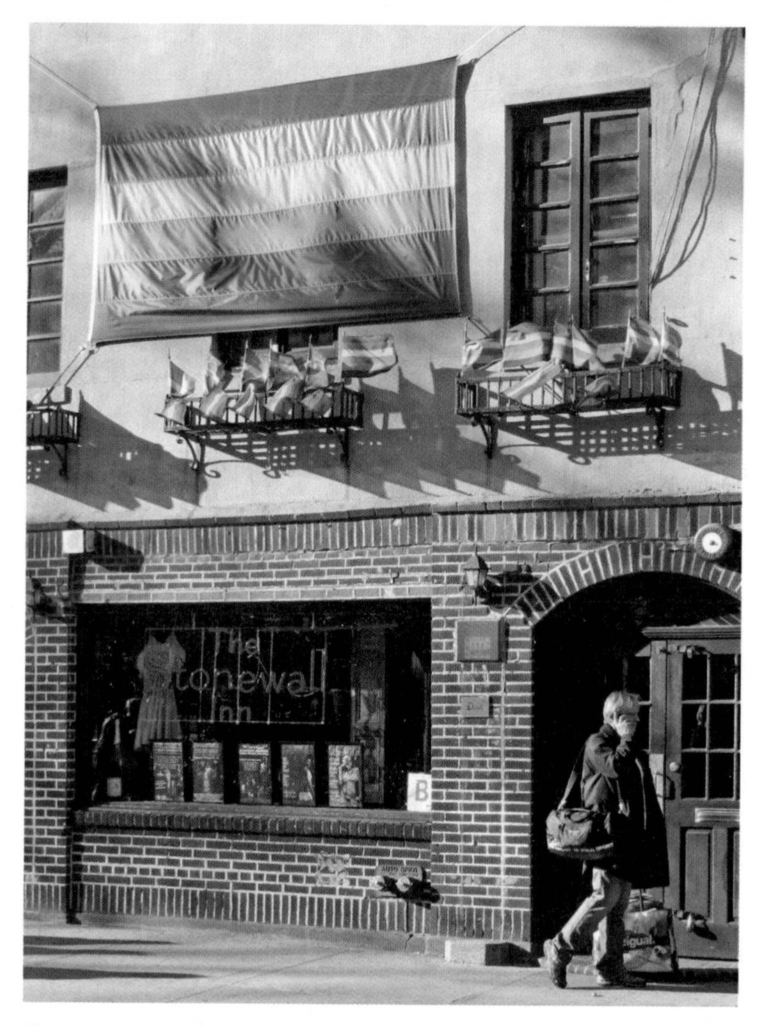

Figure 2.16 The Stonewall riots took place on June 28, 1969, outside the Stonewall Inn in New York's Greenwich Village. They are now regarded as the beginning of the gay and lesbian rights movement in the United States. *Source:* Photo by D. J. Zeigler.

Civil Service than for a highly visible queer community, the mayor and the municipal government designated a six-block section of Bank Street as the official gay village in 2011. Given its location in a medium-density commercial zone with only a few gay establishments—and its designation at a time when most gay and lesbian rights had been attained in Canada—it is doubtful that territorial visibility or commercialization are the aims of this village. In contrast, the city, prompted by a long campaign by local activists, has finally decided to create a symbolic home for the queer community in a city that has not always been welcoming to it.

Security and Urban Fortification

Threats of terrorism, whether domestic or international, have had a profound impact on many North American cities. Intense surveillance and security measures have challenged the perceived freedom of city life and altered its physical landscape. Especially since September 11, 2001, security measures have become much more visible components of urban landscapes. Given that many barricaded urban spaces are also valued public places with connections to local, regional, and national identity, this trend is of significant concern to many citizens.

The rise of security policing and new forms of surveillance are not new. Mike Davis's book, *City of Quartz*, diagnosed what he calls *fortress cities* as a response to perceived urban disorder and decay, primarily from domestic sources. He predicted that urban authorities might create fortress-style rings of steel around potential targets, creating a landscape demarcated by physical barriers such as gates, walls, and carefully hidden surveillance devices such as closed-circuit television (CCTV) cameras. This is a vision of a city that can be controlled. For some residents, security and surveillance offer reassurance in an uncertain age; for others, these measures are the architecture of paranoia.

Although cities have long had police forces and emergency plans, many did not have comprehensive security and defense strategies until the early 1990s as terrorists chose to target high-profile cities to attract global publicity. Since 9/11, it is clear that symbolic targets—such as monuments, landmark buildings, and other important urban public spaces—are increasingly at risk. Cities have responded by installing highly visible counterterrorist measures. In New York, Toronto, Los Angeles, and Philadelphia, for example, bollards, bunkers, and other barriers have been placed around selected "high-risk" targets. Since these structures often restrict access to museums, monuments, memorials, and parks, many see heightened fortification as a threat to public space.

Washington, DC, for instance, is one of the most visibly fortified cities in North America. Miles of fences, jersey barriers, and bollards surround federal buildings, monuments, and memorials throughout the city. In addition, CCTV cameras are mounted on libraries, shopping malls, banks, and even the monuments on the National Mall (Figure 2.17). Residents of all North American cities, in fact, are now more photographed and videoed than ever before. Yet, it is still unclear whether terrorism concerns merit fortification, the loss of civil liberties, inhibited public access to public space, and a new urban culture dominated by a sense of hypersecurity.

Cities have also experienced an increase in the militarization of police forces. In the 1990s, Congress authorized the Defense Department to transfer surplus military gear free of charge to federal, state, and local police departments to aid in the war on drugs. In the years after 9/11, this program (DOD 1033) accelerated. In 1990, the Pentagon gave $1 million worth of equipment to local law enforcement. In 2013, it increased to $450 million. Today, SWAT teams and police wear camouflage and military-grade body armor, carry night-vision rifle scopes, stun grenades, military assault rifles, and ride around in Humvees. There has been a proliferation of military training among municipal police departments: In some cities, local police have been transformed into small armies. Increasingly, SWAT teams are deployed to serve warrants for drug searches (as opposed to hostage situations).

Critics are alarmed by this trend because it seems to stress violence rather than working with the community to make neighborhoods safer. In fact, it runs counter to the concept of

Figure 2.17 Here, circled in blue, a security camera has been positioned atop the Jefferson Memorial in Washington, DC. What messages do surveillance cameras convey in a public space which memorializes freedom, liberty and independence? *Source:* Photo by Lisa Benton-Short.

"community policing." A 2014 American Civil Liberties Union report titled "War Comes Home: the Excessive Militarization of American Policing" noted that "our neighborhoods are not warzones, and police officers should not be treating us like wartime enemies." The report concluded that police militarization is a pervasive problem, noting that an estimated 500 enforcement agencies have received combat vehicles built to withstand armor-piercing bombs. The same report also noted that the use of paramilitary tactics primarily affected people of color; 42 percent of those impacted by a SWAT deployment were black and 12 percent Latino. And, with other hypersecurity measures, this trend has been allowed to happen in the absence of any meaningful public discussion.

Tensions rose in Ferguson, Missouri, in 2014 after a white police officer, shot an unarmed 18-year-old black man. The disputed circumstances of the shooting prompted intense debate about law enforcement's relationship with urban African American communities. For several weeks, there were both peaceful and violent protests. There were confrontations between predominately black community protesters and the nearly all-white police force. The U.S. Justice Department launched a federal investigation of the Ferguson police to determine whether officers had engaged in racial profiling or had a history of excessive force. The legacy of Ferguson started a national debate about police tactics. Less than a year later, in Baltimore, where almost a quarter of the population lives in poverty, a similar scenario played out with similar results. A young African American man was arrested on the flimsiest of charges, thrown into a van, and transported to the police department. When he exited the van, he could neither talk nor breathe; a week later he was dead. The city of Baltimore erupted, again reminding us that racial segregation, poverty, and unemployment in many cities remain an issue. For many who live in minority communities, the police are not typically seen as allies (Figure 2.18).

Rebuilding and Memorialization

Sudden events, such as those of September 11, 2001, can transform city spaces and the meaning attached to them. In the days after the collapse of the Twin Towers in New York, families and friends of missing people papered sections of New York with posters and pictures of their loved ones. In a *New York Times Magazine* article, Marshall Sella traced the evolution of these posters. Initially, they constituted a frantic effort to gain any possible information; then, the posters began to

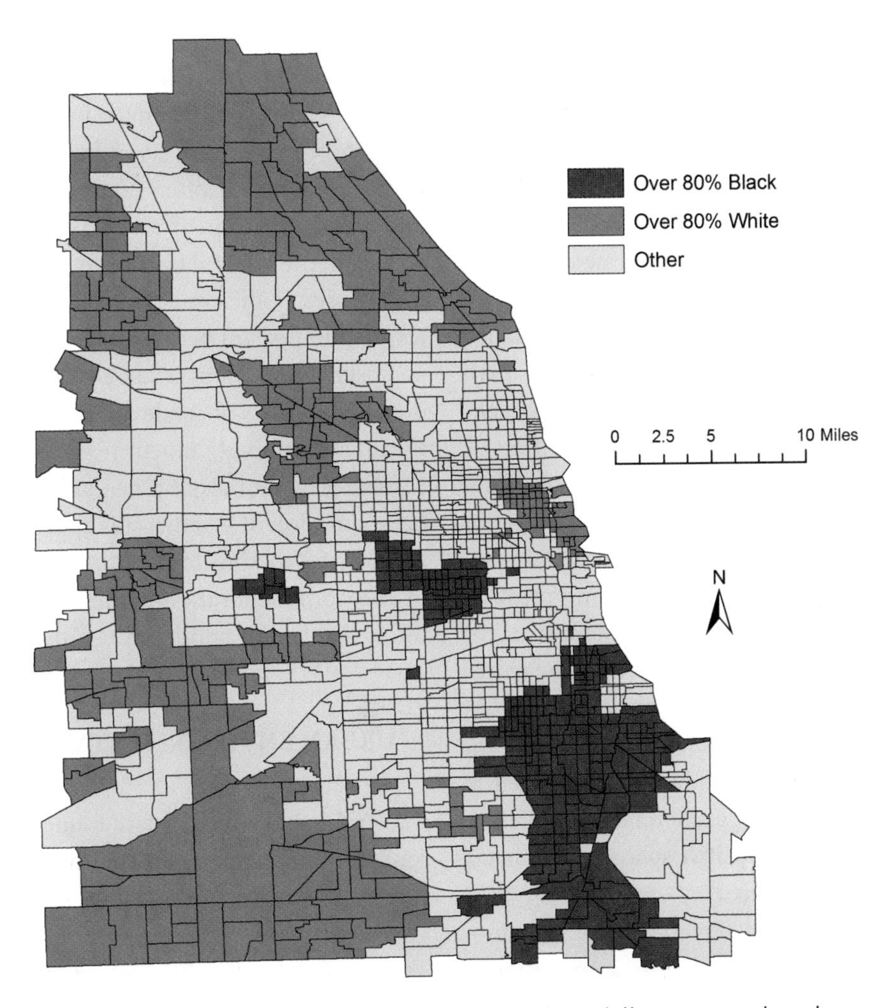

Figure 2.18 Chicago and many other cities remain racially segregated, and minorities are concerned about police profiling and violence. *Source:* US Census Bureau.

include increasingly detailed physical descriptions, apparently to make identification of the bodies possible. Most of these were hung in areas around Ground Zero. In a final evolution, posters began to address the missing people directly through "good-bye letters." The site of the World Trade Center had become sacred space not only because of tragedy, but also because it became a place of spontaneous commemoration.

In the years that followed, New York City debated what to do with the World Trade Center site. Proposals for its rebuilding were immediate, but there were also calls for memorialization. Eventually it became clear that some of the buildings would be rebuilt, but there would also be space for a memorial. Disagreements and public protests over how to design the memorial followed. Today, One World Trade Center (nicknamed Freedom Tower) soars above New York City at a symbolic 1776 feet. It is currently America's tallest building and features some 3 million square feet of office space, an observation deck, and restaurants. The building opened in 2014.

Another component of the World Trade Center site is the 9/11 Memorial and Museum. It features two reflecting pools, nearly an acre each in size, which sit within the footprints of the Twin Towers (Figure 2.19). More than 400 trees surround the reflecting pools. The names of every person who died in the attack of 2001 are inscribed into bronze panels edging the memorial's pools. The design is intended to convey a spirit of hope and renewal, while also providing a contemplative space in what many consider sacred space.

We have seen the emergence of a memorial culture. The impulse to commemorate is part of a broader social reconstruction of national or cultural identity. Seen in a positive light, proposals for memorials represent an expansion of history and identity by including the previously marginalized or ignored. In the United States, for example, the demand for new memorials is part of a larger politics of identity that seeks to make memorial spaces more reflective of a multicultural America.

Some commentators have noted, however, that contemporary society now rushes to commemorate. Historically, it was not uncommon for decades, or longer, to pass before a memorial was erected. The Washington Monument in DC, for example, was completed just prior to the 100th anniversary of Washington's death. Today, weeks after a tragic event, communities may be discussing how best to memorialize loss. The rush to commemorate, however, may prove to be more about healing and less about honoring. It often occurs when family and friends who remain try to make meaningless death meaningful by transforming victims into heroes. Some argue that allowing time to elapse is important because it allows historical perspective as well as a sense of whether the event or individual made a lasting contribution. The seemingly urgent need to plan and construct memorials also raises the more difficult question of whether it is appropriate for survivors of victims to be intimately involved in the commemoration process, or whether this is best left for another generation.

URBAN ENVIRONMENTAL ISSUES

U.S. and Canadian cities face numerous environmental problems, particularly water challenges, air quality, and climate change.

Water

A dependable supply of water is critical to sustaining life and supporting healthy

Figure 2.19 This view of the 9/11 Memorial shows one of the two reflecting pools that sit within the footprints where the Twin Towers once stood. *Source:* Photo by John Rennie Short.

communities. In every U.S. and Canadian city, there are two broad water issues: water supply and water quality, both of which rely on water infrastructure. Water supply infrastructure includes the systems of delivery (aqueducts, pipes). Water quality infrastructure includes treatment facilities and sewage systems.

Water Supply: Drinking water comes from surface sources or ground water. It goes into a water treatment facility and is purified to certain standards. In cities, an underground network of pipes delivers water to all buildings served by a public water system. The United States and Canada rank first and second with regard to water consumption per capita, compared to other highly developed countries. In part, this is due to the lack of widespread water conservation practices and water pricing that does not promote efficiency.

Each day water utilities in the United States supply nearly 34 billion gallons of water. To remove contaminants, water suppliers use a variety of treatment processes including coagulation, filtration, disinfection, ion exchange, and absorption. Both the federal government and states have responsibilities for providing safe water. The Safe Drinking Water Act was passed by the U.S. Congress in 1974 to regulate the nation's public water supply. Canada passed a similar law. While most North American cities take safe water for granted, some threats remain: contaminants such as lead, arsenic, and chromium; improperly disposed-of chemicals; animal and human wastes; wastes injected underground; and naturally occurring substances. Drinking water that is not properly treated, or that travels through an improperly maintained distribution system, may also pose a health risk. In the post-9/11 world, drinking water utilities also face new responsibilities due to concerns over water system security and threats of infrastructure terrorism.

A recent challenge facing many cities is that the existing drinking water infrastructure (treatment plants and the underground networks of pipes) was largely built during the late nineteenth century. This infrastructure is now more than 100 years old and deteriorating. In many cases, it cannot handle the volume of demand due to population growth. Recently, the EPA estimated that U.S. cities will need to invest $160 billion over a 20-year period to ensure the continued development, storage, treatment, and distribution of safe drinking water.

Water Quality: Most cities in Canada and the United States built sewage systems and wastewater treatment facilities in the nineteenth and early twentieth century. However, population growth has meant that the volume of sewage and storm water now often exceeds the processing ability of most treatment plants. This is particularly noticeable during heavy rains.

Combined Sewage Overflow (CSO) refers to the temporary direct discharge of untreated water. CSOs occur most frequently when a city has a combined sewage system that collects wastewater, sanitary wastewater, and storm-water runoff in underground pipes, which then flow into a single treatment facility (Figure 2.20). During dry weather, combined sewage systems transport wastewater directly to treatment plants. However, urban storm runoff is comingled with household and industrial wastes. When it rains, few facilities can handle the sudden increase in water volume, and the excess volume of sewage, clean water, and storm water may be discharged untreated into nearby water bodies. Forty types of disease-causing pathogens have been found in raw sewage that discharges into CSOs. CSOs are among the major sources responsible for beach closings and shellfish restrictions and the contamination of drinking water.

Municipalities in Canada and the United States have been undertaking projects to mitigate CSOs since the 1990s. For example, Ottawa has been working for many years to separate sewers from the remaining combined sewers. In southeast Michigan, prior to 1990, the quantity of untreated combined sewage discharged annually into lakes, rivers, and streams was estimated at more than 30 billion gallons per year. In 2005, it had been reduced by more than 20 billion gallons per year. Many other cities are undertaking similar projects to address CSOs. But, these are billion-dollar projects, and cities look to states and the federal government for supplemental funding.

Air Pollution

Residents in many North American cities confront the reality of air pollution. Since the 1970s, the U.S. and Canadian governments have taken steps to control emissions from automobiles and factory smokestacks. Catalytic converters capture much of the chemical pollution emitted in automobile exhaust, and vapor traps on gas pumps help prevent the escape of carbon dioxide into the air. Recent efforts to develop zero-emission vehicles (such as electric cars) are another way of using technology to alleviate air pollution. However, new sources of pollution, combined with increased use of fossil fuels, has meant that air pollution for many U.S. and Canadian cities has continued to increase despite regulatory efforts.

Smog represents the single most challenging air pollution problem in most North America cities. It is often worse in the summer months when heat and sunshine are more plentiful. Short-term exposure can cause eye irritation, wheezing, coughing, headaches, chest pain, and shortness of breath. Long-term exposure scars the lungs and worsens asthma and respiratory tract infections. Plus, it particularly affects weak and elderly residents, and those

Box 2.4 Returning to the Tap

There was a time when brands like Evian and Perrier conjured up images of purity and luxury. In 2010, Coke sold about 293 million cases of its top brand, Dasani, while rival Pepsi sold 291 million of Aquafina. The biggest player is Nestle, which sells such brands as Poland Spring, Zephyrhills, and Pure Life. But there is now a backlash. Critics of bottled water point to negative economic, regulatory, and environmental consequences of its use. For example, it costs only $10 for 1,000 gallons of tap water, while consumers spend $1,000 for 1,000 gallons of bottled water. A recent Natural Resources Defense Council study found that an estimated 25 percent or more of bottled water was really just tap water in a bottle.

The U.S. EPA requires cities to disclose drinking water conditions; no such requirement is imposed on bottled water. The reality is that tap water is actually held to more stringent quality standards than bottled water, and some brands of bottled water are just tap water in disguise. While most consumers assume that bottled water is at least as safe as tap water, there are still potential risks. Although required to meet the same safety standards as public water supplies, bottled water does not undergo the same testing and reporting as water from a treatment facility. Water that is bottled and sold in the same state may not be subject to any federal standards at all, but bottled water manufacturers encourage the perception that their products are purer and safer than tap water.

Furthermore, bottled water is wasteful, contributes to ballooning landfills, and is being marketed as a necessity by an industry making billions on what consumers used to happily get for free. Americans buy an estimated 30 billion plastic water bottles every year, nearly 90 percent of which wind up in landfills. Approximately 1.5 million barrels of oil are used to make plastic water bottles, while transporting these bottles burns even more oil. Furthermore, the growth in bottled water production has increased water extraction in areas near bottling plants, leading to water shortages that affect nearby consumers and farmers. In addition to the millions of gallons of water used in the plastic-making process, two gallons of water are wasted in the purification process for every gallon that goes into the bottles.

A growing coalition of cities and health organizations now advocates for "a return to the tap." New York-based TapIt, a nonprofit group launched in 2008, works to promote the use of tap water. They encourage restaurants to provide free refills of tap water to patrons who have their own reusable bottles. They have also worked with hundreds of colleges to install water fountains known as hydration stations so that students can refill water bottles rather than buy new ones. These fountains have taller faucets to allow tall bottles to be refilled. Hydration stations are also popping up in airports, parks, office buildings, and restaurants. TapIt is an example of a grassroots organization that leverages the power of the Web, social media, and mobile telephones to drive social change. It offers an iPhone app, mobile website, and TapIt stickers on the windows of participating restaurants.

Sources: "Tapped" http://www.tappedthemovie.com/; for more information on TapIt, go to http://www.tapitwater.com/; EPA Office of Water. 2009. "Water on Tap: What you need to know," http://water.epa.gov/drink/guide/upload/book_waterontap_full.pdf

Figure 2.20 On the Cincinnati waterfront, residents are reminded that the Ohio River is subject to combined sewer overflows that create a danger to public health. *Source:* Photo by Lisa Benton-Short.

who engage in strenuous activity. While air pollution has decreased in many urban areas due to declines in heavy manufacturing and the growing "green" movement, air quality in many cities remains poor. In 2014, the cities with the highest rates of smog were all in California (e.g., Los Angeles), followed by Houston, Dallas-Fort Worth, Atlanta, and Washington, DC. In Canada, the cities with the highest levels of ozone and particulate matter were in the Golden Horseshoe region (e.g., Toronto, Hamilton), southwestern Ontario (Windsor, Sarnia), greater Montreal (Montreal, Laval), and Atlantic Canada (Fredericton, Halifax).

Smog is one pollutant whose effects are often exacerbated by geography. Cities located in basins and valleys, such as Los Angeles, are particularly susceptible to the production of smog. Denver, the Mile High City, suffers from smog and other air pollutants that are made worse by its elevation. Because of the high altitude, Denver experiences frequent temperature inversions when warm air is trapped under cold air and cannot rise to disperse the pollutants across a wider area. As a result, smog may hover in place for days at a time, generating a "smog soup" that envelops the city. Montreal and Toronto experience another realm of "smog geography." These cities are downwind of major industrial cities in the American Midwest, meaning that many of their air pollutants originate across the border in the United States. Many of the improvements in air quality in North America's cities have been offset by population increases that drive up demand for energy and gasoline. Despite decades of regulation and good intention, cities in North America remain far from eliminating the threat of air pollution to both public health and environmental quality. Air pollution is also connected to climate change.

Climate Change

Around the world, cities today consume 75 percent of the world's energy and emit

80 percent of the world's greenhouse gases, which are the drivers of climate change. Until recently, much of the debate about climate change has been at the global or national scale. Until recently, the urban scale was often ignored. Two major effects of climate change are rising sea levels and rising global average temperatures. The major danger to human populations in cities will probably occur from extreme events such as increased storm surges (related to increasing mean sea levels) and temperature extremes (related to increasing average temperatures).

The vast and varied geographies of North America mean that cities will face different vulnerabilities with climate change. Inland cities such as Las Vegas, Denver, and Calgary may see an increase in soil erosion and a loss of water availability. Summer heat waves and droughts may affect crops; underground water sources may become stressed and overused; and competition for water may intensify. Cities in the Midwest and Northeast may also experience wetter winters with heavier snowfall. Cities located on lakes and rivers, such as Kansas City, Cincinnati, Sacramento, Winnipeg, and Quebec may see their rivers experience more flooding in the spring but reduced flows late in the summer. Finally, coastal cities everywhere may see increased flooding during storms due to sea-level rise.

CONCLUSIONS

Cities in the United States and Canada entered the twenty-first century facing complex challenges, including increasingly rapid economic, social, and environmental change. Global economic restructuring, which manifested in deindustrialization and the rise of a diverse service-sector economy, has resulted in uneven development. Cities unable to tap into global circuits of capital struggle to rejig their economies in the wake of deindustrialization. Furthermore, cities are in greater competition with each other to retain center-city populations, attract domestic and international investment, and develop diverse economies with multiple types of businesses and sectors. Yet, economic diversification is no guarantee of success. Future booms and busts in North America's regional economies will undoubtedly produce new urban development imperatives.

The demographics of North American cities are also changing, and suburban sprawl is taking on new forms. Today, not all suburbs are wealthy; some are in decline and experiencing increased poverty. Issues of immigration (both legal and undocumented) have become part of a wider public debate around citizenship, race, ethnicity, and gender. Traditional immigrant cities, such as New York and Toronto, continue to see an influx of immigrants; but cities without long histories of immigration have also begun to attract large numbers of foreign-born residents. New immigrant gateway cities are challenged to provide a range of social services (such as English as a second language in schools and translation services in hospitals). Diversity is also an issue as lesbian, gay, bisexual, transgendered, and queer communities look beyond the acquisition of same-sex marriage rights to ensure their safety and express their identities in urban space. Finally, the continuing war on terror has resulted in physical changes to the urban landscape as cities attempt to deal with safety, security, and the vulnerability of urban populations. New debates about the role of police and municipal security forces reflect concerns about racial profiling.

Box 2.5 Staying Cool in Toronto

Urban dwellers consume less electricity than their suburban counterparts—this much we know. However, the actual amount per capita in any given city can vary tremendously. The most important factor that helps to explain the difference is summer heat. Whether it is Houston, Texas or Miami, Florida, when temperatures spike in July and August, people reach for the thermostat. Even in more northern locations, such as Boston, Massachusetts and Minneapolis, Minnesota, high summer temperatures and humidity can make our lives miserable. They can also wreak havoc with monthly utility bills. Among North American cities, Vancouver, British Columbia, is often lauded for its sustainable and eco-friendly lifestyle. The reputation is well deserved. In this case, however, it is Toronto that has taken a creative and bold step in the direction of sustainability.

While many cities rely on fossil fuels to generate the electricity that cools our homes in summer, the City of Toronto has taken a different approach. In 1990, the Canadian Urban Institute, along with numerous other interested parties, began to investigate the possibility of drawing cold water from the bottom of Lake Ontario to modify air temperatures in government buildings, office high-rises, hotels, and other structures in Toronto's downtown. A unique collaboration between the Enwave Energy Corporation and the City of Toronto soon turned the dream into a reality. After an environmental assessment, the deep lake water cooling (DLWC) project was approved in 1998 and construction began. The system was officially commissioned in 2004. Today, Toronto's DLWC is the world's largest lake-source cooling system. It has been deemed a stunning success, distributing chilled water to slightly more than half the potential market in Toronto, while at the same time, lowering utility bills, attracting environmentally conscious businesses to the urban core, and significantly reducing emissions of carbon dioxide and other air pollutants associated with the burning of coal.

How does the system work? Essentially, Torontonians take advantage of a permanent reservoir of cold water that collects at the bottom of Lake Ontario. During the winter months, the lake's surface temperature cools to about 4° C. As the surface water's density increases, it begins to sink to the bottom. In summer, the situation is reversed. Water at the surface of the lake warms up, but because its density does not increase it does not sink. The result is that cold water remains trapped at the bottom year round. To take advantage of this phenomenon, Enwave sank three intake pipes along the slope to a distance of about 5 km offshore. Water is pumped to a filtration plant where it is processed and then redirected to Enwave's Energy Transfer Station. Here, an energy transfer takes place between the cold water drawn from the lake and the company's closed chilled water supply loop. Once this process is complete, the water flows to the city's potable water system. How cool is that?

Source: Geoff Buckley. 2014. "Urban Sustainability," in L. Benton-Short (ed.), *Cities of North America: Contemporary Challenges in US and Canadian Cities*. Rowman and Littlefield.

Last, environmental factors are transforming the urban landscape in many ways. The impact of Hurricane Katrina on New Orleans and the widespread economic and ecological impacts from the 2010 Gulf of Mexico oil spill are forceful reminders that many cities are vulnerable to environmental events. Cities are constantly preparing for hurricanes, earthquakes, floods, and droughts—or recovering from them. Despite the long-term effects of urban development within environmentally vulnerable areas, we continue to build homes, businesses, and roads along coasts, river valleys, deltas, and earthquake fault lines. In rapidly growing southwestern cities such as Phoenix, fresh water sources are already disappearing. Moreover, water pollution and air pollution continue to have significant health impacts on city dwellers. Climate change has emerged as a major urban issue—and cities are responding with both mitigation and adaptation efforts. In recognizing the reciprocal relationships between humans and their environments, many cities are developing sustainability plans that include new approaches to water management, air pollution, climate change, and creative reuse of previously abandoned spaces.

SUGGESTED READINGS

Anisef, Paul, and Michael Lanphier, eds. 2003. *The World in a City.* Toronto: University of Toronto Press. Analyzes the challenges of immigrants in Toronto and the municipal policies that aid in settlement and integration.

Benton-Short, L. 2014. *Cities of North America: Contemporary Challenges in U.S. and Canadian Cities.* Denver: Rowman and Littlefield. Examines critical issues including globalization, new social identities, the income gap, and environmental challenges.

Benton-Short, L., and J. R. Short. 2013. *Cities and Nature.* 2nd ed. New York: Routledge. Connects environmental processes with social and political actions, including discussion of urbanization trends and sustainability.

Bulkeley, H. 2012. *Climate Change and the City.* London and New York: Routledge. Examines how cities are responding to climate change in terms of both mitigation and adaptation.

Doan, P. 2011. *Queerying Planning: Challenging Heteronormative Assumptions and Reframing Planning Practice.* Surrey, UK, and Burlington, VT: Ashgate. Assesses how urban design strategies work to include or exclude individuals who are gay, lesbian, bisexual, and transgender.

Florida, R. 2010. *The Great Reset: How New Ways of Living and Working Drive Post-Crash Prosperity.* Toronto: Random House. Gives an overview of changes to urban economies in the wake of the global financial crisis and offers guidelines for regeneration.

Jacobs, J. 1969. *The Economy of Cities.* New York: Random House. A classic urban text that critiques 1950s and 1960s urban planning.

Melosi, M. V. 2011. *Precious Commodity: Providing Water for America's Cities.* Pittsburgh: University of Pittsburgh Press. Examines water resources in the United States and provides background on both water supply and wastewater systems.

Teixera, C., W. Li, and A. Kobayashi, eds. 2012. *Immigrant Geographies of North American Cities.* Don Mills, ON: Oxford University Press Canada. Examines the history of immigration in major gateways, challenges faced by immigrants, and specific patterns of ethnic immigration.

Zukin, S. 2011. *Naked City: The Death and Life of Authentic Urban Places.* Oxford: Oxford University Press. Explores the spaces of "authentic" urban life (art galleries, family-owned shops, etc) and how their popularity drives out residents who give neighborhoods their "authenticness."

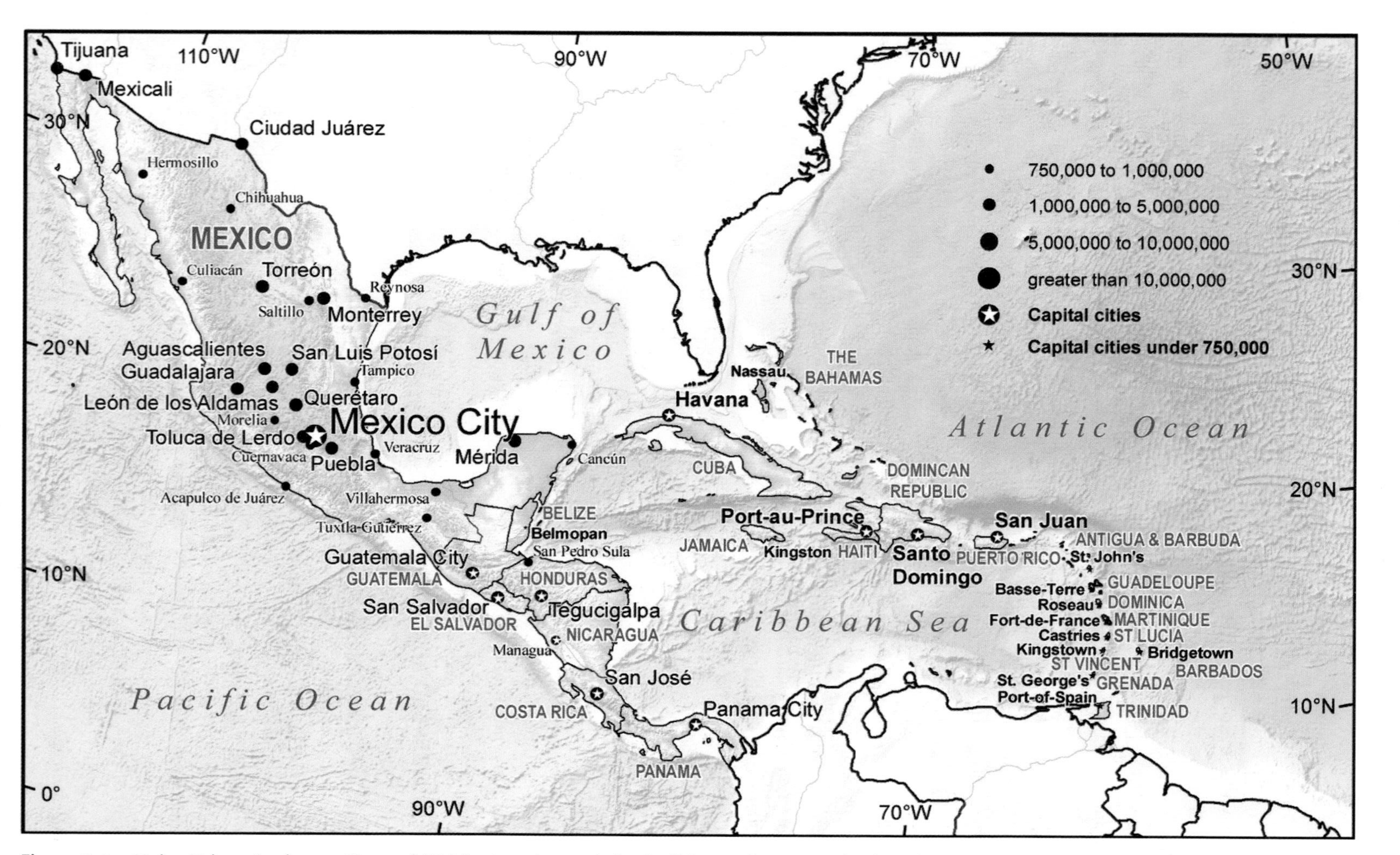

Figure 3.1 Major Urban Agglomerations of Middle America and the Caribbean. *Source:* United Nations, Department of Economic and Social Affairs, Population Division (2014), World Urbanization Prospects: 2014 Revision, http://esa.un.org/unpd/wup/.

3

Cities of Middle America and the Caribbean

ROBERTO ALBANDOZ, TIM BROTHERS, SETH DIXON, IRMA ESCAMILLA, JOSEPH L. SCARPACI, AND THOMAS SIGLER

KEY URBAN FACTS

Total Population	213 million
Percent Urban Population	73%
Total Urban Population	155 million
Most Urbanized Countries	The Bahamas (83%) Mexico (79%) Dominican Republic (78%)
Least Urbanized Countries	Trinidad and Tobago (9%) St. Lucia (19%) Antigua and Barbuda (24%)
Number of Megacities	1
Number of Cities of More Than 1 Million	22 cities
Three Largest Cities (*Metacity*)	*Mexico City* (21 m), Guadalajara (5 m), Monterrey (5 m)
Number of World Cities (Highest Ranking)	1 (Mexico City)
Emerging World Cities	Monterrey, Panama City, San José

KEY CHAPTER THEMES

1. The Mexican urban system was established in large measure by the Aztec pattern of urbanization; it was militarily subjugated by the Spanish to facilitate the colonizers' dual mission of proselytizing and mining.
2. Today, urban growth in Mexico is occurring in intermediate cities located close to large cities, in cities along the U.S.-Mexican border, and in cities far from large urban agglomerations.
3. The urban systems of Central America and the Caribbean developed under various European powers and followed an agricultural-driven model of colonial and postcolonial growth.

4. Today, Central America is over 70 percent urban, and national poverty rates are inversely proportional to urbanization rates in that the poorest countries are the least urbanized countries.
5. Social and geographic segregation has deepened in Central America's cities; crime and violence are serious problems there and in Mexico.
6. Four patterns highlight contemporary urbanization in the Caribbean: urban primacy characterizes every island; cities with 1–5 million residents have more than doubled; mid-size cities have held the same relative proportion of urban residents while smaller cities have declined; and insularity has been a key constraint on urban growth.
7. Since the mid-twentieth century, Cuba has taken the most divergent path to urban and national development with its variant form of socialist cities; improved relations with the United States may bring significant change to Cuban cities.
8. Natural disasters in the Caribbean, Central America, and Mexico compound the challenges of urban poverty.
9. Water concerns plague the region; they range from supplying potable water for expanding urban populations to droughts that threaten the agricultural base of island economies to sea-level rise that threatens low-elevation coastal cities.
10. Cross-border urbanization unfolds unevenly across the region, as, for example, the San Diego-Tijuana example contrasts with the Dominican Republic-Haiti example; however, both processes driving urbanization result from unevenly sized economies, from the demand for unskilled low-cost labor, and different commodity and retail pricing.

The cities of Middle America and the Caribbean reflect many of the historical processes that have affected the region broadly. As home to significant indigenous empires that were decimated prior to and upon the arrival of European colonists, the region's cities reflect a unique blend of cultures and political functions tied to its primary sector economies. The European conquest of the Americas and subsequent nation-state formation imposed one of the most dramatic landscape modifications in the history of the human race. The changes in the physical environment were mirrored by profound shifts in human networks and institutions as the European conquest unleashed a tragic chapter of intercontinental slavery and the annihilation of millions of Native Americans. The human drama that unfolded over the ensuing five centuries built upon and significantly transformed preexisting patterns and processes of urbanization throughout the region (Figure 3.1). This chapter shows how the Caribbean's urbanization has been shaped by the plantation system and slave trade, whereas the urban development of mainland Mexico and Central America unfolded differently. Some Spanish settlements in Central America, Mexico, and the Caribbean replaced indigenous ones (Mexico City), while others served as strategic transshipment points (Havana, Cuba, and Colón, Panama) or military outposts as part of a network of defensive safeguards (Cuba's original seven *villas*). Urban design came from Spanish "military engineers" who aimed to follow guidelines on street width and length, block size, and land use, all of which derived from versions of the colonial Laws of the Indies. Some even trace the superblocks used to build Mexico's newest city, Cancún, to the planning principles of the Spanish Indies (Box 3.1). But influences other than the uniform grid were also at work in Middle America. Some cities evolved more organically by succumbing to the demands of topography, the whims of the region's elites, or security concerns (Figure 3.3). In all instances, however, a spatially and socially segregated settlement emerged, whose irascible imprint persists more than half a millennium later.

The Mexican urban system was forged in large measure by the Aztec pattern of urbanization, which was subjugated militarily by the Spanish so that the colonizers' dual missions of proselytizing and mining could proceed. Mexico's pre-Columbian mining and agricultural system allowed the colonial and independent nation of Mexico to enter the Industrial Revolution before the rest of the region. The urban geographies of Mexico City and Monterrey highlight the relationship between these resource endowments and industrial-led urbanization.

Meanwhile, in Central America and the Caribbean, colonial and postcolonial urbanization followed an agricultural model of urban growth. Primate city functions in two capital cities—San José (Costa Rica) and Panama City (Panama)—were deepened once rail lines opened these cities and their hinterlands to world markets. Caribbean urbanization developed slowly and was restrained in large measure by limited flat terrain and tied to the fortunes of monocultural exports such as sugar, bananas, and spices. The urban geographies of Havana and San Juan highlight this region's urban development as one influenced by external dependency on trade, sugar, slavery and, in the case of San Juan, the United States.

HISTORICAL GEOGRAPHY OF MIDDLE AMERICAN AND CARIBBEAN URBANIZATION

Mexico

The modern Mexican urban system has many deep roots in the precolonial era, when the cities were originally founded. To this day, in many pre-Columbian cities, elements of the indigenous city still play a vital role in mediating social and spatial relations. Tenochtitlán (now Mexico City) is the most famous of these and its tianguis (street markets) are a testament to the enduring legacy of precolonial urbanism. At the time of the Spanish conquest in the sixteenth century, Tenochtitlán was the center of the Aztec Empire and, with a population of approximately 300,000, it was the largest settlement in the Western Hemisphere. The Aztec Empire stretched across a large swath of Meso-America that united some vital urban settlements of various ethnic groups, including the Mayan population in the Yucatán Peninsula; the Tarascos in the present-day states of Michoacán, Jalisco, Colima, and Guanajuato; and the Zapotecas and Mixtecs in the state of Oaxaca.

Two aspects of pre-Columbian settlement geography stand out. First, these large population agglomerations adopted a "city-state" model of organization, whereby a large commercial and religious settlement dominated rural communities and other smaller political-religious localities within their hinterlands. Second, the major urban centers were particularly prominent in the central region of Mexico. At the time of European contact in 1521, it is estimated that the population of this central region was 2.5 million people. This region played a historically significant role in the formation of the subsequent urban agglomerations of the Spanish and is

Box 3.1 From Cancún to Belize City

Donald J. Zeigler, Old Dominion University

Cancún was nothing in 1970. No one lived there. The eastern side of the Yucatán Peninsula, in fact, was so lightly populated it was a territory rather than a state of Mexico. Today, Cancún is one of the most well-known tourist destinations in the Americas (Figure 3.2). The history of the island, with its *zona hotelera*, and the mainland city of Cancún stand as an example of how forward-looking public sector investment can stimulate the development of thoroughly isolated locales. The first six hotels on the island had to be underwritten by the government. Today, there are well over 100 hotels on the island and twice that many on the mainland. By the mid-1990s, the city had grown to 200,000; two decades later it stands at 700,000. Cancún has become the Caribbean equivalent of Mexico's Pacific-coast tourist meccas: Acapulco, Puerto Vallarta, Mazatlán and, most recently, Cabo San Lucas. Geographer Louis Casagrande called this collection of resort enclaves one of the "five nations" of Mexico. He dubbed them "Club Mex," and called attention to the fact that they are *in* Mexico but are *not* Mexican. Rather they exist in tourist space; take their orders from the global economy; conform to international norms of orderliness, punctuality, and cleanliness; and offer more U.S.-style amenities than the Mexican cities that are actually on the U.S. border.

Figure 3.2 Over 100 hotels in the Cancun's zona hotelera offer thousands of jobs to Mexico's youth, preparing them to make a living in the service economy. Here they confront a native inhabitant of the island. *Source:* Photo by D. J. Zeigler.

The tourist trade in Cancún soon outgrew its habitation. The result was the development of Playa del Carmen (initially a fishing village) about an hour's drive south of Cancún, along with its offshore island, Cozumel. As this "tourism urbanization" spread even further south, a new name appeared to sell it as a destination of its own: the Riviera Maya. It incorporated the Mayan city, now in ruins, of Tulum, making cultural tourism an attraction, too. The Riviera Maya has not tried to mimic the glitzy intensity of Cancún Island. Rather, it has gained fame for its self-contained resort enclaves and its eco-friendly parks, *cenotes* (sinkholes that invite diving), and natural landscapes. Though vulnerable to hurricanes, neither Hurricane Wilma in 2005 nor Hurricane Dean in 2007 damped the urban expansion of either Cancún or the Riviera Maya. In fact, growth tentacles have been clawing their way south right across the international border and into Belize.

Tourism is a fickle business, and it appears that Belize's barrier reef (second only to Australia's) has been the most recent off-the-beaten path discovery. Belize City, the country's former capital, has seized the opportunity to become a gateway to the offshore islands, which are only a quick water-taxi ride away. Its airport (now linked directly to the United States by Southwest Airlines) and its cruiseport (large ships still anchor offshore) bring in tourists and the local economy makes it possible for them to enjoy island towns such as San Pedro and Caye Caulker, whose landscapes and economies have more in common with the Riviera Maya than they do with the rest of Belize. The seed planted at Cancún in 1970 has generated an elongated and now transnational arcade that may not appear very urban. But, when considered as a single coastal region, it has well over a million people, more than enough to classify it as a metropolis of its own.

today the core demographic region of Mexico. Tenochtitlán, renamed Mexico City, became the capital of the Spanish Empire.

Mining and agricultural centers constituted the first phase in the colonization of the northern region of Mexico. Spanish mining towns were founded close to important silver mines, whose indigenous settlements included Taxco, Pachuca, Zacatecas, and Guanajuato. These centers and company towns functioned as enclave economies. Bajío, in west-central Mexico, was (re-)constructed during the colonial period as a key base of the agricultural and livestock sector. Abundant natural resources in this region—its fertile plains supported food and fiber for the colonial government—were key factors in its colonization and in the establishment of conditions favorable to future urban growth.

It was not until the second half of the nineteenth century, following Mexico's independence in 1821, that new important regional centers emerged. During this period, moderate regional growth was stimulated through foreign investment and the creation of highway and railroad networks. Until the 1910 Mexican Revolution, foreign investment concentrated in railways and mining. Port development linked the railway network to maritime trade. Together, these technological and commercial links led to a proliferation of mining centers in northern Mexico, which, in

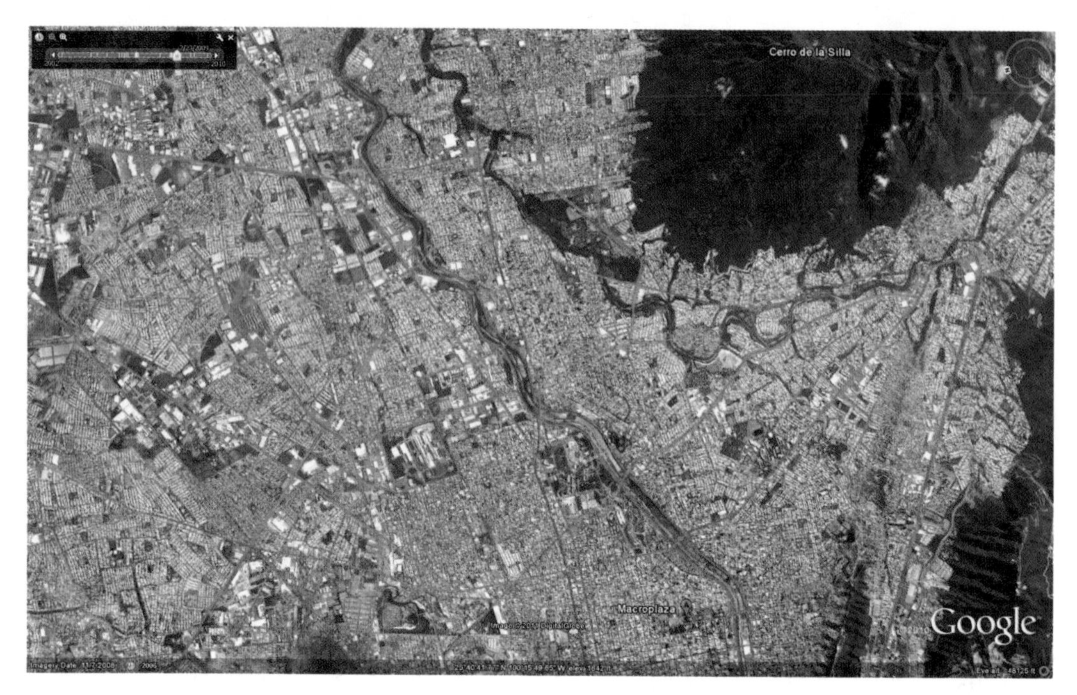

Figure 3.3 A panoramic view of Monterrey illustrates how a distinctive topographic feature, the Cerro de la Silla, can influence the shape of a metropolitan area. *Source:* Google Earth.

turn, triggered regional markets and urban growth.

Railroad expansion played a crucial role in stimulating urban growth in various cities in the central and northern regions of the country. Mérida (the hub of commercial sisal plantations on the Yucatán Peninsula) and Guadalajara, Veracruz, Monterrey, and San Luís Potosí (all with direct transport links to Tampico on the Gulf of Mexico) grew rapidly. Old mining towns in the north gave way to new cities. Monterrey, at one time known as the "Pittsburgh of Mexico," became a major center of heavy industry. Veracruz, a principal transport node, handled nearly all import and export cargo.

The economic, geographic, and political changes that took place in the latter half of the nineteenth century had long-term implications for Mexico's urban system. Although Mexico City remained the country's primate city, other regional centers provided a more diverse economic base and stimulated foreign investment. A communication network facilitated interaction between the central and northern regions of the country. High dependency on exports to the United States largely inhibited the formation of a balanced urban system; and those cities that were the largest agglomerations at the start of the twentieth century retained their economic and political dominance in the subsequent years.

Particular national and international events slowed urban growth in the first decades of the twentieth century. The revolutionary movement within Mexico of 1910–1921 and the global economic depression of the 1930s curtailed exports and funds for urban infrastructure. Nevertheless, between 1900 and 1940, the urban population grew at a rate far greater than the total population, increasing from 1.4 to 3.9 million inhabitants, with the majority of the urban growth concentrated in the largest cities of Mexico. In 1900, there were only two cities with populations greater than 100,000. Yet, these made up one-third of urban Mexico and represented 10.5 percent of the national population. By 1940, there were six cities of this size, accounting for 12 percent of the urban population and 20 percent of the total population. The population of Mexico City had reached 1.5 million, and its primacy ratio had increased; it was nearly seven times larger than the second-largest city, Guadalajara.

At the beginning of the 1970s, a shift toward metropolitan expansion emerged as a new form of urban growth in Mexico City and in some secondary cities. There was a massive rural-to-urban migration flow, with approximately 3 million migrants moving to Mexico City in the 1960s. This gave the capital an annual growth rate of 5.7 percent, which was a historic high. Eleven secondary cities experienced notable metropolitan expansion; three of these—Monterrey, Guadalajara, Puebla—had populations of over half a million. Three border cities—Tijuana and Mexicali in Baja California, and Ciudad Juárez in Chihuahua—expanded significantly and strengthened their relationships with twin cities across the border.

Mexico's border cities grew in importance when the demand for contractual migrant labor during World War II cast these cities as "staging areas" for border crossings of laborers into the United States. Between the 1940s and the early 1960s, the *bracero* workers' program (named for the day-laborers who were contracted) brought specified numbers of Mexican laborers into U.S. corporate farm operations. When the program was discontinued in the 1960s, concern grew for the service industries that had developed on the Mexican side of the "twin cities" and for potential unemployment problems. In response, *maquiladora* factories were established as part of the Border Industrialization Program. This arrangement allowed American companies to import manufacturing parts to Mexican cities, have them assembled in *maquiladora* (i.e., piecemeal assembly) plants, and reimport the finished products into the United States while paying only value-added tax. However, with the creation of the North American Free Trade Act in 1992, the relative locational advantage of being close to the United States dissipated, as trade barriers excluding the rest of Mexico fell for trade with the United States and Canada. Today, Mexican/U.S. border cities retain high levels of manufacturing and service workers and, except for the Mexican twin city of Reynosa-Tamaulipas, have even larger labor markets than their U.S. counterparts (Table 3.1). Accordingly, it is more appropriate to think of these twin cities as a single conurbation, working in similar manufacturing and service sectors, rather than as discrete cities divided by an international boundary (Box 3.2).

Between 1950 and 1970, Mexico's urban population grew at an annual rate of almost 5 percent, while the rural population (in settlements of fewer than 2,500 inhabitants) grew only 1.5 percent. From 1950 to 1970, most demographic factors signaled improvements in the quality of life. Despite this progress, there were significant gaps between urban and rural areas. Millions left the countryside in search of work in cities. Almost half of the rural migrants ended up in Mexico City, and one-fifth went to Guadalajara and Monterrey.

By the 1980s, a process of urban growth decentralization was underway, as intermediate cities in various regions began to

Table 3.1 The U.S.-Mexican Border Twin Cities Phenomenon: Population and Employment, 2009*, 2010**

City	*Population*	*Formal Employment*
El Paso, Texas*	751,296	313,882
Ciudad Juárez, Chihuahua**	1,062,913	396,911
Laredo, Texas*	241,438	79,008
Nuevo Laredo, Tamaulipas**	373,725	75,210
McAllen, Texas*	741,152	213,458
Reynosa, Tamaulipas**	589,466	191,158
Brownsville, Texas*	396,371	115,855
Matamoros, Tamaulipas**	449,815	126,458

Source: *Table 5. Estimates of Population Change for Metropolitan Statistical Areas and Rankings: July 1, 2008 to July 1, 2009 (CBSA-EST2009-05).
Source: *U.S. Census Bureau, Population Division.
Release Date: March 2010.
Source: **INEGI, XIII Censo General de Población y Vivienda, 2010 y Censo Económico 2009.

experience greater growth rates than larger cities. This process took advantage of the opportunities offered by medium-sized cities located close to large cities. Such amenities included lower costs of land and housing, newer infrastructure, more parks and open space, and less congestion.

Today, about 80 percent of Mexico's population lives in cities, but it is important to note that growth rates have dropped for cities of all sizes. And, for the past several decades, growth rates of cities in excess of 1 million residents have been consistent with overall national population growth. Today, 56 metropolitan areas located in 29 of the 32 states in Mexico, account for over half (56 percent) of the country's population and over three-quarters (79 percent) of the urban population.

Central America

Contemporary Central America consists of seven republics—Belize, Guatemala, Honduras, El Salvador, Nicaragua, Costa Rica, and Panama. Current national populations range from 340,000 in Belize to nearly 16 million in Guatemala, rendering the region's countries, and the cities within them, relatively small by global standards. Nevertheless, Central America's cities incorporate considerable ethnic diversity, including Amerindian, African, European and, to a lesser degree, East Asian and eastern Mediterranean influences. City location and population distribution conform in part to the constraints of Central America's physical geography, which is crisscrossed with extensive mountain ranges, fault lines, and volcanoes, as well as innumerable rivers, waterfalls, and lakes. Together with its weather, these environmental conditions combine to make most of the human settlements of the region vulnerable to natural disasters including earthquakes, volcanic eruptions, landslides, floods, and hurricanes. Unfortunately, most Central American countries lack the resources to prevent, prepare for, and/or manage these hazards. Yet, it was the availability of water and land resources that stimulated the growth of settlements on volcanic soil and floodplains, and which in turn

Box 3.2 Industrial Free Zones and Transnational Urbanization

In the era of globalization, even small cities become international. Goods are imported and exported across international boundaries not just in finished form but often as components of products that are truly international, whatever their apparent country of origin. Perhaps the most obvious example in the Central American and Caribbean context, as in much of the developing world, are the industrial free zones that assemble clothing, electronics, medical supplies, and other goods for shipment to the United States. Although local arrangements vary, the components for assembly are commonly imported from the United States partly processed and duty free, assembled by wage laborers in industrial enclaves subsidized by the Central American and Caribbean host countries, then re-exported—again, duty free—for sale in the United States. These industrial free zones, called *zonas francas*, *maquiladoras*, or *zones franches* in the non-English-speaking countries of the region, are often set apart from the rest of the urban landscape by walls or fences and by acres of single-storied white buildings.

Figure 3.4 Satellite image of the "sister" cities Quanaminthe (left) and Dajabón (right). The border between Haiti and the Dominican Republic follows the Massacre River in the bottom half of the image but leaves it in the top half to run more directly north. The industrial free zone, visible as the row of large white buildings near the river at the top of the image, lies in a political no man's land between the border and the river. *Source:* Google Earth.

Industrial free zones have attracted special attention along the United States-Mexico border, but they can also be found along the border between the Dominican Republic (DR) and Haiti. An industrial free zone has been established on the poorer Haitian side of the border near the sister cities of Dajabón (Dominican Republic) and Quanaminthe (Haiti) (Figure 3.4). To the casual observer, the zone seems to be on the Dominican side of the border, on the outskirts of Dajabón. In fact, the international border, which has followed the Massacre River down out of the Central Cordillera to the south, here diverges from the river to enclose a small slice of the "Dominican" side of the river in Haiti. The free zone sits on this narrow island between the river and the border, so that the Haitians who work there cross a special bridge each day to arrive at work. Visitors who enter through the main gate on the Dominican side are crossing the international border, though they might not know it. Like Mexican *maquiladoras,* this free zone takes advantage of the large pool of cheap labor on the Haitian side of the border, though here the assembly plants are partly owned by Dominican entrepreneurs, not just American companies.

Dajabón and Quanaminthe are opposite sides of a deep political divide. Their border is officially open at only a few points, all of which are well staffed by Dominican soldiers. The Massacre River was the site of the brutal 1937 massacre of thousands of Haitians at the order of the DR's dictator-president, an event still in the living memory of many Haitians in Quanaminthe. And yet the two cities are ever more closely tied by commerce. The border is opened twice a week for market days, when hundreds of Haitians cross over to buy goods for later resale in Haiti. Hundreds more Haitians cross every day to work in the free zone. Haitians have flocked to Quanaminthe from the mountains and coastal plain to seek work; and its population grew from about 7,200 persons in 1982 to about 40,000 in 2003, without any urban planning and with few city services. Lena Poschet, of the École Polytechnique Fédérale of Lausanne, found that by 2004 Quanaminthe had five times the population density of Dajabón, with one-twelfth the urban budget.

allowed the development of agriculture as the basic economic activity leading to consequent urbanization.

Human habitation of Central America may date back as far as 10,000 bce, and large-scale civilization flourished between 3,000 and 2,000 years ago. However, given the diffuse and decentralized nature of civilizations in pre-Columbian Central America, the growth of significant cities in Central America dates to the colonial era when Spain set up administrative divisions to govern the region. The Captaincy General of Guatemala—part of the larger Viceroyalty of New Spain—first used the city of Antigua, Guatemala, as its base. But, after a series of earthquakes devastated Antigua, the capital was moved to present-day Guatemala City. Many of the region's cities, in fact, began as provincial capitals under the Spanish: San Salvador in El Salvador, Comayagua in Honduras, Granada in Nicaragua, and Cártago in Costa Rica. Shortly after 1821, the year in which independence from Spain was achieved, most of these provincial capitals became national capitals.

Climatic conditions in Central America in the early colonial period proved unfavorable to agricultural development. Land on the windward Caribbean side receives more precipitation than the leeward Pacific side. At the same time, the Caribbean coast is more prone to hurricanes and is covered by jungle and swampland. Frequent attacks by pirates who roamed the Caribbean were also an important factor in locating settlements at higher elevations where the bulk of the indigenous population resided. In the highlands, Central American cities were dependent on agricultural production in their hinterlands; many settlements were created to group together dispersed agricultural producers in order to consolidate population, impose taxes, and proselytize.

The urbanization process in Central America can largely be divided into three main phases. The first period, from 1821 to 1930, includes the first century of independence from Spain and a subsequent peak of agricultural exports. The second period dates from the 1930s to the 1990s and marks a transition in both the economic model in the region and a new phase of accelerated urbanization. This era was characterized by ideologically motivated political movements which, in most cases, resulted in civil wars. The final period dates from the 1990s and is marked by the end of the Cold War, which led to the easing of tensions between "rebel" groups and national governments, both of which had for decades been co-opted by the United States and other large geopolitical players. This period is also marked by a greater incorporation of the region into the new international division of labor through trade agreements.

The era of independence after 1821 shifted hegemonic control of Central America from Spain to Great Britain and opened new external markets for the region's agricultural produce, which significantly influenced the nature of urbanization in the region. The early decades of independence marked a transition for some countries from the small-scale export of products such as indigo and cochineal (used to make blue and red dyes, respectively) to a more organized trade in cash crops. In the early nineteenth century, sugar cane, tobacco, and coffee were the three main agricultural export products, and by the end of the nineteenth century practically all of the region's countries had focused on coffee exports to satisfy growing global demand. The coffee boom consolidated the Central American capitals. This was especially apparent in Guatemala City, San Salvador in El Salvador, and San José in Costa Rica, where national governments expanded to fill this new political and economic role and city populations and physical expanse grew accordingly. This also laid the foundations for the expansion of a regional landed oligarchy, whose power would be fundamental in the urbanization process for decades to follow.

In the late nineteenth and early twentieth centuries, much of the region's export focus shifted to bananas. Enhancements in storage and sea transport, coupled with relative proximity to the United States made the sparsely populated tropical lowlands of Central America ideal for banana plantations. The industry was dominated by two large U.S. corporations, whose investments in banana production led to the development of key regional infrastructures. Extensive railway networks were developed using national funds in Guatemala and by multinationals in Honduras, which also controlled the docks and port installations of Puerto Barrios in Guatemala and Tela and La Ceiba in Honduras. Together with the coffee economy, banana production actively produced social differentiation through the need for agricultural, transport, and dock laborers in cities, ports, and hinterlands, and for salaried employees in the emerging urban centers. However, the social structures that crystallized from the agricultural export economies of Central America kept the region mainly rural until the twentieth century.

The consolidation of economic activity in agricultural export industries laid the basis for subsequent urban growth, and a series of land transformations catalyzed large-scale urbanization beginning in the 1930s and intensifying in the 1960s. As agriculture expanded in the late nineteenth century, much of the region's common lands (baldíos) were privatized, and various pieces of legislation led to forced labor, which particularly affected indigenous communities. Multinational companies and banks, in particular from the United States, held an increasing influence in local affairs, and large landowning families consolidated both power and land. The result of this was that many peasants were left landless and at the mercy not only of the family and corporate interests that subjugated them, but to global export markets that determined the price of commodities such as coffee.

By the mid-twentieth century, the region's demographic explosion led to a large-scale cityward migration. Since 1966, the population of Central America has tripled from 14 million to more than 45 million, which has been one of the dominant drivers of rapid urbanization. Rural inhabitants, particularly the poor and landless, moved to the region's cities as Central America entered the second stage of the demographic transition (steady fertility rates combined with falling mortality rates). Cities grew in size during this time, and urbanization was further accelerated by the region's civil wars, which had displacing effects on populations. In-migrants sought not only refuge from dire rural conditions, but also access to health care, education, and other social services that only cities provided to any significant degree. Economic change continued, as primary economies gave way to the development of manufacturing in urban centers. In some cases, the expectations of new urban migrants were met; in other cases, they encountered disappointment. Rapid growth left urban governments without the capacity to cope with the increase in population, and the development of shantytowns on the periphery of almost every Central American city was an inevitable outcome. Migrants not only sought better conditions in the region's cities, but a large number also migrated to the United States and elsewhere. Many families faced the difficult decision of pushing one or more family members to migrate, and it is common among Central American women, particularly from El Salvador, to make the long trip to Europe where they characteristically labor in domestic or care-giving work in an effort to support their families economically.

Since the 1990s, urbanization has continued, but with a number of significant changes. Several regional free-trade agreements (notably DR-CAFTA) spurred the expansion of export-processing zones, which feature scores of *maquilas*. This further developed industry in cities such as San Pedro Sula, Honduras, and San Salvador, El Salvador, but has only led to modest increases in wages. Though cities continue to grow in population, urbanization shows signs of slowing as demographic expansion is curbed by falling fertility and continued out-migration, particularly by younger cohorts.

By 2014, Central America was almost 60 percent urban (Table 3.2), ranging from 44 percent in Belize to 76 percent in Costa Rica. National poverty rates seem to mirror urbanization rates. Poverty is highest in Honduras

Table 3.2 Levels of Urbanization in Central America

1970–2015

	**Level of urbanization (%)*				
Country	*1970*	*1980*	*2000*	***2010*	***2015*
Panama	47.6	50.4	65.8	74.8	77.9
Costa Rica	38.8	43.1	59.0	64.3	66.9
Belize	51.0	49.4	47.7	52.7	55.3
El Salvador	39.4	44.1	58.4	61.3	63.1
Nicaragua	47.0	50.3	57.2	57.3	59.0
Guatemala	35.5	37.4	45.1	49.5	52.0
Honduras	28.9	34.9	44.4	48.8	51.4
*Total:					
Latin America & Caribbean	57.2	65.1	75.4	79.4	80.9
Central America	53.8	60.2	68.8	71.7	73.2
Caribbean	45.4	52.3	62.1	66.9	69.3

Countries are ordered by level of urbanization in 2000
*Urban population as a percentage of total population
**Based on 2009 projections, *Source:* http://esa.un.org/unup/p2k0data.asp, accessed February 11, 2011
Source: Data from United Nations, *World Urbanization Prospects, 2009 Revision* (New York: UN Population Division, 2009, http://esa.un.org/unpd/wup/index.htm).

(60 percent) and Guatemala (54 percent), and it is lowest in Panama (28 percent) and Costa Rica (21 percent). The largest and most important urban centers in Central America largely correspond to the seven countries' capital cities and their suburbs. National urban systems are supplemented by a few medium-sized cities such as San Pedro Sula (Honduras), León (Nicaragua), and Davíd (Panama), which often play a commercial function complementing the capital city's administrative role. The distribution of city sizes to some degree reflects the central-place hierarchy of agriculturally based societies, especially within countries with large amounts of pastoral land. Cities such as Santiago (Panama), Liberia (Costa Rica), and Quetzaltenango (Guatemala) play secondary roles tied to commerce and service provision. A number of port cities such as Colón (Panama), Limón and Puntarenas (Costa Rica), and Puerto Cortes (Honduras) serve as export hubs for locally manufactured goods and agricultural products and as importers of goods manufactured overseas.

Despite the economic, cultural, and political development that has occurred within and around many cities, the urban panorama in Central America is not promising. Private developers, often members of the local elite class, are increasingly responsible for suburban housing development in which service provision such as public transportation and garbage collection are minimal. Child and adolescent labor is rampant as impoverished families press their children into petty commerce, service provision, and begging. And aside from persistent poverty in the region's cities, public safety has become one of the most important concerns. On the one hand, this might be seen as a governance failure in which local governments are unable to provide

Box 3.3 Gangs: A Violent Urban Social Development

Throughout the cities of Central America, the materialization of street gangs is a consequence of many factors. Some observers hold that gangs reflect the struggle of some young people in search of an identity. Others argue that gangs are the outcome of widespread and persistent poverty and political disenfranchisement. Most observers concur that, in seeking to improve the quality of their lives and acquire what is otherwise unattainable, some youths resort to gang violence. Gangs are associated with such violent/criminal activities as organized crime, arms trafficking, forgery, gangsterism, rape, kidnapping, extortion, and the sale and consumption of drugs. Some gangs demand "taxes" from bus drivers in order to pass through their territory, while others extort protection money from small business owners who operate on their turf.

In Central America, the most notorious and violent type of gangs are known as *maras*, the most infamous of which is the *Mara Salvatrucha* or MS 13. It is made up primarily of young men between the ages of 12 and 25. Although the *Mara Salvatrucha* is dominant in El Salvador, where it represents approximately 70 percent of all youth gangs, it has spread throughout the Americas from Canada to Colombia. It has taken particular hold in the impoverished border regions of Mexico and cities of Central America where alternative sources of fulfillment are conspicuously absent. These gangs are particularly distinctive in their highly visible use of tattoos, with many gang members having identifying gang tattoos on their faces, necks, chests, and hands.

The word *mara* has become the generic term for youth gangs in Central America. *Mara Salvatrucha* was founded on the streets of Los Angeles by immigrant Salvadoran youths fleeing the Salvadoran civil war. It is alleged that *Mara Salvatrucha* was formed in response to the discrimination and victimization that Salvadoran youths experienced at the hands of ethnic gangs proliferating in Los Angeles in the 1970s. Later, other Central American immigrants were integrated into the gang. The word *Salvatrucha* refers to one who is a "shrewd Salvadoran." It is widely thought that the current proliferation of violent gangs in Central and parts of South America is related to large-scale repatriations from the United States, including many gang members who find fertile conditions in the poverty that is so prevalent in the region's cities. Gangs have come to represent (at least in the popular and political imagination) one of the most serious threats to security and democracy in the region. The formal political power vacuum created by many Central American governments enhances the power of gangs. In many cities, virulent attacks have become an issue of national security. The spread of the *maras* has undermined the authority of the police and weakened the ability of governments to protect communities.

for their citizens. Local police have resorted to military-style tactics and equipment, and armed private security guards are commonly employed to protect businesses and communities. On the other hand, however, it reflects the region's economic woes and the respective social problems that they create. Much of the crime is linked to gangs and their members, who are typically young (12–24) and tend to live in peripheral zones of large cities such as Guatemala City, Tegucigalpa, and San Salvador. Gangs' range of crimes is vast, but their source of income is almost always tied to extortion and drugs, and many have links to international crime syndicates (Box 3.3).

Ironically, Central America's major gangs were in fact developed in Los Angeles, California, by Latino youth, many of whom were displaced to the United States in the wake of civil wars in Central America. The U.S. State Department estimates that there are as many as 85,000 members of MS-13 and 18th Street—the region's primary gangs—in the "Northern Triangle" countries of Honduras, El Salvador, and Guatemala. The influence of the *maras* (gangs) has spread beyond Central America into Mexican, Spanish, and North American cities. The social and economic instability in Central America, evidenced in scarce educational and job opportunities and family disintegration, leaves many urban youths to believe that they have only two viable options: attempt to migrate to the United States or join a gang. Furthermore, many gang members are deportees from the United States with established criminal histories.

Central American countries have some of the highest rates of homicide in the world, and urban homicide rates are even higher in cities such as San Pedro Sula (Honduras), which is reputed to have the highest murder rate for any city in the world (~170 per 100,000). According to a recent article in *The Guardian*, five of the ten most violent cities on Earth are located within Central America. While violence in the region's cities remains high, a truce between El Salvador's two main gangs brokered in 2012 has had a profound effect in reducing violent crime and extortion. But, so far, most Central American countries have not developed adequate social infrastructure to curb the cycle of poverty and violence.

Caribbean

There was little urban culture in the Caribbean prior to the arrival of, and subsequent colonization by, the Spanish. The urban tradition of the region can be traced to two major influences: the Iberian Peninsula and the Caribbean island of La Española (present-day Hispaniola), more specifically the settlement of Santo Domingo. The Spanish urban tradition of the grid, utilized in the Caribbean at the time of conquest and colonization, was inherited from the Romans who, in turn, inherited it from a more basic Greek grid system. The Muslims followed the Greeks and Romans to the Iberian Peninsula and became the third group to contribute to the Spanish urban tradition.

Both Romans and Muslims gave importance to the areas where the main axes intersected, as places of religious, commercial, and social importance—something that was perpetuated in the grid system that the Spanish Crown translocated to the New World. It is worth noting that México's main square, or plaza, is commonly known as *Zócalo*, a word that seems very similar to *zoco*, an Arabic word referring to the main central market of Islamic-influenced cities. Besides the urban tradition accumulated in the Iberian Peninsula, the planning experiences acquired between 1492 and 1508 by the first colonizers in the island of La Española (later divided between the countries of Dominican Republic and Haiti) would constitute the most important and direct model of the urban forms implemented by the Spanish in the rest of the Caribbean.

At the initial stages of the colonization of La Española, Santo Domingo was not only the capital of the Indies, but also the urban planning model for other urban settlements established in the Caribbean. Many of the conquistadores were familiar with Santo Domingo's grid layout. Some plans of the sixteenth and seventeenth centuries for Santo Domingo show an almost perfect urban grid.

It was not long before the Spanish realized that the prospects of finding mineral riches (gold, silver) in the Caribbean were dissipating rapidly. This prompted a massive emigration toward Central and South America, leaving most of the Spanish Caribbean colonies in a state of semi-abandonment. Only Havana (Cuba) and San Juan (Puerto Rico), both protected by massive fortresses, remained important due to their location on Spanish treasure fleets' routes. However, the lack of attention from the Crown served to boost pirate activity in the Caribbean and to attract the attention of other European countries, particularly England, France, and the Netherlands. By the 1600s, these less powerful nations began to establish claims in the region, including western Hispaniola (now Haiti), Jamaica, and the Lesser Antilles. They were not looking for gold or silver, but for farmlands, salt, and forests (for timber).

The development of the sugar plantation system in Barbados in the 1640s accelerated, once again, the pace of colonization and settlement and caused the import of millions of slaves into the region. The Spanish colonies, where slaves and sugar had been introduced from the Old World, came late to this revolution, but by the late nineteenth century, Cuba, the Dominican Republic, and Puerto Rico had also become centers of the sugar industry.

Together, the plantation system and the slave trade established the basis for a distinctive Caribbean geography and set the fundamental settlement pattern of Antillean cities, which were first established on protected leeward harbors as trading centers. Raw sugar, molasses, and rum left these ports for Europe; European foods, machinery, and capital passed through them to the interior plantations. The early sugar ports—Bridgetown, Fort-de-France, Kingston, Port-of-Spain, and Charlotte Amalie—have remained important, even as tourism and industrial free zones supplanted sugar.

Four striking patterns highlight the contemporary urbanization and settlement patterns of the Caribbean. First, no Caribbean island is without its primate city. With the exceptions of Havana and San Juan, most primate cities are located on the leeward coast, immune from the steady trade winds and often nestled along a protected bay. These historic ports were well suited for loading sugar and unloading slaves, machinery, and provisions. Colonists built gun sites and forts on commanding hilltops and ridges to protect the locals from marauding pirates or rival European powers.

Second, Caribbean urbanization in the past half-century shows that mid-size cities (500,000–1,000,000 residents) have held the same relative proportion of urban residents, while those cities with fewer than half a million residents have declined (Figure 3.5).

Third, places with 1–5 million residents have more than doubled. These trends are particularly striking given the limited amount of low-lying land along bay fronts, coastal plains, and river valleys that can accommodate city growth. Sea-level rise, a consequence of global climate change, is a looming threat to these low-elevation cities.

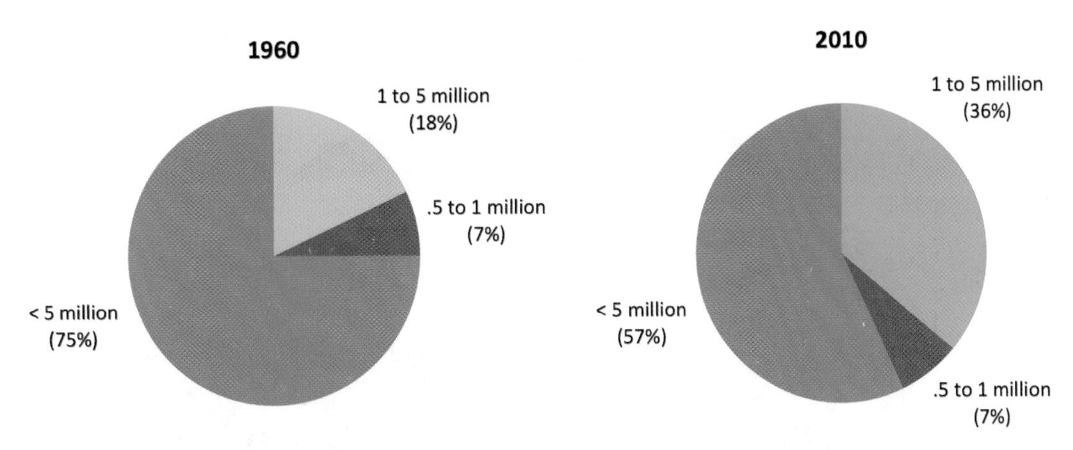

Figure 3.5 Caribbean Urbanization by City Size, 1960 and 2010. *Source:* United Nations, *World Population Prospects: The 2004 Revision and World Urbanization Prospects, 2005 Revision* (New York: United Nations Population Division, 2006, http://esa.un.org/unup).

Fourth, beyond the Greater Antilles (Cuba, Hispaniola, Puerto Rico, Jamaica), insularity is a key constraint. With the exception of Barbados and Trinidad, most of the islands are small, and many of them are restricted by mountainous topography, particularly in the volcanic Lesser Antilles. The scale of Caribbean cities therefore pales in comparison to that of other Latin American cities.

Differing Spanish and English settlement patterns provide historical backdrops to contemporary urbanization. Spanish settlements needed to defend the windward approaches into the Caribbean and major ports on the route of the treasure fleets. These locations marked early landfalls for those ships riding the trade winds, and they relied on nearby forests for shipbuilding and repair. Spanish towns in the Caribbean followed the grid-style settlement plan inherited from the Iberian Peninsula and subsequently dictated throughout the Spanish colonial empire by the Law of the Indies. Towns were centered on the main plaza, usually anchored by a government building (*cabildo*) and church at either end. Block size and street width were predetermined; locally unwanted land uses such as garbage dumps, slaughterhouses, and cemeteries were sited at the periphery of the new towns. Early Spanish Caribbean ports facilitated the extraction of mineral wealth from Mexico and other parts of the mainland, and little urban growth took place in Caribbean ports of the sixteenth century.

Non-Spanish settlements were less orthodox and more haphazard in form. In British settlements, for instance, royal favor was doled out to loyalists by the Proprietary System. Caribbean settlers from England had learned from Atlantic seaboard settlements in North America. Accordingly, their priorities entailed clearing land for timber and agriculture, constructing fortresses, and coming to terms with indigenous peoples. Although the British originally planted tobacco and cotton, they would gradually turn to sugar monoculture. In both British and Spanish settlements, little colonial architecture has survived other than a few military structures, a few churches, and some fortified (brick and stone) sugar plantations because of fire, tropical storms, and rebuilding.

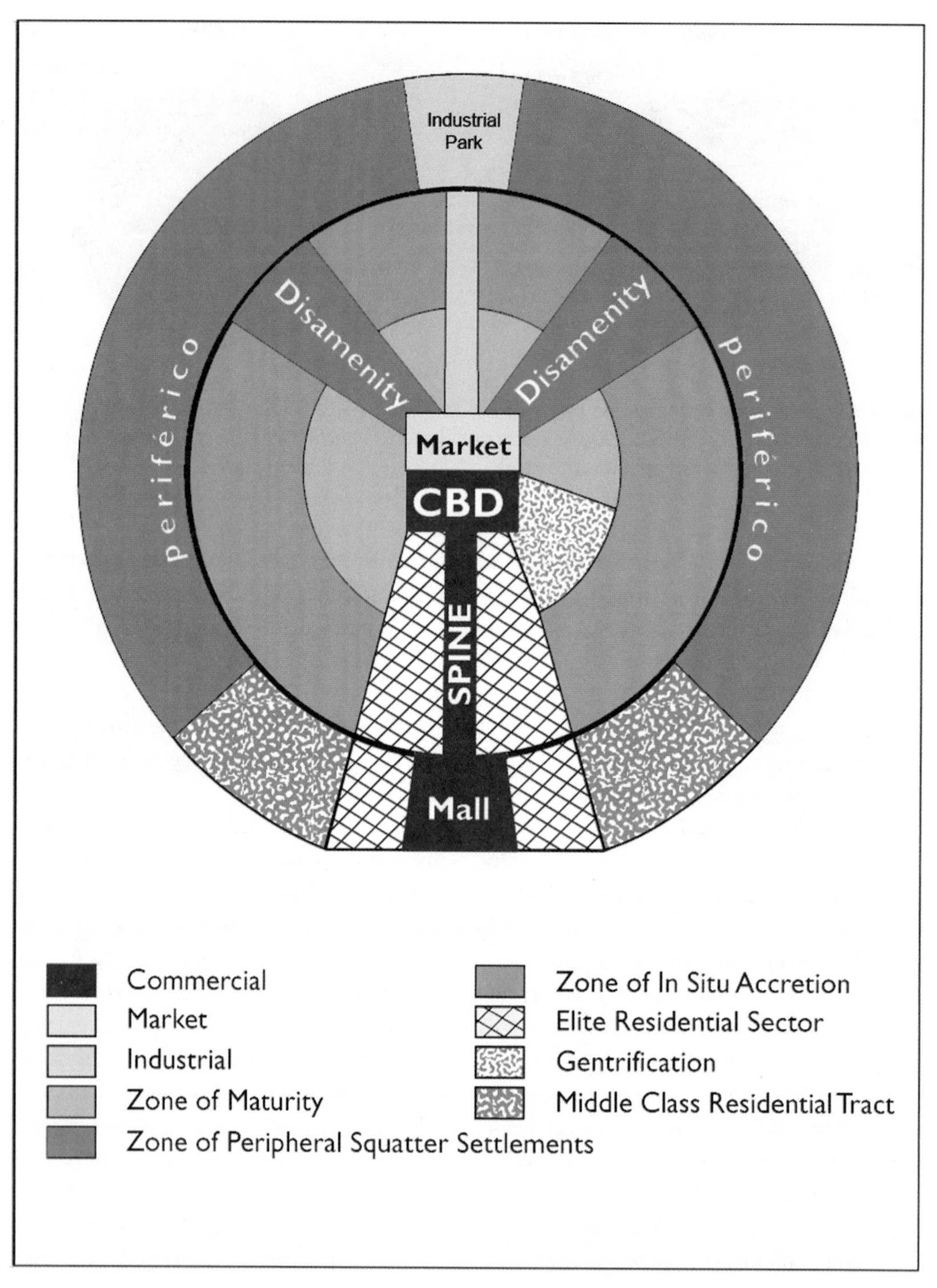

Figure 3.6 The Revised Griffin-Ford Model of Latin American City Structure. *Source:* Larry Ford, "A New Improved Model of Latin American City Structure," *Geographical Review* 86 (1996): 438. Reprinted with permission.

MODELS OF URBAN STRUCTURE

Throughout Latin America, many cities have significant similarities in city structure and urban form. Despite the diversity of histories and the local influences of economics, politics, and topography, cities across the region share such attributes as a clearly defined central business district (CBD) and, more recently, peri-urban highways connecting newer industrial estates and suburbs (Figure 3.6).

Initially proposed in 1980, Ernst Griffin and Larry Ford's model of the Latin American city serves as a valuable starting point for analyzing urban land use throughout the region. Although each city is unique, there are many commonalities—a history of colonial settlement, co-location of commercial and administrative functions at the city center, industrial development along a particular corridor, and significant areas of blight, both in the inner city and on the urban periphery. According to Griffin and Ford's model, the following characteristics are visible in many Latin American cities:

1. A geometric street pattern at the city center surrounding one major *plaza,* laid out in conformity with the Spanish "Laws of the Indies" during the colonial period. The elite resided as close to the *plaza mayor* as possible.
2. An adjacent CBD, featuring retail establishments, restaurants, hotels, and office buildings. These CBDs expanded from the 1930s onward, but their growth has been especially rapid in countries that promoted market reforms starting in the 1990s.
3. A commercial "spine" (such as Mexico City's Paseo de la Reforma) along which an expanding population of élites resides. This serves as an extension of the CBD. When it originally materialized, it contrasted with other residential zones in being well provisioned with services such as water, electricity, and trash pick-up.
4. A "zone of maturity" in which older, "filtered-down" residential areas house the working class and middle class as the urban élite move to newer communities. This zone is characterized by much mixed-use development with shops, *cocinas* (kitchens), and small industrial establishments.
5. A "zone of in situ accretion," referring to newer residential areas where a variety of housing types are in various stages of completion. These areas may still lack all or some public services such as water and electricity, and some streets may not yet be paved.
6. "Disamenities," such as rail or highway infrastructures that interrupt the urban pattern and cause noise and air pollution.
7. A "peripheral" zone of squatter settlements on the city fringe, largely housing new in-migrants from rural areas. These zones are least well connected with urban services and resemble urban "villages" in their structure, with houses that are self-built out of any available materials.

In 1996, Larry Ford published the "new and improved" model of the Latin American city to account for the fact that the region's cities had changed. Notable additions to the original model were:

1. CBD split into a more modern business precinct and a more traditional market area.
2. An "edge city" on the urban periphery, complete with shopping malls and

newer élite neighborhoods. This is often connected by a *periférico*, or ring-road highway.

3. New middle-class residential zones near the city center, often in gentrified neighborhoods that started as homes for the urban elite, filtered down to zones of maturity, and then became gentrified by an expanding middle class.

Since Ford's model was published two decades ago, suburbanization has become a more important process. Gated communities have sprung up around cities in the region, particularly as violence and domestic security become major issues for residents. Multinational companies often prefer the security of suburban office parks, which have direct highway access and often offer better access to airports and newer amenity areas such as shopping malls. San José's (Costa Rica) Escazu and Panama City's Costa Del Este are prominent examples of this newer enclave urbanism.

DISTINCTIVE CITIES

Mexico City: Ancient Aztec Capital, Contemporary Megacity

Mexico City was founded in the fourteenth century by the Aztecs and called Tenochtitlán. It was sited on a lake in the centrally located Valley of Mexico and soon became the anchor city of the largest empire in pre-Columbian Middle America. As the current capital of Mexico, it is also the country's largest urban center and serves as the nation's economic, social, educational, and political hub. With 21 million residents, Mexico City is the second-largest urban agglomeration in the Western Hemisphere, after São Paulo but larger than the New York-Newark metropolitan area. Its population in the twentieth century burgeoned from 3.4 million in 1950 to about 9 million in 1970, and was just shy of 15 million by 1990.

In local parlance, "Mexico City" (known as "day efay" for D.F., *Distrito Federal,* in Spanish) refers to the entire metropolitan area, which covers not only the Federal District but also parts of the states of Mexico and Hidalgo. It stretches over an area of more than 3,000 square miles (7,850 sq km), almost three times the size of the state of Rhode Island. Mexico City is located in a high-altitude, closed-drainage basin, which accounts for many of its ecological difficulties. At an altitude of approximately 7283 ft. (2,250 m) above sea level and hemmed in by mountains, both airborne and waterborne pollutants are concentrated and difficult to disperse.

The *Zócalo* or Main Square—now officially called the *Plaza de la Constitución*—is the traditional center of the city. On the northern side of the square, close to the ancient site of the main Aztec temple is the Metropolitan Cathedral. Spanish conquistadores frequently subjugated the Native American population by having them rebuild churches atop the ruins of indigenous temples (Figure 3.7). In fact, the square was referred to throughout the colonial era as "the pyramid" by Indians who did not want to legitimize Spanish political rule. To the east is the *Palacio de Gobierno* (main government building). Built on the ruins of the ancient Aztec emperor's palace, it is another symbolic replacement of political power. The colonial city extended in an orderly fashion for several blocks around this square, as prescribed by guidelines specified in the Law of the Indies. These orders, first issued in Spain in 1494, became the military engineer's template and mandated the location of many colonial and independence-era

Figure 3.7 The Zócalo (main square) in Mexico City is surrounded by colonial buildings, most notably the Metropolitan Cathedral and the headquarters of the Federal and Capital Governments. *Source:* Photo by Seth Dixon.

buildings in this zone. Many of the original structures and buildings of the traditional urban core—known as the *Centro Histórico*, or Historical Center—remain intact.

In general, Mexico City typifies the urbanization patterns and processes of Middle America and the Caribbean, where the historic quarters of most cities still remain somewhat intact. As a result, Mexico has dozens of World Heritage Sites that celebrate these colonial quarters. The more "modern" aspects of twentieth-century urbanization developed just beyond the *Centro Histórico*. Unlike the Anglo-American and European models of urbanization, the national elite in Spanish America placed more social value on centrality until the twentieth century when congestion and the automobile fueled the need for new suburban construction to serve middle-and upper-income residents. The poor in most Middle American and Caribbean cities tend to concentrate at the city's edge where land values are cheaper, and self-help housing develops. In the case of present-day Mexico City, the wealthy districts are concentrated in the west and various zones in the south, and in *colonias* such as Lomas de Chapultepec, Polanco, and Pedregal de San Ángel (Figure 3.8). These districts contrast sharply with the poverty in the northern zones and the illegal settlements in the eastern edges, beyond Benito Juárez International Airport, where many communities lack basic services.

Suburban retailing now challenges the traditional role of the city center as the main shopping district. Examples include the *Plaza*

Figure 3.8 The elite western corridor connecting Chapultepec Park and the Zócalo is the preeminent place to memorialize Mexican heritage and identity. Here in the Alameda is a monument honoring Benito Juárez, a Zapotec Indian, in neoclassical style. *Source:* Photo by Seth Dixon.

Satélite in the north of the city, *Perisur* in the South, and Sante Fé in the west. *La Merced,* which had been the main food market for the city since colonial times was replaced in the 1980s with a modern market in the east of the city. Nevertheless, La Merced remains the largest traditional food market in the entire city.

The import-substitution industrialization strategy implemented in the 1940s created conditions of stability and prosperity that made Mexico City the most important industrial center in the country. Today, it is responsible for 30 percent of national industrial production. In the second half of the twentieth century, encouraged by the Border Industrialization Program of 1964, heavy industry began moving from the capital to border cities of the north. Just as many U.S. manufacturing towns lost jobs to lower-wage labor in *maquiladoras,* so too did Mexico City. As a result, the under- and unemployed work in informal commerce. The growth of the border cities has somewhat stalled the growth of Mexico City.

Mexico City is also the hub of the national transportation system, a relative location that strengthens its hold on the national economy. Five main highways link the capital to the different regions of the country, as well as with Guatemala and the United States through the 75-year-old Pan American Highway. There is an extensive intra-city transport network, as well, including the Metro system that is used by over 2 million people daily (Figure 3.9), and a range of different types of buses.

As one of the most important cultural centers in the entirety of Latin America, Mexico City boasts major cultural sites, and its cinema, film, theatrical, and television industries rival those of Buenos Aires, Argentina. The *Palacio de Bellas Artes* in the center of the city is an important opera and concert venue, and

Figure 3.9 Mexico City's federally subsidized subway system is incredibly congested at key transfer stations like the Hidalgo interchange downtown. *Source:* Photo by Seth Dixon.

the Cultural Center of the National Autonomous University of Mexico (UNAM) in the south hosts the National Library, a large concert hall, and various theaters. The National Museum of Anthropology is considered one of the most important of its kind, and some monuments, such as the Chapultepec Castle and the Independence Monument, are symbols of Mexican nationhood. Mexico City is a megacity of significant history, impressive scale, and striking contrasts.

San José: Cultural Capital and Ecotourism Gateway

San José is the political and economic capital of Costa Rica. Like most Latin American cities, it is laid out in a grid pattern anchored by a series of town squares fronted by churches. San José has been declared the cultural capital of Spanish America by the Union of Spanish-American City Capitals. Costa Rica's relative economic prosperity and political stability have made its capital the safest city in the región, even if property crime (like theft) remains a problem.

San José is the primate city of Costa Rica. It is more than twice the size of Limón, Costa Rica's second-largest city located on the Caribbean coast. The semi-humid and temperate climate and fertile soils of the Central Valley favor intensive agriculture, and high-quality export products such as specialty leaf tobacco do well here. Since the colonial era, settlement and development have been concentrated in this part of Costa Rica. With time, settlement gradually spread outward toward the coastal plains, a pattern that runs counter to that experienced in most Latin American countries where settlements first took hold at navigable ports on the coasts and gradually moved inland.

The hills within the Central Valley have not curtailed San José's expansion. The metropolitan area today encompasses the adjacent communities of Alajuela, Cártago, and Heredia. This conurbation constitutes the "Central Region" and spills into adjoining valleys and mountain regions. Although the Central Region includes just 15 percent of the country's land area, it accounts for more than half of the nation's population. Wealth generated from Costa Rica's mining and agricultural sectors has historically supported business investment in San José and the subsequent expansion of its metropolitan region. Urban sprawl has overtaken small towns and outlying villages to such an extent that some peripheral zones lack basic services such as housing and schools.

Metropolitan San José, like most primate cities, contains the most important and largest industries, businesses, and residential areas of the country. This concentration implies changes in land use, private-sector investment, and the distribution of wealth. San José consists of 14 *cantones* (administrative units similar to counties in the United States). Most *cantones* are residential areas that function as bedroom communities and are distant from most places of work, retail commerce, and medical and educational facilities. There is a growing demand in the more distant *cantones* for jobs, housing, and infrastructure to accommodate the city's growth.

Continued growth reinforces San José's primacy and disadvantages other regions of Costa Rica that are less populated. Urban sprawl imposes high economic costs, necessitates the consumption of fossil fuels, and exacts human costs in the form of long and stressful commutes. A road network unable to accommodate present usage exacerbates these problems. Moreover, San José's sprawl threatens rich agricultural and protected lands in the Central Valley. In general, rapid growth and congestion threaten the sustainability of this capital city.

Taking advantage of both its physical geography and its reputation for safety and low crime rates, Costa Rica has seen the rise of a successful tourism industry over the past several decades. The country has a number of natural advantages: a variety of natural ecosystems, vertically zoned climates, classic rainforests, scenic mountains, spectacular physical features such as active volcanoes, and surfable beaches. Costa Rica has reaped the benefits of these natural advantages by promoting itself as a world-class ecotourism destination. Together, ecotourism and cultural tourism have energized the national economy as they have attracted hard currency expended by visitors from North America, Europe, and Asia. Most tourist ventures start in San José, the country's transport hub, and fan out to the interior of the country, to Arenal Volcano, and to the Pacific Coast. Although this creates economic multipliers for San José and its hinterland, it also creates economic and environmental stress. The tourism infrastructure (e.g., expansive networks of hotels, restaurants, and land and air transportation) must be maintained and upgraded continually to meet international expectations. San José experiences the financial, infrastructural, and environmental pressures that accompany Costa Rica's international notoriety as a safe, secure, and high-quality tourist destination.

Havana: The Once and Future Hub of the Caribbean?

Diego de Velázquez de Cuéllar founded San Cristóbal de Habana in 1519 as one of seven military outposts (*villas*) around the island of Cuba. Havana was originally located in 1514 on the Broa Inlet, at the Gulf of Batabanó, on the island's southern (Caribbean) side. The shallow port and the generally swampy (and unhealthy) site forced colonists to relocate to the northern side of the narrow island, where they found a deepwater harbor. The relative location of this new site was enhanced by the discovery of the Bahamian Channel, which served as a key transshipment route for goods exchanged between the Americas and Europe.

Military engineers enhanced the colonial port by building a network of fortresses over the next two and a half centuries (Figure 3.10). Flotillas carrying wealth out of the ports of Cartagena and Santa Marta in Colombia, Nombre de Dios in Panama, and Veracruz in Mexico would dock in the safe waters of Havana before crossing the Atlantic for Seville, Spain. Ranching, timber, shipping, and allied services would define the colonial city's economy. It lacked the wealth of Lima and Mexico City, but Havana served as a vital link in the Spanish colonial empire; its location on a pocket-shaped bay made it an ideal warehouse and transshipment point. In fact, the entrance to the harbor from the Florida Straits is so narrow that military officers often drew chains across it at night to entrap intruders. Located on a plain with mild marine-terrace escarpments, the city is unconstrained by topographic barriers except for the bay, which curtailed growth to the east until a tunnel was completed in 1957.

A sugar boom in the late eighteenth century brought commerce and residents to Havana, and crowding exacerbated problems within the walled city. New neighborhoods sprung up outside the walls, and the elite gradually left the walled quarters. In the early 1860s, Havana's walls were torn down, opening up a huge expanse of city blocks that were ideal for urban development.

When the United States occupied Cuba after the 1898 Spanish-American War, they found Havana to be a lackluster place, but one that

Figure 3.10 A lighthouse at Moro Castle stands at the entrance to Havana harbor, while young Cubans use the deteriorating sea wall as a recreational resource. *Source:* Photo by D. J. Zeigler.

was ripe for investment. Road building, railroad expansion, banks, customs houses, sugar and cigar-factory construction, telephone services, and the newly arrived automobile industry offered opportunities for American capitalists. The U.S. Army Corps of Engineers lent a hand, particularly in expanding, raising, and extending the seaside promenade El Malecón—a striking seaside boulevard that graces much of Havana's northern edge.

Over the course of the twentieth century, Havana became a horizontal city in the mode of Los Angeles. It developed a series of suburban enclaves west and south of the bay. Automobile commuting for a middle class of white-collar workers drove this suburbanization model and led to a scattered and deeply segregated pattern of urban growth. While a streetcar network operated until the early 1950s, automobiles and buses linked Havana's new suburban and exurban developments. When the Cuban Revolution of 1959 succeeded, about one in 20 residents were living in shantytowns of some sort (Figure 3.11). The socialist government imported models of high-rise prefabricated buildings like those used in the former Soviet Union. While only 1 million residents claimed Havana as their home in 1959, the population had barely surpassed the 2 million mark fifty years later. Over the same period, Mexico City and Lima, Peru, had increased six- and threefold, respectively.

Warfare and revolutions have given twenty-first-century Havana a unique urban morphology. It is a polycentric city that has preserved distinctive architectural designs and land uses. Colonial, Republican, new government centers, and social/cultural districts characterize this panoply of urban nodes (Figure 3.12). Havana is one of the few Latin American cities where rather benign light industry (i.e., cigar making) surrounds a city center that has served both colonial and Republican governments.

During the first three decades of Communist rule, Havana was largely a "closed"

Figure 3.11 Here are two images of a Cuba frozen in time: Che Guevara, one of the leaders of the Cuban Revolution of 1959, and a classic American sedan (one of many still on the road) that arrived prior to the Revolution. *Source:* Photo by D. J. Zeigler.

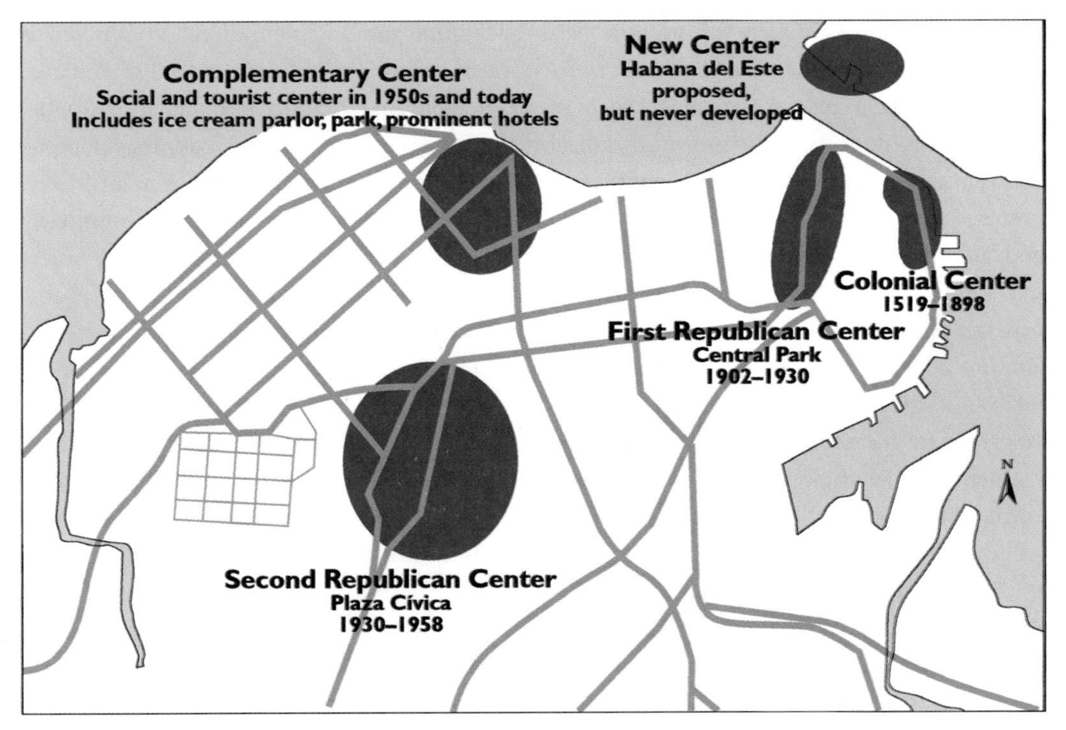

Figure 3.12 The polycentric city of Havana. *Source:* Based on J. Scarpaci, R. Segre, and M. Coyula, *Havana: Two Faces of the Antillean Metropolis* (Chapel Hill, NC: University of North Carolina Press, 2002), 87.

destination; few tourists came, and those who did hailed primarily from Soviet bloc states. Immigration to Havana from elsewhere was strictly controlled by a food-ration book (*la libreta*) and other governmental controls. However, the demise of the Soviet Union in 1991 led to a major crisis called the "Special Period in a Time of Peace." The government tightened gasoline rations as Cuba's ability to exchange sugar for Soviet oil disappeared. Thousands of un- and underemployed Cubans have migrated illegally to Havana, mainly from the eastern provinces where the dwindling sugar economy has been devastated. In typical Cuban humor, these immigrants are called *palestinos* because they hail from the east.

Other post-Soviet changes are also visible in Havana. As fuel subsidies from the USSR ended, and the relative cost of gasoline soared, bus routes were scaled back to half their number and bicycling as a means of transportation boomed. Tourism was seen as a "necessary evil" to sustain the island's economy, and the city's Old Havana district (Habana Vieja), a UNESCO World Heritage Site since 1982, became a prime destination for newfound cultural tourism. In 1993, the City Historian of Havana created a money-making corporation to address housing, hotel construction, road paving, plaza reconstruction, and urban revitalization. This firm, Habaguanex, has become one of the most powerful state enterprises in post-Soviet Havana. It has embarked on an ambitious project to rehabilitate buildings and spaces in Habana Vieja. International tourism has grown significantly from about 25,000 annual visitors to Cuba in the late 1970s to approximately 2.5 million visitors today, comparable to the growth in tourism experienced in nearby Cancún, Mexico. In 2011, the Communist Party approved radical free-market changes such as the sale of private homes (with minimum state intervention) and an increase in the number of private-sector jobs. In 2014, reversing sixty years of nonengagement, the United States and Cuba announced the reinstatement of diplomatic relations; and in 2015, the countries opened embassies in one another's capital cities. Time will tell whether these tentative steps toward entrepreneurship and U.S. rapprochement will change the face of a city that has been centrally planned for half a century.

Unlike many other Caribbean cities, Havana is home to a world-class biotechnology industry and boasts the third busiest airport in the region. It possesses the open space for more growth, either in the form of vacation homes for North Americans and returning Cuban-American expatriates or to accommodate an unfettered U.S. tourist market. In contrast to many Caribbean port capitals, Havana attracts only a few thousand cruise-ship passengers annually largely because shipping companies faced legal problems from the United States if they conducted business in Cuba. Nevertheless, the Caribbean manages some of the busiest maritime traffic in the world; approximately 50,000 ships carry 14.5 million tourists annually. Havana will be on the radar of urbanists who are interested in issues of smart growth, sustainable development, and sustainable tourism as the twenty-first century progresses.

Panama City: Child of Globalization

Of all of the cities in Central America and the Caribbean, no city has been linked to globalization and international commerce as much as Panama City, the capital of Panama. Lying on the Pacific Ocean adjacent to the southern entrance to the country's famous canal, Panama City has been a center of trade for five centuries. It is the oldest continuously inhabited city on the Pacific coast of the Americas, and is now one of the region's most prominent commercial centers, having been referred to as a "Tropical Manhattan" and a "Singapore for Central America" in the popular media.

Panama City's long history of international trade began in 1519 with the establishment of a Spanish colonial outpost from which missions to what are today Peru and Ecuador could be launched. From the outset, the city's role was trade-based, initially mediating flows of gold and silver from the New World. The city's contemporary history began with the completion of the Panama Canal in 1914, which established a U.S. presence in Panama to administer the canal and a strip of land on either side known as the Canal Zone. Panama City was located on the edge of the zone, and much of the city's infrastructure was built by the U.S. government. In 1979, the Canal Zone was abolished, and in 1999 the canal was finally turned over to Panama.

While the onset of free-market economics and globalization was actively resisted elsewhere in the region (notably Cuba and Nicaragua), Panama embraced free trade through the establishment of two large free-trade zones, the adoption of the U.S. dollar as its official currency, and the extension of tax concessions to firms relocating to the country. These days, Panama City is a prosperous national capital and one of the region's most economically bouyant cities. It is home to over 90 internationally oriented banks, serves as regional headquarters for several large multinational corporations and nongovernmental organizations (notably the UN), and is rapidly emerging into the status of a true World City (Figure 3.13). By virtue of its relative location, the city is a major shipping and logistics center, with 4 percent of global trade passing through the canal, and nearly a quarter of the world's ships registered in Panama in terms of deadweight tonnage. Panama City is also a regional shopping hub, attracting many people from surrounding countries (notably Colombia and Venezuela) to its mega malls. Albrook Mall—which was built on a former U.S. Air Force Base—is the largest mall in the Americas, and the Multiplaza Mall is stuffed with luxury boutiques. Despite the sharp social divides between the city's rich and poor, high-end development continues with the recent completion of the 70-story Trump Ocean Club and Central America's first Waldorf Astoria hotel, and the *gentrification* of the city's historic Casco Antiguo quarter (Figure 3.14). Materializing now are vast improvements to the Panama Canal that could pump even more investment capital into Panama City as longer and wider container ships are able to navigate easily through the isthmus. Two threats, however, are also materializing to challenge Panama's virtual monopoly on interoceanic commerce. First, a Hong Kong firm has been given a contract by Nicaragua to begin building a canal that would pass through that country, taking advantage of two natural lakes. Second, as global warming continues and the Arctic Ocean becomes more ice-free, the largest ships may find it profitable to avoid the canal in favor of what has been known historically as the Northwest Passage through Canadian waters.

Figure 3.13 The fishing docks and the skyscrapers of Panama City reveal traditional and emerging economic geographies. *Source:* Photo by Thomas Sigler.

Figure 3.14 The Casco Antiguo quarter in Panama City is currently undergoing the process of gentrification. *Source:* Photo by Thomas Sigler.

San Juan: American City Under Stress

San Juan, Puerto Rico, serves as capital of the world's most populous dependency. The island is not an independent state, but a U.S. territory. While most other dependencies in the Caribbean are but small islands with small populations (e.g., Montserrat), Puerto Rico has 3.5 million people, all of whom are citizens of the United States. They travel back and forth between the island and the mainland (particularly New York City and Philadelphia) at will. One would think, therefore, that the island's primate city, San Juan, would set the pace for urban well-being in the region. Instead, the island's governor in 2015 declared Puerto Rico to be essentially bankrupt, unable to pay its bills. Despite serving as the original offshore location for U.S. firms (led by Maidenform) seeking cheap labor, the island has lost its comparative advantage in the labor market to other less developed parts of the world. So far, nothing has emerged to sustain the island's economic growth, even though its pharmaceutical industry is one of the world's largest and most sophisticated. Even tourism lags behind many other Caribbean destinations and must now face new competition from Cuba.

San Juan is the hub of Puerto Rico. The city is situated astride a large bay on the northern (Atlantic) coast of the island. An elevated promontory of fresh breezes offers an ample view of sea and land. San Juan played a prime role in the geopolitical scheme of the Spanish conquest. More than a port, it was the key to the Americas' front door. The early settlement consisted of agglomerations gathered around churches and government buildings that functioned as emblematic landmarks, consistent with the Law of the Indies (Figure 3.15). Sketches from 1625 show a somewhat developed, but still incomplete, rectilinear urban outline with a rectilinear grid and central plaza. Change came during the following

Figure 3.15 The Plaza de Armas in San Juan, Puerto Rico, is now used not for drilling troops but for enhancing urban life. Fountains are common components of plazas in Spanish cities. *Source:* Photo by D. J. Zeigler.

century, when San Juan turned into a carelessly urbanized village. The poor barrios became more concentrated, first on San Juan's periphery. Eventually, as the terrain occupied by self-built housing became saturated, many poor built their humble *bohíos* (straw huts) and houses in the urban center itself.

The nineteenth century was definitive for the construction of San Juan and the island. It saw the formation of a unique Puerto Rican identity and national consciousness. San Juan reached its highest level of urbanization, becoming a neoclassic urban settlement and a center of distribution and marketing as agriculture, commerce, and industry experienced vigorous growth. San Juan benefited greatly from these economic developments; the wealthy were able to replace their wooden houses with stone ones, many of them multistoried, promoting urban densification. The improved economy attracted more people to the city, and, by the early nineteenth century, the poor barrios were disappearing to make space for new public works, for speculative residences, and for wealthier families. By the end of the century, the poor barrios inside the walls had been eliminated, private construction had become more common, and San Juan, limited by its walls, was overcrowded. The interior patios that served as orchards and gardens were occupied, and people began building on the roofs of existing buildings, beginning a process of vertical growth. Rooms on the first floors of residences were subdivided; the few remaining internal open spaces and the streets became places of domestic chores, artisanal jobs, and socialization. The city became dense, messy, unruly, and unhygienic. Poor living conditions included pests, illnesses, high rents, hunger, work insecurity, political tensions, and crime. Several wealthy families decided to move their residences outside the walls, a trend that would eventually transform San Juan. Nearby hills that had previously been considered impregnable due to their topography were occupied by new settlements. The urban profile, previously dominated by church domes and bell towers, and military structures, showed the presence of new private capital with the addition of the Territorial and Agricultural Bank. Despite other towns existing in the interior of the island, San Juan maintained its supremacy, even though it had left behind its role as Gateway to the Americas.

With the U.S. takeover of Puerto Rico in 1898, San Juan gained even more importance as a commercial, political, and economic center. New tourism developed, boosting the proliferation of small hotels. Industrial complexes took form outside the walls of the old city, opening the way for urbanization along the Central Road, which connected San Juan to the interior and also leading to the growth of poor barrios. Between 1899 and 1910, Santurce, the capital city's barrio with the most area, tripled its population.

Urban developments of the early twentieth century were pivotal in the history of San Juan and the island. Schools began to operate in English, medical centers were built on the urban fringe, and new water, sewage, and electric power systems were installed. A boom in private development accompanied the public works. The construction of modern residences for the wealthy outside the old city utilized the services of local, and recently arrived, architects. Still, despite the economic bonanza, overcrowding continued.

With the Americans also came the suburb. The first suburb in San Juan was made possible by a streetcar network and was aimed at the wealthier classes. Plots of land were sold to private individuals, and houses of several styles brought from the U.S. mainland—chalets, bungalows, and cottages—were offered as possible options. Situated on ample plots surrounded by gardens, wealthy families of Puerto Rico soon preferred this type of suburban residence. Several plots were reserved for civic, institutional, and touristic uses. This was the beginning of San Juan's, and eventually Puerto Rico's, suburban landscape.

URBAN CHALLENGES

Shifting Patterns of City Growth

Some of Middle America's and the Caribbean's largest metropolises, particularly those in Mexico, reveal slowing rates of urban growth that can be attributed to both the demographic transition and a decline in rural-to-urban migration. Nevertheless, these metropolitan expanses increasingly spread out well beyond their original limits. As a result, the urban settlement has become more dispersed and cities increasingly encroach on the adjacent countryside. Urbanization has spatially, economically, and socially incorporated many smaller towns and cities in this process. Guadalajara, Puebla, and Mexico City in Mexico, San José in Costa Rica, and Guatemala City reflect this process. Tools of urban and regional planning are needed to manage this level of urbanization. Despite a population loss or "hollowing out" of the city center in these metropolises, the capital city continues to dominate in most countries.

Mid-size cities, on the other hand, have maintained an impressive rate of population growth. They offer new promise for job creation and enhanced quality of life as opposed to the very large cities. Nevertheless, urban planners and administrators in these cities will be challenged to avoid replicating the problems that have plagued larger metropolitan areas. The viability of intermediate-sized cities will depend mainly on their economies, including the degree of integration at the global scale, the type of articulation that they maintain at the national and regional level, and the extent to which they can tap into their comparative advantages.

Social and Spatial Segregation

Social and geographic segregation in the region's cities has deepened and is a serious problem. Demand for exclusive high-income communities often leads to displacement of poor groups from targeted urban neighborhoods. Public-housing projects concentrate at the city's edge because of lower land values. In turn, this exacerbates social and spatial segregation. High-income groups increasingly isolate themselves defensively in limited-access or gated communities that feature costly houses; attractive retail, entertainment, and recreational facilities; and proximity to work, school, and other amenities. Houses of the poor households continue to occupy precarious locations on remote and marginal lands near landfills, utility plants, factories, water-treatment plants, flood plains, and on steep terrain (Figure 3.16).

Natural Disasters and Vulnerable Cities

Natural disasters compound regional poverty. Active earthquake faults run through Mexico, Central America, and the Caribbean, and active volcanoes line these fault zones, except in the Greater Antilles. Most of the region experiences periodic intense rainfall, tropical cyclones, and sometimes severe hurricanes. The steep, unstable slopes of the area's mountains encourage floods and mass landslides, especially when stripped of their natural vegetation under pressure of urbanization. Natural disasters such as hurricanes merely exacerbate these conditions.

Port-au-Prince, the capital city of Haiti, serves as an illustration. The city is clearly in harm's way. It lies in the middle of the Caribbean hurricane belt, on the edge of the Enriquillo-Plantain Garden Fault zone, and at the base of the steep Massif de la Selle. As Port-au-Prince's population exploded after World War II, shacks and poorly built cinder-block houses spread beyond the city's traditional boundaries onto rapidly growing coastal mudflats, urban washes, and the slopes of Morne l'Hôpital, the mountain that looms directly behind the city. Until 2010, flooding and landslides were the most common risks; even normal storms sometimes caused deadly floods along the city's crowded ravines.

These disasters were forgotten, however, in the massive earthquake of January 12, 2010, which killed more than 200,000 people and left hundreds of thousands homeless. Satellite images of Port-au-Prince in 2010 illustrate some of the extensive destruction and displacement caused by the 7.0 magnitude quake, which was made much worse by the instability of the urban infrastructure. The ornate presidential palace, reminiscent of French colonial times, collapsed; as did many other buildings, displacing thousands of people to tent camps like the one that took shape in the former military airport just north of the city center (Figures 3.17 and 3.18). Such camps, conspicuous for their blue tarps, popped up anywhere space was available: vacant lots, parks, roadsides, and even a golf course. One of the most conspicuous camps, perhaps symbolic of the

Figure 3.16 Two aerial views of shantytowns (*bidonvilles*) in low-lying areas just north of the Port-au-Prince, Haiti city center. Flooding occurred in these areas after Hurricane Noel struck the island of Hispaniola on October 29–31, 2007. The storm claimed at least 30 lives in the Dominican Republic and 20 in Haiti. *Source:* Photo by Joseph L. Scarpaci.

continuing plight of the refugees, is right across the street from the destroyed presidential palace.

Port-au-Prince is not unique, but rather an extreme example of how natural and human disaster work together in Latin American and Caribbean cities. Nature presents risks, but cities amplify them by destabilizing slopes, by exacerbating flood peaks, by shunting the poor onto flood plains, and by encouraging construction of high-density homes willy-nilly on every available hillside. In Middle America and the Caribbean, poverty and vulnerability are intimately intertwined with natural hazard.

Early warning systems, capable management, and institutional and political development are fundamental steps in dealing with emergency preparedness and rebuilding in cities throughout Middle America and the Caribbean. Strong political will is needed to tackle the problems affecting the daily lives of people living in cities throughout the region, in keeping with twenty-first century goals of economic development through environmental sustainability.

Figure 3.17 Former military airport north of Port-au-Prince city center, July 2009, six months before January 2010 earthquake. *Source:* Google Earth.

Figure 3.18 Tent camp at former military airport north of Port-au-Prince city center, November 2010, ten months after January 2010 earthquake. *Source:* Google Earth.

Managing Flows: Tourism and Drug Trafficking

Globalization tends to reduce barriers for cross-border trade and cash flow, uphold private property and bank secrecy, and facilitate drug trade in popular tourist destinations. The Caribbean, a region close to the Andean region of South America, where 90 percent of the world's cocaine is produced, and the United States, a major market in the region, fits the profile of a locale where tourism and drugs intersect. The geography of the region is conducive to large-scale drug trafficking. The larger islands and mainland countries of Belize, Jamaica, Guyana, Dominican Republic, and Haiti provide remote hinterlands necessary to shelter trafficking activities, whereas smaller, archipelagic states like Puerto Rico, Grenada, the U.S. Virgin Islands, St. Vincent, and the Grenadines provide numerous unguarded entry and exit points for narcotics transport. According to Interpol statistics, drug offenses are positively related not only to population density but also to visitor density. As cities develop their tourist industries, they must also guard against allowing conditions to emerge that are conducive to drug activity.

The geopolitical situation of Puerto Rico and the U.S. Virgin Islands make them ideal transshipment points for drugs to the U.S. mainland because, once inside the territories, packages do not need to clear customs. The Dominican Republic presents a complementary situation. Situated close to Puerto Rico, its coasts are poorly monitored; its interior is mountainous and underpopulated; it has significant poverty levels and poorly paid, underequipped security forces. These factors make it an attractive route to surreptitiously move narcotics from northern South America to Puerto Rico, the closest U.S. territory.

Gated Communities

Though many of the region's cities were founded within walled fortresses, modern gated communities in Middle America had their origins in the 1990s. The emergence of gated communities was motivated by a set of common reasons: fear of crime, desire to control environment, interest in private governance by the higher-income classes, and greater control over the urban development process by private developers. Increasingly, large tracts of peri-urban, family-owned land are being developed as master-planned estates, targeting a range of social classes. This pattern of urban development in the region could be characterized as "enclave urbanism"—a term indicating an urban structure in which social life is geographically fragmented. Armed conflict and everyday violence in the region have led to a proliferation of live-in gated communities, increasing the "new middle class" whose consumption preferences are modeled on those of the educated and landed elite.

Gated communities are often fenced- or walled-off with only one or two main points of entry. More affluent gated communities have armed security. In addition to houses, they usually contain community amenities such as swimming pools and playgrounds. In many cities, high-rise buildings have turned into "vertical" gated communities, with controlled access entry and parking. The same phenomenon has penetrated commercial nodes, which have over time migrated from pedestrian-access streets in city centers to privately run shopping malls. Gated communities have been widely criticized for privatizing social life, interrupting traffic flow, and dividing previously interconnected neighborhoods, but also of being instruments of socioeconomic, and even racial, segregation.

Although in most instances, this modern "enclosure movement" has been self-imposed by its inhabitants, there are cases in which the government has gated a particular community in its public-housing projects, as in the city of Ponce in south-central Puerto Rico. Four neighboring communities exist in Ponce; two are poor and two are wealthy. Of the two poor communities (Dr. Pila and Gándara), only Dr. Pila is gated. Of the two wealthier communities (Alhambra and Extensión Alhambra), only Extensión Alhambra is gated. Dr. Pila is a government-housing project whose residents did not ask to be gated but were gated by the government. They compare their community to a prison, or worse a zoo. Residents feel isolated, not only from the outside world, but also from each other because several inner sections of the project are also walled-off, which hinders community-building behavior. After further examination, the Dr. Pila project did not improve the community. It was perceived as an eyesore that would drive away tourism (an important source of revenue in the area).

In contrast, the wealthier (and private) Extensión Alhambra's gates are self-imposed; its residents agreed to it and feel it has improved their safety. The Alhambra suburb (nongated, wealthy) is one of the first suburban neighborhoods in Ponce and has been unable to cordon itself off. Its residents perceive that this has changed the character of the community since they have had to secure their individual homes. Interestingly, criminal acts against household gardens—symbols, it seems, of wealth, civilization, and class—upset these residents more than actual damage or theft.

Critics claim that the gating of public housing has converted relatively harmless, but already stigmatized, communities into ghettos making them even more stigmatized and dreaded. Gating has also disrupted the function of the city as a place where strangers can meet freely and learn from each other. It would seem that Puerto Rico's social experiment, meant to bring the poorer and the wealthier together in communities, has instead solidified and normalized urban inequality. The gate has become a symbol of social respectability to some and a palpable reminder of contempt that criminalizes others.

PROSPECTS FOR THE FUTURE

Economic Strengths and Vulnerability

Industrial development has spurred urbanization in Middle America. Initiated by the government-led industrialization policies leveraging relatively low labor costs, industry has been catalyzed most recently by free-trade agreements such as the North American Free Trade Agreement (NAFTA) and the Dominican Republic-Central America Free Trade Agreement (CAFTA-DR). Many of the poorer cities of the region are able to attract light manufacturing companies to produce textiles since the cities can offer them a large supply of low-skilled laborers, recent migrants from the countryside. A few of the more affluent metropolitan regions in politically stable countries are competing for more lucrative types of manufacturing investment than textiles. Cities in Mexico such as Monterrey, Mexico City, and Guadalajara have been able to attract high-end industrial corporations, and recently some cities in Central America and the Caribbean have been able to leverage their geographic assets to boost their economic global connections.

Costa Rica is an interesting case in point. It is famously peaceful and politically stable with the core of its labor force concentrated in the greater San José region. In 1998, after studying sites in Indonesia, Thailand, Brazil, Argentina, Chile, and Mexico, Intel chose to build a microprocessor plant in Alajuela, Costa Rica within the metropolitan area. Intel microprocessing chips are found in over 80 percent of the personal computers sold, and that volume of business became the leading driver in transforming Costa Rica's economic focus, propelling it into the globalized economy on more equal footing with the consumers of its electronic goods. Intel's investment in the San José metropolitan area created a climate that earned the trust of other investors and strengthened the overall business community. Intel's addition to Costa Rica also has affected education, business practices, and other foreign investments. Corporate call centers, U.S. medical-supply centers, and major banks are investing in Costa Rica as well.

Ironically, economic globalization creates vulnerabilities for both host countries and transnational corporations located there. In 2014, Intel announced that it would be leaving its Costa Rican microprocessor plant for lower-wage sites in Asia. With the rapid proliferation of smartphones and tablets, the demand for personal computers and the microprocessors in them has declined. Although some fear the loss of an important employer in San José, the number of workers laid off is relatively small. Costa Rica has moved on; it is investing in renewable, green energy. No fossil fuels have been used to generate electricity since 2014. Now the entire country is running on renewable, green energy and Costa Rica is garnering international attention. The infrastructure that Costa Rican cities have in place, coupled with the attention for being a leader in renewable energy, will likely attract new businesses and the country will likely survive the loss of one (albeit important) company.

While cities such as Monterrey and Guadalajara (Mexico) also attract comparable economic investments, many cities do not have the infrastructure and labor pool to attract high-end industry. Many of these cities are textile production centers such as Tegucigalpa (Honduras) and San Salvador (El Salvador). San Salvador's economic growth has been stymied by political upheaval and civil war, which has led to the loss of many manufacturing facilities, forestalling investment by many capital-intensive industries with the exception of textiles. Port-au-Prince, another city with a history of political chaos and limited opportunities for global linkages, is currently working on redeveloping its apparel export industry after the disastrous earthquake of 2010. In an effort to promote investment, the government has created a streamlined electronic process to reduce the time needed to register a limited company in Haiti to 10 days.

Cities in Middle America and the Caribbean are neither at the center of global trade nor on the periphery. As the twenty-first century unfolds, many of the region's cities will encounter possibilities that are currently unimaginable. Because Middle American cities are critically important to the national economies of the region, prospects for sustainable development may well hinge on the extent to which governments throughout the region can insulate important centers from economic vulnerability while concurrently embracing global opportunities.

SUGGESTED READINGS

Brothers, T. S., J. Wilson, and O. Dwyer. 2008. *Caribbean Landscapes: An Interpretive Atlas.* Coconut Beach, FL: Caribbean Studies Press. Surveys characteristic urban and rural landscapes of the Caribbean, using satellite imagery, ground photos, and essays.

Cravey, A. 1998. *Women and Work in Mexico's Maquiladoras*. New York: Rowman & Littlefield. Examines the relationship among economic globalization, gender, and migration in Mexican piecemeal assembly-line industries.

Hernandez, D. 2011. *Down and Delirious in Mexico City*. New York: Scribner. An observer's account of Mexico's capital from the Aztecs to the twenty-first century and from slums to subcultures.

Jaffe, R., ed. 2008. *The Caribbean City*. Leiden, Netherlands: Brill. Presents a spatial, social, and economic overview of the Caribbean's urban nodes with case studies of cities with Spanish, Dutch, French, and English histories.

McGuirk, J. 2014. *Radical Cities: Across Latin America in Search of a New Architecture*. New York: Verso. Features Rio de Janeiro, Buenos Aires, Caracas, Bogotá, Medellín, Lima, Santiago (Chile), and Tijuana, among other cities.

Scarpaci, J., and A. Portela. 2009. *Cuban Landscapes: Heritage, Memory and Place*. New York: Guilford Press. Examines the construction of sugar, slavery, heritage, and political landscapes that shape the meaning of *cubanidad* as seen from disciplines, including landscape architecture, history, popular culture, and geography.

Scarpaci, J., R. Segre, and M. Coyula. 2002. *Havana: Two Faces of the Antillean Metropolis*. Chapel Hill and London: University of North Carolina Press. Reviews five hundred years of urbanization and Havana's spatial configuration as a mirror to periods of economic development, political control, and architectural imprint.

U.N.-Habitat. 2012. *State of Latin American and Caribbean Cities 2012*. New York: United Nations-Habitat. "Presents the current situation of the region's urban world, including the demographic, economic, social, environmental, urban and institutional conditions in which cities are developing."

Ward, P. M., E. R. Jiménez Huerta, and M. M. DiVirgilio. 2015. *Housing Policy in Latin American Cities: A New Generation of Strategies and Approaches for 2016 UN Habitat III*. New York: Routledge. Considers policy choices in dealing with the "first suburbs," or squatter settlements, that came to surround Latin American cities.

West, R. C., and J. P. Augelli. 1989. *Middle America: Its Lands and Peoples*. 3rd ed. Englewood Cliffs, NJ: Prentice Hall. A highly regarded text by two prominent American geographers who worked in Middle America and the Caribbean.

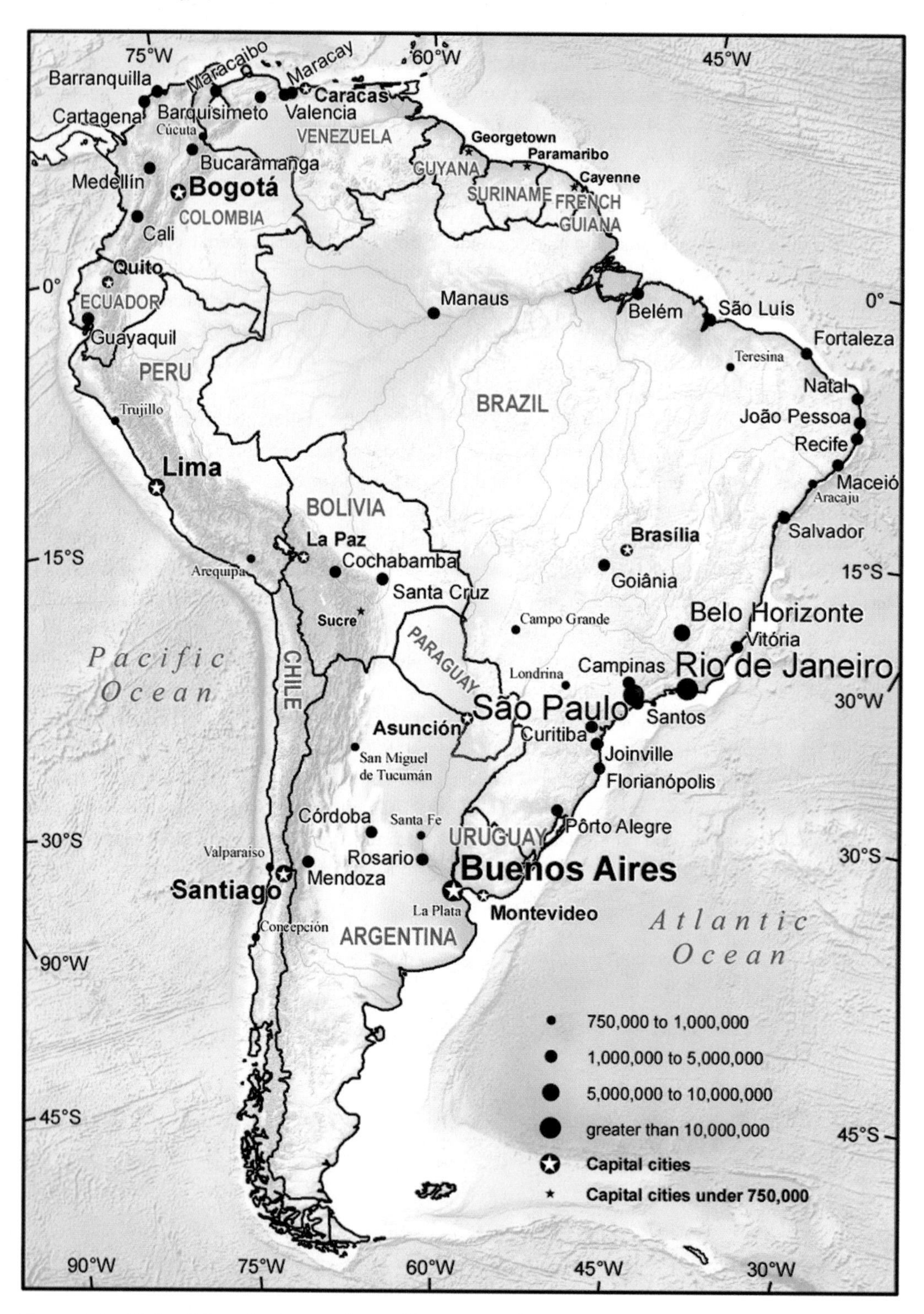

Figure 4.1 Major Urban Agglomerations of South America. *Source:* United Nations, Department of Economic and Social Affairs, Population Division (2014), World Urbanization Prospects: 2014 Revision, http://esa.un.org/unpd/wup/.

4

Cities of South America

BRIAN J. GODFREY AND MAUREEN HAYS-MITCHELL

KEY URBAN FACTS

Total Population	411 million
Percent Urban Population	83%
Total Urban Population	341 million
Most Urbanized Countries	Uruguay (95%) Argentina (92%) Chile (89%)
Least Urbanized Countries	Guyana (29%) Paraguay (59%) Ecuador (63%)
Number of Megacities	4
Number of Cities of More Than 1 Million	45 cities
Three Largest Cities (*Metacity*)	*São Paulo* (21 m), Buenos Aires (15 m), Rio de Janeiro (13 m)
Number of World Cities (Highest Ranking)	8 (São Paulo, Buenos Aires, Santiago, Lima, Bogotá, Montevideo, Caracas, Rio de Janeiro)

KEY CHAPTER THEMES

1. South America is highly urbanized, but its rate of urban growth generally has slowed in recent years.
2. The region contained four of the world's largest megacities of over 10 million people, a total of 45 cities of more than 1 million, and eight world cities in 2015.
3. Cities of Andean America reveal large indigenous and mestizo populations sharing urban space with small elite groups often of European heritage.
4. Southern Cone cities are generally heavily European in ethnic composition as well as in urban planning traditions, although recent migratory trends from Paraguay, Bolivia, and Peru have created more racial diversity.

5. Brazil's cities have a Portuguese colonial heritage and urban forms distinct from their Hispanic counterparts, including significant Afro-Brazilian cultures.
6. Most countries in South America are dominated by a primate city, often the national capital, although dynamic economic centers like Guayaquil, Medellín, and São Paulo also have arisen.
7. The cities (and countries) of South America exhibit extreme disparities in wealth, which is directly reflected in land-use patterns and the quality of life within cities.
8. Economic globalization has mainly benefited a small segment of the urban population, despite intensifying social movements, urban protests, and governmental efforts to address inequities.
9. In recent decades, a rise in urban insecurity and criminality has led to a withdrawal of elites and middle classes from many city centers, often to gated communities, shopping malls, and fortified office parks in suburban areas.
10. Rapid urbanization has caused serious environmental problems, especially air and water pollution, in many South American cities. Innovative efforts to create more inclusive and sustainable forms of urbanization have emerged in several cities, most notably such cities as Bogotá and Curtiba.

South America's cities evoke dramatic, if conflicting, mental images. The mere mention of Rio de Janeiro, Buenos Aires, Bogotá, Caracas, Lima, Quito, or Santiago conjures up images of spectacular natural settings, breathtaking vistas, cosmopolitan populations, picturesque colonial architecture, charming market streets, and impressive modern skylines. By contrast, their names also evoke images of squalid squatter settlements, intractable poverty, random violence, hapless street children, congested motorways, filthy air, and polluted waterways. To be sure, both images accurately portray urban life in South America. Just as the continent is a land of great extremes, so are its cities. Despite outward similarities, regional cities are diverse in form, environment, culture, economic structure, political governance, and quality of life (Figure 4.1).

The continent's urban centers have long participated in the world economy. Since the colonial era, cities throughout the region have served as important global producers and consumers. Today, South American cities vigorously compete for financial, manufacturing, and service-oriented multinational enterprises. Cultural currents from around the world—art, architecture, music, fashion, cuisine, athletic events, and digital technologies—flow across the continent. Both advocates and critics of globalization agree that societies are being propelled in broadly similar socioeconomic, political, and cultural directions. Is it inevitable, then, that places caught up in this process come to look and feel alike? South American cities suggest otherwise.

South American cities generally contrast with those of other world regions, given a series of shared continental characteristics: common colonial legacies of Iberian urbanism; similar histories of economic development; and recent globalization of cultural tastes, production, and technology. Despite such similar trends, the diversity of national and local experiences also stands out. Some cities originated with the Spanish conquest, while others derived from Portuguese colonization. Many are infused with indigenous ways of life,

African cultures that date from the slave trade, European immigrant influences from German and Italian settlement, and relatively small but often prominent groups descended from Asia and the Middle East (Figure 4.2). These cities exhibit dizzying social and cultural diversity, disparate urban forms, contrasting economic levels, and varying forms of governance, all spread across some of the most diverse natural environments on earth.

URBAN PATTERNS IN SOUTH AMERICA

South America's cities may be grouped into three major cultural-ecological regions: (1) Andean America (Colombia, Venezuela, Ecuador, Peru, and Bolivia), (2) the Southern Cone (Chile, Argentina, Uruguay, and Paraguay), and (3) Portuguese America (Brazil). On the continent's northern rim, the three Guianas—Guyana, Suriname, and French Guiana—may appropriately be understood in the Caribbean context (Box 4.1). Even within Spanish and Portuguese South America, despite general adherence to broad continental trends, there are significant urban and regional differences:

- Cities of the Andes reflect a strong indigenous and mixed *(mestizo)* presence, which shares urban space with small elite groups of European heritage. An "alternative economy" of the informal sector and popular markets also dominates these rapidly growing cities. As the continent's fastest-urbanizing region at present, Andean

Figure 4.2 The Pelourinho historic district, named for the "pillory" formerly used to punish slaves, indicates the strong Afro-Brazilian influence in Salvador da Bahia. *Source:* Photo by Brian Godfrey.

Box 4.1 Ethnic Geography of the Guianas

Katie MacDonald, York University

Forming a physical barrier between the Caribbean and South America, the seawall that runs along the coast of the Guianas symbolizes the rich history and culture of these three anomalous countries. Guyana, Suriname, and French Guiana together occupy an ambiguous status: although present on the South American continent, many of their social, economic, and political institutions can be linked more closely to the Caribbean, a trend which is particularly pronounced in the region's coastal cities where the vast majority of the population resides. Examining the history and culture of Georgetown's seawall, provides a lens into the unique ethnic geography of this region (Figure 4.3).

Figure 4.3 Stabroek Market is the main market in Georgetown, Guyana and always bustling with activity. *Source:* Photo by Katie MacDonald.

Designed by Dutch colonists, the seawalls were built by African slaves, imported to work on sugar plantations. Intricately connected to the drainage system of the city (built below sea level), the seawall is today seen as the first defense against flooding. Unfortunately, neither the canals nor the wall has been properly maintained, and Georgetown residents regularly suffer both dangerous canal overflow (including municipal waste and sewage) and serious seawall breaches.

Nevertheless, as a place to escape the constant tropical humidity, at low-tide the seawall is a site for gatherings, and the city's cultural diversity becomes apparent during the many ethnic festivals that occur along it. European colonists (British, Dutch, and French) brought with them Christianity, and today Easter is commemorated by flying kites along the seawall. Diwali sees a motorcade of floats celebrating light and the goddess Lakshmi, who arrived in the Guianas post-emancipation with the indentureship of Asians, primarily from India (although Suriname saw significant Javanese populations). And reminiscent of the Trinidadian Carnival, Guyana's "Mashramani" (derived from the Arawak language) is celebrated with a parade along the seawall, accompanied by the typical Caribbean sounds of steel-pan and soca. Similarly, the weekly pluralistic Sunday "lime" sees people of all ethnicities converge on the wall to talk, listen to Caribbean dancehall or chutney, and consume the wide variety of intercultural delicacies on offer, including metem (an indigenous root soup), Rastafarian "ital" (vegetarian), Creolese "cookup" (a rice-based dish), chowmein, and chicken curry, all washed down with local beer or rum.

The seawall has a darker side, however, reminiscent of the serious social problems that continue to plague the Guianas. Often the site of poverty-driven violence, the seawall is considered a dangerous place, even during festivities. Similarly, the seawall witnesses serious traffic accidents, due both to the mix of donkey/horse carts, pedestrians, and vehicular traffic that it sees, as well as to the lack of enforcement of traffic legislation, often caused by police corruption. This corruption is in part the result of the ongoing ethnic divisions of political institutions, themselves the outcome of divisive colonial practices, and continuous confirmation of these rifts manifests along the seawall during the ethnically divided political rallies that occur there. But the seawall is also the scene of activism on these social issues, and concerned citizens regularly organize to paint the wall with civic messages decrying health, social, and environmental issues.

cities operate under fiscal constraint and hence experience severe social, political, environmental, and infrastructural crises. On the other hand, such cities as Bogotá have become world renowned for their innovative programs of urban planning, mass transit, and sustainability programs.

- In the Southern Cone, cities tend to be heavily European in ethnic heritage as well as cultural traditions—apart from Paraguay, which resembles Andean countries in its significant indigenous presence and socioeconomic indicators. Argentina, Chile, and Uruguay urbanized rapidly during the early twentieth century and the largest cities—Buenos Aires, Santiago, and Montevideo—now grow relatively slowly. Human development indicators tend to suggest prosperity, although these cities have faced economic stagnation and restive middle classes.
- Brazilian cities reflect a Portuguese heritage and language, a unique popular culture, and distinctive urban forms. The Roman name for Portugal was Lusitania, so we speak of Luso-Brazilian colonial cities. Here, large Afro-Brazilian populations make black-white stratification a

key issue. With an estimated 85.7 percent of the 203 million people living in cities by 2015, Brazil has undergone a massive urban transition—even vast Amazonia is now three-quarters urbanized. Greater São Paulo now ranks as the world's third largest megacity and rivals Mexico City as the most populous megalopolis in the Western Hemisphere.

Contemporary Urban Trends

A century ago, less than 10 percent of South Americans resided in urban centers. According to 2015 estimates, 83.3 percent of South Americans now reside in urban areas and populations of all countries are highly urbanized, ranging from a low of 59.7 in Paraguay to a high of 95.3 percent in Uruguay. Five countries are now more than 80 percent urbanized: Argentina and Uruguay surpass 90 percent, with Venezuela, Chile, and Brazil not far behind. Although previously sky-high rates of rural-urban migration and natural population increase have declined somewhat in recent years, South America continues to face major problems stemming from extensive urbanization (Table 4.1).

In addition to the high levels of contemporary urbanization, the sheer size of many South American cities is noteworthy. The continent is home to massive metropolises, whose continuing expansion has transformed the human geography of South America. Throughout the continent one can travel through extensive urban areas, which relentlessly envelop surrounding rural landscapes. Forty-five urban centers contained at least 1 million people in 2015, and seven of them exceeded populations of 5 million. Four urban agglomerations are now among the world's 30 most populous metropolises: São Paulo, Buenos Aires, Rio de Janeiro, and Lima (Table 4.2).

In political-economic terms, much of the continent has industrialized and now sits on the global semi-periphery, combining characteristics both of the more developed "core" and the less developed "periphery." Financial

Table 4.1 Urbanization in South American Countries, 1850–2015

	Percentage of National Population in Urban Areas				
Country	*1850*	*1910*	*1950*	*1980*	*2015*
Argentina	12.0	28.4	65.3	82.9	91.8
Bolivia	4.0	9.2	33.8	45.5	68.5
Brazil	7.0	9.8	36.2	65.5	85.7
Chile	5.9	24.2	58.4	81.2	89.5
Colombia	3.0	7.3	32.7	62.1	76.4
Ecuador	6.0	12.0	28.3	47.0	63.7
Paraguay	4.0	17.7	34.6	41.7	59.7
Peru	5.9	5.4	41.0	64.6	78.6
Uruguay	13.0	26.0	77.9	85.4	95.3
Venezuela	7.0	9.0	46.8	79.2	89.0

Sources: David L. Clawson, *Latin America and the Caribbean: Lands and Peoples* (McGraw-Hill, 2006), 350; Population Division of the Department of Economic and Social Affairs of the United Nations Secretariat, *World Urbanization Prospects: The 2014 Revision*, http://esa.un.org/unpd/wup/unup/index_panel2.html.

circles now regard several countries as leading "emerging markets"—particularly Brazil, Chile, Argentina, and Colombia—but generally South America remains less affluent and more socially stratified than northern counterparts. While the developing countries of Africa and Asia now urbanize rapidly, South America's urban levels already approximate those of North America and Europe. Sadly, urbanization and economic growth have not been synonymous: continental cities have grown rapidly within a highly competitive global system and a regional context of poorly distributed wealth and endemic poverty. As a result, South American cities confront pressing social, economic, political, and environmental issues.

Critical Issues

Urban Primacy and Uneven Regional Development

South America has long been characterized by high rates of urban primacy, reflected in a disproportionate demographic size, political-economic power, and cultural influence of the largest cities. For instance, metropolitan São Paulo now accounts for 10 percent of the Brazilian population and 25 percent of the gross domestic product (GDP). This top-heavy style of urbanization emerged historically, as colonial centers became modern gateway cities with appeal to foreign investors, industrialists, immigrants, and internal in-migrants. The resulting demographic and political-economic dominance of leading metropolises has distorted national systems: concentration of capital and high-level functions has intensified uneven regional development, marginalized the peripheral regions, and motivated "growth-pole" campaigns of regional decentralization. Still, nearly every South American country remains dominated by one or two primate cities. The demographic concentration now often reaches astounding proportions: half of Uruguay's population lives in metropolitan Montevideo; over a third of all Argentines clusters in Greater

Table 4.2 Major Metropolitan Populations of South America, 1930–2015

Metropolitan Area, Ranked by 2015 Estimates	*Population (In Thousands)*				
	1930	*1950*	*1970*	*1990*	*2015*
1. São Paulo, Brazil	1,000	2,334	7,620	14,776	21,660
2. Buenos Aires, Argentina	2,000	5,098	8,105	10,513	15,180
3. Rio de Janeiro, Brazil	1,500	3,026	6,791	9,697	12,902
4. Lima, Peru	250	1,066	2,980	5,830	9,897
5. Bogotá, Colombia	235	630	2,383	4,740	9,765
6. Santiago, Chile	600	1,322	2,647	4,616	6,507
7. Belo Horizonte, Brazil	350	412	1,485	3,548	5,716
8. Brasília	–	36	525	1,863	4,155
9. Medellín, Colombia	–	376	1,260	2,135	3,911
10. Fortaleza, Brazil	–	264	867	2,226	3,880

Sources: Charles S. Sargent, "The Latin American city," in Brian W. Blouet and Olwyn M. Blouet (eds), *Latin America and the Caribbean: A Systematic and Regional Survey*, 188; Population Division of the Department of Economic and Social Affairs of the United Nations Secretariat, *World Urbanization Prospects: The 2014 Revision*, http://esa.un.org/unpd/wup/CD-ROM/Default.aspx

Buenos Aires, while across the Andes more than a third of Chileans reside in the vicinity of Santiago; to the north, nearly one-third of Peruvians can be found in greater Lima and one-fifth of all Colombians are in or around the capital of Bogotá. Several countries are dominated by two primate cities, including Bolivia (La Paz and Santa Cruz); Brazil (São Paulo and Rio de Janeiro); Ecuador (Quito and Guayaquil); and Venezuela (Caracas and Maracaibo) (see Figure 4.1). By contrast, the percentage of the U.S. population residing in metropolitan New York City (5.7 percent in 2015) is much lower than the corresponding figure for any South American country, as seen in Table 4.3.

Economic Polarization and Spatial Segregation

Although South America's megacities are centers of great wealth, this wealth is poorly distributed, a lingering impact of the hierarchical societies that were first implanted during Spanish and Portuguese colonization. Contemporary economic liberalization at the global scale has led to increased socioeconomic polarization nationally and locally. While elites and middle classes have benefited from globalization, and many poor residents have been lifted out of poverty, regular employment remains elusive for a large proportion of urban dwellers. While nearly all South American countries have reduced poverty rates over the last decade, international development agencies estimate that between one-quarter and one-third of the population still lives on less than US$4 a day, depending on the country and the city. This social divide can be read clearly on the urban landscape. On the one side, elite and professional districts are luxurious and boast the latest fashions, high-tech commodities, trendy shopping centers, and upscale entertainment facilities. Concerns with quality of life and security, however, have increasingly led many affluent households to live in gated communities with limited-access residential units, often located

Table 4.3 Percentage of National Population in Largest Metropolis, 1950–2015

Country	*Urban Agglomeration*	*1950*	*1980*	*2010*	*2015*
Uruguay	Montevideo	54.1	49.9	49.2	49.8
Chile	Santiago	21.7	33.2	36.6	36.3
Argentina	Buenos Aires	29.7	33.5	35.3	36.0
Paraguay	Asunción	17.5	24.1	31.6	33.5
Peru	Lima	14.0	25.6	30.6	31.8
Colombia	Bogotá	5.3	13.1	18.3	19.7
Bolivia	Santa Cruz	1.6	6.0	16.6	19.1
Ecuador	Quito	6.0	9.9	10.7	10.6
Brazil	São Paulo	4.3	9.9	10.1	10.3
Venezuela	Caracas	13.6	17.1	10.0	9.3
Brazil	Rio de Janeiro	5.6	7.2	6.3	6.3
USA	New York-Newark	7.8	6.8	5.9	5.7

Source: United Nations Population Division, *World Urbanization Prospects: The 2014 Revision*, File 16: Percentage of the Total Population Residing in Each Urban Agglomeration with 300,000 Inhabitants or More in 2014, by Country, 1950–2030.

far from the traditional elite areas. Conversely, informal settlements of self-constructed and often untitled housing cover vast areas, known as *favelas* in Brazil, *poblaciones callampas* in Chile, *villas miserias* in Argentina, *asentimientos humanos* in Perú, and so on. Here live those forced to cobble together meager livelihoods in the informal economy, laboring in such low-paying and insecure occupations as street trading, in-home manufacturing, domestic service, spot construction, itinerant transportation, and money changing. Much of this population finds itself housed in dangerous structures with poor sanitation and uncertain rights to the land on which their homes are built. As the socioeconomic divide widens in South America, violent crime, personal security, and political unrest are growing concerns.

Economic Restructuring, Structural Adjustment, and Social Movements

While some scholars trace contemporary problems of South American cities back to the socioeconomic divisions of Spanish and Portuguese colonization, others regard them as contemporary manifestations of an unfair global economy. Given ongoing programs of economic restructuring, municipal governments have been forced to curtail expenditures, payrolls, and services in neoliberal South America. Critics argue that such austerity programs of structural adjustment, imposed by international financial organizations and national governments, disadvantage the poor by cutting budgets for social services on which many depend. Such fiscal constraints complicate urban planning and management, given regional contexts of massive urbanization, widespread poverty and inadequate housing, and a host of environmental problems. Meanwhile, many urban dwellers are taking matters into their own hands by participating in self-help social movements for housing, health care, and service provision. Social movements have proliferated in recent decades, serving to fill critical needs cast off as municipal budgets have contracted. As frustrations mount, so too do urban protest, social tension, and political violence. Increasing numbers of urban residents are joining broad-based calls for economic relief, human rights, and environmental justice. In a world of instantaneous communication, their causes are garnering attention far beyond the region (Box 4.2).

Declining Infrastructures and Environmental Degradation

Given governmental budgetary constraints and privatization of many essential functions, the deterioration of urban infrastructures and the reduction of basic services put enormous stress on metropolitan systems. Furthermore, unplanned and unregulated growth exposes vulnerable populations to environmental hazards and health risks, often exacerbated by natural disasters in our era of climate change. For example, earthquakes and volcano eruptions have long been problems in Andean cities, and loss of glaciers imperils the supply of water for those cities. Meanwhile, irregular patterns of precipitation have increasingly struck Brazil and other countries. With basic services inaccessible for many, the quality of urban life steadily erodes. As household and industrial waste and traffic congestion degrade the urban environment, water and air pollution have become pervasive in cities. Of particular significance in recent years have been problems of water provision, since growing climatic

Box 4.2 Water Wars in Cochabamba, Bolivia

Andrea Marston, University of California at Berkeley

When Cochabamba's "Water War" erupted in January 2000, it caught the world's attention. Just a few months before, the city's water supply had been privatized, granting exclusive citywide water rights to the transnational consortium *Aguas del Tunari*, a subsidiary of U.S.-based Bechtel. Local residents were outraged: the new company raised water fees by as much as 150 percent, which particularly alarmed nearby rural farmers and residents of the city's impoverished *Zona Sur* (southern zone). According to contract terms, the company had exclusive rights to *all* regional water, which would in theory permit it to regulate individual and community-based wells and distribution networks. The company would even be able to regulate rainwater collection.

In response, residents formed the Coalition in Defense of Water and Life to demand that the company's contract be rescinded. Their organizing resulted in three major uprisings, finally culminating in a multiday protest in early April 2000, when thousands of farmers, workers, students, and middle-class professionals filled the central plaza. In language that artfully wove together references to human rights, trade unionism, and customary uses—the foundation of common property management in the Bolivian highlands—the coalition's leader, Óscar Olivera, transformed a local struggle around water supply into a broad-based political

Figure 4.4 Irrigators march through Cochabamba in celebration of the National Irrigators' Congress, an important milestone in the process of establishing new forms of water governance in the wake of the water war. *Source:* Photo by Tom Perreault.

movement that resisted the government's neoliberal economic program (Figure 4.4). Smaller protests ricocheted around the country, prompting then president Hugo Banzer Suárez to declare a state of siege. But protesters held strong. By the time the government had signed an agreement on April 10, 2000 to return water management to the municipal utility, at least one person had been killed by police fire.

The Water War became a potent symbol of resistance against the neoliberalization of collective resources. The addition of the phrase "y de la vida" (and of life) to the movement's name was far from incidental; rather, this signaled a refusal to divide water uses into "domestic" and "productive" spheres, and brought into view the multiple *meanings* of water beyond its market value. The coalition united a variety of social groups that had previously worked independently from one another, thereby marking a transition in Bolivian "politics as usual." In fact, the Water War was the first in a series of protests that eventually toppled the government and paved the way for the election of Evo Morales, the nation's first indigenous president.

For residents of Cochabamba's *Zona Sur*, more work remains to be done. The city's rapidly growing periphery, home primarily to rural migrants, is still largely self-provisioning when it comes to water. Many neighborhoods have formed "water committees" to manage community-owned and -operated distribution networks, but other households rely entirely on water tankers that sell water of questionable quality at exorbitant prices. While many neighborhoods have no interest in relinquishing control over their water to the government, they nevertheless seek municipal support to ensure that supply is consistent and safe. In short, they are calling for "co-governance" (*cogestión*) of water between peri-urban communities and the municipal state. But at this point, 15 years after the Water War, their demands remain unmet.

variability and more extreme weather events wreak havoc on cities. The expansion of roads, parking lots, and other impermeable surfaces inhibits the absorption of rainfall, worsens landslides from torrential rains in hilly terrain stripped of natural vegetation, and provokes flooding in low-lying urban areas. Meanwhile, water shortages have become major policy issues in many cities. While poor sectors of society have long suffered from inadequate water provision, climate variability, water pollution, and leaky pipes have exacerbated general shortages of potable water.

HISTORICAL PERSPECTIVES ON SOUTH AMERICAN CITIES

Pre-Columbian Urbanism

Urban settlements have long played an important role in South American societies. The spectacular settings and monumental beauty of the Inca cities of Cuzco and Machu Picchu spring readily to mind. The Inca, however, were only the final stage in a 4,000-year history of urban development in pre-Columbian Andean America, which stretched from present-day Colombia to Chile and Argentina. Even though the urban heritage of Andean

America has attracted most attention, settlements also flourished across a range of ecological settings in the Amazon and other lowland regions. Two notable features unite settlements in these distinctive regions: first, their successful adaptation to the challenges and opportunities of diverse habitats; and second, their nearly total destruction by invading Europeans through violence or disease and, in some cases, their reconstruction to reflect a new colonial system.

Colonial Cities: Spanish versus Portuguese America

After the fleet of Christopher Columbus first made landfall and claimed the newly discovered lands for Spain in 1492, a dispute arose with Portuguese King John II, who argued that previous treaties had given all lands south of the Canary Islands to Portugal. The Treaty of Tordesillas, signed by the two Iberian powers in 1494, attempted to resolve this conflict by dividing the new lands outside of Europe between the Portuguese and Spanish Empires along a meridian 370 leagues west of the Cape Verde islands—thereby demarcating South America along a north-south line at about the mouth of the Amazon River (see Figure 4.1). The lands to the east of this line would belong to Portugal and the lands to the west to Spain. In 1500, Pedro Cabral claimed the Atlantic coast of what is now Brazil, while Francisco Pizarro and other Spanish *conquistadores* consolidated Spanish power in the continent's Andean regions during the early sixteenth century.

After these initial voyages of discovery and conquest, both the Spanish and the Portuguese established settlements to administer, organize, and exploit their new territories. In their urban geographies, Spanish and Portuguese colonization differed in terms of site selection, general morphological characteristics, and geopolitical strategies. In both cases, however, persistent patterns of urban primacy often reflect the enduring importance of the early colonial cities. Cultural and religious landscapes of cities also echo Iberian legacies, such as the dominant Roman Catholic cathedrals and parish churches in central cities and residential neighborhoods.

The main centers of Spanish colonial power in South America lay in the Viceroyalty of Peru, centered on the former Inca Empire in the Andean highlands. The dramatic fall of the Inca Empire provided a rich source of labor, silver, and gold. Spain proceeded to extend this initial conquest with expeditions into other areas of the continent. Spain founded towns both on the coast and in highland areas. Port cities such as Callao on the Pacific, Buenos Aires on the Atlantic, and Cartagena on the Caribbean linked the new colonies to Spain. In highland areas, the Spanish conquered dense indigenous populations, along with their minerals, complex agricultural systems, and other natural resources. The colonial overlords rebuilt important indigenous centers to serve as new imperial cities. They forcibly concentrated the indigenous populations into arbitrarily created villages known as *reducciones,* rebuilt the Inca capital of Cuzco (Figure 4.5), and established such enduring Andean centers as Bogotá, Medellín, Quito, and Potosí. Town founding served as a central instrument of colonization, dominating the countryside and imposing a profoundly urban civilization.

Spain did not centrally plan its earliest colonial settlements, but new towns generally adhered to standards established during the late medieval Reconquista of southern Iberia and codified in the Discovery and Settlement

Figure 4.5 Spanish conquistadores built Mediterranean-style structures atop Inca stone walls in pre-Columbian cities such as Cuzco in present-day Peru. *Source:* Photo by Maureen Hays-Mitchell.

Ordinances of 1573. The so-called "Laws of the Indies" decreed the physical form and location of new Spanish settlements. The Spanish-American city adopted a right-angled gridiron of streets oriented around a central plaza. The imposed urban form served essentially as an effective instrument of social control: urban morphology and social geography were intertwined. Important institutions such as the Roman Catholic cathedral, the town hall *(cabildo)*, the governor's palace, and the commercial arcade bordered the central plaza. Spanish residents clustered around the urban core, often in houses built with the defensive architecture of an external wall and an enclosed inner courtyard. Indians and undesirable land uses were banished to the urban periphery, as in contemporary cities.

In Portuguese America, the Atlantic coast initially proved less alluring than Spain's Andean empire, given its rich silver and gold mines and large indigenous labor forces. Consequently, the Luso-Brazilian settlements initially tended to be smaller and less strictly planned. Most early settlements in Brazil were located at convenient coastal points of interchange between the rural areas of production and Atlantic trade routes. Except for São Paulo, all the towns established before 1600 were located directly on the coast and functioned as administrative centers and military strongholds, ports, and commercial entrepôts, as well as residential and religious centers. To strengthen these strategic footholds, the Portuguese crown began to designate captaincies, or land grants, in 1532.

The captaincy system divided Brazil's coastal strip into about a dozen fiefdoms, where favored Portuguese nobles were entitled to occupy the territory and exploit natural resources. This system combined elements of feudalism and capitalism, so as to require relatively little investment by the Portuguese crown. Yet only a few captaincies were effectively developed. As Brazil was constantly under attack from other European powers, Portugal soon created a more centralized Spanish-style system in 1549, with Salvador da Bahia as its capital. From about 1530 to 1650, sugarcane cultivation on coastal plantations became enormously profitable, powered by imported African slaves in a triangular trade with Europe, Africa, and the Americas. With a population of 100,000 by 1700, Salvador grew to become the most important early Portuguese settlement and the second largest city in the entire Portuguese realm, after Lisbon itself.

The coastal location of most early settlements underscored the importance of a good port and a defensible site, so settlers often favored hilly and topographically irregular terrain in the extensive *Serra do Mar*, the rugged mountains that stretch along much of the central Brazilian seacoast. These towns took on linear, multicentered forms. Irregular mazes of streets focused on a series of squares along the waterfront, as opposed to the more regular grid plans of the Spanish cities. Despite their apparently picturesque confusion of city streets adapted to the topography, early Portuguese settlements adhered to coherent but flexible principles of spatial order. The early colonial towns were set on defensible hilltop sites, where they prominently featured fortifications, important public buildings, churches and convents, and residential areas, all connected by a maze of winding streets and punctuated by ornate public squares. Class-segregated neighborhoods emerged, as elite mansions for rural aristocracy and urban merchant classes were set apart from slave districts. The eighteenth-century gold and diamond boom in Minas Gerais and other areas of the interior provided new wealth and stimulated urban growth, while increasing oversight by Portuguese authorities and encouraging more centrally planned and regulated cities. Late colonial Brazilian cities also witnessed a flowering of baroque art and architecture still notable in the exquisite historic districts of Ouro Preto, Salvador da Bahia, Rio de Janeiro, and other favored cities.

Neocolonial Urbanization: Political Independence, Economic Dependence

Between 1811 and 1830, independence came to each of the countries of South America (except for "the Guianas"). Colonial urban forms persisted, however, long after political independence was achieved. Until the mid-nineteenth century, when elites embarked on campaigns of economic expansion, cities remained relatively small. Thereafter, South America became increasingly integrated into the global economy through the export of primary commodities—beef, minerals, coffee, rubber—and the import of manufactured goods. Focused on trade with North America and Europe, economic expansion fostered population growth, social change, and urban morphological adaptation. Urban growth proceeded with the creation of new transportation links, rural-urban migration, urban infrastructures, and general commercial development. Leading mercantile cities diffused technological innovations and capital investments to inland centers of primary-commodity production, that is, their interior hinterlands. New urban services gave the privileged cities images of modernity and attracted migrants from the interior.

Mounting internal migration and foreign immigration contributed to South America's increasing rates of urbanization (see Table 4.1). By 1905, Buenos Aires' population surpassed 1 million and Rio de Janeiro's exceeded 800,000. Eight other South American cities—São Paulo, Santiago, Montevideo, Salvador, Lima, Recife, Bogotá, and Caracas—had between 100,000 and half a million inhabitants. Correspondingly, the percentage of the national population living in the largest city steadily rose. Commercial expansion and demographic growth led to widespread deficiencies in urban housing, transportation, sanitation, and public health, often the subjects of reform movements. The modern city emerged as entrepreneurs invested in new building projects and planners mounted ambitious public works projects to rationalize urban form. Architects and planners looked to European cities as the main sources of inspiration. For example, as urban renewal programs gentrified the center of Paris into elegant residences for elites, Latin American architects and engineers similarly reformed their own *fin-de-siècle* cities. Buenos Aires and Rio de Janeiro underwent urban renewal programs as they competed for continental leadership. This Eurocentric focus to South American city planning paralleled the continent's political-economic and cultural dependence on neocolonial powers abroad.

Twentieth Century: The Urbanizing Century

As South America moved into the twentieth century, the pace of urbanization accelerated (see Table 4.1). The metropolis, not the countryside, came to define the regional landscape. The continent's neocolonial trade status subsequently shaped the course of early industrialization and urbanization well into the twentieth century. The region's cities were promoted as poles of "modernization," defined in terms of urban-industrial infrastructure and industrial labor. In reality, cities became modern enclaves whose existence facilitated the extraction and basic processing of primary agricultural and mineral products for an export market. Their fate depended on the transfer of technology and expertise from more advanced trading partners, while the benefits of trade largely remained in the metropolitan regions and had little effect on the wider regional economies.

With the worldwide depression of the 1930s, demand for the region's primary products plummeted, unemployment soared, and poverty spread. By the early 1950s, a spirit of economic nationalism gripped most South American governments, as they intervened directly in the workings of their economies. The goal was to alter the pattern of producing primary products for export in favor of producing manufactured goods for domestic, and ultimately foreign, consumption. The development of domestic industry focused on major urban centers because they offered broad access to the national market, a concentrated pool of labor, political influence, and the infrastructure of transport and communication facilities (see Table 4.2). Investment in the urban-industrial sector predominated over the rural sector and life became increasingly untenable for small-scale agricultural producers. Cities attracted rural migrants in the hope of finding jobs, housing, education, health care, educational opportunities, and social mobility for themselves and their families (Figure 4.6).

Initially, most cities were able to accommodate the expanding populations. Rapid industrialization created manufacturing jobs as well as demand for commercial, financial, and public services. Novel building technologies, coupled with new forms of transportation, ensured that living conditions were adequate for the most part. Medical technology made cities relatively healthy places in which to live. As conditions of urban primacy intensified, however, smaller cities languished. Rapidly growing primate cities were as dependent as ever on imported technology, in the form of modern machinery and replacement parts, fostering external indebtedness and balance-of-payment deficits.

To address these shortcomings, national development shifted from an exclusive focus on nurturing domestic industries to a focus on establishing development growth poles. Growth-pole development precipitated elaborate national development plans with a range of outcomes. Chile embraced this strategy, but

Figure 4.6 At 4,000 meters above sea level, Bolivia's capital city La Paz extends throughout and beyond its crater-like valley etched into the Altiplano. The metropolitan region encompasses more than 2 million people and is the largest urban agglomeration in Bolivia. It includes El Alto, a poor and dynamic community perched on the rim of La Paz valley that, with the influx of unemployed tin miners and Aymara migrants, now surpasses La Paz city in population. *Source:* Photo by Maureen Hays-Mitchell.

it proved to reinforce preexisting patterns of industrialization and urban primacy. Brazil invoked this development model in efforts to allay the vast differences in living standards between the more prosperous and industrializing coastal southeast and the largely agrarian and impoverished north and northeast. Although growth-pole development can be credited with the expansion of northeastern industry and large-scale mining and highway projects in Amazonia, it can also be blamed for environmental degradation and enduring socioeconomic deficiencies. The most successful example of growth-pole development occurred in Venezuela, where Ciudad Guayana, founded along the Orinoco River in 1961, benefited from hydroelectric power and mineral resources to become a center of steel and heavy manufacturing.

By the mid-1970s, many growth poles were perceived to be mere enclaves of foreign capital, since investment favored export industries, which were more closely linked to northern firms than to regional or national economies. Hence, most surplus capital left the region, precluding any significant spin-off of related firms and services. Development failed to trickle down the urban hierarchy and, instead, elicited massive cityward migration and further growth of already dominant cities. Few well-paying manufacturing jobs were available to the largely underskilled rural migrants who swarmed to the cities. Most were left to seek employment at low pay and low levels of productivity, further polarizing rich and poor throughout the region (Figure 4.7).

Despite these problems, national governments continued to finance costly development—especially industrialization and infrastructure—through borrowing from foreign capital markets. Northern commercial banks aggressively courted both private and state interests in South America, as nearly every country in the region accumulated significant debt. Yet, each moved steadily along the economic and social development trajectory. Primate cities remained important (see Table 4.3). They served as national headquarters for local ruling groups and multinational enterprises and as centers for the accumulation of capital and diffusion of a globalizing consumer-based lifestyle. Moreover, they provided living space for increasing numbers of working-class and marginalized peoples.

The period between 1950 and 1980 saw consistent improvement in urban living standards. Most urban centers were characterized by an expanding middle class and active government promotion of home ownership. Mortgage systems became more accessible and urban infrastructure and services improved. Water, sanitation, education, medical care, and cultural opportunities were readily accessible. Although updated motorways and increased automobile ownership facilitated the growth of elite suburban communities, cars and mortgages were largely inaccessible to lower-income city dwellers. Consequently, cities underwent explosive growth in self-help housing—primarily squatter settlements—and related programs to service them.

By the early 1980s, however, the global economy had experienced a series of unanticipated shocks that would devastate urban life within the heavily indebted countries of

Figure 4.7 Money-changers on the streets of Lima's historic center jostle to change dollars and Euros as well as "rotos" and "deteriorados"—broken and deteriorated bills. *Source:* Photo by Maureen Hays-Mitchell.

South America. The International Monetary Fund required countries to exercise extreme fiscal restraint at every level of national life, in order to build up state revenue for debt service and eventual repayment. The debt crisis and related reforms precipitated a sustained period of deep recession and development reversal. While most countries transitioned from military to civilian rule by the 1990s, the dominant neoliberal model of privatization and deregulation increased socioeconomic polarization. Factories closed, public-sector employees were laid off, and social programs critical to the poor were slashed. Throughout the region, access to adequate shelter and public services worsened, and physical and social infrastructures deteriorated. Underemployment (the underutilization of one's skills or the inability to secure full-time employment) came to characterize a large portion of the economically active population in many cities.

The early twenty-first century has witnessed a rise of social activism and progressive democratic governments in Brazil, Argentina, Bolivia, Venezuela, and Ecuador; and more moderate-conservative tendencies in Chile, Colombia, and Peru. Creation of the South American Community of Nations (UNASUR) in December 2004 signaled increasing political-economic cooperation, despite remaining conflicts among participants. Increased commodity trade (especially oil, minerals, soy, and other agricultural products) and the rise of China's economic presence have been accompanied by a significant decline in poverty and broadening of domestic markets, most notably in Brazil, now the world's seventh largest economy. On the other hand, the distribution of income remains highly uneven and slum growth continues throughout the region.

DISTINCTIVE CITIES

The spatial structure of South American cities has been an important topic for comparative urban research, given variations in urban form. While distinctive Spanish and Portuguese urban traditions differentiated colonial cities, subsequent postcolonial influences from France, Britain, and the United States broadly affected the region during periods of rapid urbanization. Contemporary cities experience heightened degrees of internal differentiation through inner-city gentrification, affluent suburbanization, land squats by informal communities, gated communities, and peripheral commercial development of "edge cities." In larger metropolises, intense competition for available land often brings wealthy and poor populations into close proximity, while functional decentralization creates urban realms of varying socioeconomic levels, replete with shopping centers, office parks, and gated communities separate from the older CBDs (central business districts).

Although South American cities appear modern, cosmopolitan, and globally connected, they are beset by problems of poverty unparalleled in the Global North. It is tempting to speak of these urban landscapes as "dual cities" in which a modern, affluent, and progressive element has little to do with a poor, obsolete, and unseemly element. In reality, the affluent modern city and the impoverished city are intertwined aspects of the same metropolis. This urban landscape of extreme wealth and poverty epitomizes the region's enduring legacy of underdevelopment, economic polarization, and social injustice. Although every city is distinct, each also reflects the evolving urban experience of South America. Rio de Janeiro and São Paulo, joint anchors of Brazil's dominant city-region, epitomize

Luso-American urbanization, while Brasília deserves study as the most famous newly planned capital of the twentieth century. Lima epitomizes Spanish-American urbanization for Andean America, just as Buenos Aires does for the Southern Cone. Curitiba and Bogotá have played leading roles in urban sustainability planning.

Rio de Janeiro and São Paulo: Anchors of South America's Megalopolis

The vast São Paulo-Rio de Janeiro conurbation in southeastern Brazil stands alone for its urban size and scale in South America. As the twin nerve centers of a vast country and one of the world's leading emerging economies, these two megacities in southeastern Brazil have sprawled to form the joint nuclei of an integrated megalopolis with the population of a medium-sized European country. The Brazilian megalopolis now encompasses some 50 million people—one-quarter of the national population—and generates one-third of the country's GNP. Much of this huge agglomeration lies in São Paulo state, including the capital city-region (with more than 21 million in 2015) and the nearby urban areas of Campinas, Santos, and São José dos Campos (totaling another 10 million people). Altogether, this São Paulo "Expanded Metropolitan Complex" now comprises more than 31.5 million residents in 72 municipalities. Rio de Janeiro's portion of the Brazilian megalopolis includes its own capital city-region (12.9 million), along with the urbanized Paraíba Valley (2 million) and the coastal Costa Verde and Capo Frio/Búzios areas (1.5 million). The metropolis of Juiz de Fora (0.5 million) in the adjacent state of Minas Gerais also forms part of this interconnected urban region, which stretches over an area the size of Austria (Figure 4.8).

The concentration of population and economic activity in southeastern Brazil has contributed to widespread environmental problems, including air and water pollution, seasonally elevated temperatures of urban "heat islands," frequent torrential rains and flash floods, and periodic shortages of potable water. Given the widespread removal of forests, due to agricultural expansion and urbanization, regional watersheds have increasingly been unable to store and release water as reliably as in the past. During the drought of 2014–2015, the reservoirs of the major regional cities virtually went dry. While rainfall was less than normal, infrastructural neglect and pollution of major urban rivers also contributed to the overall shortage of potable water.

Coinciding with World Water Day on 22 March 2015, the environmental organization SOS Mata Atlântica Foundation released a report indicating that nearly a quarter of the 111 Brazilian rivers studied in southeastern Brazil suffered from "bad" or "extremely bad" water quality. Some of the worst levels of water pollution were found in the cities of Rio and São Paulo, which both contain dead rivers devoid of healthy biological activity and unfit for human consumption.

Despite similar environmental and socioeconomic problems, metropolitan São Paulo and Rio de Janeiro retain their own distinct identities. Residents of Rio (known as *Cariocas)* and those of São Paulo (known as *Paulistas*) are famous for their competitive, dueling dispositions. Hackneyed images of the fun-loving, easy-going *Carioca* and the intense, hardworking *Paulista* are exaggerated, but like many stereotypes it reflects a particular social history. Rio de Janeiro—famous for its spectacular seaside views and popular culture of the samba, bossa nova, and carnival celebrations—has long been an international

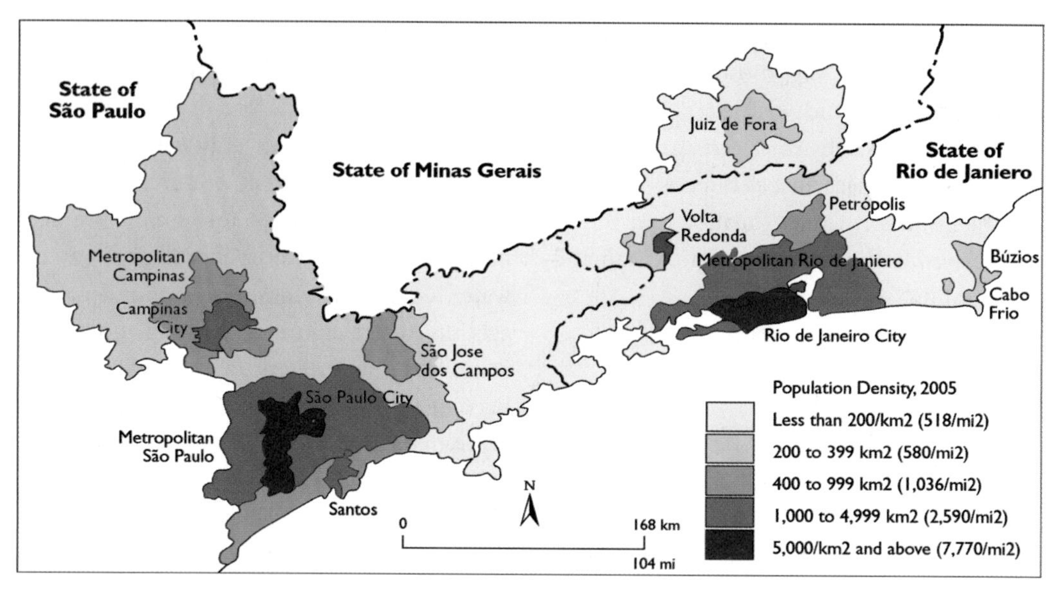

Figure 4.8 The Rio de Janeiro-São Paulo-Campinas extended metropolitan region. *Sources:* Instituto Brasileiro de Geografia e Estatistica (IBGE); Centro de Informações e Dados do Rio de Janeiro (CIDE); and the Fundação Sistema Estadual de Análise de Dados (SEADE), 2007. *Source:* Map by Brian Godfrey.

playground and beach resort. By the time Rio lost the national capital to Brasília in 1960, rival São Paulo had taken the economic and demographic lead in this rapidly modernizing country. As Rio deindustrialized and grew increasingly dependent on tourism and other urban services, São Paulo grew to become the preferred location for multinational industrial, commercial, and financial headquarters in South America. Now considered the business capital of Mercosur—the emerging common market centered on Brazil and Argentina—São Paulo is known as a fast-paced and creative metropolis with distinctly urban charms and daunting socioeconomic and environmental challenges.

Rio de Janeiro: The "Marvelous City"

On March 1, 2015, São Sebastião do Rio de Janeiro commemorated the 450th anniversary of its founding by Portuguese forces, which had just expelled French invaders. After Rio's founding in 1565, the city grew up along Guanabara Bay, one of the world's great natural harbors. With its busy port, the settlement maintained a population of several thousand before the discovery of gold and diamonds in Minas Gerais intensified growth during the eighteenth century. As a result, the colonial capital moved from Salvador da Bahia to Rio de Janeiro in 1763. After the Napoleonic invasion of Portugal, the royal family fled to Brazil and Rio served as capital of the Portuguese realm from 1808 to 1821. The imperial court's arrival stimulated building and establishment of new institutions. Extending along the bay and scaling the surrounding hills, the city acquired a linear spatial pattern (Figure 4.9).

Rio de Janeiro's status as the main seaport and capital of independent Brazil (1822–1960) secured its urban primacy. As the largest national metropolis, the port boomed, industry and commerce prospered, and cultural affairs flourished. Determined to compete with Buenos Aires as South America's most cosmopolitan city, reformers mounted extensive urban renewal programs to transform Rio into a "tropical Paris" during the early twentieth century. Using sanitation campaigns against Yellow Fever as a rallying point, authorities demolished thousands of old buildings to make way for new boulevards and high-rise structures. The port moved to modernized facilities on Guanabara Bay, while new transportation arteries encouraged real estate development in socially sorted neighborhoods during the early twentieth century. Gradually the northern zone became predominantly industrial and working class in character, while affluent populations gravitated to fashionable districts near the southern beaches.

Even as Rio grew, social-class barriers remained in place. The poor are primarily nonwhite and the middle and upper classes largely white—racial disparities that coincide with patterns of residence. While a sharp north-south split plagues Rio's social geography, settlements of self-constructed or "informal" housing known as *favelas* are highly visible on hills above fashionable southern seaside districts. Although these informal communities date from the turn of the twentieth century, they proliferated after World War II. By official figures, nearly one-quarter of Rio's population now resides in more than 600 *favelas,* scattered among the hills and lowland areas. One of Rio's largest *favelas*, Rocinha, has an official population of 70,000, but unofficial estimates range up to 180,000. Providing rent-free housing on public or disputed terrain close to employment, *favela*s have become permanent features on the cityscape, despite recurrent efforts by authorities to remove them.

By the late twentieth century, long-term governmental neglect left a power vacuum that facilitated the rise of drug-trafficking cartels, which gained control of many *favelas.* Not surprisingly, there are strong correlations between impoverished slums afflicted with drug traffic and rates of violent death, particularly among young male residents. Beginning in late 2008, as Rio prepared to host the 2014 FIFA World Cup and the 2016 Summer Olympics, special-operations forces mounted military-style "pacification" operations to rid *favelas* of drug cartels. These campaigns have so far installed units of pacifying police (UPP) in about forty of the city's *favela*s, including the "City of God" district (featured in the famous film).

Authorities have concentrated the "pacification" campaigns on favelas near the international airport, sporting venues, tourist attractions, public transportation system, and wealthy neighborhoods. These locational patterns suggest that the major strategic goal is to protect vital infrastructures, governmental institutions, foreign visitors, and affluent local residents. Governmental and NGO programs to ameliorate conditions in the favelas have focused on infrastructure improvements (e.g., street paving, provision of water and sewerage) and social services (e.g., health center, schools, and recreational facilities). With enhanced security and better services, several communities with UPPs have begun to experience real estate appreciation and even gentrification. On the other hand, violent episodes have poisoned police-community relations and led to protests in several communities. Still, the more accessible *favelas* of the city's southern zone have become tourist attractions for curious foreign visitors, and bed-and-breakfast lodging has sprung up here for adventuresome youthful travelers (Figure 4.10).

Figure 4.9 This panoramic view of Rio de Janeiro includes Sugarloaf Mountain (Pão de Açúcar) at the entrance to Guanabara Bay, Corcovado Mountain with its majestic statue, Rodrigo de Freitas Lake, and the lush forests of Tijuca National Park. *Source:* Photo by Brian Godfrey.

Rio's environmental problems have mounted with urbanization. Torrential summer storms often devastate precariously perched *favelas* and flood low-lying streets below. Fifty years ago, thick hillside vegetation absorbed most of the rainfall, but now the water runs off impermeable urbanized surfaces, dislodges unstable structures, and blocks transportation arteries. Water pollution is another major environmental problem. The state's environmental agency admits that only a third of the city's sewage receives treatment, while most spills raw into rivers, bays, and coastline. A study by the SOS Mata Atlantica NGO, released in March 2015, found that, of fifteen rivers sampled in Rio, ten suffered from "bad" water quality and only five had "normal" levels of contamination. Not a single river studied was considered in "good" or "excellent" condition.

Rio's Guanabara Bay, one of the world's greatest natural harbors, has served as a dumping ground for centuries. While the enclosed, 148-square-mile bay shelters the port from Atlantic storms, the narrow entrance also inhibits the ocean's natural flushing action, particularly with mounting landfill, pollution, and sedimentation. The highly urbanized waterfront now contains large populations, industries, oil refineries, two major airports, the seaport and naval base, and the federal university. For decades, the beaches within the bay have been unsuitable for swimming. About 8 million people live within the bay's watershed, many in precarious housing conditions with little sanitation. Trash and untreated sewage flow into the bay from 55 rivers and streams, most of them highly contaminated. An ambitious clean-up program, launched after the Rio "Earth Summit" of 1992, failed to

Figure 4.10 A view of the Cantagalo district, located on steep hillsides between Copacabana and Ipanema beaches, illustrates the informal, adaptive geography of Rio's favelas. *Source:* Photo by Brian Godfrey.

make significant progress in cleaning up the bay. Recently, concern has mounted that the bay and other local bodies of water may not be clean enough, despite assurances, to hold the sailing, rowing, and other aquatic events for the 2016 Summer Olympics.

Recognition of the city's fragile urban environment came in 2012, when the UNESCO World Heritage Program inscribed a new property, "Rio de Janeiro: Carioca Landscapes between the Mountain and the Sea." The site comprises the key natural elements that have shaped and inspired the city's development, from the peaks of the Tijuca National Park down to the scenic coastal areas. It also includes such cultural attractions as the Botanical Gardens, Corcovado Mountain with its famous statue of Christ the Redeemer, the historic hills around Guanabara Bay, and the designed landscapes along Copacabana Beach. UNESCO has emphasized the need to address the issues of water pollution and environmental conservation. While ambitious conservation and management plans have been approved, the key will lie in the implementation, monitoring, and enforcement of the guidelines. These issues have gained international visibility as a result of hosting international athletic mega-events, which have provoked heated debates about the city's development priorities, environmental problems, and displacement of low-income populations (Box 4.3).

São Paulo: The Making of a Megacity

São Paulo's distinctive origins began with its inland site, which contrasted with the coastal locations of most other early Luso-Brazilian cities. Jesuits founded São Paulo de Piratininga in 1554 on the gently rolling hills of an inland plateau, strategically located at a critical transportation juncture between the coast and interior. Lacking valuable resources or lucrative plantations, the village remained small for three centuries. São Paulo's locational advantage became more apparent during the mid-nineteenth century, when the city became the center of a prosperous coffee-growing region, favored by fertile soils and mild subtropical climate. With railroads financed by British capital, São Paulo became the chief point of transshipment for the lucrative new cash crop. As a result, turn-of-the-century São Paulo grew rapidly, increasingly populated by Italian and Japanese immigrants after the abolition of slavery in 1888 led to a shortage of labor in the coffee fields. Profits from the coffee trade were invested in urban commerce, industry, and real estate development. Enterprising immigrant families made fortunes in food processing, textiles, and other early industries. By the 1920s, São Paulo overtook Rio de Janeiro as the principal industrial center of Brazil.

São Paulo's dizzying growth, sometimes called "three cities in a century," occurred in successive urban layers. During the late nineteenth century, São Paulo rapidly changed from an agricultural boomtown with colonial features, largely constructed of mud and thatch, to a modern industrial and commercial hub. A dense, concentrated city with a high-rise core began to take shape through early programs of urban renewal, as in Rio de Janeiro, during the early twentieth century. Large-scale demolition, redevelopment, and new transportation lines facilitated the growth of a burgeoning office and commercial district downtown; in outlying areas served by trains and streetcars, real estate speculation encouraged housing development in socially sorted districts. Working-class districts emerged in run-down central slums and near industry in

Box 4.3 Mega-Events: The 2014 FIFA World Cup and the 2016 Olympics in Brazil

Like the world fairs of earlier eras, contemporary mega-events have become important strategies for urban and national development. With the Olympic games and the FIFA World Cup for soccer every four years, advances in transportation and media coverage have globalized international athletic spectacles as never before. Given the high costs of such mega-events, however, their socioeconomic, ecological, and political impacts have come under increasing scrutiny. Such concerns became apparent in planning the 2014 FIFA World Cup in 12 Brazilian cities, and similar issues remain for the upcoming 2016 Rio Olympic Games.

Although an upbeat developmentalist perspective on these mega-events initially prevailed, Brazilians soon became critical of apparent corruption and misplaced governmental priorities. This shift in outlook came as a surprise, since Brazilians are so passionate about soccer *(futbol)* and sports generally. In a context of slowing economic growth and rising inflation, large demonstrations rippled across the country a year before the World Cup. Energized by young people, networked by social media, protests began in São Paulo over hikes in bus fares and then spread nationally. On June 17, 2013, an estimated 100,000 people marched through downtown Rio against political corruption and in favor of expenditures on health, education, and other services. While generally peaceful, such protests turned violent as fringe groups engaged in vandalism and the police responded forcefully.

Even before the national team's humiliating loss to Germany in the quarter-finals, public disillusionment became widely apparent. Brazil spent about US$4 billion—80 percent of its public funding—on 12 new or renovated stadiums for the World Cup. FIFA required only eight stadiums, but organizers decided to build four more than needed to satisfy regional interests. Several of the host cities did not have top-level professional soccer teams, raising questions about the long-term value of investments. Total spending on World Cup preparations ballooned to $15 billion, swallowing entire regional development budgets. Brazilians demanded "FIFA-quality" hospitals and schools, but those projects often did not materialize as cost overruns mounted.

In Rio, development for the spectacles resulted in record real-estate values and inflated consumer prices, along with cost overruns and the displacement of low-income populations. A 2014 report documented the displacement of 3,507 families—12,275 people in 24 communities—due to projects for the two mega-events. The areas of greatest displacement included the communities near the Maracanã stadium and adjacent to the Olympic Village in the city's western area. The latter project largely removed Vila Autódromo, where a close-knit community of 3,000 residents dwindled under pressure from authorities to accept monetary compensation or replacement housing.

These two sports extravaganzas in Brazil provide cautionary tales for other countries. Some previous mega-events, such as the Barcelona Olympics of 1992, have been widely praised for transformative investments in infrastructure and urban revitalization. More often, host countries have suffered from the "Olympic curse" of high costs and little public gain. Certainly the World Cup of 2014, even after a massive public investment, did not fulfill promises of widespread economic growth and improved public services. On the contrary, evidence suggests that elites succeeded in socializing the costs while privatizing the profits. Overall, it appears unlikely that the political-economic benefits will outweigh the financial and human costs of hosting such international mega-events, at least in developing countries with a free press, such as Brazil.

the low-lying river basins and railroad corridors. Generally, the wealthy sought higher terrain in the city's southwestern districts, distant from industry, along the Avenida Paulista (Figure 4.11). Programs of import-substitution industrialization, initiated during the depression of the 1930s, consolidated São Paulo's industrial dominance.

After World War II, São Paulo metastasized into a sprawling, dispersed metropolis with a center-periphery geography. Along the Avenida Paulista, the town houses of coffee barons and business leaders gave way to the headquarters of banks and corporations. After the 1956 Development Plan designated São Paulo as site of the foreign-led automobile industry, Volkswagen established the country's first automobile assembly plant here. Subsequent investments by other Brazilian and multinational firms expanded the industrial plant. Construction of São Paulo's modern freeway and subway systems encouraged new areas of urban expansion in peripheral areas. Since the 1950s, metropolitan transportation policy has favored individual automobile travel by the middle and upper classes through a massive investment in new arterial roads, while the poorer sectors of society are underserved by the city's inadequate public transportation system. Working-class areas and peripheral shantytowns often depend on tortuous, unreliable bus service.

During the late twentieth century, metropolitan São Paulo experienced complex processes of economic restructuring, deindustrialization, and decentralization. The metropolis shifted from a modernist center-periphery model to a more diverse and fragmented urban geography, marked by relatively greater proximity of social classes and guarded by heightened forms of security, surveillance, and militarized space. Discourses of violent crime now often incorporate racial and class-based referents, which intensify social polarization in the huge metropolis. Shopping malls increasingly draw customers away from the old downtown. The suburban industrial region faces cutbacks and job loss as industries move away to neighboring states that offer attractive tax breaks to lure automobile assembly plants. Meanwhile, outlying satellites are known for their universities and high-technology sectors.

São Paulo also faces severe problems of environmental degradation and related health concerns, accumulated during years of explosive growth. Given its inland location and concentration of heavy industry, motor vehicles, and informal peripheral growth, São Paulo endures heavy air and water contamination. Air pollution worsens particularly in the winter, when temperature inversions trap pollutants and prevent contaminants from blowing away. While state agencies monitor

Figure 4.11 Once lined by elite mansions, the Avenida Paulista became the city's corporate "Miracle Mile" after World War II. *Source:* Photo by Brian Godfrey.

pollution and impose penalties on the offending industries, it is difficult to regulate more than 4 million cars and buses since vehicular emissions are the concern of federal authorities. Sewage and waste treatment systems also remain inadequate, particularly in the informal *favela* settlements, where wastewater often pollutes surrounding areas. Fiscal problems have hindered ambitious clean-up programs for befouled rivers that snake through the metropolitan area, and the Billings Reservoir remains a heavily polluted cesspool on the southern metropolitan fringe. Such continuing pollution of metropolitan rivers has eliminated potential sources of potable water.

São Paulo also found itself at the center of a severe regional drought in 2014 and 2015. In fact, shortages of potable water affected all of southeastern Brazil, including Rio de Janeiro, Belo Horizonte, and other metropolises. The region received only about half the normal rainfall during a prolonged dry period, and Brazilian experts voiced concern that such climatic variability may be increasing due to a combination of global, regional, and local environmental factors. As reservoirs ran dry, a regional "water war" erupted between São Paulo and Rio de Janeiro, while agricultural, industrial, and residential users also battled over scarce supplies of potable water. Authorities considered imposing mandatory water rationing in São Paulo, but fearing the public response they simply lowered water pressure to limit consumption of the scarce resources.

Some observers have dismissed this regional water shortage as a temporary fluke of nature, or blamed the privatization of the state's water and waste management company, SABESP. Environmentalists point to long-term ecological problems. Massive urbanization has destroyed regional forests that once absorbed, filtered, and gradually released water into local rivers, while São Paulo has created urban "heat islands" that often divert rainfall from rural catchment areas and reservoirs. Another cause of the regional water shortfall may derive from deforestation of the Amazon Basin, due to the expansion of agriculture, ranching, forestry, and other activities. When intact, the dense Amazon forests generate rainfall by circulating water in hydrological cycles that benefit local and even more distant ecosystems. The continuing deforestation of the Amazon, many experts fear, disrupts the "floating rivers" of humidity that typically circulate over central and southern Brazil.

No easy solutions are in sight for what appears to be a long-term struggle with scarce water resources in populous southeastern Brazil. While authorities have proposed expensive infrastructural solutions to capture and transfer more regional water supplies, conservation measures have not received adequate attention. An estimated 25–30 percent of São Paulo's water is lost in leaks, as opposed to about 10 percent in New York City. In addition, watershed protection remains a daunting challenge, given the array of vested development interests in opposition to environmental protection. Without a concerted regional effort, Brazil's most populous and economically productive southeastern region faces major questions about its long-term ecological sustainability.

After a century of rapid growth, Brazil's two leading metropolitan areas now face the challenges of deteriorating physical and social infrastructures, traffic congestion, air and water pollution, fear of crime, housing scarcity, and saturated job markets. While the twin anchors of the Brazilian megalopolis are unlikely to lose their global and national prominence, metropolitan decentralization, economic restructuring, and environmental degradation have created increasing problems of social inequality and urban livability.

Brasília: Continental Geopolitics and Planned Cities

Urbanization has now spread to South America's long-forsaken interior, including the Brazilian central plateau *(planalto)*, the Amazon Basin, and other inland areas. The founding of new inland cities has presented a prime opportunity for modern urban planning and industrial development, as in Ciudad Guyana of Venezuela and, in Brazil, Goiânia, Belo Horizonte and, most famous of all, Brasília. The transfer of the federal capital from Rio de Janeiro to Brasília in 1960 served as a dramatic notice of the determination to redistribute the population from the coast to preconceived cities of the interior. Under Juscelino Kubitschek, president of Brazil from 1956 to 1961, construction of the new capital played an important part of an ambitious program of national urban-industrial development. The new capital's spectacular modern design and rigorous land-use controls were meant to contrast with more spontaneous earlier cities, seen to be plagued by irregular urban growth (Figure 4.12).

Brasília's construction began in 1957 on a barren site in the state of Goiás, about 600 miles (970 km) from the coast. Brazilian architect and planner Lúcio Costa designed the new capital's visionary land-use morphology, while his colleague Oscar Niemeyer designed the city's most impressive modernist buildings, such as the National Cathedral, Senate and Chamber of Deputies complex, the Itamaraty Palace of the Foreign Relations

Figure 4.12 The spectacular modern architecture of Brasília, designed by Brazilian architect Oscar Niemeyer, highlights the federal buildings located along the Monumental Axis (Eixo Monumental). Here we see the Ministry of Justice in the foreground with the iconic congressional complex in the distance. *Source:* Photo by Brian Godfrey.

Ministry, the Planalto Palace executive building, and the Alvorada Palace of the president. Costa's highly symbolic "Pilot Plan" of Brasília features two great intersecting axes, one governmental and the other residential, which observers have likened to the outline of a bird or an airplane with the wings outstretched. Costa himself saw the new capital's ground plan to be "born of the primary gesture of one who marks or takes possession of a place: two axes crossing at right-angles—the very sign of the Cross" (Figure 4.13).

In functional terms, federal government buildings cluster at the eastern end of the plan's monumental axis, centered on the Plaza of the Three Powers, which joins the executive, legislative, and judicial branches of government. At a central intersection of major boulevards sit the bus terminal, stores, hotels, and cultural institutions. Farther west is the local governmental complex of the Federal District, along with a sports arena and recreational facilities. Residential areas, which extend north and south along the "wings" of the plan, comprise groups of six-story apartment buildings to house government functionaries and their families. Each "superblock" of apartments contains a school, playground, shops, theaters, and so on. On the eastern side of the Pilot Plan lies the scenic Lake Paranoá, where expensive private residences have been built, especially in the exclusive Lago Sul ("South Lake") sector (Figure 4.13).

In his critique of Brasília, James Holston suggests that "the modernist strategy of defamiliarization intends to make the city strange." The Pilot Plan imposed a new order at odds with prior expectations of urban life.

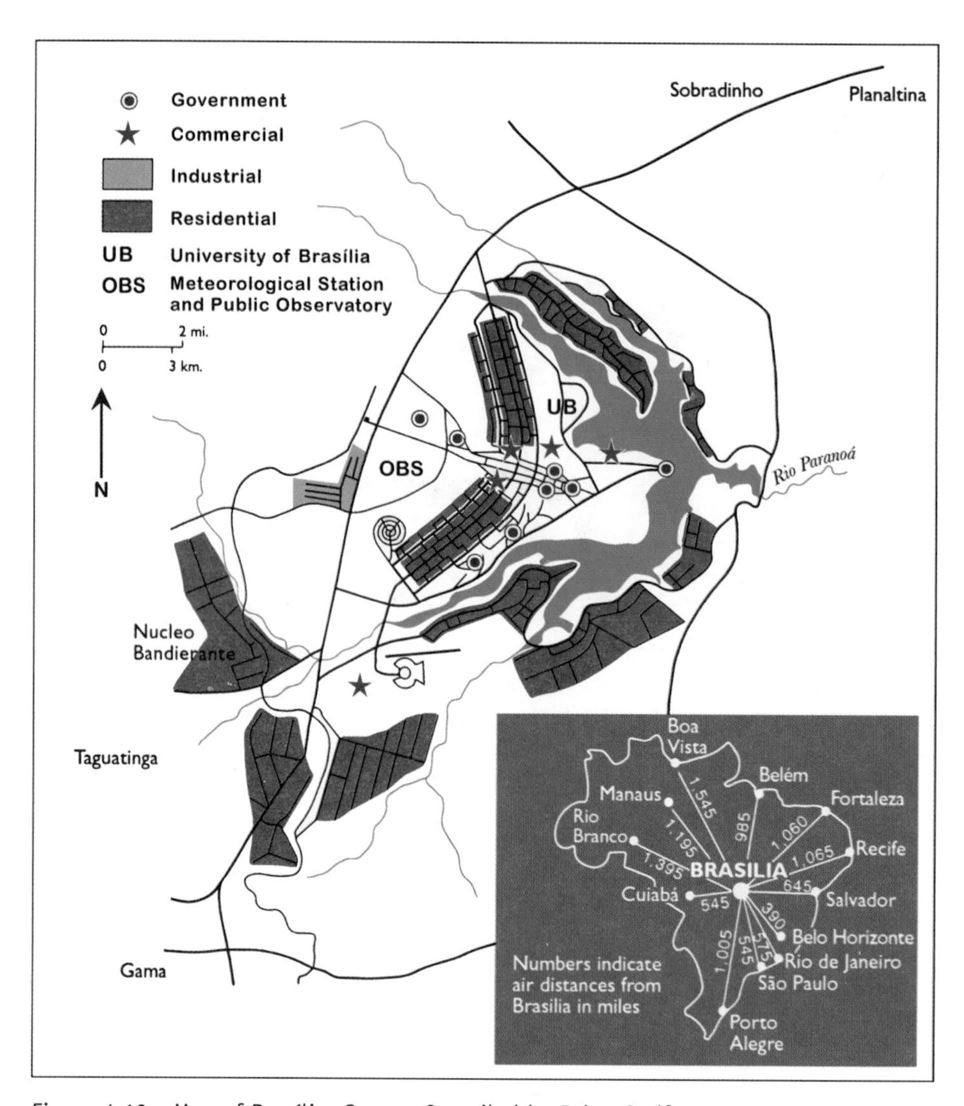

Figure 4.13 Map of Brasília. *Source:* Compiled by Brian Godfrey.

Indeed, early residents and architectural critics often found Brasília sterile and monotonous, lacking the vibrant street life of other Brazilian cities. Many government officials initially maintained homes in Rio, the former capital. In time, however, Brasília filled in with upscale businesses, diverse services, appealing residences and, along with the new amenities, the capital developed a certain character. Residents have adapted public spaces to their uses, such as the informal craft markets held regularly in the central mall. Brasília certainly has become an effective symbol of national identity, symbolized by the exuberant free-form architecture of the National Congress, the National Cathedral, and other modernist monuments.

UNESCO designated the Federal District's central planned area—the Pilot Plan—as a World Heritage site in 1987. The organization's International Committee on Monuments and Sites (ICOMOS) found that "the creation of

Brasília is unquestionably a major feat in the history of urbanism," although it also cautioned that the "new capital of Brazil encountered serious problems which, even today, have not been totally overcome." This decision to recognize Brasília included a precautionary warning that "minimal guarantees of protection" must "ensure the preservation of the urban creation of Costa and Niemeyer." That the central Pilot Plan of the modernist capital of Brazil would be historically preserved, less than 30 years after its founding, reflects more than admiration of an architectural icon; it also speaks of widespread concerns over the rapid and largely unplanned urbanization of the rest of the Federal District.

Away from the central Pilot Plan of the new capital, informal settlements quickly emerged in what were called the "satellite cities"—out of sight but within commuting distance of the city center. Housing was not provided for the construction crews, other workers, and their families. So, a series of spontaneous suburbs some distance from the attractive residential "superblocks" of the city center were built by and for the migrant laborers and their kin. These unplanned communities were composed mainly of low-rise, self-constructed wooden homes and initially exhibited a ramshackle frontier atmosphere. Several of the early settlements, like Taguatinga, in time became established centers with public services, while other more recent areas are still in rudimentary conditions. The majority of the population—in 2015, 2.85 million in the Federal District and 4.2 million in the metropolitan region—lives outside the Pilot Plan in what are now preferably called "surrounding cities" *(cidades do entorno)*. Despite the widespread early criticism of Brasília, the Federal District's steady growth suggests a successful pole of in-migration. Yet the inability to plan effectively the entire Federal District, the symbol of a modernizing regime, underscores the persistence of familiar social problems, such as widespread poverty, self-constructed housing, and the informal sector. The experience of Brasília speaks to the challenges of centralized planning in a developing country beset by high levels of income concentration and a dearth of basic public services.

Lima: Tempering Hyperurbanization on South America's Pacific Rim

Historical and modern, cosmopolitan and deprived, luxurious and squalid, problem plagued and splendid—this is Lima, capital of Peru. Lima and its port Callao are centrally located on South America's Pacific coast, squeezed into a narrow coastal desert between the Pacific Ocean and the Andes Mountains. Initially serving as a point of contact between Spain and its colonial empire in South America, Lima quickly evolved into a transshipment point for the mineral, agricultural, and textile wealth extracted from the Andean interior as well as the unrivaled capital of Spanish-American high culture (Figure 4.14).

Although Lima was founded before the "Laws of the Indies," its founding anticipated them; the city was laid out in a grid pattern, with streets radiating from a central plaza in a regular east-west and north-south pattern. Urban development took hold along a set of axes, each of which had a distinctive character. The area north-westward to the port of Callao would become the city's industrial corridor; the seacoast to the southwest would develop into an elite residential zone; and small industry would intermingle with working-class housing to the east. By the mid-twentieth century, the areas radiating from the old Lima center to the Pacific coast were fully urbanized.

Figure 4.14 Lima's central plaza, known as the Plaza de Armas, dates to the city's founding and served as the central point from which streets extended in the four cardinal directions consistent with the Laws of the Indies. *Source:* Photo by Maureen Hays-Mitchell.

Soon, shantytowns would be commonplace in the desert regions to the north and south of the city (Figure 4.15).

Today in downtown Lima, ornate colonial architecture contrasts sharply with the modern high-rise buildings that accommodate government ministries, banks, law firms, and businesses. The enclosed wooden balconies that typified the colonial city have become a point of interest for preservation; UNESCO designated much of central Lima a World Heritage site in 1991 (Figure 4.14). Notwithstanding, many private-sector businesses and international agencies have moved their offices to less congested and more secure suburbs, and deteriorating colonial mansions have transitioned to slum housing. Yet, the most defining feature of Lima is its expansive shantytowns, which have been euphemistically renamed *pueblos jóvenes* (young towns) or *asentimientos humanos* (human settlements). Shanties have been constructed on the barren slopes that rise above the red-tiled roofs of the inner suburbs and on the flat desert benches that encircle Lima (Figure 4.16). Approximately forty percent of the city's population is estimated to reside in *asentimientos humanos.*

Population growth, agricultural stagnation, economic injustice, and armed violence in rural Peru are responsible for successive waves of cityward migration. Until mid-century, mostly rural elites and people from nearby provinces migrated to Lima. In the two to three decades following World War II, Peruvians from all regions, lured by new industry, found their way to the city. The political violence and economic crisis of the 1980s and 1990s brought an influx of poorly prepared and traumatized displaced persons, primarily from the southern highlands, seeking safety and refuge. In relatively short order, provincial migrants and their offspring transformed Lima from a bastion of elitist creole culture

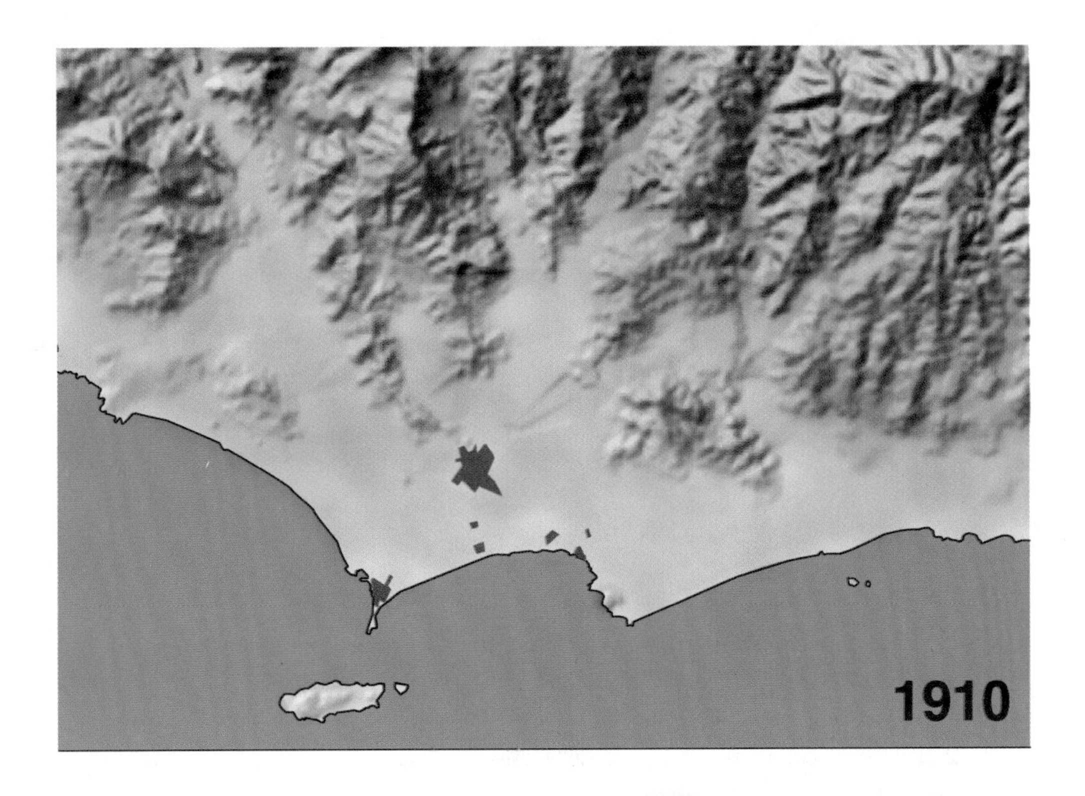

Figure 4.15 Growth of Lima, 1910–2000. *Source:* Centro de Promoción de la Cartografía en el Perú, Avda. Arequipa 2625, Lima 14, Peru. (Continued on next page)

Figure 4.15 (Continued)

Figure 4.16 Villa El Salvador is among the oldest and most well-known shantytowns (asentimientos humanos). Established as a land invasion south of Lima by migrant families from the Andean highlands in 1970, it epitomizes the self-help housing movement. It was awarded formal status as a district within metropolitan Lima in 1983. Today, it is home to some 400,000 people and hundreds of businesses. The pink buildings are schools. *Source:* Photo by Alex Pustelnyk.

(European culture within America) into a microcosm of contemporary Peru. Today, food, music, dance, artisanry, accents, dress, and festivals from every region of Peru are found in Lima.

The social fabric of present-day Lima is more complex than ever. Race, ethnicity, and class defy easy classification. As Lima's population shifts, Andean, and to a lesser extent Amazonian, culture infuses its streets and public spaces. Pressure to assimilate is less today as migrants and their offspring assert their cultural heritage and their claim to Lima as a multicultural city. In response, wealthy Limeños pick up the process begun centuries ago of distancing themselves from the poor. Now they are moving not simply to the traditionally more elite western districts of the city, but also beyond to quasi-rural settings to the east as well as the more distant seaside communities to the north and south (Box 4.4).

Lima dominates all aspects of national life. Seventy-nine percent of the national population currently resides in cities, with nearly one-third of that population, approximately 10 million, living in greater Lima. The city's primate status is the cause and effect of growth. The concentration of political influence, capital, industry, communications, workforce, consumers, and the most prestigious institutions of research, learning, and culture induces further concentration of all these activities and reinforces Lima's primacy. In recent years, the Peruvian government has undertaken a decentralization process intended to stimulate political and economic

Box 4.4 Street Working Children in the Andes

Kate Swanson, San Diego State University

Working children are a common sight on the streets of Andean cities. They juggle at busy intersections, beg on crowded sidewalks, sell in public markets, and sing on city buses. For some, it can be quite jarring to see an eight-year-old girl carrying her two-year-old brother on her back while she sells packages of gum to passers-by. For those from the Global North, it is at odds with common understandings of how children should behave. Isn't childhood supposed to be a time for play, innocence, and fun? Why do children in so many parts of the world have to work? The reality is that not so long ago, children everywhere had to work. The modern Western construction of childhood—an understanding that prioritizes play, education, and innocence—stems from a place of wealth and privilege. As nations in the Global North developed and grew in prosperity, children became valued more for their intrinsic worth than their economic worth. This means that families no longer needed children's labor to survive, as they could make ends meet through adult wages alone. Yet in the Andes, there remains tremendous poverty, which forces families to enlist each family member into paid labor. This poverty is highly racialized and disproportionately affects those of indigenous and Afro-Latino descent. In some Andean indigenous communities, they joke that children get four years to live for free. After four, they have to start working in order to contribute to their family's economic well-being. As a result, on weekends, after school, and on school holidays, children head to the streets to sell goods, shine shoes, sing, perform, and beg in the hopes of earning a little extra money (Figure 4.17).

Figure 4.17 Three young girls find time for fun as they assist their mothers who labor as ambulantes (street vendors) in the informal economy of Huancayo, a city in the Peruvian central Andes. *Source:* Photo by Maureen Hays-Mitchell.

Of course, some argue that the streets are no place for a child. In Ecuador and Peru, there are state and municipal-level anti-child labor campaigns designed to remove children from the streets. They argue that the streets expose children to many dangers, such as crime, air pollution, youth gangs, and traffic, among others. There is certainly truth to these arguments. On the other hand, some argue that children's work is empowering and builds many entrepreneurial skills to help them succeed into adulthood. Some advocate for children's right to work and push back against organizations that try to remove them from the streets. Others suggest that children's income is critical as it helps them pay for school supplies, such as uniforms, textbooks, and school lunches. In fact, many children say that they enjoy working on the streets as it gives them a great deal of personal freedom—something that is lacking for young people in the United States and Canada these days due to helicopter parenting, municipal curfews, and high-tech surveillance systems. Ultimately, it is poverty that forces so many Andean children to work on city streets. To address this issue, nations must tackle the deeply entrenched structural inequalities that plague the region.

development in rural regions, while mitigating Lima's growth and dominance.

As Lima emerges from the severe economic and political crisis of the late twentieth century, poverty rates are declining and its middle class is expanding; construction is booming; roads are being paved, and public spaces illuminated. Yet, the city confronts problems of unprecedented proportion and complexity. Not only is prosperity highly concentrated there, but it is unequally distributed within the city; and the gap between Limeños who benefit from improved economic conditions and those who do not remains wide. The city's rapid and unplanned growth has also caused severe environmental degradation, especially of the city's water and air. Lima is a megacity in a desert. The rapid loss of Andean glaciers threatens Lima's sources of water. The very rivers that gave rise to human settlement here some 4,000 years ago are shrinking and contaminated by mining and agricultural runoff as well as residential and industrial waste. Traffic congestion and unregulated industries pollute the air. Urban sprawl has eaten away at green space and consumed wetlands, reducing biodiversity and affecting microclimates within the metropolitan region.

The UNEP (United Nations Environmental Program), among others, has identified water as the most critical environmental problem in the Lima-Callao conurbation. More than 1 million Limeños, primarily in poor districts, lack access to potable water and sewer service, and more than forty percent of the city's water supply is lost to leakage. Efforts are underway to address this. "Water for All," a $715 million public-awareness campaign designed to increase efficient water use, has constructed water treatment facilities and initiated projects to bring potable water to underserved districts. AguaFondo, or Lima Water Fund, works with local communities in water conservation by stabilizing slopes, recovering lagoons, and reforesting watersheds. The private sector is also contributing to green innovations. In 2012, an advertising agency partnered with Lima's University of Engineering and Technology to create a billboard that captures moisture from air and converts it to filtered drinking water.

Most Limeños, however, consider air pollution the most pressing environmental issue. Limeños who frequent the city center and/or reside in poorer districts inhale large quantities of airborne particulates and other pollutants. A study conducted between 2007 and 2011 found that the high pollution rate caused over 5,000 deaths, 80 percent of which were directly attributable to pollution from public transport. Government-led initiatives, such as the Clean Air Initiative and innovations in public transport, seem to be having impact as Lima's air quality gradually improves. In 2010, Lima launched a high-capacity transportation system of rapid buses—"El Metropolitano"—that run on natural gas and link the north and south corridors of the city. The system moves nearly 1 million passengers per day along a 26 kilometer corridor. In 2014, construction began on Lima's first subway system. When complete, the train line will link the port area of Callao on the west with Ate in the east through a 35 kilometer tunnel, and cut travel time from 2 hours to 45 minutes. To increase the share of green space, 49 districts in the metropolitan area have signed the "Green Lima and Callao Pact." The program prioritizes building large recreational areas and parks in the city's poorest neighborhoods, as well as revamping public squares and green areas along main avenues in the city center. Green spaces have increased by over 50 percent in the past decade, enhancing the health and quality of urban life while creating a shared sense of civic identity for many Limeños.

Lima has long served as the gateway between the outside world and the rest of the country. Indeed, in hosting the UN Framework Convention on Climate Change (COP20) in December 2014, Lima—and by extension Peru—sought to establish its reputation as a forward-thinking society concerned with critical issues of local, national, regional, and global magnitude. Notwithstanding, serious problems continue to plague this national capital and primate city. Recent initiatives to address Lima's seemingly intractable problems—social, economic, political, and environmental—may be cause for cautious optimism not only among Limeños but all Peruvians.

Buenos Aires: Global City of the Southern Cone

Long regarded as one of Latin America's greatest cities, Buenos Aires stands as the most visible symbol of Argentina's history and identity. Once a minor colonial outpost of Spain, Buenos Aires grew rapidly as a center of immigration, urban design, and modernism from roughly 1880 to 1930. While the country emerged as an agricultural and industrial power, the Argentine capital became known as the "Paris of South America"—an elegant city of broad boulevards, graceful public squares, and impressive public buildings. Monumental Buenos Aires has long served as the stage for national political movements, as dramatized by the famous scenes of Juan and Eva Perón addressing the multitudes from the balcony of the Casa Rosada, the presidential palace. More recently, Mothers of Plaza de Mayo have continued to demonstrate to protest the "disappearance" of their children during the "dirty war" of the military regime. Despite a contemporary decline in regional importance vis-à-vis São Paulo, Greater Buenos Aires remains a vital metropolis with a 2015 population of 15 million and a high degree of national primacy. The city's residents, known as *porteños* (port-dwellers), continue to be trendsetters. On the other hand, the Argentine metropolis now faces growing problems of socioeconomic inequality, popular discontent and insecurity, spatial segregation, and environmental degradation.

Colonial Buenos Aires followed the characteristic Spanish-American urban form. The central plaza (later named the Plaza de Mayo) served as the core of the colonial settlement, surrounded by the important governmental, religious, and commercial structures. The city council, or Cabildo, sat across from the Cathedral and a commercial arcade lined much of the plaza. While colonial Buenos Aires typified a Spanish "Laws of the Indies" town, the postcolonial city's design increasingly reflected French and British influences. To resolve prolonged centralist-federalist conflicts, Buenos Aires was federalized and removed from the dominant Buenos Aires Province in 1880 and developed rapidly. British-financed railroads fanned out into the pampas, opening up an agricultural breadbasket to world trade, while the development of refrigeration and improved port facilities allowed export of Argentine beef to Europe.

An emergent country needed a world-class capital city, graced by monuments and public buildings, worthy of Argentina's new wealth and aspirations. The Avenida de Mayo, completed in 1894, provided a striking visual corridor, reminiscent of the Champs-Élysées in Paris, linking the executive seat of government and the national capitol. The Avenida 9 de Julio, one of the world's widest avenues, is centered on an Obelisk visible from various vantage points downtown. With the opening of additional boulevards in the early twentieth century, the city's reputation as the "Paris of South America" was sealed (Figure 4.18).

As Argentina became a postcolonial beacon of order and prosperity, European immigrants poured in, and 30 percent of the Argentine population was foreign born by 1914. As the federal capital, transportation hub, commercial center, cultural mecca, and immigrant port of entry, Buenos Aires experienced a high degree of urban primacy in the national city-system. The metropolitan population now represents roughly a third of the country's total, more than 15 million residents (see Tables 4.2 and 4.3). With suburbanization, however, only about a fifth of the metropolitan population resides in the capital city itself.

Unimpeded by physical barriers, districts called *barrios* covered the Federal District by 1930. New immigrants first settled in central barrios near the port; the local Italian-Spanish dialect known as "Lunfardo" emerged here, along with the Argentine "Tango" dance. The city's southeastern areas generally became industrial, working-class districts. In contrast, elegant upper-class neighborhoods emerged on the northwestern side of Buenos Aires. The two socially sorted residential sectors—generally more affluent to the northwest of downtown, more working class toward the southeast—continued their historic trajectories in contemporary metropolitan growth beyond the Federal District. A massive influx of impoverished migrants from the Argentine interior, Bolivia, and Paraguay has created extensive shantytowns or *villas miserias* ("towns of misery"). An estimated 640 *villas miserias* encompass up to a million people in the suburbs in Greater Buenos Aires, and studies suggest that urban slums now grow ten times faster than the national population.

Argentine society has long been regarded as affluent—given middle-class living standards, high levels of education, and good public health—but economic restructuring and neoliberal reforms shattered such assumptions during the 1990s. The country grew economically but suffered a contraction of government

Figure 4.18 The Diagonal Norte (Northern Diagonal Boulevard), officially the Avenida Presidente Rouge Saenz Pena, highlights the imposing Obelisk monument in downtown Buenos Aires. *Source:* Photo by Brian Godfrey.

services, privatization of state enterprises, and widespread deindustrialization. While elites prospered, much of the population suffered from increasing unemployment and poverty. An economic recession began in 1998 and culminated in the crisis of 2001–2002, when Argentina defaulted on international debt obligations and devalued the peso. With growing public protests came new social movements, such as the *piqueteros,* unemployed workers who blocked roads, bridges, and buildings. Unemployed workers organized into cooperative markets and businesses, and neighborhood-based assemblies *(asambleas populares)* arose.

As elsewhere, Buenos Aires has witnessed a proliferation of gated communities, characterized by low-density residential complexes guarded by defensive enclosure and private security. These affluent enclaves cluster primarily in suburban areas with good highway access to the city center and, paradoxically, in poor localities with relaxed land-use laws. The clustering of exclusive gated communities in low-income jurisdictions has deepened social polarization by juxtaposing wealthy and poor households.

The Argentine metropolis also faces growing environmental problems. Given the city's low-lying coastal location on the edge of the humid pampas, water management has long played an important role in urban development. Buenos Aires suffers increasingly from flooding, wastewater disposal, and industrial pollution. The city is also at risk from climate change and sea-level rise. Due to rapid urbanization and the concentration of impervious surfaces, the drainage system cannot adequately handle storm runoff. Since the 1980s, the contamination of aquifers has led to increased reliance on surface water supplies.

As part of neoliberal restructuring, the French Suez Company won the metropolitan concession for water management and sanitation in 1992. While controversial, privatization initially resulted in improvements to water service and a 27 percent reduction in customer bills. The Suez profit margin remained low, however, and a renegotiation of contract terms in 1997 led to widespread protests. Suez lost the concession in 2006, after which a conglomerate of companies, *Aguas del Gran Buenos Aires*, has managed metropolitan water and sewer provision. Currently, four wastewater treatment plants treat only about 5 percent of wastewater before discharging it into the Rio de La Plata.

Another major challenge comes from high levels of industrial discharge. The most polluted area is the Matanza-Riachuelo River basin, home to Argentina's largest concentration of urban poor. Among the 3.5 million inhabitants in the basin, one-third live below the poverty level and an estimated 10 percent reside in flood-prone informal communities that suffer from contact with both untreated organic waste and toxic industrial chemicals. In 2004, a group of residents sued the national government, the Province of Buenos Aires, the Federal District of Buenos Aires, and 44 businesses for damages suffered from pollution of the Matanza-Riachuelo River. In a 2008 landmark decision, the Argentine Supreme Court ruled in favor of the residents and determined the defendants to be liable for ecological restoration of the river basin.

Overall, Buenos Aires now finds itself in the midst of swirling currents of change that are restructuring the metropolis. While the emergence of suburban shopping centers, office parks, and informal and gated communities challenges the supremacy of the traditional urban core, contemporary redevelopment projects suggest a continuing concern for the central city, as evidenced in the renovation of the abandoned downtown piers at Puerto Madero (Figure 4.19). While Buenos Aires retains a cosmopolitan air and cultural status, the contemporary intensification of socioeconomic inequality and spatial segregation raises troubling questions for the future. Long thought to be different from other South American megacities, Buenos Aires now converges with them in terms of growing urban socioeconomic and environmental problems.

Curitiba and Bogotá: Planning For Sustainable Urban Development

Written with Andrés E. Guhl, Universidad de los Andes, Bogotá

Given high rates of urbanization and environmental degradation, many South American cities now pursue policies of sustainable urban development. Two world-famous examples are Curitiba, Brazil and Bogotá, Colombia. In these continental trendsetters, governments have adopted innovative policies intended to encourage compact, livable, and environmentally friendly urbanism. The cities both implemented cost-effective bus rapid transit (BRT), which inspired similar express-bus programs in other cities. In their efforts to combat the pressures of suburban sprawl, urban planners have endeavored to implement pedestrian streets, preserve historic centers, concentrate growth along commercial corridors, promote ecological design and green buildings, and encourage parks and open spaces, among other progressive measures.

In Curitiba, capital of the state of Paraná, sustainability planning began as urbanization rates exceeded five percent in the 1960s. Long a regional center of agriculture and timber production, industrialization and rural-urban migration accelerated after World War II. Fears that rapid urban growth threatened the quality of life prompted development of a 1965

Figure 4.19 Recent renovation of Puerto Madero, long a deteriorated inner harbor, created a revitalized waterfront district adjacent to the downtown of Buenos Aires. *Source:* Photo by Brian Godfrey.

Preliminary Plan by a team under architect Jaime Lerner, who later served as mayor and governor. The Institute for Urban Research and Planning of Curitiba developed a Master Plan *(Plano Diretor)* officially adopted in 1966. This plan proposed to minimize traffic congestion, control urban sprawl, preserve the historic city center, provide parks and open space, and develop an efficient public transit system. Implementation began dramatically in 1972, when planners converted one of the major downtown thoroughfares, November 15 Street, into a pedestrian street. Although disgruntled motorists threatened to ignore the traffic ban, local authorities dissuaded them with an act of public theater, which featured the unfolding of large sheets of paper for school children to paint on the street.

Subsequent zoning regulations promoted development along arterial corridors, revitalization of the commercial core, and maintenance of peripheral open space. A "Trinary" road system, consisting of five traffic arterials that converge downtown, separates automobile traffic in two outer lanes, going opposing directions, from central lanes reserved for express buses. "Tube stations" feature elevated passenger shelters to facilitate fare collection, rapid entry to, and exit from express buses. As of mid-2014, there were more than 350 tube stations in the metropolitan area. The Integrated Transport Network permits transit between any points in the city with a unified fare. Curitiba also promotes design-with-nature principles of urban ecology; low-lying areas subject to flooding are reserved for parks. Despite continuing problems of poverty and service provision in peripheral shantytowns, the city's program of "Faróis de Saber" (Lighthouses of Knowledge) offers free educational centers, including libraries, internet access, and other social and cultural resources.

These successes, however, have generated new problems. With an official city population of 1.75 million and a metropolitan population of 3.5 million in 2010, Curitiba is a major political and economic center of southern Brazil. The metropolitan demographic growth rate of 3.19 percent between 2005 and 2010 remained among the country's highest, as were per-capita incomes and rates of automobile ownership. This relative prosperity has added to growth pressures. Although express buses continue to be heavily used, development of the transit corridors has encouraged metropolitan sprawl. Whether planners can build on Curitiba's innovative record of transit-oriented development and environmental conservation to meet these new challenges remains to be seen.

The larger metropolis of Bogotá has faced even more daunting growth pressures. The Colombian capital's rate of urbanization reached a dizzying seven percent between 1950 and 1965, gradually dropping to a more manageable 2.9 percent in 2005–2010. And from 1950 to 2010, the metropolitan area grew from 630,000 to 8.5 million residents. Fortunately, governmental reforms in the early 1990s facilitated local innovation which, in turn, has gained momentum under a series of progressive mayors. As a result, *Bogotanos* have witnessed significant improvements in urban transportation, utilities, and public space. Additionally, the city's health services and libraries have expanded, and virtually all children have gained access to public schools. In 2000, Bogotá adopted a master plan *(Plan de Ordenamiento Territorial* or *POT),* revised in 2004, which has helped prioritize resources and improve quality of life. The plan sets clear zoning patterns to regulate land use, promote transit-oriented development, and restore the ecological assets of the city.

The most dramatic change has occurred in public transit with the implementation in 2000 of the *Transmilenio,* a network of express-bus lanes based on the model of Curitiba. This BRT system is part of an integrated transportation plan designed to provide subway and train service throughout Bogotá and surrounding municipalities. In late 2014, Transmilenio supplied about 40 percent of the city's transportation needs, with 70 miles (113 km) of exclusive bus lanes and 411 miles (663 km) of bus routes feeding into the system. Yet, the system has developed just one-third of the originally planned BRT lines; it has not expanded since 2012 and is overwhelmed by escalating demand.

Transit mobility throughout the city is worsening with a steady increase in car and motorcycle usage. The city now plans to build its first subway line, which is likely to take decades to complete. To mitigate congestion, Bogotá has encouraged the use of bicycles by building exclusive bike lanes that provide a safe and environmentally friendly way to move around the city. As of 2015, there were 233 miles (376 km) of bike lanes in the city that served roughly 14 percent of the population. The number of people using bicycles has grown more than 20 percent in the last two years. Every Sunday, main thoroughfares turn into recreational space in the *Ciclovia* program, as about 2 million residents flock to the streets on bicycles, roller skates, and other means of recreation.

The *POT* master plan further seeks to restore ecosystems lost to ill-conceived twentieth-century policies. It encourages the restoration of natural assets and has engaged the private sector as providers of valuable ecosystem services, ecological design, and green building. The city has invested heavily in parks, recreational facilities, sidewalks, and public spaces. For example, Avenida Jiménez, a central boulevard in the historic center, has been transformed into a leisurely walkway (Figure 4.20). This urban intervention restored an important stream by transforming it into a popular linear park known as *Eje Ambiental* (Environmental Axis). However, much remains to be done. Air pollution is a persistent problem, and sewage continues to be dumped largely untreated into the metropolitan watershed. Ecological restoration has been difficult due to limited resources and lack of environmental consciousness on the part of many Bogotanos.

As in Curitiba, Bogotá has improved the quality of life of its citizens through coordinated urban planning, economic development, and environmental protection. Many challenges remain, however. The city currently is undergoing a governance crisis that pits the mayor and city council at odds. The early success of the Transmilenio BRT and ecological restoration now requires systematic and far-reaching efforts to address emerging problems. While both Bogotá and Curitiba have clearly moved along the path toward sustainable urbanism, they must consolidate this trend through careful planning

Figure 4.20 Eje Ambiental in historic Bogotá, where a dechannelized stream is part of a linear park along Avenida Jiménez. *Source:* Photo by Andrés Guhl.

to ensure inclusive, equitable, and environmentally friendly urbanism.

URBAN CHALLENGES AND PROSPECTS

The Urban Economy and Social Justice

Recent trends in economic globalization are benefiting some countries, most notably Chile and Brazil where middle classes are growing and poverty rates declining somewhat. Most countries in South America are not as fortunate. Throughout the continent, the proportion of poverty households remains relatively high. In cities, long-standing conditions of socioeconomic inequality and social and environmental injustice endure. Issues of employment, housing, and environmental degradation affect the poor more severely than they do other sectors of urban society. It is a sad fact that the areas of cities where life expectancy is lower than citywide averages are low-income communities with high levels of contamination.

It is not uncommon for many urban residents to spend more than half of their cash income on food—only to barely meet nutritional needs. In the absence of unemployment insurance or an adequate social security system, many South Americans cannot afford to be unemployed and are forced to turn to their own resourcefulness. Research on urban labor markets in South America indicates that, although participation within the paid workforce has improved, participation in the informal economic sector has increased. This is especially true among lower-income groups and the more vulnerable, such as poor women and children.

Despite indicators of stabilization, even expansion, at the macro level (e.g., growth in gross national income), socioeconomic polarization persists in South American cities as the benefits of economic development accrue unevenly. When such conditions are concentrated among certain social groups or regions, they can generate restive conditions that challenge the cohesion of a society and the stability of a government. The rise of indigenous politics and social protest in Bolivian cities is a fascinating example that is playing out on the streets of La Paz, the national capital, Cochabamba, site of the infamous water wars, and Santa Cruz, where a secession movement is underway (see Box 4.2).

Defensive Urbanism and Self-Help Housing

South America's cities reveal a curious sociospatial pattern of segregation that often juxtaposes those with wealth in secure high-rises or gated communities alongside those without in *favelas, asentamientos humanos, villas miserias* (shantytowns). Indeed, large-scale urbanization has spawned "defensive urbanism." The fear of crime has led the urban elite to retreat into protected areas in luxury apartment buildings or suburban communities, where security is enforced by walls and armed guards, and children are chauffeured to private schools. New security infrastructures—video surveillance, remote-controlled gates—are proliferating in cities across the continent (Box 4.5).

Today, one- to two-thirds of the population of any given city resides in informal-sector housing. Similar to its employment counterpart, the informal housing sector exists outside the bounds of "officialdom" in that it ignores building codes, zoning restrictions, property rights, and infrastructure standards. In South America, informal-sector housing is commonly known as "self-help" housing,

Box 4.5 Urban Security and Human Rights

Increasing concerns with violent crime now plague South American cities. Fears of violence have been fed by vivid accounts in the news media, tourist guides, and popular films. Such acclaimed recent films as "City of God" (Brazil, 2002) or "Our Lady of the Assassins" (Colombia, 1999) feature racy stories full of sex, drugs, violence, and armed conflict in urban slums. Such representations sensationalize violence and serve to stigmatize the urban poor, who are disproportionately of indigenous or African racial origins. The preoccupation with urban insecurity has created a culture of fear, which Teresa Caldeira relates to "the increase in violence, the failure of institutions of order (especially the police and the justice system), the privatization of security and justice, and the continuous walling and segregation of cities. . . ." Widespread concern over crime has served to maintain class and racial boundaries, despite the expansion of formal democratic rights.

Official statistics often underreport crime, since distrust of the police discourages residents from reporting incidents. Rates of homicide (murder and manslaughter) represent the most reliable data, given compulsory death registrations. In 1980, national homicide rates in Brazil and the United States were about the same (about 10 per 100,000 residents), but the Brazilian rates were twice as high by the late 1990s. Of course, violent crime tends to be worse in large cities than rural areas, especially related to urban drug traffic, gang wars, and police brutality. Even so, while North American crime rates dropped dramatically in New York and many other U.S. cities, Brazilian cities like São Paulo, Rio de Janeiro, and Recife became steadily more violent, although their homicide rates have tended to drop somewhat in the last decade.

São Paulo, for example, experienced a dramatic rise in homicide rates between 1980 and 2000, usually involving firearms. Most victims have been young men (15–29 years old) suffering from both poverty and drug-trafficking activities. Other factors commonly include socio-spatial segregation, high unemployment, and widening income inequality. On a positive note, São Paulo's murder rate fell to 14/100,000 in 2007, which researchers attributed to more effective policing methods and better enforcement of gun-control legislation, despite the persistence of socioeconomic problems. Subsequently, the cities of both Rio de Janeiro and São Paulo both experienced a significant decrease in homicides from 2007 to 2011. Despite the decline in the Brazilian southeast, violent crime has risen in other parts of the country, particularly the north and northeast. The booming Amazon metropolis of Marabá in the state of Pará, for example, is now one of Brazil's most violent cities with homicide rates as high as 125 deaths per 100,000 residents, according to recent federal statistics.

Such high levels of urban violence reflect issues of human rights. Community initiatives now feature programs to prevent violence, particularly among young people in poor communities. Nongovernmental organizations (NGOs) began in the 1990s to offer programs to reduce firearm injuries, promote social justice, and provide vocational training for young people in poor communities. For example, the Mangueira Social Project, located in one of the city's *favelas,* provides after-school programs for local youth who demonstrate regular school attendance. Such grassroots campaigns attempt to change the perception of communities through the Internet, media outreach, and partnerships with the government, universities, and the private sector.

a term that carries a double meaning. Most commonly, self-help refers to the characteristics of the homes and the process through which they are built. Self-help housing tends to be built by the inhabitants themselves, using simple—often hazardous—materials that the owner-builder-occupier has accumulated over time. Additionally, the term conjures up images of impoverished, yet well-intentioned, urban dwellers "helping themselves" to unoccupied land—in the absence of a more viable option. Self-help housing communities are commonly considered shantytowns (see Figures 4.10 and 4.16). Many settlements lack basic services, such as running water, sewerage, electricity, and garbage removal. They are constructed of scrap materials that often do not provide adequate protection from inclement weather, have limited access to services, are overcrowded, and lack the security of tenure (i.e., title to the land). Shantytowns—or self-help communities—are marginal in terms of both their location on the urban periphery and the quality of the land occupied, which tends to be undesirable and often unhealthy and dangerous. They may be constructed on toxic "brownfield" sites, alongside noxious landfills, on steep hillsides, or in polluted wetlands. Their overcrowded conditions are ideal for the transmission of disease. Shanties are the first structures to collapse in mudslides and the first to be carried away in floods, and they easily go up in flames.

Under favorable conditions, self-help communities strengthen and improve over time. After the initial land invasion, settlements can evolve into consolidated and well-organized communities. Structures are steadily improved and basic services are addressed in one way or another. With time, municipal governments officially recognize the communities and extend urban infrastructure, supplying water and electricity, paving roads, extending public transportation lines, providing garbage removal, building schools, and staffing clinics. El Alto, perched above La Paz, and Villa El Salvador, outside Lima, are cases in point (see Figures 4.6 and 4.16). Despite the celebration of the self-help movement in many circles, it is nevertheless an inadequate proxy for regulated housing and urban services.

Spatial Segregation, Land Use, and Environmental Injustices

Although South American cities have long been highly segregated, the pattern of segregation is more complex today. Population expansion and variegated topography are bringing distinct social groups into closer contact. As intervening land is occupied, self-help communities and elite developments often exist side by side. An interesting phenomenon is the proliferation of affluent enclaves within low-income districts, where relaxed land-use laws attract real estate developers. There is little indication that residential segregation is abating and, ironically, proximity accentuates class tensions. Indeed, South American cities are characterized by greater polarization in lifestyle. Glass-fronted skyscrapers and shopping malls characterize business districts and elite neighborhoods, while peripheral shantytowns are built of scrap materials and lack basic services.

Metropolitan expansion and decentralization have eroded the relative dominance of the traditional city center. Employment in the center is decreasing as industrial activity shifts to peripheral or nearby rural locations, and government and professional offices move to affluent suburbs that are less plagued by traffic congestion and crime. Although historic preservation and heritage sites in traditional downtowns have encouraged tourism, there is little evidence of residential gentrification and high-end commercial revitalization. Affluent residents now prefer suburban locations with their amenities and security infrastructure. Indeed, urban elites are more likely to enjoy the advantages and to escape the disadvantages of urban living. Affluent business and residential districts tend to be better serviced with running water, sewerage, electricity, garbage service, public transportation, paved streets, sidewalks, and public parks. In contrast, low-income districts are characterized by inadequate urban services and infrastructure.

A differentiated urban landscape is also evident in terms of environmental justice. Air pollution in some cities commonly surpasses safe levels as established by the World Health Organization. The wealthy can more readily escape these negative externalities as they listen to car stereos while waiting out traffic jams in air-conditioned cars. Meanwhile, the less affluent are crowded onto hot, noisy, diesel-spewing buses. The discharge of untreated urban sewage into rivers and streams occurs more regularly in low-income districts. Children who live in shantytowns are especially vulnerable to gastrointestinal and respiratory illnesses, due to the poor water, inadequate sanitation, contaminants, open garbage, and burning refuse that characterize their living spaces. In contrast, the better-off reside in less polluted areas, are more able to control some aspects of their living environment, and are more able to escape to country clubs and vacation homes. Indeed, evidence clearly suggests that vulnerability to environmental hazards parallels income and status in South American cities.

Widespread efforts are now underway, however, to reconcile urbanization and environmental quality in South America. Municipal governments and grassroots organizations now promote urban sustainability, environmental justice, and the greening of cityscapes around the continent. Inspired by innovative programs in such role models as Bogotá, Colombia, and Curitiba, Brazil, many additional cities now endeavor to provide more socially inclusive, equitable, and environmentally friendly forms of urbanism. Historic preservation, express buses, bicycling, pedestrian spaces, ecological restoration, community gardens, recycling programs, and tree-planting campaigns—the residents of South America's cities are paragons of creativity and resourcefulness.

AN EYE TOWARD THE FUTURE

The cities of South America have long played crucial roles in a global urban network and capitalist economy. Economic, political, social, and cultural currents from around the world have flowed through the region's cities since the arrival of Iberian conquistadores. Today, as in the past, these global forces and the region's cities continue to shape and influence one another. Indeed, the escalating reach of globalization is adding new urban dimensions to long-standing problems of uneven development, regional shifts of industry, environmental degradation, socioeconomic polarization, urban insecurity and violence, and spatial and environmental injustice. Such urban problems are stimulating economic decentralization and rising growth rates of small and intermediate-sized cities in South America. In cities large and small, the region's intractable social divide can be read on its urban landscape, which is at once magnificent and tragic. South America's cities contain a disproportionate concentration of regional wealth and power, as well as a disproportionate concentration of marginalized people who are undeterred in laying claim to their cities. Contemporary democratization has facilitated the rise of social movements and political activism throughout the continent, often shifting the balance of power to new groups. Although the future remains uncertain, it is being debated, contested, and enacted now.

SUGGESTED READINGS

Browder, J., and B. Godfrey. 1997. *Rainforest Cities: Urbanization, Development, and Globalization of the Brazilian Amazon.* New York: Columbia University Press. Comparative study of urbanization in Amazônia.

Goldstein, D. 2012. *Outlawed: Between Security and Rights in a Bolivian City.* Durham, NC: Duke University Press. Reveals how indigenous residents of marginal neighborhoods in Cochabamba, Bolivia balance security with rights through "community justice."

Goldstein, D. 2003. *Laughter Out of Place: Race, Class, Violence, and Sexuality in a Rio Shantytown,* Berkeley: University of California Press. Paints an intimate ethnographic portrait of women in Rio's *favelas* who use black-humor storytelling to deal with tragedy.

Holston, J. 2009. *Insurgent Citizenship: Disjunctions of Democracy and Modernity in Brazil.* Princeton, NJ: Princeton University Press. Explores struggles for homeownership and service provision of residents of São Paulo's peripheries.

Kohl, B., and L. Farthing. 2012. *From the Mines to the Streets.* Austin, TX: University of Texas Press. Draws on the life of a Bolivian political activist to convey how profoundly political systems can affect individual life.

Mann, C. 2006. *1491: New Revelations of the Americas before Columbus.* New York: Knopf. Reveals that large indigenous populations actively shaped their environments through early agriculture, trade, and urbanization.

McGuirk, J. 2014. *Radical Cities: Across Latin America in Search of a New Architecture.* New York: Verso. Features Rio de Janeiro, Buenos Aires, Caracas, Bogotá, Medellín, Lima, Santiago (Chile), and Tijuana, among other cities.

Perlman, J. 2010. *Favela: Four Decades of Living on the Edge in Rio de Janeiro.* Oxford: Oxford University Press. Restudy of author's earlier work finds that most *favela* residents are more pessimistic about prospects for social mobility and fearful of violence in their communities.

Scarpaci, J. 2005. *Plazas and Barrios: Heritage Tourism and Globalization in the Latin American Centro Histórico.* Tucson: University of Arizona Press. Examines local experiences of historic preservation and heritage tourism in nine Latin American cities.

Ward, P., E. Jiménez Huerta, and M. DiVirgilio. 2015. *Housing Policy in Latin American Cities: A New Generation of Strategies and Approaches for 2016 UN Habitat III.* New York: Routledge. Considers policy choices in dealing with the "first suburbs," or squatter settlements, that came to surround Latin American cities.

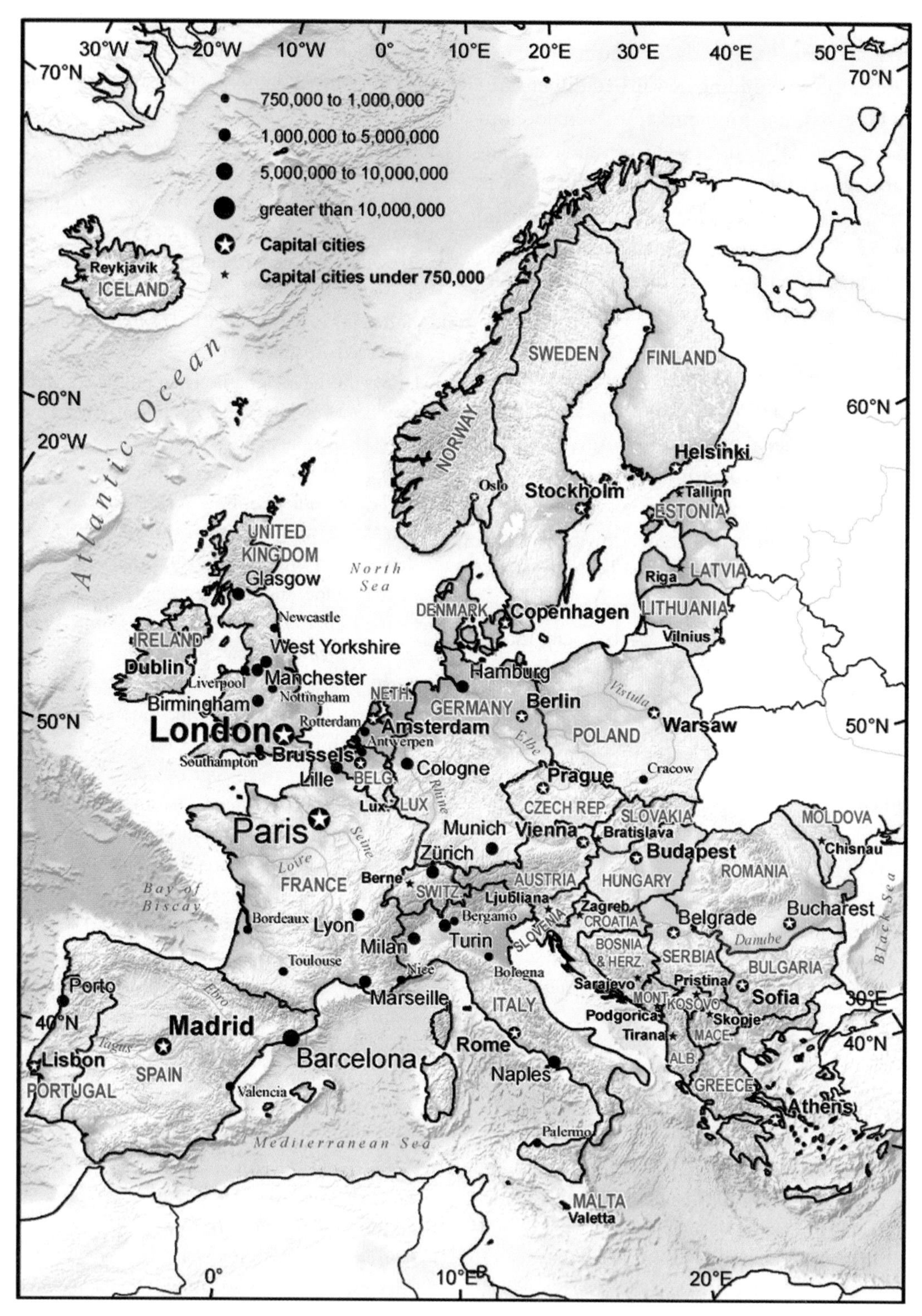

Figure 5.1 Major Urban Agglomerations of Europe. *Source:* United Nations, Department of Economic and Social Affairs, Population Division (2014), World Urbanization Prospects: 2014 Revision, http://esa.un.org/unpd/wup/.

5

Cities of Europe

LINDA MCCARTHY AND COREY JOHNSON

KEY URBAN FACTS

Total Population	547 million
Percent Urban Population	73%
Total Urban Population	402 million
Most Urbanized Countries (not including microstates)	Belgium (98%) Netherlands (90%) Luxembourg (90%)
Least Urbanized Countries	Bosnia-Herzegovina (40%) Moldova (45%) Slovenia (50%)
Number of Megacities	2
Number of Cities of More Than 1 Million	41
Three Largest Cities	Paris (11 m), London (10 m), Madrid (6 m)
Number of World Cities (Highest Ranking)	30 (London, Paris, Milan, Frankfurt, Madrid, Amsterdam, Brussels)
Global City	London

KEY CHAPTER THEMES

1. Europe is integral to the study of urban development because of its long history and the extraordinary impact of European urban influences worldwide.
2. Europe's urban system is dominated by the world cities of London and Paris, the result of their dominance within former vast empires.
3. Europe has quite a few cities larger than 1 million people, but in general these cities are growing more slowly than those in other world regions.
4. European cities exhibit great diversity in style and form, the result of a long history and complex mix of people and cultures.

5. The demand for low-wage labor in western Europe has meant that immigration has gradually produced new cultural mixes in the largest cities.
6. Complex land-use patterns within European cities have certain similarities but also important differences when compared with U.S. cities.
7. Cities within the European Union form part of an international trading bloc that contains more than half a billion people with a combined gross national income greater than that of the United States.
8. Since the end of the Cold War, Communist-era cities have undergone radical transformations, bringing them closer to their western European counterparts.
9. Europe became the birthplace of modern city planning as it reacted to the drawbacks of uncontrolled growth during the industrial period.
10. Sustainable urban management, including a search for best practices in water management, is increasingly becoming a priority in Europe.

Europe is a vital focus in the study of cities for a number of reasons (Figure 5.1). First, European cities are interesting in their own right; indeed, the great ones, like London, Paris, and Rome, come to mind when we think about Europe or plan a trip there. Second, because European cities are quite old, they reflect the history of many different economic, political, social, and technological changes. Third, a study of European urbanization then is made all the more exciting by the tumultuous changes since the fall of Communism. Fourth, as a hearth area of urban design, European cities are essential to understand the urban landscapes elsewhere. Fifth, European cities are part of global networks that they directly impact and are impacted by. A global city, such as London, is an important node in networks of capital investment, corporate decision making, and transportation infrastructure.

History, even recent history, has strongly conditioned the character of European cities. During the period of Soviet-style totalitarian governments after World War II, for instance, cities in much of central and eastern Europe diverged in form and function from their western European counterparts. In the early twenty-first century, the once-pronounced legacies of Communism and totalitarianism on urban landscapes have in many places given way to gleaming skyscrapers and subdivisions, although in others, socialist-era apartment blocks and factories still predominate. Today, cities within the European Union (EU) fall under a single economic and political framework that affects the living and working conditions of urban residents. The predominantly urban European Union contains more than half a billion people with a combined gross national income greater than that of the United States. Yet, while cities across Europe share many characteristics, including bustling city centers and compact form, differences remain, in land-use patterns, quality of urban infrastructure, and city planning and architectural design, to name a few.

Europe is more than 70 percent urban, but there is no continent-wide definition for a city. National definitions range from a minimum population of 200 in Norway and Sweden to 20,000 in Greece and Spain. Nevertheless, most Europeans live and work in urban areas, and Europe's urban population of about 400 million represents approximately 10 percent of the world's urban population.

Just as there is disagreement over what constitutes an urban population, there is also disagreement about how to define Europe as a region. A look at a world map shows that the region labeled "Europe" is, physically speaking, a peninsula of a much larger region, Eurasia. As sensitive debates illustrate, including political disagreements in Brussels about the accession of Turkey to the EU or Georgia to NATO, a definition of Europe as a region is not settled. Geographic labels such as those in the chapter titles of this book are more convenience than objective truth. Indeed, Europe is much more a cultural idea than a neatly bounded region on a map.

HISTORICAL PERSPECTIVES ON URBAN DEVELOPMENT

One of the exciting things about studying cities is learning to decipher the landscapes of bygone eras—their streets, buildings, and monuments. An historical perspective is necessary for understanding the evolution of the European urban system because the same forces that modify the built environment of individual cities also determined where cities were initially located and how they flourished or declined.

Classical Period: 800 BCE to 450 CE

In the ancient Greek realm, independent city-states were located along coastlines, reflecting their sea-faring culture, and on easily defendable hill sites, reflecting the need for security in turbulent times. As cities like Athens, Sparta, and Corinth grew, bands of colonists left to establish cities around the Aegean and Black Seas, along the Adriatic Sea, and as far west as present-day Spain.

Greek towns shared common traits. At the center was the *acropolis*, or high city, which contained temples and municipal buildings. Below the high city, in the "sub-urbs," were the *agora* (market place), more government buildings, temples, military quarters, and residential neighborhoods. These cities were laid out in a north-south grid pattern and surrounded by defensive walls. Greek cities, though, remained quite small by today's standards. Although Athens probably reached a population of about 150,000, most cities ranged from 10,000 to 15,000, while the majority had only a few thousand people.

Greek civilization was displaced during the second and first centuries BCE by the expanding Roman Empire. Although the structure of Roman cities (like Pompeii) was similar to their Greek predecessors—including the grid system, central market place (*forum*), and defensive walls—there were important differences. Roman cities were established mainly inland and operated as command-and-control centers. They functioned within a well-organized empire and were designed along hierarchical lines, reflecting the rigid Roman class system. By the second century CE, the Roman Empire extended over the southern half of Europe (Figure 5.2). Roman cities, though, remained fairly small. Although Rome's population probably reached the million mark by 100 CE, large Roman towns contained only about 15,000–30,000 inhabitants, while most had fewer than 5,000.

The vacuum created by the collapse of the Roman Empire in the fifth century was filled by various tribes who greatly disrupted urban life. Most urban centers became depopulated, and their crumbling buildings were a source of building materials for rural residents. At the

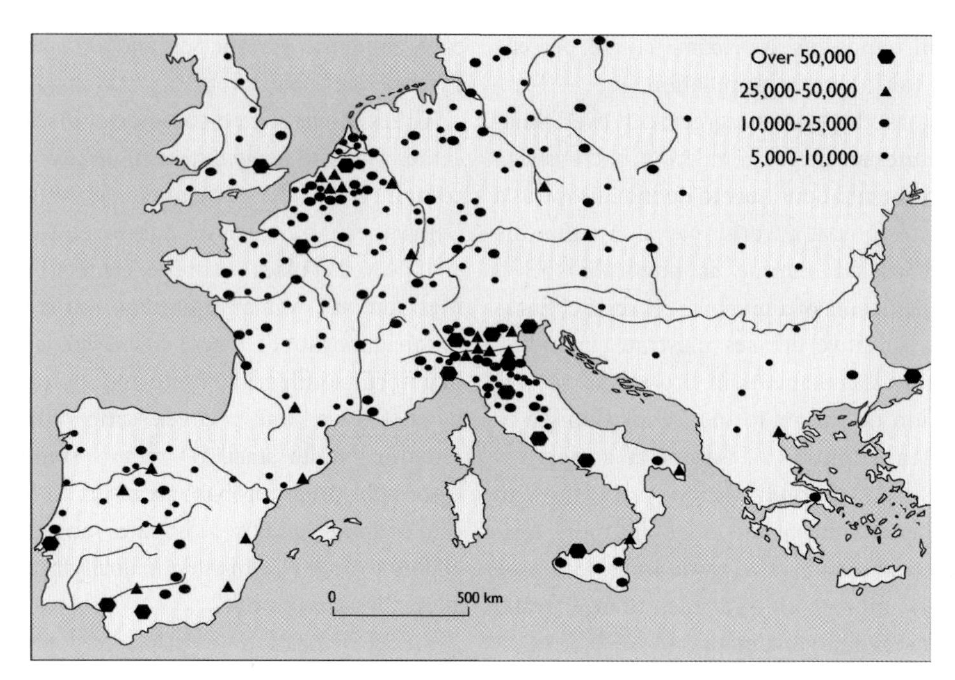

Figure 5.2 Roman Cities in Europe, second century CE. *Source:* Adapted from N. J. G. Pounds, *An Historical Geography of Europe* (Cambridge: Cambridge University Press, 1990), 56. Reprinted with permission.

same time, the constant threat of attack spurred the construction of castles and other fortifications, even in some parts of Europe that had previously seen more limited urban development.

Medieval Period: 450–1300 CE

Feudalism curtailed the development of cities during the early medieval period because its highly structured nature favored the self-sufficient country manor as the basic building block of settlement. The only urban places to thrive or even survive were religious, trade, or defensive centers. With the resumption of long-distance trade after 1000, many medieval towns grew along commercial routes that crisscrossed Europe (Figure 5.3).

At the center of the typical medieval city was the town square. In larger cities, this market square was surrounded by the main cathedral or church, town hall, guildhalls, palaces, and houses of prominent citizens. Close to the center were streets or districts that specialized in particular functions, such as banking, furniture, or metalwork. The streets and alleys were quite narrow. The enclosing walls often had water-filled moats to enhance defensive capability. Finally, medieval towns were decidedly unhygienic. Given the cramped conditions, lack of air circulation, poor sanitation, and absence of waste treatment, it is little wonder that the Black Death (1347–1351) progressed so easily, killing one-third of the people in urban Europe.

Most development during the medieval period was in the western and southern parts of Europe that had a Roman heritage of city-building. Urban development was impeded in southeastern Europe where the Byzantine Empire was in control, whereas much of

Figure 5.3 Ljubljana, Slovenia, took advantage of the collapse of Communist rule to bring out the medieval elements of the city's center, including the Dragon Bridge and St. Nicholas Cathedral. *Source:* Photo by D. J. Zeigler.

western and northern Europe remained in a pre-urban state. Conversely, the Moors, who spread into Iberia in the early 700s, founded or restored many cities and elevated urban culture in what would become Spain. At the close of the medieval period, Europe had about 3,000 cities, most of which had fewer than 2,000 people; only Milan, Venice, Genoa, Florence, Paris, Córdoba, and Constantinople (now Istanbul) had more than 50,000.

Renaissance and Baroque Periods: 1300–1760 CE

The Renaissance, beginning ca. 1300, was marked by significant changes: in the economy (from feudalism to merchant capitalism), in politics (rise of the nation-state), and in art and philosophy. Beginning in Florence in the 1300s, these changes spread throughout western Europe; conversely, feudalism was still strong in eastern parts of Europe; southeastern Europe fell under the grip of the Ottoman Empire; while much of northern Europe remained outside the progressive influences of the Renaissance.

Spurred by heightened demand for such luxury goods as spices and silks introduced during the time of the Crusades (1095–1291 CE), merchants greatly expanded the trade functions of Mediterranean cities. Later, the economic center of gravity shifted to the port towns along the North and Baltic Seas in conjunction with the Hanseatic League, an association of towns with the goal of promoting trade. Remarkably, these networks of trading cities presaged contemporary trading patterns amid globalization. At the height of the Hanseatic League, cities such as Lübeck, Hamburg, and Visby were highly integrated with each other through mercantile trade. Today, a global city such as London is more functionally networked with distant world cities, such as New York and Tokyo, than it is with much closer cities with dissimilar economic and political profiles.

Changes in the political system, particularly the growth of nation-states, had an impact on European urbanization. Best exemplified by Paris and Madrid, the central location of these capitals aided the process of political consolidation; in turn, both cities were given further impetus for growth by their administrative functions and enhanced status at the vortex of social, economic, and political change. Similarly, regional centers and seats of county government emerged to fill out expanding national urban networks.

The overall appearance and structure of cities changed because of the new forms of art, architecture, and urban planning. Especially in capital cities, flourishing artistic and architectural expression brought about greater use of sculpture in public areas, other urban beautification such as fountains, and embellishment on monumental buildings, which reached a peak during the baroque period (1550–1760).

Changes during this period were not just cosmetic. After the introduction of gunpowder, massive city walls became obsolete. In many cities, the walls were removed to make space for wide boulevards that were becoming fashionable. Also, the accumulation of great wealth by the nobility led to opulent palaces being built in many cities, notably Vienna and Paris, and to replanning parts of cities. Beginning in Paris, many districts containing narrow medieval streets were torn down to make way for wide boulevards that radiated outward and connected the various palaces and formal gardens laid out for the nobility. The emphasis on the control of visual perspective and the rediscovery of classical models of design marked a significant departure from medieval times. Overall, the urban network remained largely unchanged, although individual cities had grown larger.

Industrial Period: 1760–1945 CE

Large-scale manufacturing began in the English Midlands in the mid-1700s and spread to Belgium, France, and Germany, reaching Hungary by the 1870s. New factories, making a range of products from textiles to machine tools, changed the structure of cities and led to massive rural-to-urban-migration (Figure 5.4).

In many cities, whole districts of factories emerged, easily identified by their belching smokestacks, deafening machinery, and general hustle-and-bustle of industrial activity. By the mid-nineteenth century, trains transported much of the industrial inputs and products, so new tracks, stations, and rail traffic began to play a significant role in urban development. Public transportation—trolleys and subway systems—also modified the look and functioning of cities. Large tracts of often cramped worker housing were constructed. The industrial period also heralded the development of the CBD with its office buildings and corporate headquarters.

The growth of cities closely mirrored the spread of industrialization. By the mid-1800s, industrial towns in the English Midlands

Figure 5.4 Much of the coal that fired the industrialization of cities came through the Welsh port of Cardiff. That era is commemorated with public art on the reclaimed waterfront, along with one of the chimerical animals from a Bob Dylan poem. *Source:* Photo by D. J. Zeigler.

(notably Birmingham) and Scotland (notably Glasgow) had grown to more than 100,000 inhabitants and the proportion of the population living in cities larger than 10,000 had risen to 30 percent. The growth of industrial cities in France, Belgium, and Germany reflected the same pattern. In contrast, expansion of the industrial sectors in southeastern Europe did not occur until the early- or mid-twentieth century.

URBAN PATTERNS ACROSS EUROPE

A glance at the map of Europe shows the impact of central place theory on the size and spacing of urban places. In southern Germany, the largest metropolitan areas—Frankfurt, Munich, and Stuttgart—are spaced about the same distance apart. In Hungary, centrally located Budapest is ringed by the regional centers of Debrecen, Miskolc, Szeged, and Pécs. Of course, political, economic, cultural, environmental, technological, and other changes can alter the role and rank of a place within an urban hierarchy. Still, the empirical observation of rank-size distribution holds for Belgium, Germany, Italy, Norway, and Switzerland. In other national urban systems, a deviation occurs at the top of the hierarchy to create primacy. Primate cities that are also national capitals include Athens, Budapest, Dublin, London, Paris, Reykjavik, Sofia, and Vienna.

Historically, rural-to-urban migration has been the most important component of urban growth, especially during the industrial period. This form of migration within Europe, though, has largely ceased. And as birth rates have fallen considerably in recent decades, European cities are among the slowest growing in the world, averaging just 0.2 percent a year, with most of this growth accounted for by immigration from abroad.

The cities of Europe, however, have coalesced into conurbations in tandem with the transportation and communications infrastructure. Europe now contains about 50 conurbations with more than a million inhabitants. The metropolitan regions between London and Newcastle form an area of extensive urbanization in England. Germany's Rhine-Ruhr conurbation has a diameter of about 70 miles (110 km) and runs from Düsseldorf and Duisburg in the west to Dortmund in the east (Figure 5.5). Of similar diameter is the Randstad, a densely populated horseshoe-shaped region in the Netherlands that runs from Utrecht and Amsterdam in the north through The Hague and Rotterdam in the west, to Dordrecht in the southeast (Figure 5.6). Only about 60 miles (100 km) apart, these two conurbations may eventually coalesce to become a dominant European metropolitan core.

Postwar Divergence and Convergence

Western Europe

After World War II, separate urban systems developed on either side of the Iron Curtain—the boundary that divided Europe into a capitalist west and Communist east. Cities in western Europe cultivated connections with the capitalist world, especially the United

Figure 5.5 The Rhine-Ruhr Conurbation in Germany. *Source:* Compiled from various sources.

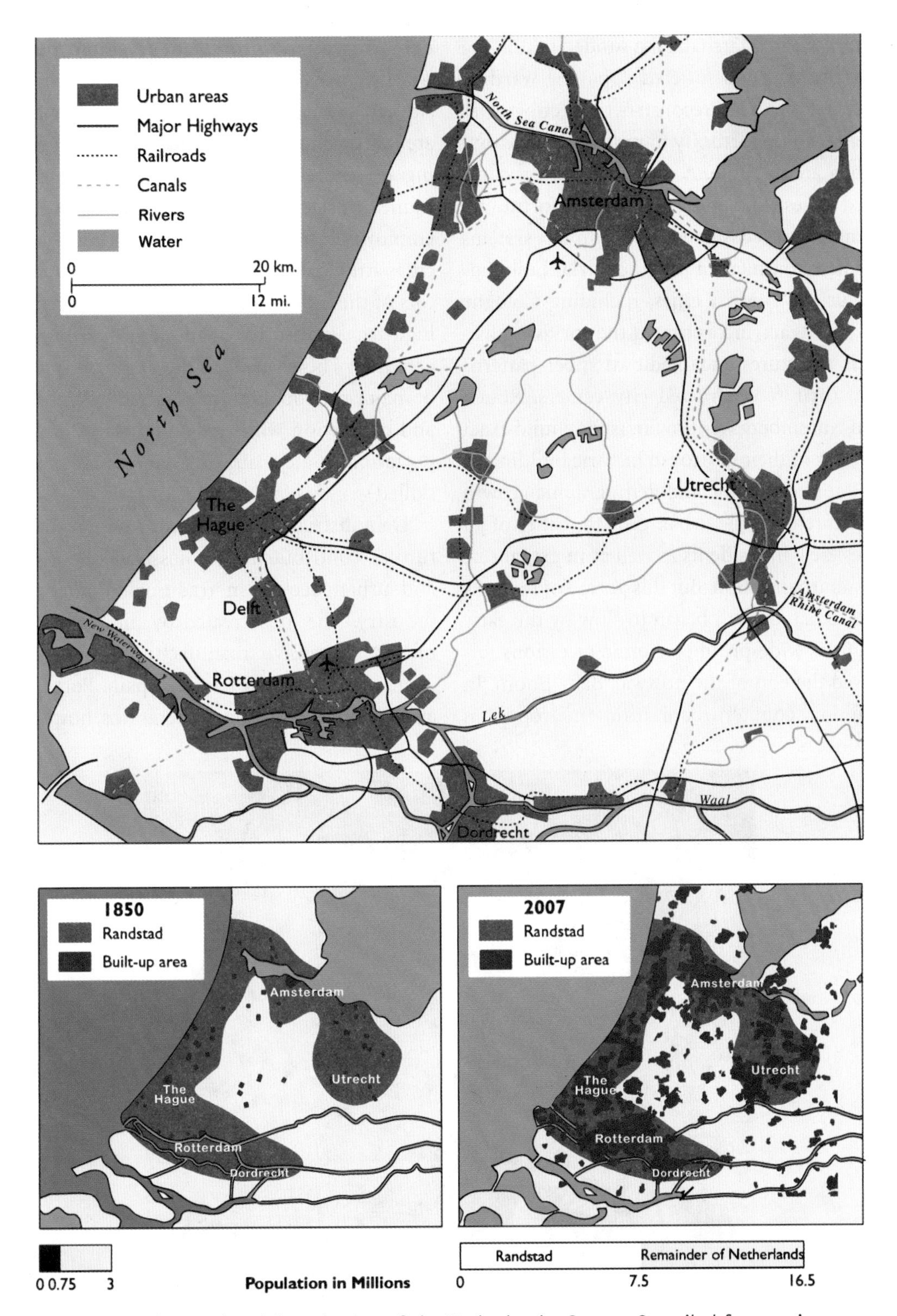

Figure 5.6 The Randstad Conurbation of the Netherlands. *Source:* Compiled from various sources.

States, whose Marshall Plan funded rebuilding in cities that had suffered appalling wartime destruction. The reconstruction effort was seen as an opportunity to replan bombed-out urban areas. Some of the most heavily damaged cities, like Rotterdam and Dortmund, completely redesigned their street systems for new commercial and industrial buildings (Figure 5.7). Most cities, including Cologne and Stuttgart, incorporated the surviving historic structures and medieval street patterns into their reconstructed city centers. Rouen and Nuremberg went so far as to rebuild exact replicas of their destroyed historic buildings.

Rapid economic and demographic growth fueled remarkable urban growth. Cities of all sizes grew in residents as well as in extent due to suburbanization. But this period of remarkable urban growth began to slow by the early 1970s as widespread economic recessions followed the end of the postwar Baby Boom. In addition, counterurbanization (metropolitan decentralization) promoted development in nearby towns and rural areas, while growth slowed toward the urban core. Peripheral areas attracted residents and businesses looking for more space and less pollution and crime. Most city centers lost retail and office employment to outlying areas. Medium-sized cities attracted employment in expanding sectors of the economy like information services, high-tech industries, or modern distribution activities. These smaller cities at the periphery of major metropolitan centers had lower rents and congestion while enjoying nearby transportation routes, airports, universities, and skilled workers.

Deindustrialization and corporate restructuring contributed to massive job losses and urban decline in traditional centers of industry. The jobs created by the relocation of labor-intensive manufacturing benefited some urban areas in Ireland, Spain, Portugal, and Greece. Branch plant operations, however,

Figure 5.7 Nation building is a function of every capital city's landscape. In Amsterdam, a statue says thank you to Queen Wilhelmina, who gave her subjects hope during World War II. Next to the Dutch flag is the U.S. flag. *Source:* Photo by D. J. Zeigler.

are vulnerable to decisions made outside the area and to company relocations when government tax incentives expire. Since the early 1990s, competition for investment has come from central and eastern Europe where production costs are lower.

During the last few decades, the city centers have seen quite significant changes as employment has shifted to professional and business services such as banking and insurance. New developments include shiny new high-rise offices, luxury condos and apartments, and gentrified neighborhoods with expensive restaurants, bars, and boutique stores. In world cities like London and Paris, the most visible group of people on the streets of the CBDs are young professionals chatting on their cell phones and wearing the latest fashions.

Socialist Urbanization

Following World War II, cities behind the Iron Curtain developed independently of their western counterparts. Totalitarian governments engaged in sweeping reforms that led to considerable changes to their national urban systems, which evolved in response to centralized planning rather than market forces.

Communist governments had to contend with the pressing need to rebuild cities left in ruins after the war. The damage sustained by these cities, particularly Dresden, Berlin, and Warsaw, was more severe than in western Europe, but subsidies were not available to Communist bloc cities through the Marshall Plan. The earliest stage of postwar economic development involved rapid expansion of heavy industry, particularly iron and steel, chemicals, and machinery. Coupled with collectivization and increased mechanization in agriculture, this *extensive* industrial development soon led to unprecedented rural-to-urban migration. Levels of urbanization rose quickly in conjunction with rising levels of primacy, severe housing shortages, insufficient social infrastructure and basic services, and environmental degradation.

So, beginning in the mid-1970s, Communist governments set out to erase the difference between city and village life by emphasizing light industry and services, decentralizing production from capital and larger cities to smaller ones, developing transportation networks, and increasing levels of public infrastructure and housing in cities. Rural-to-urban migration slowed significantly. Despite these efforts, by the fall of Communism, the urban network in much of central and eastern Europe was still less developed than in the West.

Post-Socialist Changes

In the late 1980s and early 1990s, Communist governments were toppled, Germany was reunited, and Czechoslovakia, Yugoslavia, and the USSR were broken up. Central economic planning was abandoned in favor of democratization and transformation from socialist to market economies. The demise of the Soviet Union opened the way for western investment to move in and people to move out. Many countries have since joined the western political, economic, and military alliances, including the North Atlantic Treaty Organization (NATO) and the European Union.

These changes impacted cities and urban systems in a number of ways. First, city names that were inspired by revolutionaries, such as Leninváros (Lenin City) in Hungary (now Tiszaujvaros) and Karl-Marx-Stadt in Germany (now Chemnitz) were changed back to their prewar names or to honor individuals or events associated with the 1989 revolutions. Statues of Communist and Soviet leaders were removed and some were later put on display in statue parks or museums like Memento Park in Budapest. Second, foreign direct investment flooded in, targeted mainly at capital cities. This boosted the transition to capitalism and fueled speculative construction booms and gleaming, Western-style commercial and residential developments. Warsaw's skyline, for example, was once dominated by the Stalinesque Palace of Culture and Science, the tallest building in the Eastern Bloc outside of Moscow. Today, its unmistakable "wedding cake" architectural style building is dwarfed by newer steel-and-glass skyscrapers (Figure 5.8). Third, decentralization gave more authority to city planners.

More color and neon lights now characterize the cities of central and eastern Europe. Advertising has replaced Communist slogans on billboards. Shabby, old department stores have been renovated or replaced by boutiques and shopping malls. Beggars have appeared, as well as casinos and night clubs; crime has risen and congestion is ubiquitous. Social differentiation in housing has increased tremendously. Democratization and market economies are erasing the Communist legacy and bringing these cities closer to their western European counterparts.

Core-Periphery Model

A core-periphery model is often used to describe urban patterns in Europe (Figure 5.9). The dominance of cities and conurbations at the European core is based on their superior endowment of factors influencing the location of economic activity, such as accessibility to markets. The largest cities are connected by the most advanced transportation and communications systems. Labor force quality and government policies make the core the most attractive area for modern companies. French geographer, Roger Brunet, identified the "Blue Banana," a curving urban corridor of high-tech industry and services that includes London, the Randstad and Rhine-Ruhr conurbation, and Milan.

Cities in the core and periphery are linked in a symbiotic if unequal relationship. Core cities prosper and maintain their economic dominance at the expense of the periphery by capturing flows of migrants, taxes, and investment in cutting-edge industries such as high-tech manufacturing and in command-and-control functions like the headquarters of transnational corporations (TNCs). At the other extreme, peripheral cities have more limited potential for economic development, and attract tourists and investment in branch plants from core locations.

The European core, however, has been shifting to the south and east, to areas of high-tech industrial growth. Newer core cities include Munich in Germany, Zürich in Switzerland, Milan in Italy, and Lyon in France. This southeastward shift intensified after the fall of the Iron Curtain; people and companies have been attracted to cities such as Bratislava and Budapest because of the surge in economic activity, as well as to the region as a whole due to its relatively low production costs. London and Paris, however, have retained their historic importance because of their size and established positions as major national and international cities. The continued economic strength of the core is reinforced by the considerable political control that comes with the role of the largest cities as major centers of international decision making.

European cities are also part of a global core-periphery model of urbanization. World cities, such as London, Paris, Frankfurt, and

Figure 5.8 Warsaw's skyline, once dominated by the Stalinesque Palace of Culture and Science's "wedding cake" architectural style, and the tallest building in the Eastern Bloc outside of Moscow, is today dwarfed by newer steel-and-glass skyscrapers. *Source:* Photo by Linda McCarthy.

others, are vital nodes in international networks of investment capital, business decision making, and transportation infrastructure, while other once important centers have melted into relative oblivion, at least as far as global processes are concerned. One measure of this is rents: central Frankfurt's commercial real estate is in a league with such cities as Tokyo, Hong Kong, and New York, given its status as a banking center, while in comparably

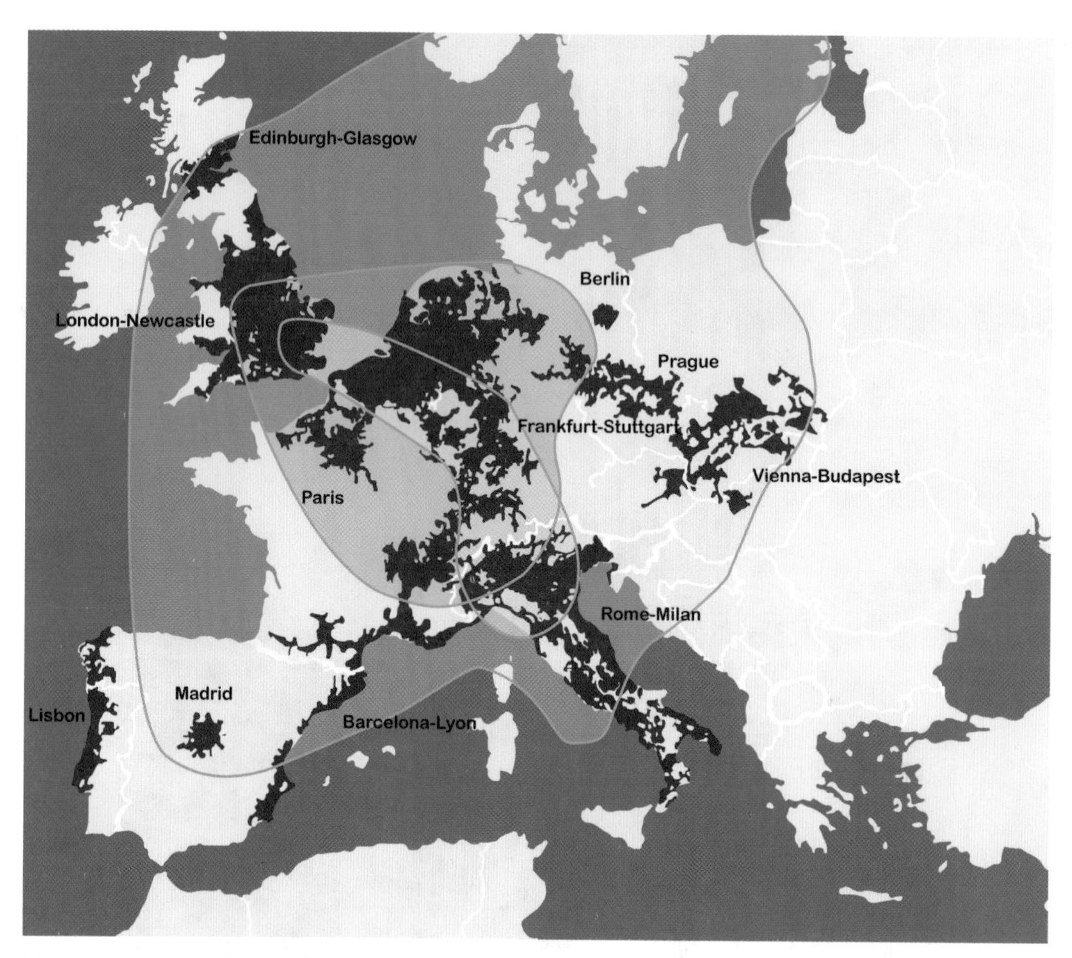

Figure 5.9 Europe's conurbations within the context of Europe's "Blue Banana" and core-periphery conceptualizations. *Source:* Photo by Linda McCarthy.

sized cities such as Essen and Dortmund, rents are dramatically lower.

IMMIGRATION, GLOBALIZATION, AND PLANNING

The Challenge of Integrating Immigrants

The rebuilding of western Europe's urban infrastructure and industry after World War II generated strong demand for labor, especially in the more prosperous countries. In the 1950s and 1960s, rural-to-urban migration fueled growth, especially in the largest cities. In addition, foreign guest workers were brought in to fill low-wage assembly-line and service-sector jobs that the more skilled domestic labor force would not take. Guest workers came from Mediterranean Europe and former colonies. Then West Germany attracted immigrants from Turkey and Yugoslavia; France brought in workers from northern and western Africa; and Britain drew on Commonwealth citizens from the Caribbean, India, and Pakistan.

In the European Union today, about 7 percent—about 35 million—of the people were

Figure 5.10 The salon de thé (tea house) is a common element of urban landscapes in French-speaking North Africa. As Arab immigrants arrive in Brussels, they bring with them their preferences for particular tastes and social settings. *Source:* Photo by D. J. Zeigler.

born outside its 28 member countries (Figure 5.10). More than one-third of the foreign-born immigrants in France are concentrated in the Paris region, where they represent over 15 percent of the population. Foreign-born residents from outside the European Union comprise 15–25 percent of the population in German cities such as Frankfurt, Stuttgart, and Munich. More than half the population of Amsterdam is non-Dutch, being strongly represented by individuals from Morocco, Turkey, and Indonesia. In addition to the demographic data, the presence of recent immigrant arrivals is reflected in other data such as the list of most popular baby names for cities like Brussels or London (Table 5.1) or in the most popular ethnic food in different cities across Europe (Table 5.2).

Europe's aging population coupled with demand for low-wage labor means that some jobs will continue to be filled by immigrants. But there has been an anti-immigrant backlash

Table 5.1 Top 10 Boys' and Girls' Names in London

	Boys	*Girls*
1.	Muhammad	Amelia
2.	Daniel	Olivia
3.	Alexander	Sophia
4.	Mohammed	Isabella
5.	Joshua	Mia
6.	Oliver	Emily
7.	Harry	Jessica
8.	Samuel	Sophie
9.	Thomas	Ava
10.	James	Chloe

Source: London Evening Standard, 2014, http://www.standard.co.uk

by some people, often xenophobic and racist, against newcomers as well as minorities, such as the Roma, who have lived in Europe for centuries. They have been the target of discrimination and persecution, including by the Nazis who murdered hundreds of thousands

Table 5.2 Popular Ethnic Food in European Cities

City	*Country*	*Ethnic Food*	*Description*	*Original Influence*
Amsterdam	Netherlands	loempi	spring roll	Indonesia (former Dutch colony)
Berlin	Germany	döner kebab	"rotating" grilled meat on a spit (aka gyros (Greek)), served sliced, typically in flatbread shawarma (Arabic) (e.g., pita)	Turkish immigrants
Bucharest	Romania	covrigi	pretzels	Italian monks; German bakers later
London	United Kingdom	chicken tikka masala	chicken curry	India (former British colony)
Paris	France	falafel	deep-fried chickpeas +/or fava beans, typically served in flatbread (e.g., pita)	Middle Eastern immigrants
Sarajevo	Bosnia & Herzegovina	burek	filled baked phyllo dough pastries	Ottoman Empire

of Roma. More recently, in a 2009 referendum, the Swiss voted to ban the construction of new minarets. Since then, France and Belgium have passed legislation banning the wearing of face-covering veils and other conspicuous religious symbols in public. Most countries have enacted immigration restrictions. The contradiction between rising demand for low-wage labor and an unwillingness to accept nonnationals has proven quite costly and dangerous, as the weeks of rioting in the largely Muslim working-class suburbs of Paris amply demonstrated in 2005.

European and Global Linkages

European cities are part of urban networks that operate at different spatial scales. Since 1989, cities on either side of the former Iron Curtain have become more interconnected. Increasing EU economic and political integration has influenced the development of the European urban system. For example, the removal of national barriers to trade within the European Union, with the internationalization of the European economy, has encouraged population increase along certain border regions. Urban growth zones straddle the boundaries between the Netherlands and Germany, Italy and Switzerland, and the southern Rhine regions of France, Germany, and Switzerland.

European cities are linked through trade and other mechanisms to major urban areas throughout the world. A select group of cities contain the headquarters of major international agencies, many of which were founded after World War II to promote economic, political, or military cooperation. Geneva is the main European center for the United Nations. Paris is the headquarters for the Organization for Economic Cooperation and Development (OECD) and the European Space Agency. Vienna is the headquarters for the Organization of Petroleum Exporting Countries (OPEC).

Important decision-making functions are located in the EU's "capital cities": Brussels (both the Council and the European Commission), Strasbourg (Parliament), and Luxembourg (Court of Justice). Brussels is also the headquarters of NATO and Strasbourg additionally serves as the headquarters of the Council of Europe, an organization of nearly 50 countries that promotes European unity, human rights, and social and economic progress.

The major centers of international banking and finance in Europe have been London and Paris, but now include Frankfurt and Luxembourg. Frankfurt hosts the Bundesbank, Germany's influential central bank, as well as the European Central Bank that manages the euro, making Frankfurt the financial capital of the European Union. Luxembourg, on the other hand, is the headquarters of the European Investment Bank and over 150 other banks serving an international clientele.

London and Paris rank among the select number of world cities that contain the headquarters of some of the most powerful TNCs in the world. London contains about 21 of the 500 largest global companies (78 percent of the United Kingdom's total), including BP, HSBC, and Lloyds. Paris has even more—28 of these companies (90 percent of France's total), including BNP Paribas, Christian Dior, and Vivendi. In addition to housing 3 of the 500 largest global companies, Rome contains Vatican City, the seat of the Roman Catholic Church. Paris and Milan are major centers of fashion and design, while London is the premier insurance center. In addition, despite the proliferation of fast-food restaurants such as McDonald's, a city such as Paris can still be recognized as the global capital of haute cuisine and is the European city with the most restaurants with Michelin stars for fine dining. Adding to Parisians' cultural pride and identity, the United Nations cultural organization, UNESCO, added France's multicourse gastronomic meal to the world's "intangible cultural heritage" list in 2010.

Accessibility via the latest transportation and communications technologies allows some cities to strengthen their international positions. High-speed trains reinforce the dominance of London, Paris, Brussels, Amsterdam, and Frankfurt. The cities with the busiest airports are London, Paris, Frankfurt, and Amsterdam. These four cities form one of the world's major clusters of airline hubs in a global network of air travel. As such, these airports are important not only as final destinations for passengers and cargo, but also as transit hubs where people and goods change planes along their route.

At the mouth of the Rhine, Rotterdam is Europe's largest port and one of the largest in the world. Its annual turnover of more than 400 million metric tons of cargo is third only to Shanghai and Singapore. Rotterdam's water and pipeline connections with the Ruhr in Germany make it the main oil distribution and refining center in Europe. Antwerp, Marseille, and Hamburg are other major ports.

Trucking is the most important mode of ground transportation for freight. Nearly 2,000 billion tons of goods are transported by road annually in the European Union alone compared to about 400 billion by rail. There are close to 500 passenger cars for every 1,000 people in the European Union. With about 600 cars per 1,000 inhabitants, German and French levels already match those of Canada, but even Italy's higher rate of nearly 700 is still far below the U.S. level of more than 800. Steadily increasing automobile ownership, with the distance traveled tripling since 1970, has overwhelmed existing and new highway capacity and led to traffic congestion within and between cities.

Formerly Communist areas of Europe still lag behind in terms of the extent and efficiency of their transportation systems, though this is rapidly changing. In Germany, rail and road links abandoned during 40 years of Communism that divided the country into West and East have been rebuilt. Even though Communist states were allied with each other and the Soviet Union, this did not ensure a high-performing network of connecting infrastructure. Traveling between Budapest and Warsaw by land, for example, entailed multiple border crossings, transit visa requirements, and long waits. Travel was slowed by narrow, dangerous, often prewar roads. In part due to large investments by the European Union, new multilane highways are being built, but it will be years before transport linkages resemble those in the West.

Urban Policy and Planning

Europe became the birthplace of modern city planning as it reacted to uncontrolled growth during the industrial period. Planning to address urban problems now pervades European city life. After World War II in western Europe, national policies promoted regional decentralization. Industry, commercial activity, and population were redirected from the large congested cities to new towns, as in the case of Abercrombie's Plan for Greater London. Postwar planning for growth ended, however, in the early 1970s. Declining population growth rates and widespread economic recessions forced governments to reconsider large-scale publicly funded projects and the need for new towns.

Given the dissatisfaction with alienating high-rise buildings and open spaces, policy shifted to planning for conservation and restructuring, and combating urban decline. This reappraisal has had two, often conflicting, components: budgetary constraints forcing governments to seek private-sector investment in revitalization projects, and growing concern for social equity, citizen participation, environmental protection, and aesthetic quality.

A shift to neoliberal policy and planning reflected factors such as the severity of decline in the central parts of larger cities and their importance as national engines of growth in a global economy. The government of the United Kingdom established Urban Development Corporations to attract businesses to declining industrial and port areas in cities like London and Liverpool. In recent years, most countries in western Europe have decentralized power and responsibility for urban planning to local governments. In addition, smaller units of local government have been consolidated into larger regional ones to achieve economies of scale. These policy and administrative changes have set the scene for more coordinated regional planning. The Dutch "compact city" policy in the Randstad endeavors to curb counterurbanization by concentrating new development within existing major cities in an effort to maintain their economic competitiveness.

The urban revitalization policies in the older industrial cities of western Europe have promoted economic restructuring away from traditional manufacturing. Cities use local, national, and EU funds to attract private-sector investment in high-tech and service industries. The transition toward K-economies (knowledge economies) has favored diversified metropolitan economies with highly skilled workforces, major universities, and good quality of life. Traditionally in southern Europe, cities in lower-cost production areas attracted labor-intensive branch plant industries. More recently, cities like Montpellier in France, Bari in Italy, and Valencia in Spain have focused on providing attractive environments for high-tech industries.

In Communist central and eastern Europe after World War II, government planning was guided by the basic tenets of Marxist-Leninist ideology: to remove the "contradiction" between living standards in urban and rural areas and to create a classless society. Urban planners sought to avoid excessive population concentration in large cities and to achieve a balanced urban infrastructure. These social goals, however, often clashed with economic directives, especially the development of heavy industry.

In an attempt to increase overall industrial capacity and provide urban functions to underserved areas, governments implemented a program of new town construction away from existing cities. These towns were developed around a large industrial facility, typically an iron and steel mill or chemical processing plant. New towns included Eisenhüttenstadt in East Germany and Nova-Huta in Poland. By the 1970s and 1980s, Communist planners had turned their attention away from promoting large-scale industry and new towns to developing light industries and filling out the national urban systems. In many countries, central place theory became an explicit guide as planners tried to create multitiered urban hierarchies that provided goods and services to particular regions according to their size and function.

Since the early 1990s, central and eastern European cities have experienced dramatic changes because the transition to a market economy involved rapid, large-scale privatization of state-owned housing, industry, and services. Similar to their western counterparts, Communist-era cities like Dresden, Budapest, and Warsaw, now work to attract new commercial and industrial investment. National policies have evolved to address urban problems that were unknown in the former socialist states, such as unemployment, crime, poverty, and homelessness. EU integration has helped alleviate somewhat some of the problems through a number of urban redevelopment projects.

In fact, Europe is the scene of significant international urban planning and management initiatives. The Council of Europe and the European Union, for example, celebrate Europe's cultural heritage through European Heritage Days. Cultural events are planned in cities and towns across the European Union that are aimed at bringing European citizens together through highlighting local traditions, skills, and works of art and architecture. The European Union also selects two cities every year as European Capitals of Culture, with Donostia-San Sebastian in Spain and Wroclaw in the Poland selected for 2016.

The European Union established the European Green Capital Award to promote local government efforts to improve their urban environment, and to showcase best practice. The first award winner in 2010 was Stockholm; Copenhagen won in 2014; and the first eastern European city to win was Ljubljana, Slovenia, in 2016. In terms of more measureable environmental quality, Copenhagen and Stockholm also lead the rankings on the European Green City Index (Table 5.3).

EU integration efforts have led to unprecedented achievements in international policy and planning. The publication of the "Green Paper on the Urban Environment" in 1990 reflected the need for EU policies to address specifically urban issues. European cities are generally predisposed to being "green" because

Table 5.3 European Green City Index: Top 10 Cities

Rank	*City*	*Score**
1	Copenhagen	87.31
2	Stockholm	86.65
3	Oslo	83.98
4	Vienna	83.34
5	Amsterdam	83.03
6	Zurich	82.31
7	Helsinki	79.29
8	Berlin	79.01
9	Brussels	78.01
10	Paris	73.21

*Out of a possible 100, based on eight categories (CO2, energy, buildings, transport, water, waste and land use, air quality, environmental governance) using 30 indicators, conducted by the Economist Intelligence Unit, sponsored by Siemens, 2012.

of their high density and compact form and associated walkability and high usage of public mass transit. Even so, European cities are not exempt from environmental problems associated with issues such as climate change and sea-level rise (Box 5.1) or traffic congestion and associated high levels of airborne particulate matter and unhealthy ozone levels. Over the years, the European Union has adopted policies that have provided funding for innovative environmental management projects, including the Sustainable Cities project involving research, information exchange, and networking in conjunction with the implementation of Local Agenda 21, the TRUST (Transitions to the Urban Water Services of Tomorrow) project involving sharing best practice for urban water cycle solutions, support for Mayors Adapt, which involves adaptation to climate change in cities, and Eurocities, a network of well over 100 major cities across more than 30 countries that provides a platform for best practice exchange.

CHARACTERISTIC FEATURES WITHIN CITIES

"Our cities are like historical monuments to which every generation, every century, every civilization has contributed a stone" (Ildefons Cerdà, Spanish town planner, 1867). The landscape of European cities today represents an incomplete catalog of urban development and redevelopment over time. Typical historic and contemporary features include:

Town Squares

The town square, the heart of Greek, Roman, and medieval towns, has often survived as an important open space. Some medieval town squares boast a continuous tradition of open-air markets. In central and eastern Europe, the large open square, typical of socialist cities, was used for political rallies. Today, like their western European counterparts, many central squares and their historic buildings contain modern commercial functions, such as tourist offices and fashionable restaurants and cafés.

Major Landmarks

Historic landmarks in western European city centers have become symbols of religious, political, military, educational, and cultural identity. Many cathedrals, churches, and statues serve their original purpose and some still dominate the skyline. Town halls, royal palaces, and artisan guildhalls have been converted into libraries, art galleries, and museums. Medieval castles and city walls are tourist attractions. Today, of course, the major landmarks are expressions of economic power—offices of TNCs and sports stadiums, for instance.

Box 5.1 Venice and the Challenges of Climate Change

The stark reality of climate change and the resulting sea-level rise raise significant challenges for European cities. An extreme case is Venice, Italy. Situated in the 210-square-mile (550 km^2) Venetian Lagoon on Italy's northern Adriatic coast, Venice is famous for an iconic urban landscape that dates from the thirteenth to seventeenth centuries, when it had a merchant empire of its own. The city itself sits atop a human-made foundation of wooden piles that are now hundreds of years old. Its location at the interface of land and water served it well in the past, but Venice is now facing multiple threats to its very existence from climate change and environmental degradation.

The health of the Venetian Lagoon is vital to the well-being of the city. Apart from serving as home to humans and wildlife alike, the lagoon is also a source of economic prosperity in the form of fishing and, of course, tourism, which is the biggest economic engine of the region. The natural processes that initially created the lagoon have been altered by Venetians since the city's founding. These environmental alterations often come at the expense of the lagoon's ecosystem. Sea-level rise means that the balance between saltwater and freshwater is changing in the estuarine environment, which impacts everything from fisheries to the iconic summertime odors of the city's canals.

A far more serious challenge is flooding. As a city built on canals, Venice is barely above sea level and experiences frequent tidal floods in the fall and winter months during the *acqua alta* (high water) season. This regular flooding has become more severe recently because of the rising sea level. Photos of the iconic Piazza San Marco submerged under more than a foot of water are becoming more frequent. If action is not taken, flooding will become more severe.

The problem of sea-level rise and flooding is exacerbated by the sinking and shifting of the lagoon's silt floor upon which Venice rests. Centuries ago, in-flowing river sediment was seen as a threat to the watery advantages of the lagoon, so rivers flowing into the lagoon were diverted to the Adriatic Sea to halt the silting process. But now this loss of sediment is threatening the city. The city's subsidence is a natural phenomenon, but the rate of sinking has been increased due to urban development, fresh water and natural gas extraction from beneath the city, and pollution of the lagoon's water. While urban growth and underground extraction have created pressure differentials, pollution from industrial dumping and inadequate sewage infrastructure has substantially changed the ability of native plant species to thrive. Yet the lagoon plants are essential to preventing faster erosion rates of the lagoon floor.

The Italian government is close to completing the MOSE (MOdulo Sperimentale Elettromeccanico), a massive engineering project designed to protect Venice from the sea. MOSE consists of a system of gates located on the lagoon floor near the inlets that connect the lagoon to the Adriatic. The system is designed to protect the city from the extreme flooding events during high tide that have been exacerbated by subsidence and sea-level rise. As a city uniquely tied to water, Venice provides an example of why the needs of the human communities and the natural environment must be balanced in order to maintain a sustainable relationship between the two.

In central and eastern Europe, in addition to prewar landmarks, the hallmarks of socialist cities included massive buildings in "wedding cake" style, red stars, and "heroic" statues. Since the late 1980s, socialist political symbols have been replaced by billboards advertising the trappings of consumer culture.

Complex Street Pattern

The narrow streets and alleys of the medieval core developed in the pre-automobile era (Figure 5.11). During the medieval period, suburban areas grew around long-distance roads that radiated outward from the city gates. In the nineteenth century, cities like Munich, Marseille, and Madrid made radial or tangential boulevards the axes of their planned suburbs.

Figure 5.11 Here on Ludgate Hill in the City of London, a new immigrant from Bangladesh directs people to the nearest McDonald's. In medieval times, this area would have been a shadowy tangle of narrow alleys that passed for streets. *Source:* Photo by D. J. Zeigler.

High Density and Compact Form

The constraints of city walls kept population density high during medieval times. Several factors maintained the compact form that is now characteristic of many large cities in Europe. A long tradition of planning that restricts low-density urban sprawl dates back to strict city-building regulations in the earliest suburbs. Compact urban form also reflects the relatively late introduction of the automobile, as well as high gasoline prices.

Bustling City Centers

Their high density and compact nature create city centers that bustle with activity (Figure 5.12). Heavily used public transportation systems of buses, subways, and trains converge on the core, and central train stations figure prominently.

In larger cities, distinct functions dominate particular districts. Institutional districts house government offices and universities. Financial and office districts contain banks and insurance companies. A pedestrianized retail zone often leads to the train station. Cultural districts offer museums and art galleries. Entertainment areas include theater and "red light" districts.

Many buildings in the city center have multiple uses. Apartments are found above shops, offices, and restaurants. Large department stores, such as Harrods in London and Kaufhaus des Westens in Berlin, are prominent features in most city centers. Modern quite centrally located malls include Westfield

Figure 5.12 Busy, pedestrianized shopping streets, such as this one in the heart of Dublin, are typical of the European city centers. *Source:* Photo by Linda McCarthy.

London (more than 300 stores) and Prague's Palác Flóra, both accessible by mass transit.

Suburban malls are becoming prevalent. Many cities on a coast or a river have also refurbished old port and industrial buildings to house mixed-use waterfront developments like Ķīpsala in Riga and HafenCity in Hamburg. Other cities have renovated obsolete historic structures, such as London's Covent Garden, as festival marketplaces with specialized shops, restaurants, and street performers.

Low-Rise Skylines

For North American visitors, the most striking aspect of the older parts of many European cities is the general absence of skyscraper offices and high-rise apartments. City centers were developed long before reinforced steel construction and the elevator made high-rises feasible. Building codes designed to minimize the spread of fire maintained building heights between three and five stories during the industrial period. Paris fixed the building height at 65 feet (20 m) in 1795, while other large cities introduced height restrictions in the nineteenth century. Still regulated today, high-rises are found in some cities only in redevelopment areas or on land at the periphery of the city, such as La Défense in Paris. Skyscrapers have also been built in the central financial districts of some of the very largest cities, including London.

Neighborhood Stability and Change

Western European cities historically have enjoyed remarkable neighborhood stability. Europeans change residences much less frequently than North Americans. As a result, some older neighborhoods at or near the center of large cities enjoy remarkably long lives, despite suburbanization.

Some districts of handsome mansions built by speculative developers for wealthy families in the seventeenth and eighteenth centuries remain stable, high-income neighborhoods, such as Belgravia and Mayfair in central London. High-income suburban neighborhoods developed in the western parts of older industrial cities, upwind of factory smokestacks and residential chimneys.

Wealthy residents, in fact, have situated themselves at or near the city center in western Europe since before the Industrial Revolution. Higher taxes on city land until the late nineteenth century kept the poorest people outside the city walls. Beginning in Paris in the mid-nineteenth century, this tradition was strengthened by the replacement of slums and former city walls with wide boulevards and imposing apartments.

Since the eighteenth century, however, urban growth has also spread to suburban areas and even enveloped freestanding villages and towns. These separate urban centers became distinct quarters within the expanding city as they maintained their long-established social and economic characteristics and major landmarks and links to the city center by public transit. During the second half of the nineteenth century, annexations of these suburban areas produced distinctive city districts with their own shopping streets and government institutions.

In the past few decades, city governments funded urban renewal projects designed to attract higher-income residents to the revitalized parts of central areas. The success of these large-scale redevelopments has given rise to gentrification in the surrounding area. Demand for housing that can be renovated for higher-income occupants, however, has raised property values in certain areas and pushed out lower-income residents.

The presence of recent immigrant arrivals is evident in the cultural diversity reflected in the names of stores and restaurants in some neighborhoods. Immigrants typically live in poor quality suburban high-rise apartments or inner-city enclaves left vacant through suburbanization. Each enclave is dominated by a particular ethnic group. Enclaves in Frankfurt and Vienna are home to mostly Turks, while in Paris and Marseille, they house Algerians and Tunisians. In large British cities, in contrast, there is significant mixing of different ethnic groups. Within each neighborhood, however, the ethnic groups are highly segregated from each other. And although there are large numbers of Asians and West Indians, the foreign-born population represents only 15–20 percent of the population within most neighborhoods.

In addition to outright discrimination, the labor and housing markets help create inner-city enclaves. Low wages force immigrants to rent lodgings in deteriorating inner-city locations. The internal cohesiveness of the ethnic groups also contributes to residential segregation. Existing residents are more likely to share information about vacancies in their neighborhood with members of their own ethnic group.

Housing

Apartment living is common in Europe. Apartments are a good land-use choice when space is at a premium and land values are high. Instead of growing outward, cities grew upward, to the limit of the height regulations.

The multistory apartment building originated in northern Italy to accommodate the wealthy during the Renaissance. By the early eighteenth century, apartment buildings had spread to the larger cities in continental Europe and Scotland. Until the invention of the elevator, social stratification within individual buildings was vertical: wealthier families occupied the lower floors; poorer residents lived in smaller units above. Horizontal social stratification also developed within apartment blocks. Larger expensive units faced the front; small low-rent units faced the rear. By the late eighteenth century, as the Industrial Revolution spurred increasing urbanization, apartment blocks had spread to medium-size cities. Speculators built large-scale standardized tenements for middle-income occupants and barracks for low-income residents.

The two-story, single-family row houses with small gardens are found in England, Wales, and Ireland (Box 5.2). This tradition

Box 5.2 Growing Power: Urban Agriculture in Europe

With growing concerns over environmental sustainability and food security, attention in European cities has turned to the possibilities for urban agriculture. Liberalized laws mean that apartment dwellers in London and Rome, for instance, can keep chickens. Rooftops and garden patios now double as vegetable gardens, while old allotments are seeing new life as people want food security and quality assurance about the food they eat.

Urban agriculture is not new. Rapidly industrializing Germany became a pioneer in the nineteenth century as socially conscious lords and later city governments sought to offer relief to migrants who lived in squalid conditions in cities such as Berlin, Munich, and Leipzig. Urban community gardens had their origins as part of larger projects of providing low-income residents with a means to feed themselves. Allotment gardens, in Germany called *Kleingärten* (small gardens) or *Schrebergärten* (after a physician who promoted gardening as a means for urbanites to escape the ills of city life), were usually located on low value land, such as railroad rights-of-way. In Scotland, the 1892 Allotments Act provided a legal means for working-class people to petition for an allotment garden. During wartime in the twentieth century, allotments served a crucial role as sources of food. "Dig for Victory" was a rallying cry for Britons during World War II in the same way as victory gardens sprouted across the United States. After the war, when critical shortages of food in central Europe caused widespread malnutrition in cities, small plots provided much needed vegetables. As Europe became more and more prosperous after World War II, though, increasingly the population's food requirements were met by an increasingly industrial, large-scale system of agriculture. This lengthening of the food chain meant that the food for most people in European cities was coming from an ever more complicated web of suppliers spanning the world.

Urban agriculture's recent strong comeback across Europe is not about fear of starvation, but about a trend toward a "return to roots." The slow food movement, which originated in Italy and spread throughout the world, emphasizes that overall physical and psychological health is tied to healthy eating and locally produced food. Trends such as the slow food movement can be seen as a desire to shorten food chains—making them closer to what they had been historically—in order to gain environmental and human-health benefits. The EU's common agricultural policy (CAP), which constitutes the largest share of the massive EU budget, has historically privileged large, rural farms in its payment schemes, but recent initiatives have called upon the CAP to fund more urban agriculture.

The geography of urban agriculture reflects general shifts in urban morphology during the last century. Where market gardens once occupied peripheral lands around European cities, those areas have long been overtaken by suburbanization. Since urban land commands premium prices, gardeners have found novel locations for farming. Old tourist boats that once plied Amsterdam's canals have found new life as floating greenhouse gardens. Sections of the former "no-man's land" of the Berlin Wall are now community gardens. Paris rooftops hum with beehives, providing honey to kitchens and restaurants. And the nineteenth century allotment gardens are abuzz with a rejuvenated agricultural economy centered on the sustainable local provision of food.

Resources on the Web: (a) COST—Action Urban Agriculture Europe, and (b) European Federation of City Farms.

can be traced back to efforts to restrict congestion in London in the late 1500s that made it illegal for more than one family to rent a new building.

The serious housing shortage that started with the economic recession of the 1930s was exacerbated by the lack of construction and significant destruction during both world wars. The public-housing programs that began in Vienna in the early 1920s were stepped up after World War II across western Europe. Modern architecture and urban design were combined with low-cost factory production. Many war-damaged historic houses and dilapidated nineteenth-century tenements were replaced by monotonous high-rise apartments after World War II.

In the 1950s and 1960s, most governments adopted a policy of metropolitan decentralization. Massive modern high-rise apartment blocks were concentrated in large peripheral housing estates known by their French name—*grands ensembles*. The amount of public housing was highest in cities with serious housing shortages and liberal municipal governments such as Edinburgh and Glasgow in Scotland, where the number of public units grew to well over half the housing stock. Traditionally, public housing comprised 25 percent of the total in England, France, and Germany, and 10 percent in Italy. Public housing represents only 5 percent or less of the housing in the more affluent and conservative Swiss cities. Since the 1970s, however, dependence on public housing has declined significantly due to government cost cutting and privatization programs.

In contrast, cities that developed under socialism were less spatially segregated. Certainly mansions, the prewar residences of the social elites, were used for political purposes to house party officials, foreign delegations, or institutes. But housing was viewed as a right, not a commodity, and each family was entitled to its own apartment at reasonable cost.

In the face of the tremendous housing shortfalls following World War II, as well as the needs of rapid industrialization, Communist governments built massive housing estates. Prefabricated multistory apartment blocks were constructed in groups to form a *neighborhood unit*, with shops, green space, and play areas for children at the center. Individual apartments were small. The housing estates were typically built in large clusters, forming massive concrete curtains, on land near the edge of cities. As a result, urban population densities could actually *increase* near the urban periphery.

MODELS OF THE EUROPEAN CITY

The concentric zone model, with concentric circles of increasing socioeconomic status with distance from the center, is most applicable to

British cities. In contrast, Mediterranean cities, as in Latin America, exhibit an inverse concentric zone pattern. There, the elite typically concentrate in central areas near major transportation arteries, while the poor live in inadequately serviced parts of the periphery. In Europe, the number of people per household usually increases with distance from the city center.

The sector model explains the pattern of socioeconomic status in which different income groups congregate in sectors radiating outward from the city center. The wealthy may prefer to locate along monumental boulevards or upwind of pollution sources. Poorer residents are left with unattractive sectors along railway lines or strips of heavy industry. Finally, the multiple nuclei model describes the pattern of ethnic differentiation in which different groups are concentrated in ethnic neighborhoods within the inner city or in high-rise public housing near the periphery.

Northwestern European City Structure

The preindustrial city center contains the town square and historic structures such as a medieval cathedral and town hall (Figure 5.13). Apartment buildings host upper- and middle-income residents above shops and offices. Narrow, winding streets extend out about a third of a mile. Some wider streets may radiate out from the square to form a pedestrianized corridor that runs to the train station and contains major department stores, restaurants, and hotels. Skyscrapers are concentrated in the commercial and financial district. There are downtown shopping malls or festival marketplaces in refurbished historic buildings. Some old industrial and port areas may have been recycled into new retail, commercial, and residential waterfront developments.

Encircling the core are some zones in transition. The area of the former wall is a circular zone of nineteenth-century redevelopment. Some of the deteriorated middle-income housing has been gentrified, while other sections provide low-rent accommodation for students and poor immigrants.

Surrounding this area is another zone in transition—an old industrial zone with disused railway lines. In the 1950s and 1960s, new industrial plants (e.g., light engineering, food processing) replaced many of the derelict old factories and warehouses. Low-income renters and owners live in run-down nineteenth-century housing. Some houses have been refurbished or replaced. Certain neighborhoods are quite distinctive because they house foreign immigrants who often live above their exotically painted stores and restaurants.

Beyond this inner area is a zone of "workingmen's homes": a stable, lower middle-income zone dating from the early twentieth century. These *streetcar suburbs* contain apartment blocks and houses without garages, and are typically anchored by small shopping areas, community centers, libraries, and schools. Beyond these areas are middle-income automobile suburbs containing apartments and single-family homes with garages that correspond with the zone of better residences in the concentric zone model. Farther out are clusters of the most exclusive neighborhoods.

The multiple nuclei model best explains the estates of public high-rise apartments and new middle-income "starter" homes at the urban periphery that lack basic amenities like shops and banks. The periphery also contains commercial and industrial activities, such as shopping malls, business and science parks, and high-tech manufacturing.

Beginning in the early twentieth century, cities like London established a greenbelt at

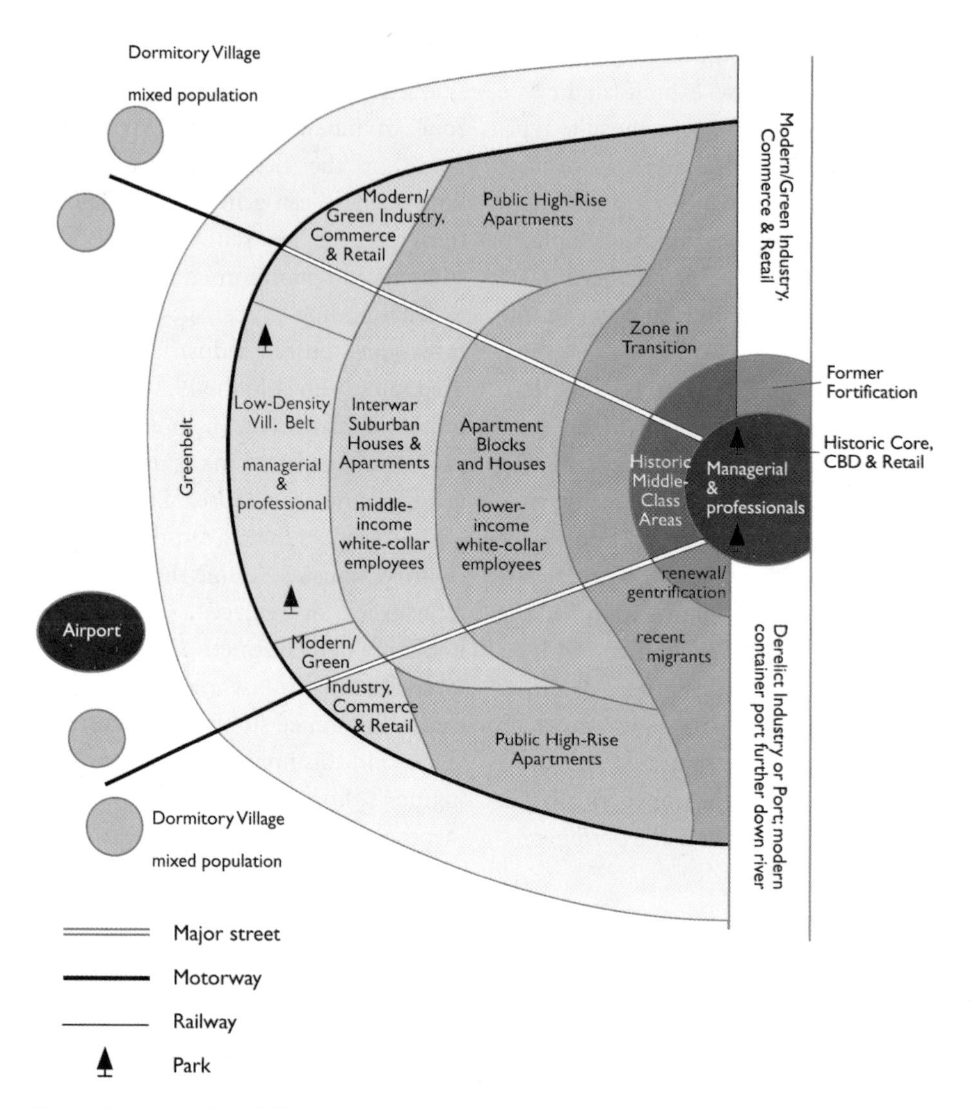

Figure 5.13 Model of Northwestern European City Structure. *Source:* Linda McCarthy.

the edge of the built-up area where development was prohibited. The greenbelt was intended to prevent urban sprawl and provide recreational space. Commuters live outside the greenbelt in dormitory villages and small towns that correspond with the commuters' zone in the concentric zone model. Airport and related activities, such as hotels and modern factories, are located farther out on major freeways.

Mediterranean City Structure

The structure of the preindustrial core reflects the history of each city (Figure 5.14). In Greece and Italy, the historic core can show traces of the grid pattern of streets from the first walled enclosure of Greek or Roman origin. In Spain and Portugal, remnants of narrow alleys of the Arab quarters date back to Moorish control. The central town square is home to markets and festivals; and in Spain, bullfights. The

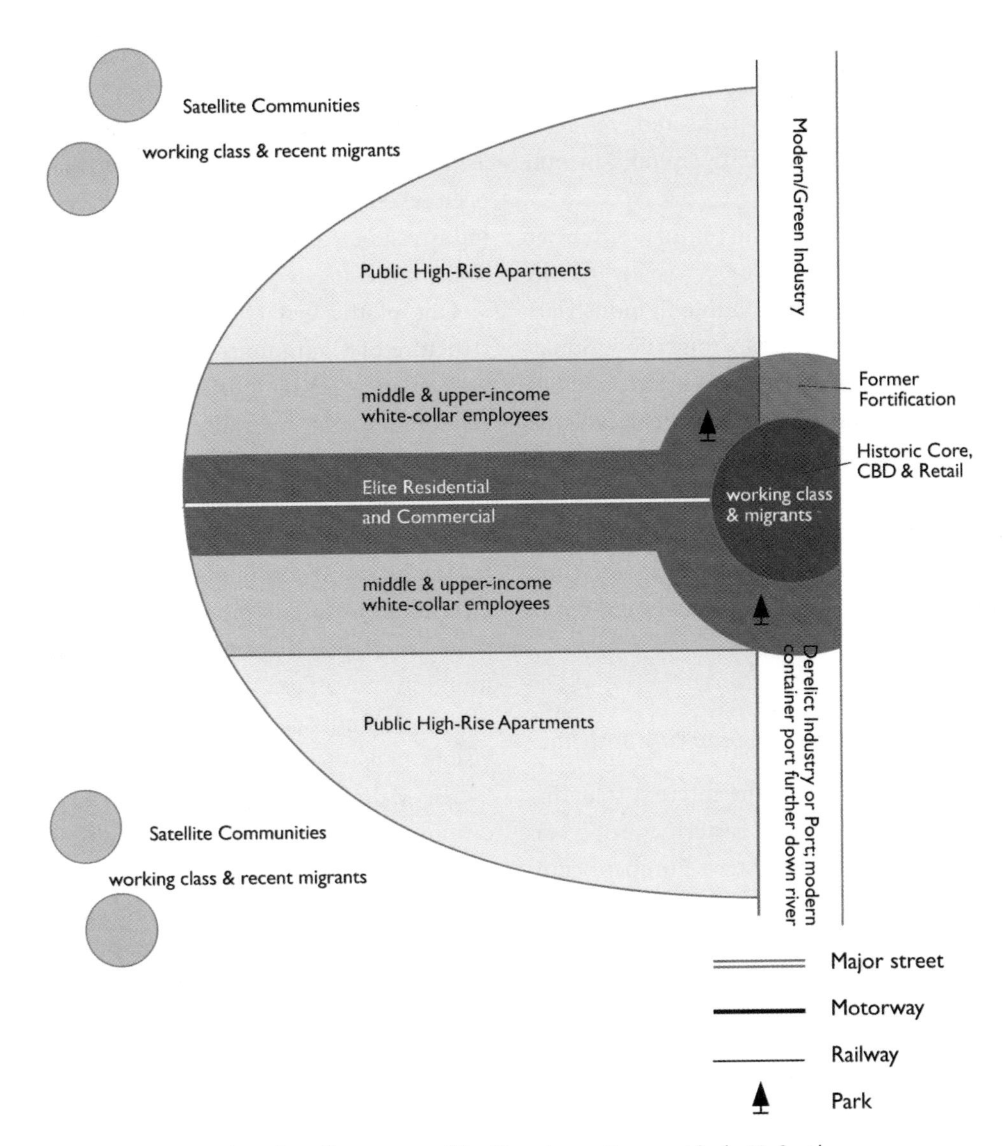

Figure 5.14 Model of Mediterranean City Structure. *Source:* Linda McCarthy.

area around the town square contains the cathedral, town hall, and the narrow streets of the walled medieval city. Lower-income residents live at high densities above street-level shops and offices. A retail corridor runs from this old commercial core to the train station. The high-rise offices of the modern CBD are nearby. As in the multiple nuclei and sector models, new industries are found in former old industrial sites and in locations well served by the Mediterranean region's generally more limited transportation infrastructure.

Until the nineteenth century, urban growth was absorbed in increasing densities within the medieval city. Larger cities like Barcelona that removed their medieval walls in the nineteenth century laid out new monumental districts. A grand new thoroughfare lined with public works such as statues and fountains was extended out from the city. This area attracted commercial development and wealthy residents, as suggested by the sector model. These elite residential areas of parks and tree-lined boulevards were flanked by middle-income neighborhoods.

In the early twentieth century, suburban sprawl began to be an issue, especially in cities experiencing rapid growth due to industrialization and rural-to-urban migration. Squatter settlements encircled the outskirts of cities. After World War II, these were replaced with low-cost, high-rise public housing that today contains low-income households. Farther out, near a natural resource or industrial plant, are the remote, poorly serviced satellite communities for low-income residents and recent immigrants.

Central and Eastern European City Structure

Prior to World War II, the internal structure of cities in central and eastern Europe was much the same as in western Europe. Beginning in the late 1940s, however, the imposition of socialist planning set "Eastern Bloc" cities on a different trajectory, resulting in a set of features that typified Communist-era cities. These cities did not conform to Western models of urban structure because land use was based more on government decisions than economic forces.

A typical socialist city contained a central square for political gatherings. Following the imposition of socialism, former mansions were converted to government use, religious establishments were used for other purposes, and statues of revolutionary heroes dotted the cityscape. Clusters of housing estates and neighborhood units were interspersed with factories, transportation hubs, and retail establishments. Not all cities exhibited these features to the same degree. Few socialist elements are evident in central Prague, which escaped major destruction during World War II. The socialist city model was most clearly achieved in cities such as Warsaw that had been severely damaged during the war, in the industrial new towns, and in countries where socialist ideology was especially strong.

One of the first changes to occur in the structure of Communist-era cities after 1989 was an increase in tourist facilities—hotels, restaurants, and entertainment—to cater to foreign visitors. A building boom, especially in the capitals, larger cities, and tourist centers, resulted in foreign-financed office buildings, trade centers, and shopping malls becoming a common feature of cities like Berlin, Budapest, and Prague (Figure 5.15). These are also found in brownfield redevelopments. Corporate logos and billboards have become very visible signs of change. Suburbanization has increased dramatically, but relatively strict traditions of planning regulation often mean that a suburban development will have a core of a historic village and be linked to the city center by public transit. Nevertheless, many cities are becoming increasingly oriented toward the automobile, as witnessed by large commercial and office park developments at the periphery along ring roads.

DISTINCTIVE CITIES

London: Europe's Global City

As hub of the British Empire, London became the center of global economic and political power in the nineteenth century. Today, London enjoys global city status shared by New York and Tokyo. Greater London has a population of over 8.5 million, and, its metropolitan area boasts more than 14 million people. At the head of navigation on the Thames

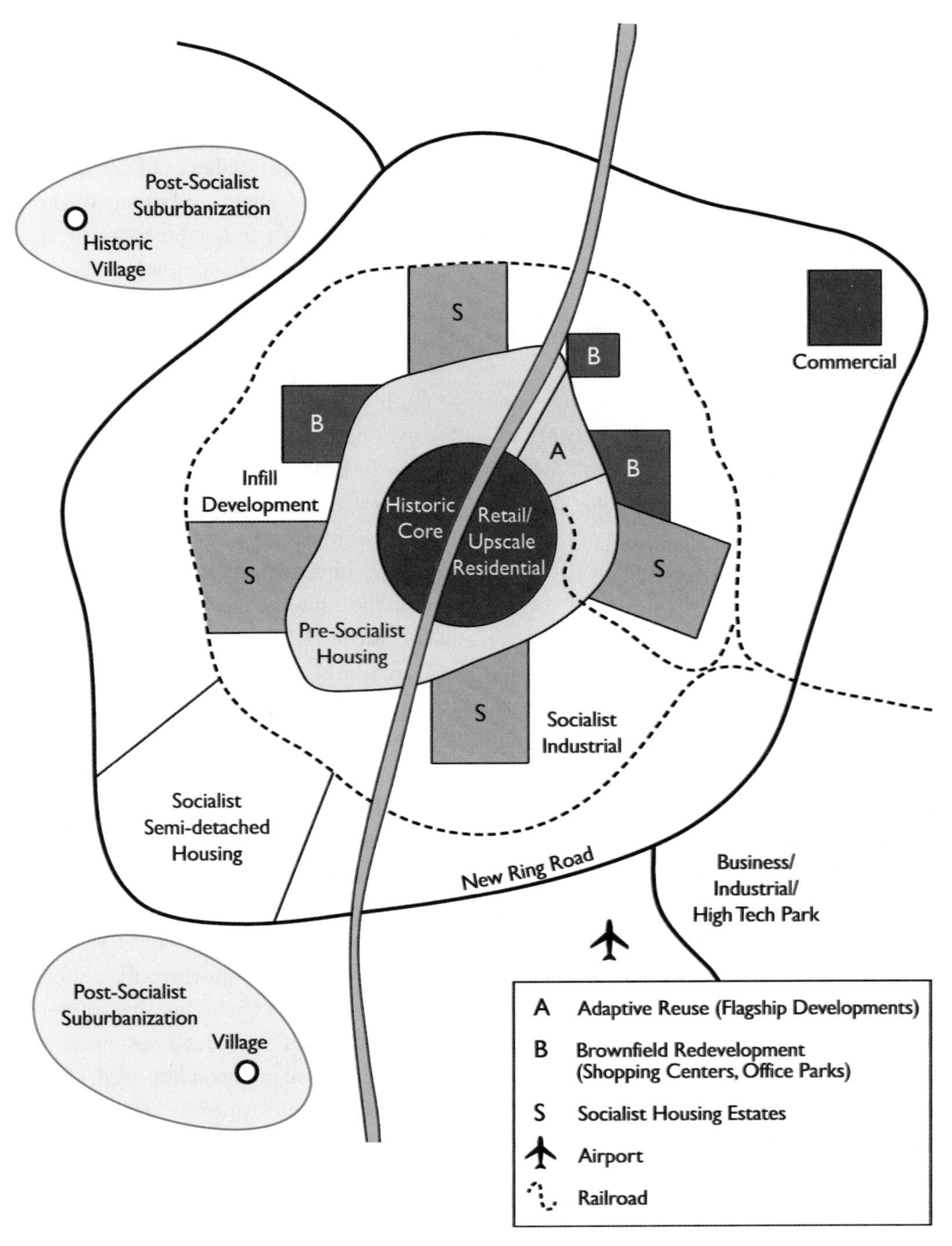

Figure 5.15 Model of Central and Eastern European City Structure. *Source:* Corey Johnson.

River, London dominates the United Kingdom from southeast England. It is the seat of national government, core of the English legal system, headquarters for TNCs, and a leading center for banking, insurance, advertising, and publishing.

In 1666, a fire destroyed virtually the entire city, setting off an immediate building boom. Many historic structures survived until bombed in World War II. London's old nickname, "The Smoke," recalls the days when a haze of pollution from industrial and domestic chimneys hung over the city. London has since undergone deindustrialization and a shift to services and modern manufacturing. Despite these changes, London remains firmly rooted in its historic past.

Central London grew around two core areas, both along the Thames: the City of London (port and commercial hub) and the City of Westminster (government and religious hub). The former developed from a Roman fort, Londinium, which became the fifth-largest city north of the Alps with trade networks extending as far as the Baltic and Mediterranean. In the medieval period, London's protected inland site and strategic location for North Sea and Baltic trade allowed port and commercial activities to thrive. The docks spread from the Tower of London into the East End. Specialized market areas developed near St. Paul's Cathedral in the original square mile of the Roman city. This "City of London" is now the financial precinct, containing the offices of the world's largest banks and insurance companies. The City also houses powerful institutions such as the Stock Exchange and the Bank of England.

About two miles (3.2 km) upriver, the "City of Westminster" developed around Westminster Abbey to become a second core during the medieval period. The present Houses of Parliament were built in the mid-nineteenth century and Queen Victoria made Buckingham Palace the monarch's residence in 1837. This institutional core grew eastward toward the commercial core along Whitehall, where government offices include 10 Downing Street, the prime minister's residence (Box 5.3). The royal hunting grounds in the west became St. James's, Green, Hyde, and Regent's Parks. The area attracted mansions of the nobility, centers of culture such as the National Gallery, and exclusive shops. In the seventeenth and eighteenth centuries, large-scale townhouse developments were speculatively built for the aristocracy. Belgravia, the last of these West End developments, has survived as an affluent neighborhood.

In the nineteenth century, major retailing axes developed along Oxford and Regent Streets. In addition to the The City, the inner city (12 of London's 32 boroughs) comprises a ring of nineteenth- and early-twentieth-century suburbanization. From the early 1840s, the railways allowed wealthier families to move farther out. The higher-density Victorian and Edwardian housing nearer to the center included middle-income detached and row houses such as those in Islington and laborers' cottages in the East End. With increasing industrialization and the incredible growth of new docks, the East End became home to the poorest immigrants.

Much of the original housing in the East End is gone—destroyed in World War II air raids or replaced by high-rise public housing, now deteriorating. Other housing, dispersed among old factories, warehouses, docks, and railway yards, is in poor condition too. Many of the decaying middle-income residences have been subdivided into low-rent apartments. Within the inner city, however, residents are differentiated into neighborhoods, each with its own high street, socioeconomic and ethnic mix, and political and sporting allegiances.

Since the early 1980s, an extensive area of London, the Docklands, has been revitalized through public and private investment. The British government established an Urban

Box 5.3 Security and Surveillance in London

Cities—especially world cities such as London—are a preferred location for terrorist attacks, for several reasons. First, they have symbolic value. They are not only dense concentrations of people and buildings but also symbols of national prestige and military, political, and financial power. A bomb in London's Underground (subway) arouses international alarm and is communicated instantly to a world audience. Second, the assets of cities—densely-packed with a large mix of industrial and commercial infrastructure—make them rich targets for terrorists. Third, cities are nodes in vast international networks of communications—reflecting not only their power but also their vulnerability. A well-placed explosion can cause enormous reverberations by triggering fear and economic dislocation. Finally, word gets around quickly in high-density localities. These kinds of environments can be a source of recruits for terrorist organizations.

Central London has attempted to reduce the real and perceived threat of terrorist attacks. Physical and increasingly technological approaches to security have been adopted at ever more expanded scales. In 1989, the prime minister installed iron security gates at the entrance to Downing Street to control public access (Figure 5.16). In 1993, a security cordon was set up to secure all entrances to the financial zone of the City of London (the "Square Mile").

Figure 5.16 The iron security gates at the entrance to Downing Street in London prevent the public from getting close to the official residence of the Prime Minister. *Source:* Photo by Linda McCarthy.

The 30 entrances to the City were reduced to seven, with road-checks manned by armed police. Over time, the scale of this security cordon was increased to cover 7 percent of the "Square Mile" (Figure 5.17).

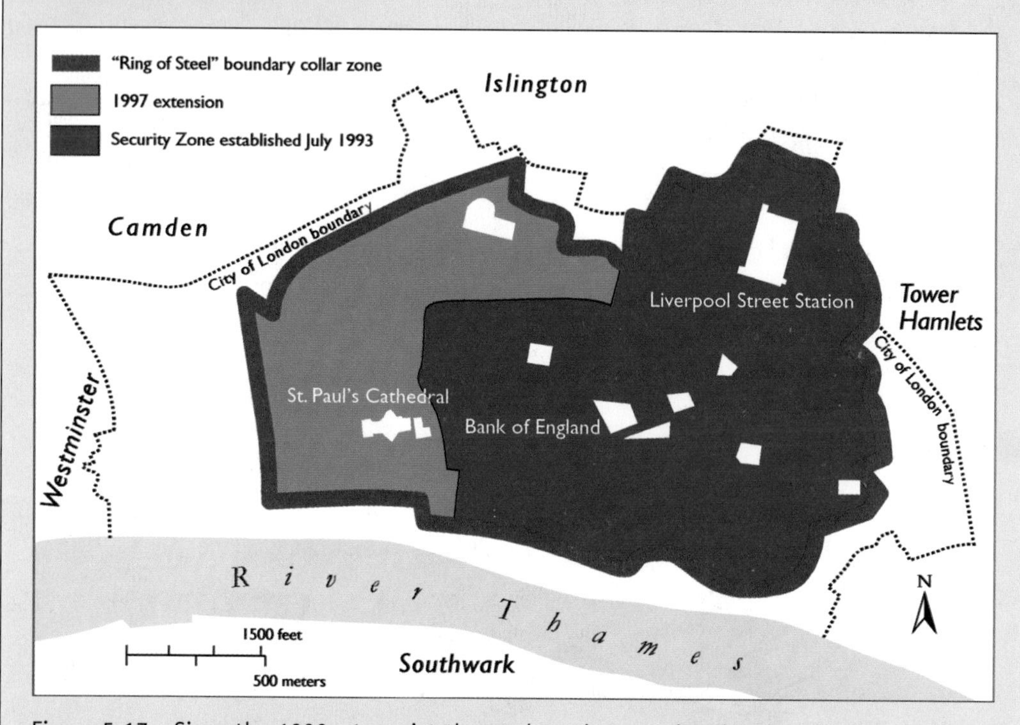

Figure 5.17 Since the 1990s, terrorist threats have increased and so has the security zone in London's financial district, "The City." *Source:* Adapted from J. Coaffee, "Rings of Steel, Rings of Concrete and Rings *of Confidence," International Journal of Urban and* Regional Research 28 (2004): 204.

As a territorial approach to security, this cordon was augmented by enhanced surveillance, especially by retrofitting the closed circuit TV (CCTV) system. The police, through its "CameraWatch" partnership effort, encouraged private companies, such as retail establishments, offices, and warehouses, to install CCTV. At the seven entrances to the security cordon, 24-hour Automated Number Plate Recording (ANPR) cameras, linked to police databases, were installed. The City of London is now the most surveilled space in the United Kingdom, and perhaps the world, with some estimates of the number of private and public CCTV cameras as high as nearly half a million. This level of surveillance raises important questions about how best to balance the benefits of surveillance associated with crime prevention and detection against the drawbacks for urban residents associated with the loss of privacy and threats to civil liberties.

Sources: J. Coaffee, "Rings of Steel, Rings of Concrete and Rings of Confidence," *International Journal of Urban and Regional Research* 28 (2004): 201–11; H. V. Savitch with G. Ardashev, "Does Terror Have an Urban Future?" *Urban Studies* 38 (2001): 2515–33.

Development Corporation that used public funds to stimulate private development. Until the Shard was built at London Bridge, the city's tallest building was a 50-story tower containing offices and specialty stores built at Canary Wharf. Dockland revitalization projects such as this now extend as far west as upscale Saint Katharine Docks (just east of the Tower of London). These dockland developments have attracted higher-income occupants and promoted gentrification.

Outer London is a lower-density belt of interwar housing with some shopping streets and industrial parks. These outer suburbs comprise the remaining 20 of London's 32 boroughs. Between 1918 and 1939, the expansion of the London Underground (subway) and private automobile use promoted suburbanization. Middle-income residents live in well-maintained houses with gardens. Neighborhood stability is strong. Second-generation immigrants have moved into pockets of older housing. An innovative approach to London's traffic congestion was the introduction of a daily congestion charge for motorists driving into the most heavily congested zone of the city. The estimated benefits have been a more than 20 percent reduction in traffic entering the zone, and funds to reinvest in public transportation.

The outer suburbs end abruptly at a 5–10 mile (8–16 km) wide greenbelt within which development is restricted to prevent sprawl and provide recreational space. Villages and small market towns remain much as they were when the greenbelt was established in 1939. Growth pressures are evident only in the rural dwellings that have been gentrified by newer wealthy residents. Prohibiting development within the greenbelt has forced growth into either the existing built-up area or farther out. Eight new towns were built beyond the greenbelt to house London's overspill population and migrants from the rest of the United Kingdom. This metropolitan fringe extends more than 50 miles (80 km) from the city center and includes large towns like Guildford, Reading, and Luton. The relatively strong economy, the removal of trade barriers within the European Union, and business from the Channel Tunnel have put pressures on housing, government infrastructure, and transportation services. The high-speed rail line that includes the Channel Tunnel (or "Chunnel") passes under the English Channel and connects London to Paris and Brussels.

Paris: France's Primate City Par Excellence

The city of Paris has a population of over 2 million people, but with more than 12 million, it is Europe's second-largest metropolitan area. As France's primate city, Paris dominates the national urban system and the country's economy, politics, and culture. In the wake of deindustrialization, Paris has become a major international center for modern industry and finance. The outer suburbs contain high-tech plants and research and development companies. Inner-city workshops produce *haute couture* and jewelry.

Since World War II, Paris has grown almost continuously due to migration from the rest of France and the former French empire, and the city's high proportion of young adults of childbearing age. Much of this growth has been concentrated in the outer suburbs. The city center and inner suburbs are losing population.

The original site of Paris was an island in the Seine, today called Île de la Cité (Figure 5.18). The Romans seized the island in 52 CE from the Parisii, a Gallic tribe. They built a temple and a palace for the city's governor, and the island

Figure 5.18 Paris evolved around an island in the Seine River: Île de la Cité. Today, it is most famous for the cathedral of Notre Dame, whose spire is barely visible here. *Source:* Photo by D. J. Zeigler.

settlement attracted convents and churches. The magnificent Gothic cathedral of Notre Dame was begun in the twelfth century and took more than 170 years to complete.

As a royal center, the grandeur of its architecture and planning made Paris an intensely monumental city. The "Royal Axis" is the imposing entry to the city. It runs from the Louvre (a royal palace, now national art gallery) and Tuileries Gardens across the Place de la Concorde, along the Champs-Élysées, to the Arc de Triomphe. The nearby Eiffel Tower was erected for the Paris exposition of 1889. The tallest structure in Paris, the Eiffel Tower is one of the most recognizable monuments in the world. Paris still produces imposing architecture. Initially, controversial structures include the sleek glass pyramid in the nineteenth-century forecourt of the Louvre and the Pompidou Center, the national museum of modern art, nicknamed the "arty oil refinery" for its multicolored exterior ventilation and steel-and-glass escalators.

The Paris region, Île de France, comprises eight administrative units (*départements*) that date from the French Revolution. Most familiar to tourists, the innermost coincides with the historic City of Paris. This high-density area developed within the confines of the medieval wall. Its distinct quarters include Île de la Cité. Facing downstream, the "right bank" of the Seine has become the economic heart of Paris. It contains offices, fashionable shops, hotels, restaurants, and high- and middle-income apartments. The "left bank," the seat of intellectual and cultural life, is dominated by its oldest part, the Latin Quarter, with the Sorbonne University, bookshops, theaters, and middle- and low-income apartments. Unlike London, there are few large parks. Paris gets its feeling of openness and greenery from the wide boulevards and tree-lined river walkways.

The outer parts of Paris include the "little ring" (*petite couronne*) of inner suburbs that extends out about 15 miles (24 km) from the center. It developed between the late 1800s and World War II. Interwar speculative developments of single-family homes were built on prime sites. Public high-rise apartments were erected later on the less marketable land. The big ring (*grande couronne*) of outer suburbs spreads out another 10–15 miles (16–24 km) and contains the postwar *grands ensembles* of poorly serviced public high-rise apartments. These poor suburban communities contain the highest proportions of Muslims in France. Three groups of immigrants—from North Africa, Sub-Saharan Africa, and Turkey—together comprise about 50 percent of the immigrant population in the Ile-de-France region. As many as one-third of suburban residents—increasingly not only newly arrived but also second- and third-generation—live below the poverty line. Although France prides itself on being a progressive society, discrimination against many of these suburban residents continues to fly in the face of the ideals of liberty, equality, and fraternity.

Beginning in the late 1940s after the publication of Jean-François Gravier's book, *Paris and the French Desert*, planners began to focus on counteracting the extraordinary economic and demographic primacy of Paris. National decentralization policies attempted to limit growth and congestion problems within the Paris region, while promoting development in the eight *métropoles d'équilibre* of Lille-Roubaix-Tourcoing, Metz-Nancy, Strasbourg, Lyon, Marseille, Toulouse, Bordeaux, and Nantes-St. Nazaire. Five new towns (St. Quentin-en-Yvelines, Evry, Melun-Senart, Cergy-Pontoise, and Marne-la-Vallée) were built along two east-west axes of growth to the north and south of Paris. These new towns grew as extensions to the city, however, and became middle-income dormitory communities for some of the more than 1 million daily commuters to central Paris.

Complementing the new towns are four suburban employment centers. The largest and most successful is La Défense. It boasts high-rise offices containing the headquarters of TNCs, shops, public buildings, and housing. Its modern Grand Arche is visible from the Arc de Triomphe along Avenue Charles-de-Gaulle—the modern extension of the "Royal Axis."

Barcelona: Capital of Catalonia

With more than 1.6 million people, Barcelona is Spain's second-largest city after the capital, Madrid. On the northeastern coast of Spain, Barcelona is the country's largest port and leading industrial, commercial, and cultural center. Housing the seat of the Catalan government, this regional capital of Catalonia is a bilingual city: Spanish and Catalan are widely spoken official languages (Figure 5.19).

The Phoenicians founded Barcelona more than 2,000 years ago. The street plan reflects its three main phases of growth—its ancient and medieval origins, nineteenth-century additions, and late-twentieth and early twenty-first-century suburbs. The old town is the symbolic and administrative center of the city. Remnants of the Roman wall and grid pattern of streets are overlain by the narrow streets of the medieval core. Here, residents and tourists alike stroll along the famous *Ramblas*. Barcelona is the most popular tourist port

Figure 5.19 Throughout Catalonia, signs of Catalan nationalism—and separatism—are to be found. This banner, in Girona, speaks to the world in English. *Source:* Photo by D. J. Zeigler.

in the Mediterranean with more than 7 million annual visits, including 2.5 million from cruise ships.

In 1859, Ildefons Cerdà drew up a plan of expansion into the area of the former medieval wall. His pioneering design was based on a grid pattern with wide, straight boulevards and unique 8-sided city blocks containing parks surrounded by apartment houses. Largely ignored during the nineteenth- and early-twentieth-century era of speculative growth, Cerdà's plan was fully realized only in the *Eixample* district, a new precinct that was built just north of the old city. This new development also contains the high-rise offices and apartments of the modern CBD.

At the end of the Spanish Civil War in 1939, when the country was still a dictatorship, Barcelona's Catalan culture was suppressed and the city experienced uncontrolled speculative development without adequate public infrastructure and services. Massive rural-to-urban migration fueled rapid population growth. Tens of thousands of illegal squatters ended up in shantytowns at the sprawling edge of the city. In the 1960s and 1970s, several hundred thousand poorly designed and serviced peripheral high-rise public apartments were built to address the acute housing shortage.

Since the mid-1970s and the establishment of Spain's parliamentary democracy, the increased autonomy of Barcelona's elected local governments contributed to a rebirth of planning as well as growing prosperity. Barcelona's urban renewal program benefited from funding for infrastructure from the European Union. The construction of the 1992 Olympic village helped rejuvenate an area of derelict docks into waterfront redevelopments (Box 5.4). Popular World Heritage sites include a park by Antoni Gaudí—Park Güell—and his unfinished church—Sagrada Familia—financed by private donations since 1882!

At the same time, a continued influx of poor residents puts pressure on housing, infrastructure, and services. These poor migrants become socially, economically, and locationally polarized in the poorest inner-city

Box 5.4 Making the Spectacular Happen: Mega-events in European Cities

Mega-events such as the Olympics, FIFA's World Cup, and world expos, involve huge public expenditures, massive construction projects in already dense built environments, and logistical and security nightmares. Why, then, do cities across Europe continue to covet them? Some people see mega-events as a way of putting their city on the map. The spectacle and intense scrutiny that a mega-event brings are seen as potentially associated with investment and tourism long after the event is over. Whether the legacies of mega-events justify their incredible upfront costs remains open to debate.

Nevertheless, these legacies do have long-lasting impacts on urban landscapes across Europe. Barcelona, Spain, hosted the 1992 Summer Olympic Games, and it is often touted as a city that was transformed positively by hosting this mega-event. Local and national governments invested in infrastructure projects, such as demolishing waterfront factories and warehouses and building parks and an Olympic village in their place. Nearly 50 miles (80 km) of new roads were constructed (15 percent more than before), the airport was renovated, and hotel capacity was dramatically increased. Barcelona now ranks among Europe's most popular tourist destinations, and this in part can be attributed to the city's reputation shifting after 1992 from a run-down industrial city to a vibrant, culturally rich, and beautiful city on the sea. Other European cities to host Summer Olympic Games recently include Athens (2004) and London (2012).

London's 2012 Olympics illustrate some of the challenges of organizing mega-events in big cities. Coming in the wake of terrorist bombings in the city several years before and generalized fears about terrorists targeting spectacles with spectacular attacks, the London games were a massive security operation. By some accounts they were the largest peacetime security operation in the country's history. During the games, 13,500 military troops—in addition to tens of thousands of police—were deployed to help keep order, an aircraft carrier docked on the Thames, and surface-to-air missiles were stationed on apartment rooftops. Protestors protested, and debates raged in the London newspapers about displacing low-income people to build venues, banning displays of brand names that were not official sponsors, and costs to the public in the wake of the financial crisis and government belt tightening. At the conclusion of the games, the success seemed to be measured by the world press in large part by the fact that nothing truly disastrous or embarrassing occurred.

There is growing debate in Europe and elsewhere as to whether the production costs of hosting mega-events are worth the potential benefits. Certainly investments in infrastructure carry benefits for host cities long after the visitors leave, but specialized stadiums and other venues often simply fall into disuse, as the case of Athens illustrates. The swimming pavilion, beach volleyball arena, and softball arena built for the 2004 Olympics are crumbling, causing many Athenians to ask themselves if the $15 billion dollars to host the games might have been better spent.

Source: Ferran Brunet y Cid. *The economic impact of the Barcelona Olympic Games, 1986–2004.* Barcelona: Centre d'Estudis Olímpics UAB, 2005.

neighborhoods and peripheral public apartment blocks. In contrast, higher-income residents live in nicer central districts or well-serviced lower-density parts of the suburbs.

Oslo: Low-Key Capital of Norway

Oslo is the largest urban center in Norway as measured by both the more than 600,000 people that inhabit the city and its metropolitan area population of about 1.5 million. At the mouth of the Oslofjorden (Oslo Fjord), this city is Norway's capital, main port, and leading commercial, communications, and manufacturing center.

Oslo was founded around 1000 CE and became the national capital in 1299. After a devastating fire in 1624, King Christian IV of Denmark designated another site for the town nearer Akershus Castle on the east side of the inner fjord. The new town was named Christiania (later Kristiania). It was planned with a grid system of spacious streets, a square located between the town and castle, and ramparts protecting its northern flanks. For fire resistance, buildings were required to be constructed of brick or stone; soon, however, extensive tracts of wooden houses were built on the outskirts of the built-up area. The town grew slowly: in 1661 only around 5,000 residents had made Christiana their home; by 1800 the population had risen to only 10,000.

During the mid-1800s, the administrative function of the city was augmented by industry, based mainly on textiles and wood processing. Many landmarks, such as the university, royal palace, parliament, national theater, and stock exchange, were built. The city expanded in a largely unplanned manner as the population swelled to 28,000 by 1850 and 228,000 by 1900. In 1925, the city reverted to its original name, Oslo. After World War II, Oslo's outward expansion continued, largely as a result of public policies that subsidized owner-occupied housing.

Oslo has 40 islands and 343 lakes; about two-thirds of the city comprises protected natural areas, which give it a picturesque appearance. While most of the surrounding forests and lakes are private, the public is strongly against developing them. As is common throughout northern Europe, the city extends around the port, is flanked by the centrally located train station, has a royal palace overlooking the historic core, and pedestrianized shopping streets. Oslo is where the Nobel Peace Prize is awarded because Alfred Nobel decided that this prize—awarded to the International Campaign to Ban Landmines, Doctors without Borders, and the European Union, to name a few—was to be awarded by a Norwegian committee (while the other four prizes were to be awarded by Swedish committees). Despite being Scandinavia's oldest capital, Oslo today is a modern, though low-key, city.

Berlin: The Past Always Present in Germany's Capital

Around every corner, across every bridge, and in nearly every U-Bahn (subway) station, Berlin offers tantalizing morsels from its fascinating past. Relatively unimportant for most of its 800-year history, Berlin found itself in the middle of many important political struggles of the nineteenth and twentieth centuries. More recently, as the capital of reunified Germany and a key political node in the European Union, Berlin continues to undergo profound changes.

Berlin had rather humble beginnings in the thirteenth century on the flat, glaciated marshlands of the North European Plain at a convenient crossing point on the Spree River. The growth of Berlin reflected its political fortunes as the center of what would become the Kingdom of Prussia. The architectural heyday of the city came after 1701 as the official capital of Prussia. The royalty sought to give the capital an impressive built environment worthy of long-established capital cities such as Paris, Vienna, and London. Walking along Unter den Linden, the city's most important axis, you see the bravado of the Prussian ruling family, the Hohenzollerns, who built up the boulevard during the early nineteenth century to project their pride after helping to defeat Napoleon's armies. The street culminates at the Brandenburg Gate, a monument symbolic of Prussian power, German Imperial pretension, Cold-War division, and since 1990, a reunified Germany. As people walk through the Brandenburg Gate, they are confronted by more history: to the south, the Memorial to the Murdered Jews of Europe; to the north, the center of Germany's government in the renovated Reichstag building with its huge glass dome symbolizing the transparency of the Federal Parliament. Perhaps more than any other spot in the city, modern and old, painful and joyous coexist in an almost surreal urban assemblage.

The growing power of Prussia during the 1800s was accompanied by industrialization. Large companies such as Siemens and AEG were founded, while Berlin-based insurance firms (e.g., Allianz) and banks (e.g., Deutsche Bank) served the booming industrial economy. Reflecting shifting political fortunes, none of these TNCs remain. But the evidence of Berlin's status as one of Europe's major industrial cities can still be seen. Working-class apartment houses from its industrial heyday are found in neighborhoods such as Wedding and Kreuzberg, which encircle the historic core. The city's old industrial breweries have found new life in post-unification Germany as cultural and arts centers. The Schultheiss brewery in Prenzlauer Berg is now a cultural center and shopping area called *Kulturbrauerei*, and the Kindl brewery in Neukölln houses artist studios and apartments. Berlin also became a major transportation hub during the nineteenth and early twentieth centuries. The major trunk rail line winds its way through the city center, stopping at iconic stations like Zoologischer Garten, Lehrter Station (the main station, the largest in Europe), Friedrichstrasse, and Alexanderplatz.

Berlin's economic and demographic peak came in about 1940 when it was the industrial, transportation, and government center of the Third Reich. Its population was over 4 million in 1939 on the eve of Germany's invasion of Poland, compared with just over 3.5 million today. The Nazis left their indelible mark on Berlin, although not entirely as planned. Hitler's planned Germania, a megalomaniacal rebuilding of Berlin's core as the capital of the German Empire, was never realized. The most readily apparent legacy of the war is destruction. The core of the city was up to 95 percent destroyed during bombing raids by the Allies and the bloody campaign of the Soviet Army during the last days of the war.

After World War II, when Berlin was divided into four sectors by the Allies (France, Britain, United States, Soviet Union), the priority was constructing housing for the remaining people and large refugee population in both East and West Berlin. The results were often more functional than architecturally appealing. Though bland housing blocks are still apparent, some Cold-War prestige projects remain. The Stalinallee (later Karl-Marx-Allee) in East Berlin was designed as a showcase of Communist architecture, and it offers remarkable insights into the aesthetic ideals of the Communist regime. In West Berlin, projects such as the Cultural Forum, home to the concert hall of the Berlin Philharmonic Orchestra, is a modernist icon, which was designed to show off the merits of West Germany's social market economic system.

Since reunification in 1990, Berlin has been the site of one of the largest urban reconstruction efforts of all time. Potsdamer Platz, a prewar buzz of activity, was in the area of the Berlin Wall during 1961–1989. Shortly after reunification, it was rebuilt in the then popular steel-and-glass style. Friedrichstrasse, the major north-south axis that the Wall once cut in two, has blossomed as Berlin's most fashionable shopping and office address. Meanwhile, Kurfürstendamm, an icon of West German consumerism, appears frozen in 1980's time. One of the few new airports to be constructed in Europe in the last 20 years is on the outskirts of Berlin. The massive project, designed to replace two airports close to city center, was completed in 2011, but embarrassing planning mistakes and faulty construction, including a fire suppression system that was not up to standard, delayed the airport's opening by several years.

Berlin's fascinating history is also the subject of frequent debates over appropriate land uses, and nearly every major building decision is accompanied by sensitive emotional excavations of the more sordid periods of that past. In recent years, this is perhaps best illustrated by the reconstruction of the *Stadtschloss* (city palace). The damaged nineteenth-century original was destroyed after the war by the East German regime, and on its site a modernistic Palace of the Republic was constructed, a building intended to distance the site from a militaristic, Prussian history. Demolished in 2008, the Parliament voted to rebuild an exact replica of the city palace that would be called the Humboldtforum and house the Humboldt collection and gallery of non-European art. Reconstruction began in 2013 and is scheduled to be completed in 2019.

Bucharest: A New Paris of the East?

On the Romanian Plain between the Carpathian Mountains to the north and the Danube River lowlands to the south, Bucharest is by far Romania's largest city, with a population of nearly 2 million. In addition to being the capital, it is the country's most important economic and industrial city. Like Berlin, Bucharest bears the marks of its Communist past, in the form of monumentalist Stalinist architecture, large Communist-era apartment blocks, and a somewhat run-down prewar built environment. Prior to World War II and the postwar Communist era, however, its elegant architecture and social elite made it the "Paris of the East." Since Romania's accession to the European Union in 2007, there are signs that Bucharest would like to reclaim that title.

Bucharest is relatively young by European standards: the first references to the city date to 1459. During the 1800s, Bucharest became an important transportation hub, acquired a manufacturing base, and became Romania's capital when the country was formed in 1862. By the end of the century, Bucharest boasted a tram system and the world's first electric streetlights. The city enjoyed continued growth until World War II. The population rose from about 60,000 in 1830 to slightly more than 1 million at the end of World War II. During the 1930s, a master plan for the city sought to vastly change the compact medieval core and envisioned expansion like Berlin and Paris with their wide boulevards, parks, and grand public buildings. This laid the groundwork for what would come later under postwar Communist rule.

Bucharest suffered heavy damage during World War II from Allied and Nazi bombing. After the war, socialist planning guided development: industrial capacity was greatly expanded; new housing was constructed; and former villas were converted into government offices and foreign embassies. The population increased to 1.4 million by 1966, the year Ceauçescu came to power. Perhaps no city in Europe bears the personal imprint of an individual to the same extent as Bucharest at the hands of this deposed leader, who was executed in 1989 along with his wife after a hasty trial amid a democratic revolution. Romania's capital allows fascinating insights into the impact of a megalomaniac personality. Ceauçescu's program for urban redevelopment and systematization led to single-family houses in some suburbs being replaced with apartment blocks. He greatly expanded housing estate construction within existing districts and redesigned certain boulevards as impressive entryways to represent the revolutionary aesthetics of socialism.

After a 1977 earthquake caused significant damage, Ceauçescu turned his attention to central Bucharest. By 1989, approximately 25 percent of the historic central area had been bulldozed for a new civic center. An estimated 40,000 families were evicted practically overnight, and those with dogs were forced to let them go (giving rise to Bucharest's intractable stray dog problem). At the heart of the construction scheme was The House of the Republic, which was to be the new seat of government. This grandiose structure became one of the world's largest buildings. Over a thousand acres of neighborhoods were torn down to make room for the complex. This building consumed virtually all the country's marble during construction and featured hand-carved wood paneling and crystal chandeliers. The second component of the scheme was the construction of the Victory of Socialism Boulevard. This finely appointed ceremonial route was intended to be longer and grander than the Champs-Élysées in Paris and eventually was adorned by a lavish fountain and lined by Bucharest's finest apartments. A rather bizarre element of Ceauçescu's modifications involved churches. He simply did not like them; but rather than being accused of ordering their destruction, he had many moved behind other buildings.

For much of the 1990s, Bucharest suffered the aftermath of Ceauçescu's experimental planning. The city was run-down and known to visitors more for its stray dogs than as the "Paris of the East." The completed House of the Republic became the seat of Romania's democratic parliament, symbolizing a coming to terms with the past and a more forward orientation for the country and its capital city. More recently, as Romania has benefited economically from its integration into Europe, including accession to NATO and the European Union, Bucharest is again blossoming. Wide boulevards that were once used to showcase a regime's hold on power are now monuments to consumerism. Bucharest still has its stray dogs—40,000 by recent estimates—but the number is declining due to a controversial municipal "killing law" that has resulted in the extermination of 10,000 strays.

URBAN CHALLENGES

Compared with the problems faced by cities in many other regions of the world, European cities are fairly well off. They do, however, face challenges similar to those in other more developed parts of the world.

As the earliest place to industrialize, Europe was also the first to suffer deindustrialization. Rising long-term unemployment among inner-city residents has concentrated poverty and a wide range of social problems in some neighborhoods of older industrial cities. These neighborhoods also contain the city's oldest and most deteriorated housing and urban infrastructure. Privatization of public housing by governments attempting to cut back on expenditures has exacerbated the shortage of decent affordable housing.

As private automobile ownership has risen, traffic congestion and air pollution, especially in the medieval cores, have reached critical levels. Transportation policies in western Europe typically shifted from investment in freeways and central parking facilities to transportation demand management involving ride sharing and public transit. Many cities built or extended their subway and light-rail systems.

The inadequate road system throughout central and eastern parts of Europe, particularly given the dramatic increase in car ownership rates since 1990, has placed considerable strain on roads and parking facilities. Before the fall of Communism, most people could not afford a car, waiting periods for car orders were long, and restrictions on ownership applied in some countries. A major challenge continues to be to improve the transportation infrastructure as part of a European-wide transportation network.

The presence of significant numbers of foreign workers and their families has generated problems in some parts of Europe. Language differences create difficulties for the educational system in countries with large numbers of children born to foreign workers. These students represent more than 10 percent of the school population in some French, German, and Swiss cities. During times of recession and rising unemployment, existing prejudices can be intensified based on stereotyping associated with differences in religion, language, culture, or race. Anti-foreigner sentiment (xenophobia) has contributed to some governments, as in France, banning headscarves worn in public schools and universities. Vicious attacks on migrants by violent elements such as skinheads have occurred, especially in some German and French cities. Discrimination against African immigrants, as well as high unemployment and lack of opportunities in France's poorest immigrant suburbs (*banlieue*), has sparked riots.

More recently, the considerable economic and social changes in central and eastern Europe have increased the opportunities for criminal activities. The incidence of petty crime, such as pickpocketing and graffiti, has increased (Box 5.5). Organized crime has also grown. Besides the more typical drugs, gambling, and prostitution, "Mafia"-type crime organizations are common in some cities.

Pollution is a problem in nearly every city of the world, though many European cities rate more favorably than most. Indeed, Swiss, Austrian, and Scandinavian cities are practically sanitized daily and citizens are conscious not to litter. Many street frontages in countries such as France, Italy, and Spain are routinely washed down by proprietors before they open for business. Levels of pollution in most of the former heavy industrial regions have fallen. Indeed, air quality in the English Midlands has improved since polluting industries have closed or relocated, and even the Ruhr area boasts clear skies and clean lakes.

Formerly Communist parts of Europe are still tackling the legacy of weak Soviet-era environmental standards, use of higher-risk

Box 5.5 Urban Graffiti: Is the Writing on the Wall?

Art or eyesore? Free expression or vandalism? European urban landscapes bear the marks of urban graffiti. On a stroll through just about any neighborhood in Brussels, Paris, Prague, or Warsaw, you will see spray paint adorning buildings, signs, buses, and trains. An aesthetic issue for many, graffiti in European cities is also often a political issue tied up with ethnic, socioeconomic, or generational conflict.

Urban graffiti is nothing new. Ancient Greek and Roman cities had graffiti, including a caricature of a politician etched on an outdoor wall in the excavated Roman city of Pompeii from around 79 CE. More recently, graffiti has played an important role in some of the ethno-political struggles in Europe. The Basque separatist group ETA and sympathizers with their cause have used graffiti as a means of protesting the lack of autonomy within Spain and France. During Spain's 40-year dictatorship under Francisco Franco, which lasted until 1975, graffiti was one of the few forms of protest that the ETA could get away with. Since then graffiti as political statement has continued alongside more publicized acts of violence. Similarly, Irish Republican Army (IRA) markings are common in contested Northern Ireland cities, while politically motivated graffiti by Bosnians and Serbians can be found in Sarajevo nearly two decades after the war that broke up Yugoslavia.

In recent years, socially and politically conscious works of art have sprung up in Athens, Greece, a city whose middle class has been hit exceptionally hard by pressures brought about during the recent financial crisis. The city's younger residents, many of whom are unemployed, have taken to the street and used graffiti to express disenchantment with economic uncertainty and ineffective government institutions. Some have suggested that the frustration has fueled the production of graffiti to such an extent that Athens has become a "contemporary mecca" for street art in Europe.

With a rise in graffiti, new forms of policing have emerged to combat it. Also, public officials have commissioned street artists to paint murals in designated public spaces thereby giving street art an air of public approval. Such attempts by the city governments to combat unsanctioned graffiti while encouraging "official" murals are viewed by many street artists in Athens as an attempt by the authorities to control or neutralize the power of graffiti as a form of political expression.

An example of more drastic efforts taken against European street artists can be found in Berlin, Germany. Berlin's cold war history and its numerous potential spaces for art have made it a blank canvas for street artists within the city and beyond. In 2013, the operator of the city's commuter railway, Deutsche Bahn, proposed using drones to monitor trains in an effort to discourage graffiti—the company's plans to use tiny remote-controlled helicopters equipped with infra-red cameras to patrol the company's stations and train yards. While Deutsche Bahn claims that the drones will not target the traveling public in open spaces, the plan has met with much opposition. In a country still highly sensitive to the surveillance of law-abiding citizens, given a history of dictatorial Nazi and Communist rule, such opposition is not surprising.

Sources: L. Alderman, Across Athens, Graffiti Worth a Thousand Words of Malaise, *New York Times*, April 15, 2014; M. Eddy, Some Germans Balk at Plan to Use Drones to Fight Graffiti, *New York Times*, May 28, 2013; T. Moreau and D. Alderman, Graffiti Hurts and the Eradication of Alternative Landscape Expression, *Geographical Review*, 2011, pp. 106–24; A. Tzortzis, "Bombing" Berlin, the graffiti capital of Europe, *New York Times*, March 3, 2008.

Figure 5.20 Communism brought extensive industrial development (evident in the background) and isolation to Plovdiv, but post-Communist cell phone networks now connect a new generation of Bulgarians to the world. *Source:* Photo by D. J. Zeigler.

industrial processes, and greater reliance on aging Soviet-type nuclear reactors. The EU's stricter environmental regulations have closed most of central and eastern Europe's iconic smoke-belching plants (Figure 5.20).

Since the end of the Communist era, improved air, rail, and road transportation linkages connecting the major urban centers across Europe have been laying the foundation for the complete reintegration of the urban system. Businesses and city governments in the eastern European countries that joined the European Union in the 2000s (Bulgaria, Croatia, Czech Republic, Estonia, Hungary, Latvia, Lithuania, Poland, Romania, Slovakia, and Slovenia) have already developed stronger ties with their counterparts in the preexisting member states.

Membership in the European Union has enhanced the opportunities for former socialist cities to address their pressing social and economic problems. Certainly, the future and prosperity of Europe as a whole in the global economy depend on creating a more economically and socially equitable situation for all European urban residents.

As the common currency, the euro, is gradually extended to more countries, the urban system will undergo more profound changes as governments, businesses, and people in cities across Europe and elsewhere reorient their activities to take advantage of the changing economic environment.

At some point in the future, serious consideration will be given to moving some of the administrative functions of the European Union to cities farther east. While London, Paris, Berlin, Frankfurt, Brussels, and Milan will continue to dominate as major financial, political, and cultural centers, Warsaw, Prague, Budapest, and Sofia will surely shift the center of gravity of the core-periphery model farther east as the twenty-first century progresses.

SUGGESTED READINGS

Beatley, T., ed. 2012. *Green Cities of Europe: Global Lessons on Green Urbanism.* Washington, DC: Island Press. Examines some of the world's best examples of urban sustainability to show how cities can be green and livable.

Hall, P., and R. Pain. 2009. *The Polycentric Metropolis: Learning from Mega-City Regions in Europe.* London: Routledge. Examines eight networked, polycentric megacity regions in northwest Europe.

Hamilton, F. E. I., K. D. Andrews, and N. Pichler-Milanovic, eds. 2005. *Transformation of Cities in Central and Eastern Europe: Towards Globalization.* New York: United Nations University Press. An overview with rich examples of major cities on the road to globalization and European integration.

Herrschel, T. 2014. *Cities, State and Globalisation: City-Regional Governance in Europe and North America.* New York: Routledge. Examines how city-regions are governed by comparing European and North American examples.

Kazepov, Y., ed. 2005. *Cities of Europe: Changing Contexts, Local Arrangements, and the Challenge to Urban Cohesion.* Malden, MA: Wiley-Blackwell. Chapters focus on issues such as segregation, gentrification, and poverty.

Kresl, P. K. 2007. *Planning Cities for the Future: The Successes and Failures of Urban Economic Strategies in Europe.* Cheltenham, UK: Edward Elgar. Examines the relationship between competitiveness and economic strategic planning for 10 internationally networked cities.

Murphy, A. B., T. G. Joran-Bychkov, and B. Bychkova Jordan. 2014. *The European Culture Area: A Systematic Geography.* 6th ed. Latham, MD: Rowman and Littlefield. A major text with chapters on cities, culture, the EU, and the environment.

Ostergren, R., and M. Le Bossé. 2011. *The Europeans: A Geography of People, Culture, and Environment.* 2nd ed. New York: Guildford. A comprehensive view of Europe with two chapters on towns and cities.

Penninx, R., K. Kraal, M. Martiniello, and S. Vertovec, eds. 2004. *Citizenship in European Cities: Immigrants, Local Politics, and Integration Policies.* Aldershot, UK: Ashgate. Examines citizenship in European cities with a focus on immigration policies and immigrant participation in civil society.

van den Berg, L., P. M. J. Pol, W. van Winden, and P. Woets. 2005. *European Cities in the Knowledge Economy.* Aldershot, UK: Ashgate. Examines the knowledge economy using case studies of Amsterdam, Dortmund, Eindhoven, Helsinki, Manchester, Munich, Munster, Rotterdam, and Zaragoza.

Figure 6.1 Major Urban Agglomerations of Russia. *Source:* United Nations, Department of Economic and Social Affairs, Population Division (2014), World Urbanization Prospects: 2014 Revision, http://esa.un.org/unpd/wup/.

6

Cities of Russia

JESSICA K. GRAYBILL AND MEGAN DIXON

KEY URBAN FACTS

Total Population	143 million
Percent Urban	74%
Total Urban Population	105 million
Annual Urban Growth Rate (2010–2015)	0.1%
Number of Megacities (> 5 million)	1
Number of Cities > 1 million	
Populations of Megacities (> 5 million)	Moscow (12 million)
Fastest Annual Urban Growth Rate (2010–2015)	0.1%
Largest Urban Agglomerations	
World Cities	Moscow

KEY CHAPTER THEMES

1. Russia's urban development reflects the impact of three distinct eras in the country's history: tsarist, Soviet (communist), and post-Soviet.
2. Russia's cities experienced two reconstruction phases in the twentieth century, one after the creation of the Soviet Union in 1917 and the other when the Soviet Union collapsed in 1991.
3. The main pattern of the urban system, with its strong reflections of European urban planning characteristics, was established in the tsarist era.
4. Russia's rapid urbanization in the early twentieth century accelerated the country's historic patterns of urban growth and contraction over the last thousand years.
5. As a result of the disintegration of the Soviet-era socialist support system, crime and corruption have hindered the emergence of a democratic post-Soviet governance and civil society, especially in cities with increasing in-migration.
6. Environmental issues in Russia's urban centers are increasingly recognized as severe and have become an important issue requiring attention from post-Soviet city leaders.

7. The need to overhaul and redesign urban places and urban governance raises new questions about the roles of government and citizens in the post-Soviet era.
8. Changing demographics and shifting cultural and religious identities have reinvigorated questions about tolerance and acceptance of multiculturalism in post-Soviet cities.
9. In the post-Soviet period, cities are no longer subsidized by the central government; many have experienced economic recession, significant population loss and, at least, seasonal deurbanization or ruralization.
10. Cities that are prospering are those with superior locations, strong historic roots, or attractive environments for foreign investment and economic growth.

The urban landscape of the Russian Federation, commonly known as Russia, is today characterized by ornate tsarist-era buildings and monuments (palaces, churches, museums) standing alongside utilitarian, concrete-and-steel structures of the Soviet era (office buildings, communal apartments, community centers) and the newly erected European-style, elite apartments and shopping centers of the post-Soviet era. This landscape reflects the impacts of urban development during three distinct periods in the country's history: tsarist, Soviet (communist), and post-Soviet.

Diverse ethnic groups have inhabited this region of Eurasia for at least a thousand years, continually contributing to Russia's diverse population. A blending of many cultures, religions and histories across the European and Asian realms of Russia for several centuries resulted in multicultural settlements that eventually grew into towns and cities during the Tsarist Russian Empire (1721–1917). Under the Union of Soviet Socialist Republics (the USSR, or the Soviet Union) from 1917 to 1991, that multiculturalism was celebrated but also tempered by communist universalism; standardized Soviet urban forms spread across a Eurasian territory, either rebuilding existing cities or creating new ones (Box 6.1). Across Russia, but especially in cities, the collapse of the Soviet Union was symbolically marked on December 25, 1991, as the Soviet hammer-and-sickle flag lowered, replaced by the red, white, and blue flag of the Russian Federation, now one of 15 independent post-Soviet nations.

Russia began the twentieth century with less than one-fifth of its population classified as urban; by 1989, 74 percent of Russia's population lived in urban places. The average percentage leveled off in the late 1990s at about 73 percent, a figure that remains stable today. Although it has been two decades since the end of the Soviet period, the imprint of Soviet-era urban policy and form still profoundly affects the larger territory—urban and rural—of the Russian Federation (Figure 6.1) and other post-Soviet states. The Soviet attempt to provide greater cultural, educational, employment, and housing opportunities in cities produced specific urban landscapes; many of these remain as Russia continues to undergo a series of socioeconomic, political, and cultural transformations, shaping urban trends that contradict some expectations common in the West. For example, the severity of the economic collapse following the end of the Soviet Union and Russia's abrupt confrontation with the global economy prompted many urban workers to fall back on *dacha* settlements (rural areas established in the Soviet era) to practice subsistence farming, a process known as *ruralization*.

Box 6.1 Where does Soviet Influence Begin or End?

Cities evolving under post-socialist regimes are breaking away from the urban plans so strictly enforced by communist/socialist governments. Socialism's largely compact, comprehensively planned cities were structured internally and regionally to be self-sufficient, but this is changing today as individuals and businesses make their own decisions about where to locate residences and businesses in freer market economies. The rate of change in urban planning for the post-socialist city depends on many factors, including a city's location and involvement in a regional or global economy, and the visions of individual city leaders.

Three growth trends alter the urban form, function, and internal spatial structures of post-socialist cities, reforming socioeconomic and political processes in addition to the built environment. First, emerging land markets and commercial real estate spaces transform the urban fabric as new housing, shopping, and sometimes industrial developments are created within city limits and in suburban or exurban locations. Second, increased automobile and cargo truck ownership causes new kinds of movements in and around cities, causing congestion in the city but the possibility of movement into suburban locations by businesses and individuals. Third, as suburban growth develops, there is a tendency for previously compact post-socialist cities, often radial or quadrangular in form, to become linear in form, as economic activities occur along arterial routes out of cities into the surrounding countryside.

While city centers, urban peripheries, and suburban locations are all foci for redevelopment and new growth, often-outward development is chronological. In early post-socialist city development, the most important growth region is the inner city, where densification of the urban fabric occurs as new residences and small entrepreneurs appear. In later post-socialist city development, inner city development continues but is joined by peripheral urban and suburban growth, where small-scale development is replaced by mega-shopping complexes and gated residential communities. As development continues, land outside cities but along transportation routes is redeveloped first because it has the greatest access points to customers and infrastructure. Thus, post-socialist city growth is both vertical and horizontal, integrating commercial and residential development in new ways in city centers and peripheries.

Not surprisingly, in a country nearly twice the size of the United States, the effects of "wild" capitalism and its visual imprint on the landscape are unevenly distributed across Russian cities. In some Russian cities, the built environment has changed so dramatically since 1991 that many cities are nearly unrecognizable to those accustomed to quiet, somber Soviet landscapes. Commercial retailers, private transportation, and new housing construction have altered Russian social and cultural urban landscapes (Figure 6.2). For example, Moscow's Red Square is no longer a nearly deserted public space awaiting military parades; instead, this central-city landmark area abuts a bustling high-end retail and

Figure 6.2 New construction in cities around Russia (Vladivostok is pictured) relegates Soviet urban landscapes to the background as new commercial and residential buildings vie for valuable real estate locations. *Source:* Photo by Sergei Domashenko.

tourist space (Figure 6.3). This new socioeconomic landscape changes how people use the built environment, creating new cultural spaces and practices; luxury shopping and café lifestyles have become daily activities for Muscovites and tourists alike.

Construction of tsarist and Soviet cities emphasized urban planning principles such as pedestrian walkways and mass transit, thus ignoring or minimizing the needs of automobiles. Indeed, across the territory of the former USSR, over 20 metros (subway systems) were developed, far more than any other country in the world. The Soviet vision of accessible transportation for all urban citizens thus predates many sustainability-minded mass-transit projects now becoming popular in the West. However, an exponential increase in private and commercial vehicles has occurred in post-Soviet Russia, and cities were neither built nor have been redeveloped to accommodate them. The concept of rush hour has great meaning now, and for many Russian cities, especially Moscow, rush hour begins mid-afternoon and extends through the evening (Figure 6.4).

The typical Soviet rings of monolithic apartment complexes on city outskirts are increasingly mixed: new elite apartment buildings are constructed alongside Western-style suburban developments of "cottages" and gated communities (Figure 6.5). Soviet neighborhoods were often ethnically and socioeconomically intermixed, but today, depending

Figure 6.3 Renovations in GUM shopping center on Red Square make it a top destination for tourists and Russia's elite seeking high-end shopping experiences. *Source:* Photo by Jessica Graybill.

on the city, stratification by socioeconomic class and sometimes by ethnicity is beginning to occur, as upwardly mobile residents choose to live in newly constructed, high-security apartments or McMansions in suburban or exurban locations. Cultural and social change is noted in the replacement of Communist Party billboards (formerly present in every city and town) with brilliant neon and banner-type commercial advertisements for consumer goods and services along major urban thoroughfares.

Post-Soviet Russia still grapples with the legacy of the spatial framework created by Soviet urban development. The Soviet planning regime often located settlements near natural resources in isolated, inhospitable environments, resulting in far-flung, potentially unsustainable urban growth. In the post-Soviet period, capitalist notions of efficiency have made such cities' locations and industrial operations unprofitable, resulting in economic decline and population outflow. This type of development has been called "archipelago urbanization," because Soviet cities arose like urban islands in a vast rural Eurasian hinterland that remained—and remains—seemingly unchanged culturally or economically for centuries. Cities' position as islands within Russia's vast territory persists due to a lack of transportation infrastructure connecting them or, today, the lack of affordable transportation to any destination but Moscow. For example, the price of a round-trip plane ticket from Petropavlovsk-Kamchatsky to Magadan (a 1-hour flight) is usually greater than a round-trip ticket from either of these cities to Moscow (a 9-hour flight), indicating the centralized power that Moscow still wields over individual Russian regions.

In both tsarist and Soviet Russia, many cities owed their existence and location to questions of national security, but military-industrial complexes now play a lesser role in determining the location of urban investment and growth. Today, cities previously favored by Soviet urban and economic policies are undergoing economic restructuring processes not dissimilar to the restructuring experienced by major North American and European cities during deindustrialization, beginning in the 1970s. For example, cities such as Ekaterinburg in western Siberia are becoming service-based transportation and corporate centers for European businesses, and other gateway cities near the Chinese border (e.g., Vladivostok and Khabarovsk in eastern Siberia) are transforming Russian-Asian business relations.

Figure 6.4 Since the fall of communism, automobile ownership in Moscow has soared, and with it has come urban gridlock. *Source:* Photo by Alexei Domashenko.

Figure 6.5 New microrayon developments, with varied architectural styles and imposing gates and fences, are rapidly changing the face of Russia's suburbs. This picture is from Balakovo. *Source:* Photo by Jessica Graybill.

The early post-Soviet period exposed many cities to increasing poverty, economic collapse, and restructuring, with large inflows of refugees from more troubled parts of the former Soviet Union (Table 6.1). Many people from Ukraine sought residence in Russia in 2015 due to conflict in and near Crimea. Today, rather than buying subsidized goods and services from the state, urban governments are challenged to transform into self-sufficient capitalist entities responsible for self-promotion in an economic climate marked by rapid and widespread changes in the distribution of development both within cities and between them. Urban in-migration and development has led to increased growth of larger cities (e.g., 100,000 or more) in western and southern Russia (Figure 6.6). Just as throughout Russian history, harsh climate, a poorly developed (and frequently impassable) network of roads, and immense distances still exacerbate the fragmentation of the Russian urban system. Russian cities will continue seeking successful solutions to these challenges well into the twenty-first century.

HISTORICAL EVOLUTION OF THE RUSSIAN URBAN SYSTEM

The Pre-Soviet Period: Birth of the Urban System

Historical settlement patterns have depended on access to water, transportation, and the location of military and economic outposts. The eastward spread of Russia's urban population dates to the first Slavic cities that appeared in the Valdai Highlands of the Russian plain at the end of the ninth century. A vast river network provided connectivity through this region, often called Rus, creating vital trade routes between Scandinavia, Russia, and the eastern Mediterranean regions. The Vikings established a set of city-principalities, at once both military outposts and trading centers, where they collected tolls from merchants traveling through the region. Kiev (or Kyiv, the capital of independent Ukraine), Novgorod, and Smolensk were among the earliest urban settlements of this period.

The region gradually began to function independently from the Viking settlers, and

Table 6.1 Percent Urban Population in Each Federal Okrug

Federal Okrug	*1926*	*1939*	*1970*	*1989*	*2002*	*2010*	*Percent change, 1926–1989*	*Percent change, 1989–2010*
Central	19	34.2	64.3	78	79.1	81.3	310.5	4.3
Northwest	29.2	48	73.3	82.2	81.9	83.5	181.5	1.6
Southern	19.2	31	52.1	60	57.3	62.4	212.5	4.1
North Caucasus	–	–	–	–	–	49.1	–	–
Privolzhskaya	12.1	23.8	56.1	70.8	70.8	70.8	485.1	0.0
Ural	21	45.4	71.3	80.2	80.2	79.9	281.9	−0.3
Siberia	13.3	32.6	62.5	72.9	70.5	72.0	448.1	−1.3
Far East	23.4	46.5	71.5	75.8	76	74.8	223.9	−1.4
Russian Federation Total	17.7	33.5	62.3	73.6	73	73.7	315.8	0.1

Sources: Percent urban population in each federal okrug. The North Caucasian Federal District was split from Southern Federal District on January 19, 2010. Percent urban population in each federal okrug. The North Caucasian Federal District was split from Southern Federal District on January 19, 2010. 2010 Census Data; www.perepis-2010.ru.

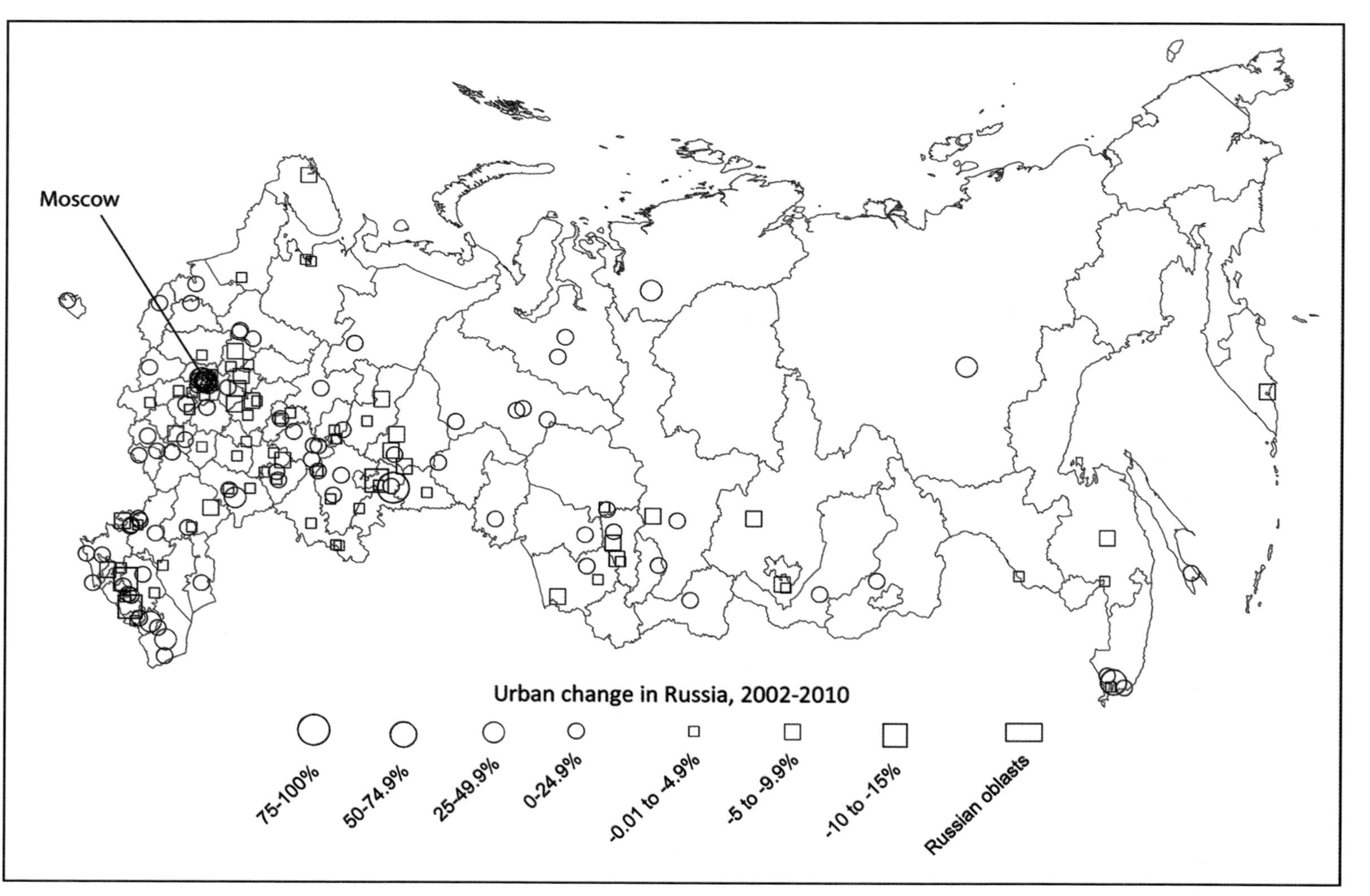

Figure 6.6 Population Change in Russian Cities, 2002–2010. *Source:* Map by Jessica Graybill using Russian Census 2010 data; www.perepis.ru.

Kiev became the focal point for Slavic political and economic development because of its location on the navigable Dnieper River with access to the Black Sea and Constantinople. In 988, Orthodox Christianity became the official religion, constraining the open practice and tolerance of other beliefs (e.g., Islam, Judaism, paganism, and pantheism). Most cities in Kievan Rus were located along rivers and were originally established as *kremlins*, or forts, because of constant conflict among the settlements as well as for protection against raids by Mongols and, later, Mongolian Tatars. The importance of hills for defense and rivers for communication lines during this period explains common features of many city centers. Kremlins were always located on high riverbanks and streets were radially planned, to facilitate rapid dispatch of troops. Many of these cities have survived today with their kremlins still intact. The famous Golden Ring cities around Moscow (e.g., Yaroslavl, Suzdal, Vladimir) have origins in Kievan Rus and remain important centers of Russian Orthodox Christianity.

After an important victory against the Tatars in 1480, a new polity called Muscovy arose and aided in the development of a new type of urban network that developed east of Rus. The growing city of Moscow dominated this new region of settlement from its location at the center of another river system, allowing cultural and economic growth in new directions. Access to the Volga and its tributaries aided eastward expansion; the Western Dvina led to the Baltic; and the Don and Dnieper rivers led to the Black Sea. Theologians who envisioned Moscow as the 'Third Rome' provided Muscovy with a vision and self-proclaimed mission to build a new Russian empire firmly rooted in Christian missionary traditions with expansionist intentions.

Indeed, Russian settlements expanded eastward across the Ural Mountains into Siberia, encountering little resistance after final defeat of the Tatars at Kazan in the middle of the sixteenth century by Tsar Ivan the Terrible. New settlements, such as Tobolsk and Yakutsk, began as military outposts; they remained isolated frontier towns until Soviet expansion. Trappers plundered Siberia for furs; explorers and scientists who sought to map the territories to the East and South also brought back tales of ethnic groups and raw resources in Siberia, the Far East, and Central Asia.

As the seventeenth century ended, Russia's network of cities had become landlocked. Seeking access to the sea, Tsar Peter the Great founded St. Petersburg in 1703, touching off spectacular urban transformations within Russia. Consistent with Russian urban history, St. Petersburg was built for cultural, economic, and security reasons, to be a showcase naval and commercial port with crucial access to maritime routes (Figure 6.7). New to Russian urban development, however, was the cultural purpose of becoming the country's "window on the West"; the city was designed according to European planning principles. Peter the Great also expected adoption of western cultural norms. For example, he required men in cities to cut off their long beards or pay an annual beard tax, thus hoping to shear away old Muscovite customs and traditions in favor of new styles and habits of living.

As the new national capital, St. Petersburg quickly supplanted Moscow. The urban focus moved westward physically and culturally. Reforms undertaken by Tsar Peter revitalized both local and long-distance trade, encouraging growth in new market centers as well as in more established ones. The creation of this new, Western-oriented city fueled social and spatial tension between those who believed in modernizing the country and those who emphasized Russia's traditional Slavic origins. Current debates about Russia's direction of development mirror these earlier ones; some

Figure 6.7 The Church of Our Savior on the Spilled Blood, in St. Petersburg, was built on the spot where Emperor Alexander II was assassinated in March 1881. Built from 1883 to 1907, the Romanov family provided funds for this glamorous cathedral. *Source:* Photo by Jared Boone.

look to the West or East for support, and some look inward for purely Russian inspiration and solutions. The development of St. Petersburg in originally inhospitable swampy land was also a precursor to the Soviet belief that humans can conquer nature in the name of economic progress. By the end of the nineteenth century, about 16 percent of Russia's population lived in urban areas. Factories in the region around Moscow and several nearby centers (e.g., Tver, Vladimir, Ivanovo, Kostroma) fueled economic and urban development.

The Soviet Period: New Urban Patterns

After the Russian Revolution (1917) and ensuing civil war, the Communist Party took steps to consolidate its political power and reshape the economy, establishing a political-economic and urban system unlike any other worldwide. In 1918, the leadership moved the capital from St. Petersburg (renamed Leningrad in 1924) back to Moscow. The move was both symbolic and strategic: the relatively recent capital built by the tsars as a window on the West was replaced by an older capital (Moscow) in the country's heartland, which would be easier to defend. It also made a statement that the country's gaze was no longer to the West but to the East and within empire.

Soviet doctrine privileged urban life over rural as the proper environment for communist man, drawing on British and French models and experiments in worker housing. Urbanization was seen as necessary to create an industrialized working class that would embrace communist ideals and thus became synonymous with the construction of communism. Russia rapidly urbanized after the communist era began in 1917 and significant levels of urbanization continued in every region of Russia throughout the Soviet period, bringing electricity and indoor plumbing to many regions. Even in predominantly agricultural regions, more than half of the population lived in urban places by 1979.

The Communist Party established a new economic system guided by communist and socialist principles instead of market forces, called the command economy because a group of central planners located in Moscow made all decisions. Central planners allocated all investment resources and set standards for urban development, privileging national over local needs. This meant that cities had little influence over local economic development, urban growth, and internal city structure.

Private property was abolished. To provide immediate housing for the crush of population moving into cities to supply industrial labor, private apartments were appropriated and subdivided to create *kommunalka*, or communal apartments, which provided immediate housing in the era of urban industrialization. Multiple households had to share spaces that formerly accommodated a single family. At first, such lack of privacy was tolerated as part of the excitement of building a new communist society. Later, however, *kommunalka* became slum-like dwellings for the urban poor where sharing was largely practiced out of economic necessity, not out of idealism about a better future.

Obeying another ideological principle, Soviet planners attempted to distribute urban settlements evenly across the Soviet expanse, even into harsh, inhospitable regions, obeying the injunction of Friedrich Engels to distribute large-scale industry equally across the country. Using algorithms based on European and North American urbanization models, Soviet planners chose so-called optimal locations for industrial development and built cities around them. This led to the construction of new cities with predetermined sizes (e.g., less than 50,000 people, more than 100,000 people) in previously lesser developed and less populated regions. This approach created a seemingly irrational pattern of economic flows between quite distant cities; the locations of suppliers, intermediate producers, markets, and managerial bureaucracies were of little concern in a system where transportation and energy costs were state subsidized and therefore perceived to be virtually free.

The artificiality of this urban planning policy is especially visible in the rapid urbanization and settlement of Siberia, the Far East, and the Far North. Soviet planners regarded mastery of these regions as an ideological necessity and as a challenge to their technological ability to tame harsh environments, such as permafrost or steppe regions. Prior to the Soviet period, small, autonomous villages and indigenous settlements dotted the vast territory of the Russian Far North and East. Modernization of lifestyles in these regions in the early Soviet period was achieved by pushing people off their native lands and into regional towns and collective farms (*kolkhozi*). Ostensibly undertaken to ease central management and regional planning for cities and towns, resettlement of nomadic and semi-nomadic indigenous peoples from small villages across the former Soviet territories ultimately enlarged the Soviet industrial workforce but greatly altered residential patterns and traditional ways of life.

High rates of urbanization in the Arctic and Siberian regions existed as early as 1959 and persist today (Table 6.1). Even into the late twentieth century, the population of the Far North (the Arctic region) remained nearly 80 percent urban, well above the Russian average of 73 percent. The ideologically based subsidies that enabled this process included higher salaries offered to persuade people to join the social and physical construction of communism in the new industrial settlements. This practice highlights a significant mismatch between the location of labor resources, markets, and urban-industrial power in the western portion of the country, and the location of natural resources, including energy, in the eastern and northern portions of the country.

After World War II, national security needs also prompted the creation of a vast urban network connected to the military-industrial complex (MIC). Closed to outside visitors, these cities grew in economic importance and population only because of their attachment to the MIC. Decades of defense-related investment in these cities' industrial bases, housing stocks, roads, schools, and other urban infrastructure influenced their urban geographies in ways impossible in capitalist economies (in Figure 6.20, note the enclave of Zelenograd near Moscow but distinct from it).

Subsidization of transportation included spectacular megaprojects that aimed to increase connectivity in the urban system. Joseph Stalin envisioned the construction of a canal network across northern European Russia to promote trade. While never completed, it is remembered for the use of prisoners to construct it, especially the White Sea-Baltic Canal (Belomorkanal) portion. In the 1970s, Soviet planners began constructing a second Siberian rail route, the Baykal-Amur Mainline (BAM), to supplement the capacity of the Trans-Siberian Railroad. The BAM facilitated new natural resource exploitation and the transportation of goods across Russia's vast expanse, providing lifelines to cities located thousands of miles from central Russia. In the Far North, cities primarily depended on boats using the Northern Sea Route along the Arctic coastline or Siberian rivers. Even today, frozen rivers are used as winter roads until the ice breaks. Crucially, however, the USSR never developed a network of highways such as those in North America or Europe, thus greatly hampering circulation between cities.

To help carry out political and economic agendas, as well as to reflect the new ideology, the communist leadership established a hierarchical urban administrative system that located all power in Moscow. Administrative centers in the *oblasts* (political units comparable to states or provinces) were subordinate to central planners in Moscow, who controlled resource allocation and use in each region. Not surprisingly, administrative centers benefited disproportionately from central investment decisions. *Oblast* centers became the locations for massive industrial investment and grew rapidly. Historic industrial centers such as Moscow, Yaroslavl, and Kazan were joined by administrative/industrial centers in Siberia and the Far East (e.g., Omsk, Novosibirsk, Krasnoyarsk, Irkutsk, and Vladivostok). Many *oblast* centers still function like primate cities in other world regions, where investment, services, and labor are concentrated in one city, creating uneven regional development.

Planners also used investments to develop a system of secondary industrial cities focused on heavy industry (e.g., the automotive industry in Tolliatti, aluminum production in Bratsk) or natural resource exploitation (e.g., nickel in Norilsk, oil near Surgut). Thus, a new system of large cities (i.e., cities of more than 50,000 people) developed in Russia. In a country where bigger was seen as better, planners and politicians spoke glowingly of cities with more than 1 million inhabitants. Indeed, many Russians have a cultural urban bias; they consider it more prestigious and advantageous to live in cities and, despite urban hardships, prefer not to leave them for rural settings.

While planners directed investment resources to specific cities, they simultaneously pursued a contradictory policy: limiting population growth in many of the same cities through formal control mechanisms such as the *propiska* (legal permission to live in a specific city). Many individuals found legal ways around the system, such as marrying someone who had a *propiska* or finding employment and having the employer secure a *propiska.* Ultimately, pressure for ever-greater production made investments in established sites more economically rational. But additional production created demand for increased labor. This had the dual effect of increasing city sizes beyond intended targets and intensifying industrial production (and thus pollution and waste) inside city boundaries.

Urban and Regional Planning in the Soviet Period

Central planners also influenced the internal spatial structure of Soviet cities. To create cities consistent with socialist ideals, planners adopted specific principles to guide urban planning that included adopting urban growth boundaries in order to constrain city sizes, distributing consumer and cultural goods and services equitably to the population, minimizing journeys to work and providing public transportation for spatial mobility, and segregating urban land uses. Interestingly, some of these Soviet principles, such as urban growth boundaries and reduced commuting, have been abandoned in Russia today but are propounded in the West as "smart" growth.

The basic building block of Soviet cities was the *microrayon.* Constructed near industry and other places of work to minimize journeys to work, *microrayons* housed 8,000–12,000 people in living areas designed as integrated units of high-rise apartment buildings, stores, and schools, providing residents with cultural and educational services required by Soviet norms. In this urban planning scheme, all daily life activities (e.g., education, shopping, and the use of city services like post offices and utilities payments) could be conducted without leaving the *microrayon,* thus influencing how children, workers, the elderly and others moved through the city. People living in close proximity in the same city, but not in the same *microrayon,* might never meet on the street because of the highly structured nature of urban life in these neighborhoods.

Built using standardized plans irrespective of local environmental conditions, *microrayons* are numbingly similar whether in Novosibirsk, Vorkuta, or Moscow, and large tracts of identical or similar multistory apartment buildings still ring all Russian cities. For example, similar construction materials and designs were used to build *microrayons* located in diverse physical geographic regions found across the former Soviet Union, such as in earthquake hazard zones (e.g., Almaty in Kazakhstan); in cold, damp climates (e.g., Petropavlovsk-Kamchatsky); in flood hazard regions (e.g., St. Petersburg); or on the semi-arid steppe (e.g., Barnaul). *Microrayon* locations were set by the so-called General Plans, which determined the location of *microrayons* within cities, as well as all other land uses. General plans were so detailed that a milk store could not be built legally on a site designated for a bread store. General plans were intended to complement the shorter-term five-year economic plans, which determined what would be made, how it would be made, who would make it, who would receive the final product, and at what price.

The Urban Environment in the Soviet Period

Absent from Soviet planning principles were concerns about the impact of industrial or urban development on the environment or about ecosystem limits. Planners, and the Soviet system in general, believed in and practiced technological control over nature. This practice, combined with the zeal to reach economic goals, resulted in almost complete disregard for the ecology in and near Russia's cities. Teams of planners, geo-engineers, and economic geographers choreographed large-scale development projects to modernize society, especially in large urban areas. For example, dams, hydroelectric power stations, and industrial complexes were constructed in and near cities. "Progress" was narrowly conceptualized as industrialization at all costs, and nature was society's tool to create the new socialist reality.

For example, near the city of Okha on Sakhalin Island, onshore oil deposits have been exploited since the early 1900s. Exploitation increased in the Soviet period, and the evidence of poor environmental standards for extraction remains today. On a road out of town, adjacent to local residents' summer homes (*dachas*) and within sight of high-rise apartment buildings, numerous rusting and leaking oil pumps stand in pools of stagnant water mixed with leaked oil. Although signs posted in this suburban oil field warn pedestrians of the toxins in the area, they are often illegible or half buried in oil muck. This mixture runs into local creeks, which in turn empties into the Sea of Okhotsk, where discharge from runoff pipes disrupts ecologies in nearshore bays and coastlines. This environmental and human-health hazard was—and remains—less important than the economic bottom line.

Examples like these abound in and around Russian cities and can be understood as examples of environmental and social injustice. Although many urban residents were and are aware of urban environmental issues, Soviet newspapers and scientific-engineering literature remained silent about the growing environmental problems across most of Russia's industry-driven cities until the late Soviet period, when the extent of environmental degradation began to be publicized. Only in the late 1980s did people openly begin to express concern about environmental health issues after nationwide reporting of air, water, and land pollution; environmental degradation with economic consequences (e.g., decreased fishing catches in lakes and rivers); and human-health issues (e.g., asthma, kidney diseases, lung diseases). Many urban dwellers found a silver lining in the industrial decline of the 1990s—the spiraling decline of urban environments was temporarily halted until massive increases in automobile use caused air pollution to rebound. Disregard for the impact of economic development on ecology shaped investment practices that continue today.

The Soviet history of urban and regional planning has left an indelible mark on Russia's built and natural environments, precisely because buildings and industries were located without reference to market forces and environmental conditions. In the command economy, land was not bought and sold in Soviet cities, but allocated roughly in accordance with the socialist ideology and planning principles outlined above. The absence of a free market meant that land was not recycled for other purposes, as would have been the case in market economies. As a result, new construction tended to continue moving outward from the city center. Compare, for instance, the population density of Paris and Moscow as it varies with distance from the city center (Figure 6.8). In Paris, market forces mean that valuable land near the city center is more densely populated than less valuable land on the city outskirts. In Moscow, just the opposite was true. The most densely populated parts of the city were on land far from the city center, which was often reserved for culture and political symbolism.

Clearly visible historic rings of development remain. Beginning in the 1930s, huge factories were erected outside tsarist-era city cores as part of the industrialization drive. In subsequent years, especially after the late 1950s, a near catastrophic housing shortage and renewed determination to improve people's living conditions began decades of construction of *microrayons* on city outskirts while historic buildings crumbled in city centers (Figure 6.9). Instead of developing technology to build skyscrapers on valuable land in the city center, central planners focused on building high-rise apartment buildings on the outskirts, creating a characteristic "bowl" skyline in many cities. Ironically, this undervaluing of

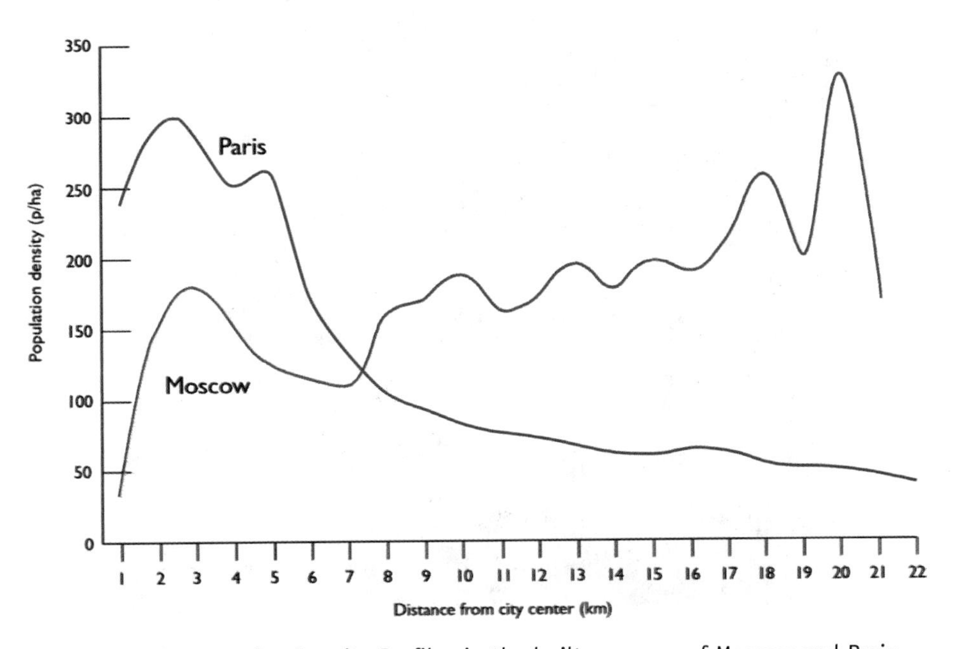

Figure 6.8 Comparative Density Profiles in the built-up areas of Moscow and Paris. *Source:* Beth Mitchneck and Ellen Hamilton.

centrally located land preserved many older buildings up until the post-Soviet period.

Late Soviet Period: The Beginning of Change

Noticeable urban restructuring preceded the final economic and political collapse of the Soviet system in 1991. Starting in the 1980s, a phenomenon of "disappearing cities" revealed a pattern of manufacturing and industrial decline. Towns in European Russia lost so many people that they no longer appeared in Soviet statistical accounts of urban places. While distributed throughout Russia, more than half of these towns were in the industrial core regions around Moscow and St. Petersburg and in the Urals. By the mid-1990s, however, Siberian cities accounted for a larger proportion of shrinking urban areas. The exodus of residents from shrinking towns, combined with new migration patterns westward, contributed to the worsening situation of overburdened and decaying urban infrastructure, including housing and utilities, as

Figure 6.9 Historic buildings in Vladivostok's urban core crumble today from neglect in the maritime climate of this port city. *Source:* Photo by Jessica Graybill.

population became concentrated in many large Russian cities. Thus, despite natural population decline (a greater number of deaths than births), many Russian cities—especially in the south and west—are growing rapidly even as others shrink (Figure 6.6).

Analysis of the profile of disappearing and shrinking towns reveals the impact of economic restructuring on the urban system. Because of the boom-bust economies surrounding mineral and other resource exploitation, mining towns account for a large proportion of declining towns (roughly a quarter by 2000). Urban centers related to the MIC comprise another large proportion of declining cities. Previously, MIC cities received special attention from the central government, including better-than-average access to goods and services as well as higher salaries. This special treatment does not continue today. For example, in the border zone of Kamchatka in the Far East, MICs have become near ghost towns due to a perceived end in the need for border control in this region.

An interesting counterpoint to the disappearance of many towns is the appearance of previously unacknowledged cities, known as *secret cities* or, in Soviet parlance, "closed administrative-territorial formations." Secret cities never appeared on maps, and some estimates place the number of secret cities around 40. Located throughout Russia, but with clusters in Murmansk Oblast, the Far East, the Urals and Moscow Oblast, employment was focused on highly classified military production including nuclear research and missile production (Figure 6.10).

CONTEMPORARY RUSSIA: RECONFIGURING THE URBAN SYSTEM

In the post-Soviet period (1992–present), heavy industry was especially hard hit by the

Figure 6.10 A submarine in Kaliningrad, a former secret military city in the former Soviet Union, is now used as a tourist attraction. *Source:* Photo by Annina Ala-Outinen.

reintroduction of market forces and the reorientation of the economy away from defense. As Russia's borders opened, it was flooded with cheaper and better consumer goods, and demand for locally produced goods—everything from steel to planes—dried up. Cities experienced economic restructuring processes not dissimilar to those in major North American and European cities during deindustrialization in the 1970s. At the same time, regions began scrambling for new investment capital to replace the funds that used to flow in from Moscow. Economic restructuring away from manufacturing to increased raw resource extraction and export means that cities dependent on manufacturing now struggle to cope with high unemployment rates and few opportunities for new development. It is precisely these deindustrializing cities that are losing population in the current period.

In the post-Soviet period, the places with the highest rates of gross regional product today are found in just a few natural resource-rich regions in Siberia, the Far East, and in Moscow. As a direct result of the change from Soviet central planning to market-driven processes, there has been significant movement out of Siberian and Arctic cities (as well as from cities in the newly independent countries in Central Asia and the Caucasus) back to European Russia. The introduction of market forces in cities in harsh and inaccessible places, such as Norilsk and Surgut, caused rapid increases in costs for energy and transportation, food, housing, and industrial production. As subsidies dropped sharply, urban-industrial complexes closed and unemployment surged. Many who could simply pulled up stakes and moved. One unique urban phenomenon in the former Soviet nation of Kazakhstan, is the creation of a new national capital (Astana), ostensibly to replace the heavily Russified older capital (Almaty) and renew economic growth and sociocultural importance of this place by and for Kazakh citizens (see Box 6.2).

The uneven spatial distribution of the benefits of economic reform has created vast differences among individual cities in Russia's urban network. Some cities are clearly thriving in the transition to a market economy, but those whose geographic locations are not conducive to taking part in transformation are struggling. However, many people (including the elderly and the poor) stay in these cities because of strong social or kinship ties forged in the urban archipelago, or a strong belief in the Soviet system. For those who stay behind, the "new" traditional economy depends on activities like hunting, gathering, fishing, and domestic agricultural production. For example, people have intensified agricultural crops and livestock husbandry on household plots and at *dachas* (Figure 6.11). In some places, foresters report increased gathering of communal forest resources such as mushrooms, berries, or herbs. In this way, traditional agricultural activities are woven into everyday life in shrinking urban places across Russia.

Cities in regions with growing economies and growing populations (like Moscow, Yakutsk, Vladivostok, and Kazan) have become attractive destinations for migrants from more depressed areas of Russia and other parts of the former Soviet Union. This phenomenon has rearranged hierarchical relationships in the urban system by increasing the population and economic clout of previously minor cities. For example, migrants from other former Soviet republics are attracted to economic growth and jobs associated with natural resource development in booming northern locations such as Sakhalin Island and Norilsk (Figure 6.12). Cities in the Southern Federal

Box 6.2 New Capital Cities in the Post-Soviet Sphere: Astana's Amazing Growth

Natalie Koch, Syracuse University

"The goal is to have people talk about Astana like Dubai," explained the Master Plan chief of Astana in 2011. Some years prior, Kazakhstan's president Nursultan Nazarbayev tasked urban planners in the capital with using the capital city's development to raise the international prestige of this young, oil-rich Soviet successor state. This is a task that Dubai is frequently understood to exemplify. But long before the city's spectacular rise, capital cities have frequently been treated as the "face" of a nation, or a "shop-window" to broadcast its modernity and importance to the global community. Nowhere is the continued salience of this practice better illustrated than in Astana, Kazakhstan.

Astana, which simply means "capital" in Kazakh, became Kazakhstan's official capital in 1997. Astana has served many ideological agendas of the authoritarian Nazarbayev regime, which has dominated Kazakhstan since the collapse of the Soviet Union. Through the city's fast-paced development, the leadership has promoted the idea that the entire country is rapidly modernizing. Explicitly treated as the "business card" of the developing country, Astana's rise is used to advertise the country's new market orientation and to entice foreign investors. What has this process looked like in Astana?

In their efforts to raise Kazakhstan's international prestige, state and urban planners use many of the same urban development strategies that have helped to raise the international profile of Dubai. Specifically, these have included developing Special Economic Zones and technology parks; hosting mega-events such as international conferences, festivals, sporting events; introducing state-of-the-art new educational facilities, and developing a hypermodern image for the built landscape, through prioritizing a pastiche of iconic architecture by internationally renowned architectural firms like Foster + Partners.

While Astana's transformation has been remarkable (and remarkably costly), it is not based on popular demand. Rather, like numerous other boom cities in Asia, Astana's explosive growth is justified on the basis of the "build it and they will come" cliché. That is, planners claim that building in itself will create the demand. As many scholars have shown, this logic of "urban boosterism" routinely fails to deliver on its promised benefits. Indeed, many of the iconic new towers, stadia, and palaces that have redefined the urban landscapes across Eurasia stand largely empty and underused. But the development continues apace. Why?

In Kazakhstan, there are several reasons. First, construction contracts are an important way of distributing state oil revenue to political elites. Many are connected to companies headquartered abroad, which are typically overpaid for their work. Skimming money off the top, elites can thus transfer funds abroad—making the continued building boom quite attractive to top decision makers. Second, the authoritarian political system is such that ordinary people cannot hold their government accountable for using state funds in a more socially responsible manner. There are no free elections, opposition figures are systemically persecuted, and popular protests are put down with swift and severe action. But lastly, many ordinary people actually believe that a beautiful capital city is important and truly makes them feel proud—especially at a time when everyone in the world is laughing at Borat's imagined homeland.

Figure 6.11 Space around many Russian homes, such as this one near Moscow, and apartment buildings is devoted to subsistence agriculture during the short summer season. *Source:* Photo by Nancy Ries.

Okrug, such as Krasnodar, Stavropol, Vladikavkaz, and Novorossisk, are also growing, but mostly as a result of large influxes of migrants from conflict-ridden areas of the North Caucasus and the former Soviet republics. This trend, also in the capital city of Moscow, has also brought about high populations of non-Russians and with them, a variety of languages and other belief systems (e.g., Islam).

Significant population losses nationwide and depopulation of many urbanized regions in the post-Soviet period (Table 6.1) are clear signs of the failure of the communist approach to city planning. In the early twenty-first century, Russia is experiencing a dramatic restructuring of urban centers. Whereas the centralized economic authorities used to designate the political and economic importance of any particular city as well as the location of residences, individual Russians can now choose from a widening array of housing options. A growing trend is suburbanization and new housing developed outside city limits (Figure 6.5). Many new housing developments are video monitored, guarded by security officers, and approachable only by

Figure 6.12 Opened in 2010, "City Mall" in Yuzhno-Sakhalinsk is the largest shopping mall in the Russian Far East and boasts a microbrewery for beer and loudspeaker announcements in Russian and English. *Source:* Photo by Jessica Graybill.

automated private transportation; although this contributes to increased traffic woes, it satisfies a desire for private space for those who can afford this lifestyle. While there are no systematic data on suburbanization processes across Russia, it is clear that this trend is especially prevalent in the European portion of the country. Yet like past urban processes, it is spreading eastward.

Political Urban Transformation

Democratization and political decentralization have also influenced post-Soviet urban geographies. The first democratic elections took place in cities throughout Russia shortly before the end of the Soviet Union. For the first time, local politicians, at least in theory, became accountable to local populations instead of to higher-level government officials. During the 1990s, this accountability had important new implications for the spatial structure of cities, as urban geographies began to reflect local economic needs and social desires instead of national ones. For example, one result of local autonomy in the 1990s was the popularity of renaming cities and streets; names associated with prominent Soviet leaders have been replaced with historic names of the tsarist past. For example, Leningrad reverted to St. Petersburg and Sverdlovsk (named after a local communist leader) reverted to Ekaterinburg (literally, Catherine's City, after Tsar Catherine the Great). Similarly, streets were renamed: Moscow's Gorky Street, named after the Soviet writer, was renamed

Tverskaya Street. The changes contributed to a feeling among Russians of reclaiming their cities and neighborhoods.

Following the Soviet era, it became increasingly possible to express political visions for cities and the state that departed from the official line, and many groups chose public urban space as a venue. For example, since the early 2000s, extremist nationalist groups whose views were long kept quiet by Soviet multiculturalism have held gatherings and targeted non-Slavic populations (e.g., Africans, African-Russians, ethnic peoples of the former Soviet Union, especially from the southwest), sometimes resulting in violence. Under Putin, these possibilities in public space have steadily decreased. Dissident political groups strove to hold parades on Nevsky Avenue in Petersburg and on Red Square in Moscow, but have been increasingly diverted to less central spaces in the city. While a series of visible anti-Kremlin political demonstrations took place throughout the 2000s, particularly from 2006 to 2013, the murder of prominent politician Boris Nemtsov in February 2015 just one day before a planned demonstration against Putin's government chilled a climate already hostile to alternate viewpoints. The arrest of several members of Pussy Riot, a feminist protest group, after a guerilla performance of a song in Moscow's Cathedral of Christ the Savior and the house arrest of Alexei Navalny, candidate for mayor of Moscow and critic of Putin, have both been taken as warnings to political groups that criticize the Kremlin. Groups seeking to hold a public meeting must follow restrictive protocols and obtain permission from city authorities. A gradual return to a more open urban political scene may come from the work of groups like Open Russia—with branches in cities including St. Petersburg, Kaliningrad, Voronezh, and Ekaterinburg—organized by former oil oligarch and political prisoner Mikhail Khodorkovsky, released in late 2013.

Changing Urban Structure and Function

Notable changes to urban structure include new kinds of infill within the city, suburbanization, and "slumification." Important changes to urban function include increased finance and retail commerce at multiple scales. Processes absent during most of the twentieth century govern these changes—market forces and the active participation of municipal and regional governments.

New infill appears in city cores as old factories on land surrounding historic city centers are increasingly torn down and the land reused for other purposes, such as housing (apartment buildings) or retail. Existing buildings in poor condition but in good locations are purchased, upgraded, and converted into office space or upscale, gated apartments (Figure 6.13). In Moscow and some other Russian cities, this has led to gentrification and displacement of long-time residents from city cores.

Suburbanization is another visible change in urban form that results from the development of real estate markets in cities where they had been prohibited for most of the twentieth century (Figure 6.5). Carefully guarded single-family housing has appeared seemingly overnight in what has become a new ring of housing developments referred to as cottages (*kottedgi*) surrounding the older Soviet and new post-Soviet multifamily high-rises.

"Slumification" of parts of Russian cities is a result of transition from a command to a market economy. Run-down high-rise apartment buildings far from the city center are located on nearly worthless and often polluted land near former industrial production sites.

Figure 6.13 Tsarist-era buildings in Vladivostok's urban core are being revitalized in the post-Soviet era. *Source:* Photo by Sergei Domashenko.

These high-rises, and sometimes entire *microrayons*, are deteriorating rapidly as better-off tenants move to superior locations, leaving only poorer residents behind in what will likely become vertical slums.

Finance and banking—particularly international banking—is increasingly a feature of larger post-Soviet Russian cities. Just as in deindustrializing North American and European cities, the economic function of Russian cities and the new labor market are more oriented toward services in general, and toward financial and retail services in particular. Members of the international financial and banking sector have added Russian locations near manufacturing or natural resources. For example, European and Japanese banks can be found in Yuzhno-Sakhalinsk today, which marks a change in the function of this city from a small regional capital to a globalizing city involved in oil and other natural resource exploitation. Other Russian cities actively seek foreign investment to restructure their cities via partnerships among government and local and international businesses.

Retail commerce, powered by market forces, has also visibly changed the economic geography of Russian cities. Previously, retail trade occurred in state-owned stores or in a limited number of farmers' markets. Now, the spatial structure of urban retail has been altered dramatically. Transportation hubs (subways, rail stations) are multiscalar centers of retail trade where peddlers vend wares and where retail centers, such as malls, have been built (Figure 6.14). Some prerevolutionary shopping centers have regained their functions; for example, Moscow's famous GUM department store was remodeled as a high-end shopping mall. The urban periphery has turned into a new retail environment; megastores, such as IKEA, are opening outside of traditional retail centers, extending urban retail spaces.

Sociocultural Urban Transformation

Notable social transformations include changing labor and leisure structures. There is a growing class with extra money to spend, resulting in increased consumerism and more availability of goods. A revived entertainment sector has also sparked a service industry catering to 20- and 30-year-olds who have cash to spend. However, goods, services, and entertainment remain expensive; many people cannot afford a high quality of life in the new Russia, or cannot afford it after leaving home or college, as the costs of buying and furnishing apartments are out of reach for many. Recent devaluation of the ruble and economic sanctions against the Russian economy following support for military aggression by Russian separatists against (see Box 6.3), particularly, the annexation of Crimea has also constrained the growth of this new consumer sector.

Although the *propiska* no longer exists, the current registration system exasperates people moving around the country for jobs. Nonpermanent residents working away from their hometowns must purchase temporary urban registration in a semi-legal system, thus increasing their cost of living and jeopardizing their ability to succeed in Russia's changing spatial economy. This type of registration system relegates them to second-class citizenship in their city of employment, often putting them and their children last in line to receive government services, such as socialized health care, education, or employment services.

Transition from the Soviet to the post-Soviet era was difficult for those who were raised, trained, and already employed in the Soviet system, as it led to the disappearance of many jobs, the depletion of pension funds, and an unknown future. Many turned to the countryside to survive (ruralization), but others (considered "victims" of economic transition) turned to alcohol, theft, and/or prostitution. Because of Russia's slow response to providing social services to people in need, a new and growing class of the very poor, the homeless, and the disenfranchised exists in cities today.

Urban governments have meager funds to allocate for social services; in Moscow, only 1,600 beds in 2005 gave shelter to the entire homeless population (with homeless population estimates ranging from 30,000 to 1 million). Official sources reported in 2006 up to 55,000 homeless children in Moscow alone and 16,000 in St. Petersburg, but the problem occurs across Russia. These figures include children—many teenagers but some younger—living without parents on the streets in Russian cities.

Many of Russia's homeless children are escaping domestic situations that include

Figure 6.14 Street peddlers hawk a variety of fresh goods along the railroad tracks across eastern Sakhalin Island. *Source:* Photo by Jessica Graybill.

extreme poverty, alcoholism, domestic violence, and neglect. Research suggests that these social ills result from the post-Soviet economic restructuring and are experienced at the household level. Other causes include forced migration as a result of civil unrest in particular regions of Russia and a migration process in the 1990s that brought people from former Soviet republics to Russian cities, where they often failed to find housing or employment. These forms of displacement especially affect children: street children often do not attend school, experience harmful health impacts (e.g., contraction of infectious diseases), and

Box 6.3 Russia in Ukraine: Understanding the Annexation of Crimea

Michael Gentile, University of Helsinki

Crimea is a strategically located peninsula extending from the Ukrainian mainland into the Black Sea. It has the legal status of Autonomous Republic within Ukraine, providing residents with a degree of self-determination. On February 23, 2014, "little green men" started appearing near strategic objects in Crimea; the Russian Federation had annexed the entire territory from Ukraine in a move for which the international community was unprepared. The official "reunification," as the landgrab is known in Russia, took place on March 18, 2014, and Russian president Vladimir Putin's declining popularity rebounded almost overnight to unprecedentedly high levels.

While annexation was unexpected, it transpired against a background of long-term, but mostly low-grade and decreasing, political unrest with ethnic undertones within Crimea. This unrest is related to the complex nature of Crimea's cultural geographies, to the contested political status of the region, and to the peninsula's role within the demised Soviet and, before that, Tsarist empires. Crimea was under Ottoman rule between 1475 and 1774, after which a short but turbulent period of independence, under the Crimean khanate, followed. In 1783, it was annexed by Russia under the rule of Catherine the Great. The territory was subject to rapid colonization, which particularly influenced the cities. With the advent of the Soviet Union, it was included in the Russian Republic, but was transferred to the Ukrainian Republic in 1954. When Ukraine became independent in 1991, Crimea became its constituent part, backed by the support of the majority of its population in the immediately preceding independence referendum.

Despite Vladimir Putin's recent attempt to "sacralize" Russia's connection to Crimea, suggesting that the two are united by an unchallengeable historical affinity, the nature of this relation is debatable.

The territory retains an important regional identity, which relates, in part, to the region's ethnic composition, characterized by a majority share of Russians (about 58 percent at the time of the 2001 census, but decreasing since then), followed by large minorities of Ukrainians (24 percent), Crimean Tatars (12 percent), and smaller communities of Armenians, Bulgarians, Greeks, Germans, Jewish, and others. However, the relative shares of these groups have differed dramatically over the centuries, and the Crimean Tatars generally view themselves as the peninsula's true indigenous population. They were the majority ethnic group until 1783, but declined to about 35 percent in 1897—still the region's largest ethnic population—and 26 percent in 1921. In 1944, the entire remaining Crimean Tatar population was deported to Central Asia by decree of Soviet leader Joseph Stalin, who suspected the group of collaborating with Germany's Nazi forces. In 1989, the Crimean Tatars were allowed to return, and by 1993 they composed about 11 percent of the population, which has since stabilized at around this level.

While fears that the region was a potential powder keg of ethnic strife have long existed, political forces on the peninsula never succeeded in fully mobilizing ethnic particularism to create strong divisions within the community, not least because the Russians and Ukrainians of Crimea consistently express similar political (including geopolitical) preferences. Therefore, the 2014 annexation by Russia had nothing to do with inter-ethnic tensions in the region and neither did the doctored 96.7 percent referendum vote in favor of reunification with Russia. Rather, many consider the "reunification" to be, instead, a landgrab by Russia that has created, rather than soothed, ethnic and cultural tensions in a region already struggling with socioeconomic and political rebuilding in the post-Soviet era.

become targets for illegal activities. Some work in slave-like conditions, whereas others engage in child prostitution or drug trading.

The post-Soviet period has also brought new kinds of residents to Russia's cities. For example, the influx of Asian migrants into Russia creates fairly large groups of Chinese, Indian, and Japanese in cities across Russia, especially in border cities of the Far East but also in Moscow, complete with their own banks and social organizations. The Soviet-era representative populations of Central Asians and Caucasian Muslims—praised by the multicultural rhetoric—have been swelled by economic migrants who do much of the unskilled labor in urban construction and seek to establish or expand their places of worship (Box 6.4). These migrants bring new worldviews, cuisines, religions, and lifestyles to Russia, which is increasingly curious about these people and their homelands, as noted in the increasing interest and ability to travel abroad.

Twenty-first-Century Environmental Concerns

First raised openly in the late 1980s, environmental concerns are now growing across Russia. Ever-increasing publications relate the harmful environmental legacy of Soviet urban development, suggesting an energized engagement with socio-environmental issues today. For example, one prominent issue plaguing urban areas is garbage. Soviet goods were often wrapped only in paper and string; larger amounts of waste in the post-Soviet era from imported packaged goods have not led to increased infrastructure to contain or remove garbage from urban centers. This results in garbage accumulating in public spaces (Figure 6.15), increasing environmentally hazardous conditions, and contention between citizens and city governments.

Urban environmental concerns, such as motor vehicle emissions and chemical poisoning (e.g., lead) from contaminated water supplies, remain largely unaddressed in many cities. Urban environmental problems are largely understudied and misunderstood in the post-Soviet era. Many people feel that solving environmental woes is the government's responsibility and that they individually cannot respond because socioeconomic and political issues are currently more pressing. In some cases, where urban and regional growth infringes upon land valued by environmentalists, conflict arises when citizens join forces to criticize or block urban growth. Protests around the development of a road

Box 6.4 Islam, Language, and Space in Moscow

Meagann Todd, University of Colorado—Boulder

Every Friday around 1 p.m., the Zamoskvorechnye neighborhood in central Moscow fills with men of differing nationalities—Russian, Kyrgyz, Tatar, Kazakh, Chechen, and more. Devout Islamic practitioners, the men are heading to Historical Mosque to attend Jum'uah, a weekly Muslim prayer service. Many wear tubeteikas, traditional caps made of velvet designed after the yurt. Others wear t-shirts or business suits. This scene emerges every Friday and becomes part of the everyday landscape of this historically Tatar neighborhood.

Best known for onion domes and Stalin's gothic spires, Moscow's architectural iconography does not represent on-the-ground realities. A center of capital and business, Moscow attracts skilled and unskilled immigrants. Officially its population is 11.5 million, but when undocumented workers are included, the population is estimated to be between 13 and 17 million. Many of these immigrants are from areas of the former Soviet Union with traditions of Islam—primarily from Uzbekistan and Kyrgyzstan. Without including the illegal migrant Muslim population, Moscow has an estimated two million Muslims, leading to overcrowding at the city's four mosques.

Tensions exist between the devout and city dwellers, and among Moscow's diverse Islamic communities. This is best illustrated by police presence at the Historic Mosque. The riot police serve many purposes: they protect mosque attendees from violence, check documents of illegal migrants, and help conduct traffic. However, many attendees consider their presence as a holdover of Soviet monitoring of religions, impeding their freedom to worship.

Because Moscow has been a site of terrorist attacks by the Caucasian Emirate, many locals express conflicting values with migrants and often refer to them with the derogatory term *chyorni*, or black. While these workers are valued for their cheap labor, they often do not speak Russian and have difficulties communicating with other Muscovites or even among themselves, as they come from numerous other places and religious or cultural backgrounds that were formerly included in the Soviet melting pot. Indeed, not all of them are Islamic. Many live in enclaves or on the outskirts of the city.

Despite recent unrest, Moscow is also a traditional homeland of some Muslim communities. For example, a Tatar ethnic group has lived and practiced Islam in Moscow for over 600 years. Members of this group dominate Islamic religious clergy in Moscow's mosques, where prayers are conducted in Arabic, Tatar, and Russian. One solution enacted by mosque leaders to ease tensions between Moscow's Islamic migrants and Muscovites is language instruction. For example, free Russian language classes for migrants are offered at mosques and many Russian Orthodox churches. Mosques also offer Tatar language lessons so that Tatar attendees can learn their ethnic language.

After state-sponsored atheism, Russia's Islamic community is growing. One effect is the growing number of Russians and Muslims studying Arabic and the increase in Arabic language offerings at universities and mosques. Moscow is a global city where many nationalities, cultures and religions converge, providing places in the city where worshippers grow in faith and connect in this megacity.

Figure 6.15 Increasing consumption and lagging public services are reflected in the garbage-strewn landscapes surrounding many Russian apartment buildings. *Source:* Photo by Sergei Domashenko.

through the Khimki Forest near Moscow is an example. Such instances, however, are not the norm in today's Russia, where most citizens are not well versed in opposing the government and instead see the state as responsible for protecting its people and the environment. Antidevelopment movements have succeeded in a few cases where they have been linked to protection of sites with a particular role in local urban culture, such as the successful effort to block the Gazprom/Okhta-Center skyscraper in St. Petersburg.

At the federal and international levels, however, environmental discussions since the mid-1990s have included the concept of sustainability. In creating new environmental policy directives, Russian policymakers invoke tsarist-era Russian and Western ideas about living in harmony with the biosphere as a foundation for creating sustainable development. Indeed, recent legislation that recognizes the need to address anthropogenic climate change provides hope that achieving economic growth while balancing social and environmental goals is a real concern among politicians and will increasingly become so among citizens in Russia today.

DISTINCTIVE CITIES

Moscow: Russia's Past Meets Russia's Future

Perhaps no city captures Russia's long history as vividly as Moscow. Modern Moscow is a chaotic blend of brash and unfettered capitalism, seen in its casino lights and chic boutiques; monolithic apartment blocks from the Soviet period (where most residents live);

new construction of glass skyscrapers and gated communities; and buildings renovated in the old Russian style. New Russian Orthodox churches join places of worship of many faiths (including other forms of Christianity and Islam) on the urban landscape. Museums and theaters have undergone much revitalization and reconstruction in the post-Soviet era, bringing new cultural capital to the city.

In Moscow, Russia's past lives alongside its future (Figure 6.16). Founded over 850 years ago in the declining years of Kievan Rus, the city grew rapidly in importance until Peter the Great moved the capital to St. Petersburg. The 1917 Russian Revolution returned the seat of power to Moscow. On December 25, 1991, resignation of the last Soviet leader, Mikhail Gorbachev, once again brought Moscow into the international limelight and launched massive socioeconomic and cultural changes.

After the collapse of the Soviet Union, Moscow exploded with the signs of capitalism. Foreign investment flooded the city, and new foreign and Russian capital created business centers, real estate companies, and a new retail sector. Once-empty avenues filled with cars—seemingly overnight. In the early 1990s, kiosks appeared everywhere stocked with an improbable mix of everything from candy bars to vodka, socks, and toys. Now, more permanent stores selling food, liquor, clothing, and toys, as well as every other possible consumer good, largely replace kiosks. Mega-scale shopping malls in the city outskirts also have replaced hastily built and remodeled stores. Historically the political, economic, media, educational, and cultural center of the country, restructuring of Moscow's new economy has added importance for the city's existing functions, especially as it rises to be the nation's banking and consumer capital.

Moscow is vastly richer than most other parts of Russia, partially as a result of inheriting immensely valuable real estate from the Soviet government and Communist Party. But the introduction of capitalism has still resulted in a highly fractured city. City residents experience life in vastly disparate ways. For the expatriate community, continuously growing as a result of foreign investment, Moscow is

Figure 6.16 Iconic Moscow River and Kremlin view at night. *Source:* Photo by Ian Helfant.

ranked as one of the most expensive cities in the world depending on choices for housing, dining, and other forms of consumption. For "New Russians," the city is a 24-hour shopping, dining, and business extravaganza.

For the vast majority of Muscovites, however, life in the city is, as they say, *normal'no* (normal). In addition to choking traffic jams, overcrowded metro commutes during morning and evening rush hours, and a decrease in environmental quality along major roadways due to the explosion of automobile ownership in the city, *normal'no* also includes a dizzying array of consumer services, fulfilling every desire, such as 24-hour gyms and sushi and coffee bars, businesses offering extended educational opportunities (e.g., computer science, foreign language training), rental agencies, and much more. In Moscow, it is no longer a question of seeking out something new to do or become; rather the challenge is to choose among the many options.

Along with growth in consumer options, the city is slowly expanding beyond its urban growth boundary, a road that rings the city 20 kilometers outside the city center. Delineated by urban planners in the Soviet era, land outside the boundary was meant to be a green zone for leisure purposes (e.g., hiking, picnicking, camping). With the development of megashopping centers and the single-family housing market, the green zone is slowly diminishing as the city creeps outwards radially, especially along major transportation routes.

Always a multiethnic city, Moscow is now increasingly so because of a large, constant flow of labor migrants from other parts of Russia and former Soviet republics. While migrants arrive from all regions of Russia—urban and rural alike—perceived increasing numbers of ethnic migrants from the south (e.g., Georgia, Chechnya, Uzbekistan) causes inter-ethnic conflict in Moscow and other major Russian cities (see Box 6.1). While some ethnic migrants are actually Russian citizens, some ethnically Slavic portions of the population feel that increased numbers of non-ethnically Russians threaten the economic and cultural futures of ethnic Russians in Russian cities. Largely, however, Moscow's long-standing tradition of being a multicultural locale remains, and the city is highly cosmopolitan and vibrant. In the post-Soviet era, Moscow now also ranks as a world city, the first in Russia.

St. Petersburg: Window on the West—Again?

Peter the Great founded St. Petersburg in 1703 as Russia's "window on the West," making it Russia's capital city; although thousands of Russians lost their lives while building the city, its ornate palaces and bridges over winding canals came to symbolize Russia's effort to join the European community. In 1917, the battleship *Aurora*, now a museum, fired shots at the tsars' Winter Palace, thus signaling the start of the Russian Revolution. When victorious Soviet leaders centralized economic and political resources in Moscow, again making it the capital, St. Petersburg was freed of its tsarist-era bureaucratic atmosphere and was renamed Leningrad in 1924. Aggressive neglect by Stalin ironically preserved the city's many parks and its architectural heritage.

While Leningrad did not grow as quickly as Moscow, the population doubled from 1917 to reach a high of about 5 million in 1989. In the post-Soviet period, its official population size declined to about 4.6 million in 2006 but the 2010 census reported it as 4.85 million. It remains Russia's second largest city. Well-known worldwide as the home of a unique cultural legacy, including the renowned

Hermitage Museum, Leningrad gained its prestige in Soviet times from its many prestigious universities and research institutions. The city's economy depended on educational and research activities, particularly defense-related industries in the Soviet military-industrial complex.

After 1991, the city's original name was restored and it moved to capitalize on its distinctive past. A deteriorating economic and financial situation caused by the near collapse of government support for education, culture, and especially defense signaled a need for significant restructuring. In the mid-1990s, as a result of leadership from Mayor Anatoly Sobchak, a former lawyer active in the political transition, the city embarked on a strategic planning process—the first city in Russia to do so. In contrast to top-down Soviet central planning, this plan aimed to be participatory, building on a partnership between local government, private businesses, and citizens to analyze possible scenarios for the city's development. Extensive discussions with private businesses, residents, and local organizations in St. Petersburg pinpointed the city's highly educated population; its ranking as the largest-capacity port in Russia; and its favorable location, not far from Finland and the rest of Europe, with excellent access to major railroads and highways. An initial concept document laid the foundation for the subsequent Master Plan and new legislation to standardize the construction and development process as well as to modernize city zoning. Under City Governor Valentina Matvienko (2003–2010), the city embarked on numerous high-profile projects to improve its appearance and infrastructure, including a ring road to divert industrial traffic away from the city center. It remains to be seen whether the government of St. Petersburg can improve living standards and promote long-term restructuring. Halting attempts to increase hotel capacity and urban amenities suggest that tourism has not grown as expected. Meanwhile, road congestion and air pollution have increased dramatically because per-capita car ownership has more than doubled recently; the city struggles to provide adequate traffic flow and parking throughout the historic area.

St. Petersburg touted its success at attracting investment, including plants for General Motors, Ford, Caterpillar, and Hewlett-Packard. However, due to lowered oil prices and other economic factors, the car market, for example, has contracted in Russia (according to *Fortune* magazine, car sales fell 10 percent in 2014 and are expected to fall further in 2015). General Motors closed a plant in early 2015 and Ford started to operate one in the oblast on reduced shifts. In March 2015, the CEO of Caterpillar, Inc., which has a plant in Tosno outside St. Petersburg, even suggested that regime change will be needed to secure increased foreign investment. As long as the Russian economy relies primarily on its petroleum exports to generate capital, a slowdown in foreign investment will probably slow related developments in Petersburg.

The protection of St. Petersburg's architectural fabric has also weakened since the late 2000s. Although the city's culture of participatory urban development has produced vigorous public outcry over aggressive land-use choices in the historic center, protests against these have increasingly failed, aside from certain highly visible examples. For example, from 2006 to 2010, a struggle took place over plans by the state-owned natural gas corporation, Gazprom, to build a new skyscraper in St. Petersburg called Okhta-Center. Fueled by residents' pride in the traditional low-rise city skyline punctuated by cathedral spires, resistance to construction of a central skyscraper became an international cause célèbre. Since the entire historic city center is on the list of World Heritage sites, UNESCO asked the city of St. Petersburg to halt the project in order to study potential impact on the city's historical monuments. Protest against the plan was a central theme in a series of unprecedented street demonstrations in 2007 and 2008; several prominent cultural figures in St. Petersburg and eventually Moscow joined the opposition. While plans for Okhta-Center were eventually scrapped, other less high-profile buildings are evidence of the power that business interests wield in the city administration and have quietly shaped a new skyline. These include the Stock Exchange building on Vasilievsky Island, erected in 2007, and the new Stockmann shopping center behind historic Ploshchad Vosstaniya.

Yuzhno-Sakhalinsk: The International Power of Oil

Yuzhno-Sakhalinsk, located on Sakhalin Island in Russia's Far East, is an oil boomtown fueled by multinational investment. This city, long a small urban hub for the military and natural resource exports (coal, oil, fish, timber), is ushering in the era of globalization in the Russian Far East because of its proximity to offshore oil and gas reserves in the Sea of Okhotsk. The oblast is a leading destination for foreign direct investment—second only to the city of Moscow—indicating the importance of natural resources and regional centers to Russia as a whole. Located 4,000 miles (6,500 km) from Moscow and only 110 miles (175 km) from Japan, Yuzhno-Sakhalinsk is sited on a tsarist-era settlement for exiled prisoners (Vladimirovka) that later became a Japanese village (Toyohara) during Japanese rule of southern Sakhalin Island from 1905–1945. After World War II, the USSR reclaimed the entire island, and Yuzhno-Sakhalinsk was created as the new oblast capital for Sakhalin and the Kuril Islands.

Yuzhno-Sakhalinsk embodies the multiethnic character of Soviet-era cities. Home to about 175,000 people, the urban population is comprised of ethnic Russians, Ukrainians, and other Slavs; native peoples of Sakhalin (Nivkh, Evenk, Orok); and ethnic Koreans. Koreans, the last "newcomers," were brought to Sakhalin during World War II to work the coal mines. Expatriates associated with the hydrocarbon industry have arrived since 1995, and it is common to find workers from Europe, North America, and Russia's neighboring states residing semi-permanently in city hotels. The city's multiethnic history is noted in the mélange of architectural styles in the city. The few remaining traditional Japanese structures stand next to tsarist-era frontier houses (turn-of-the-twentieth-century wooden multifamily dwellings with rudimentary utilities), Soviet-era five-story apartment buildings, post-Soviet suburban kottedgi on the city's outskirts, and gleaming Western-style offices and houses occupied by expatriate executives. Previously a small urban center connected to the MIC, the large and increasing presence of foreigners in Yuzhno-Sakhalinsk is a big change for this formerly closed city. Offshore hydrocarbon sites lie further north than Yuzhno-Sakhalinsk, but smaller settlements there lack the infrastructure and political capital necessary to accommodate international companies. Hence, as the oblast center, Yuzhno-Sakhalinsk has become the bustling urban hub for the hydrocarbon industry. New service industries associated with hydrocarbons are changing the labor market for the city, but not for the entire island. Current spatial patterns of economic growth mimic the Soviet urban settlement pattern where large urban centers were favored. The economic boom occurring in this city provides hope for future regional economic growth and is a refreshing change from the Soviet period when the government had to send people to work on Sakhalin using the propiska system. Today, it is a destination city both for younger generations from the Russian Far East and for international migrants associated with the hydrocarbon industry, each seeking promising jobs.

While Yuzhno-Sakhalinsk enjoys international investment and a relatively high standard of living, many residents wonder when they will see benefits from oil extraction promised to them by the regional government and multinational companies. Many fear that the promises made in the mid-1990s to develop the island will remain unfulfilled. For example, infrastructure projects (e.g., transportation, education, health facilities.) promised in return for allowing hydrocarbon extraction have not actualized, and residents remain saddled with decrepit Soviet-era dwellings, transportation systems, and services. Gated and gleaming buildings for expatriate workers taunt neighboring buildings lacking decently operating heat, hot water, or electricity. This disparity raises questions about the strength of Yuzhno-Sakhalinsk as an emerging hub in Russia's globalizing economy, as well as the preparedness of urban and regional governance to secure even economic growth for all citizens in the post-Soviet era. Many residents of Yuzhno-Sakhalinsk struggle to understand their rapidly changing socioeconomic and cultural place in the post-Soviet era.

Norilsk: The Legacy of Heavy Industry

Planned, developed, and federally subsidized to house over 100,000 people above the Arctic Circle, Norilsk is the northernmost large city in the world (population 175,300; 2010 Census). Temperatures in the city can reach –58°C, and snow cover lasts 70 percent of the year. Gulag laborers (prisoners in the Soviet penal system of forced labor camps) worked from the 1920s until the mid-1950s to construct many of the city's buildings, mines, and smelting facilities. Norilsk remains a closed city, ostensibly maintaining security of nationally valued metallurgic operations by restricting travel and residency of nonresident Russians and foreigners. It is the world's largest producer of nickel and palladium and one of the world's largest producers of platinum, rhodium, copper, and cobalt.

The city was carved onto the tundra homeland of the indigenous peoples of the Taimyr Autonomous Okrug, bringing modernization and traditionalism into close spatial contact; it is not surprising to see native Evenk driving caribou-drawn sleighs through the city, vividly juxtaposed with the city's modern transportation and high-rise buildings. As a Soviet creation, Norilsk's urban landscape consists of high-rise, concrete-panel *microrayons* intermixed with and surrounded by mining and metallurgical industrial facilities (see Figure 6.7). At 69° N, the city lies in the continuous permafrost (soil at or below the freezing point of water) zone. As a result of the urban heat island effect and anthropogenic climate change, the permafrost warms under the city's foundations, compromising the city's transportation networks and building foundations.

Although Norilsk's history as an "urban gulag" is legendary, the longest lasting legacy of

Norilsk may be as Russia's most polluted city. One percent of global emissions of sulfur dioxide are estimated to come from Norilsk. Air, soil, and water pollution degrade the physical environment and health of all living inhabitants. For example, in the process of ore smelting, sulfur dioxide (SO_2) is emitted. Acid rain forms and precipitates back on the city, releasing contaminants (acid, heavy metals) into the soil and water supply. The legacy of environmental damage to vegetation and waters is noticeable at a regional scale (Figure 6.17) but is highlighted in the city, where the surfaces of buildings and monuments decay where

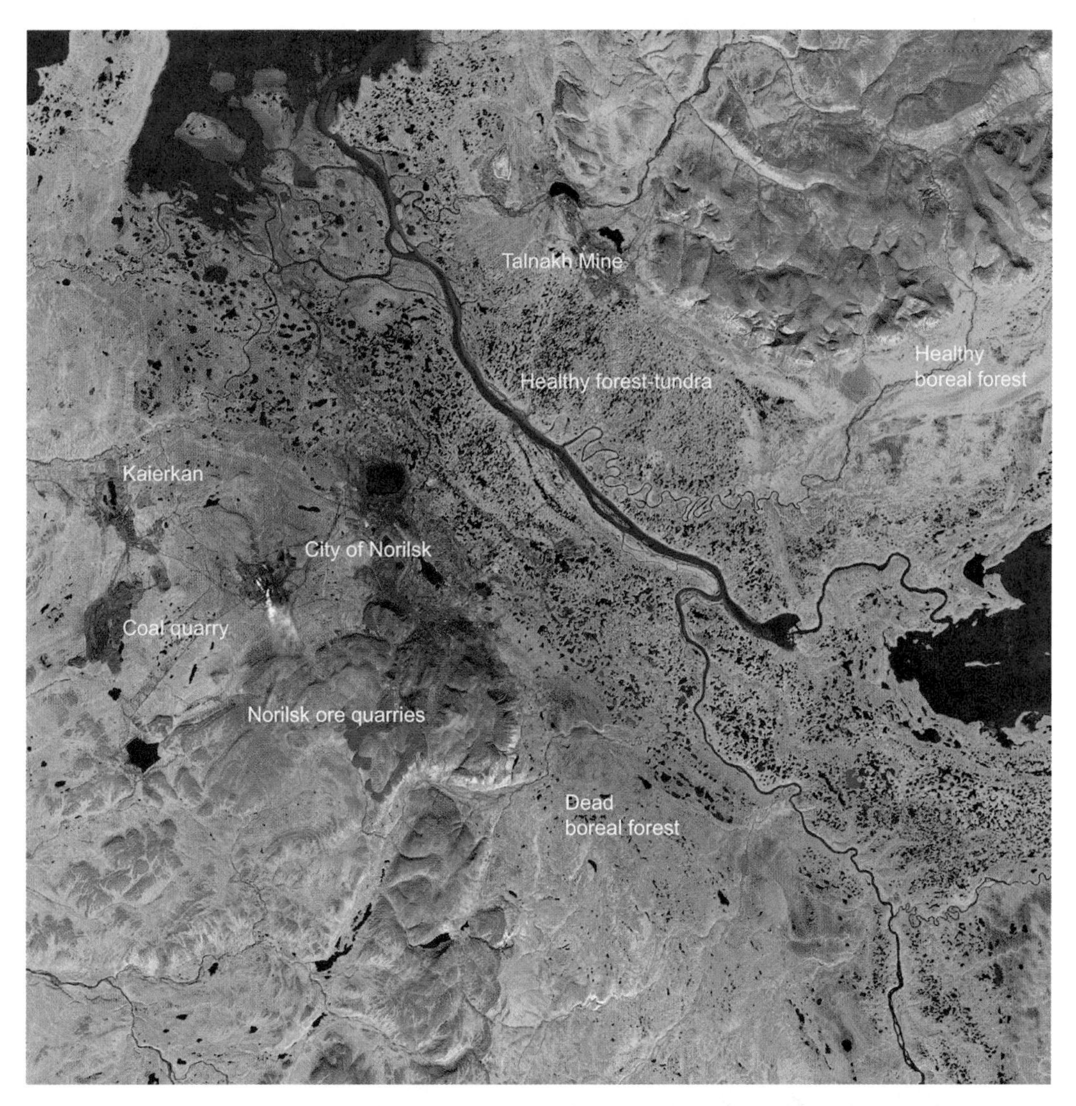

Figure 6.17 False-color image of Norilsk. Shades of pink and purple indicate bare ground (e.g., rock formations, cities, quarries) where vegetation is damaged from heavy pollution. Brilliant greens show mostly healthy tundra-boreal forest. South and southwest of the city are moderately to severely damaged ecosystems, and ecosystems northeast of the river and away from the city and industrial centers are healthier. *Source:* NASA.

acid rain eats away at the stone or cement. Some businesses propose mining Norilsk's urban soils, because the proportion of metals in them is economically viable. As in many industrial Russian cities, illnesses (e.g., lung cancer, asthma,) affect the young and old alike. Despite the risks, many residents remain because of the importance of resource extraction and industry in Russia, noted in high wages (up to four times the national average) and because of the strong historical roots or social networks linking people to place. Once state owned but now in private hands, the smelting operations of Norilsk Nickel, the primary regional employer, drive the continued existence of the city. Ongoing urban pollution problems remain largely unchecked by Norilsk Nickel's corporate headquarters in Moscow (a remnant of command economic planning).

The combination of being situated physically in the environmentally sensitive far north and economically as crucial to Russia's resource industry makes addressing Norilsk's environmental issues both timely and necessary. As awareness of Norilsk's power to alter regional and possible global environments (through global climate change) becomes increasingly known in Russia and abroad, industrial managers and the government are beginning to implement environmental management systems, providing environmental and social impact statements of urban-industrial activity. Some industry leaders also desire to streamline Russian environmental legislation to meet international environmental safety standards, hopefully reducing regional and global pollution derived from Norilsk. In this way, Norilsk is poised to become a leader in addressing urban environmental concerns of the north, especially regarding resource extraction, industrial pollution, and local and indigenous concerns.

Kazan: Volga Port in Tatarstan

Kazan, the capital of the Republic of Tatarstan, is the seventh largest city in Russia (1.1 million) and one of the largest outside of European Russia. Kazan is distinctive for many reasons, not least of which is the political struggle of the 1990s that resulted in the establishment of the republic as one of the leaders of the independence movement within the Russian Federation. Two recent events have drawn worldwide attention to this city along the Volga River. In 2000, UNESCO included the historical city center on its World Heritage List, which denotes places of universal value to the world community. And in 2005, the city celebrated its 1,000th anniversary.

While most cities in Russia are largely multiethnic, few have as large and powerful a non-Russian population as this city and republic. Tatars, whose origins are in the Central Asian steppes, make up the largest ethnic group in the republic. Russian settlement and domination of the city, however, goes back to the sixteenth century when Ivan the Terrible invaded the region. Even today, despite a plurality of the Tatar population in the republic, the city has a slight majority of Russians—about 50 percent relative to about 42 percent Tatar. Other non-Russian populations from the Volga region also live in the city including the Chuvash, a Turkic group, and the Maris. Migration in the 1990s brought new ethnic groups to the city, namely those from the Caucasus and Central Asia.

The Tatar population is traditionally Islamic. Kazan's history as a prominent Muslim city extends back to the fourteenth century. Today, the city is home to a new school training Russians in the Islamic religion, the Islamic University. Since the early 1990s, at least 40 new mosques have been built in the city. The city government helped construct a new mosque on the grounds of the historic kremlin. The political and social significance of both the site and the leading role of the city government should be recognized as a symbol of cultural as well as political independence from Russia. The city's cultural independence from Russia is also seen in the historic forms of architecture that make up the urban built environment. Buildings in the city combine many architectural styles, ranging from Baroque to Moorish. Bas-reliefs created by traditional Tatar stone workers embellish buildings in the city, and minarets dot the skyline.

Kazan and the region have major economic significance to the Russian economy. The city is a major port on the Volga River, the main water route through European Russia. The city and the region have been a transportation gateway for centuries. Currently, the European Union is helping to modernize its port facilities, and in 2005, it became one of only a few cities in Russia to receive a direct loan from the World Bank. The city's economy is also strongly tied to the production of transportation equipment. It produces military transport equipment, including helicopters, and is home to KamAZ, still a gigantic automotive production firm.

Vladivostok: Russia's Pacific Capital?

Home to the Russian Navy's Pacific Fleet, Vladivostok from 1958 to 1991 was a closed city where even Soviet citizens needed permission to enter. Before this time, Vladivostok had been an international city. Valuable for its port facilities and proximity to Asian markets, the city drew a diverse international population, including Chinese, Koreans, Japanese, and Americans. Since 1991, these populations are back, as the city is now a gateway into Russia, largely for Asian tourists and businesses. Many Pacific Rim nations operate consulates in Vladivostok.

Founded in 1860, Vladivostok is the capital of Primorsky Krai and is the Russian Far East's largest city, with nearly 600,000 residents. It is the largest Russian port on the Pacific Ocean and historically has been an important regional industrial center for shipping and fishing. Located 3,800 miles (6,430 km) from Moscow, it is the eastern terminus of the Trans-Siberian Railroad. The great distance to European Russia feeds the imagination of Vladivostok as a gateway to exotic, Asian Russia. Historically, it led Vladivostok residents to be self-reliant and to expect little from Moscow; indeed, the first Soviet leader to visit Vladivostok was Nikita Khrushchev in 1954. Also headquarters of the Far Eastern Division of the Russian Academy of Sciences, the city hosts many academic and research institutions. These factors have fueled hopes that Vladivostok could become an urban economic hub for a range of businesses. For example, entrepreneurs dream of Vladivostok-based regional ecotourism, and port facilities are a strong asset in rebuilding a strong import-export city on the Pacific. The Asia-Pacific Economic Cooperation Summit was held here in 2012, for which the federal government provided funds for urban improvements. Stunning views of the Golden Horn Bay from the hilly city also provide incentive for restoration of the historic center and its tsarist-era buildings and monuments, many of which could be refurbished to rival those in St. Petersburg. As the intellectual capital of the Russian Far East, Vladivostok has much to offer post-Soviet Russia, despite the need for crackdowns on polluters and numerous illegal activities.

Because much international trade with Asia is funneled through this Russian city, organized crime in Vladivostok and Primorsky Krai since 1991 purportedly involves not only Russian mafia but numerous mafia-like groups from former Soviet republics in Central Asia. It is unfortunate that, since the early post-Soviet years, many elected government officials work with, instead of against, organized crime. Illegal trade includes marine resources from China, cars from Japan, heroin from Central Asia, and timber exports from Russia to China and the Republic of Korea. The rise of informal and mafia-driven economies has hindered the growth of legitimate, tax-paying businesses, which slows overall development. The city is also an aviation gateway to Asian cities, since its airport is one of the few that handle international flights. When Russians from across the Far Eastern region return from destinations across Asia via Vladivostok, they often bring commodities to sell in Russia.

In addition to economic difficulties, Vladivostok suffers from severe air, water, and soil pollution. Ecologists consider much of the city to be hazardously polluted by heavy metals and industrial (cadmium, mercury, arsenic) and agricultural (nitrates, phosphates) waste. Despite the city's location adjacent to the Pacific Ocean, local wind and water circulation patterns in the Amursky Bay do not remove pollutants from densely populated or intra-urban-industrial areas. Unchecked, pollutants have built up over time in the soil and nearshore environments, and their detrimental effects on ecosystem and human health are only slowly being recognized.

Despite its distance from European Russia and from the federal center, Vladivostok (and other Siberian and Far Eastern cities) remains in the cultural, political, and economic orbit of Moscow. While the rich economies of Japan and the Republic of Korea and the rapidly developing economy of China are far closer neighbors than Western Russia, Vladivostok is a city settled by ethnic Slavs (largely Russian and Ukrainian), who remain the largest percent of the population today. Indeed, many Ukrainian refugees moved to Vladivostok in 2014 because of unrest in Crimea, and long-term Ukrainian heritage in the region creates mixed emotions about Russia's involvement in Ukraine today. Generally, Russians are suspicious of the Chinese and their interests in Russia's vast raw and land resources. The facts that the Russia-Chinese border is largely unmanned and many Chinese enter Russia illegally to reap resource rewards are not unnoticed by Russians; indeed, many in this region deeply distrust the Chinese. Thus, cultural allegiance to European Russia is stronger than new Asian economic ties. The loyalty of Vladivostok's political elites to the center is being rewarded by massive capital investment in Vladivostok by the federal government. Recent construction in this city, including new infrastructure—sewage treatment plants, underground freshwater reservoirs, highways, bridges, dams, and airport terminals—are positioning this city to become a center of international cooperation in the Asian Pacific region (Figure 6.18). This can only aid Russia in stabilizing the population and growing the economy in the Far Eastern region.

PROSPECTS FOR THE FUTURE

Cities in Russia today are the products of tsarist, Soviet, and post-Soviet Russian societies, as well as of the many different ethnic groups and cultures that have inhabited the region for at least 1,000 years. The blending of this diverse set of cultures and histories results in cities with varied built and social landscapes.

Figure 6.18 New urban infrastructure (bridges, roads) in Vladivostok, built for the 2012 Asia-Pacific Economic Cooperation Summit, revitalizes this regional capital and port city in the Far East. *Source:* Photo by Sergei Domashenko.

The natural environment, however, has always wielded an important influence over the location of urban settlements, irrespective of the time period or the dominant ethnic group. The harsh Siberian landscape originally posed barriers to Russian expansion and settlement, but widespread urban settlement of Siberia became a great accomplishment of Soviet central planners—in spite of environmental degradation wrought by settlement and the immense social and cultural costs of stranding people in isolated and inhospitable places along with the immeasurable financial implications. Cities in the post-Soviet period continue to struggle with the consequences of harnessing nature.

The federal and municipal governments are attempting to integrate economically, politically, and geographically disparate cities into a larger geopolitical and economic framework, with implications for Russian transportation and communications systems. For example, should cities closer to Tokyo or Beijing than to Moscow rely primarily upon trade within Russia for economic direction, or should they look to Asia for new markets and influence? What will happen to cities constructed within the Soviet system but now functioning under another? How will existing structures such as factories, housing, roads, school, and other buildings be adapted for new uses in a market economy? How (and

Figure 6.19 Suburban development on the fringes of compact Soviet-era cities, such as Balakovo, brings socioeconomic division and expansion into agricultural zones to previously mixed and compact urban settings across Russia. *Source:* Photo by Jessica Graybill.

by whom) should pollution in urban environments be addressed? How sustainable are cities in extreme environments, such as the Arctic and Siberia? How can cities quite distant from one another remain connected both economically and politically? Perhaps most importantly, what will happen to the people who live and work in Russian cities? Should the Russian government promote integration into the European Union? How far down the path of destroying the Soviet housing system should Russian cities go? How best should Russian cities manage land-use change, especially in the face of increasing suburbanization (Figure 6.19)? Will the nascent increase in birth rates stem the overall population decline in Russia? If not, what will the national declining population mean for the regeneration of many Russian cities? Answers to these and many other questions confront the people of Russia today as they continue the process of reinventing their cities.

SUGGESTED READINGS

Axenov, K, I. Brade, and E. Bondarchuk. 2006. *The Transformation of Urban Space in Post-Soviet Russia.* London, New York: Routledge. Focuses on the transition from socialism and communism to democracy and capitalism in urban areas of post-Soviet Russia and Eastern Europe.

Barnes, I. 2015. *Restless Empire: A Historical Atlas of Russia.* Cambridge, MA: Belknap Press. A graphical explanation of the cultural, political, economic, and military developments of Russia's past including up-to-date coverage of current land claims in Ukraine.

Bassin, M., and C. Kelly, eds. 2012. *Soviet and Post-Soviet Identities.* Cambridge: Cambridge University Press. Provides an overview of issues of identity and culture in post-Soviet Russia.

Blinnikov, M. 2010. *A Geography of Russia and Its Neighbors.* New York: The Guilford Press. Thorough coverage of Russia and cultural, economic, and geographic relations to former Soviet states.

Chernetsky, V. 2007. *Mapping Postcommunist Cultures: Russia and Ukraine in the Context of Globalization.* Montreal: McGill-Queen's University Press. Focuses on post-Soviet cultural developments, puts them in a global context, and suggests that Russia and Ukraine form the basis of post-Soviet culture.

Clowes, E. 2011. *Russia on the Edge: Imagined Geographies and Post-Soviet Identity.* Ithaca: Cornell University. A discussion of what it means to be Russian today using examples from popular culture and literature.

Figes, O. *Natasha's Dance: A Cultural History of Russia*. New York: Picador Press, 2002. A survey of European Russian culture from the beginning of the tsarist era through the Soviet era, especially focusing on the roles of multiculturalism, Europe, peasant society, and expansionism in the creation of the arts and lives of citizens in the tsarist and Soviet empires.

Hill, F., and C. G. Gaddy. 2003. *The Siberian Curse: How Communist Planners Left Russia Out in the Cold*. Washington, DC: Brookings Institution Press. Traces the failed attempt to establish an industrial base in Siberia, and argues for abandoning the eastern territories because of their economic instability.

Hiro, D. 2009 *Inside Central Asia: A Political and Cultural History of Uzbekistan, Turkmenistan, Kazakhstan, Kyrgyzstan, Tajikistan, Turkey, and Iran*. New York: Overlook Duckworth.

Oldfield, J., and Denis Shaw. 2015. *The Development of Russian Environmental Thought: Scientific and Geographical Perspectives on the Natural Environment*. London: Routledge. A comprehensive overview of the rich thinking about environmental issues that has grown up in Russia since the nineteenth century.

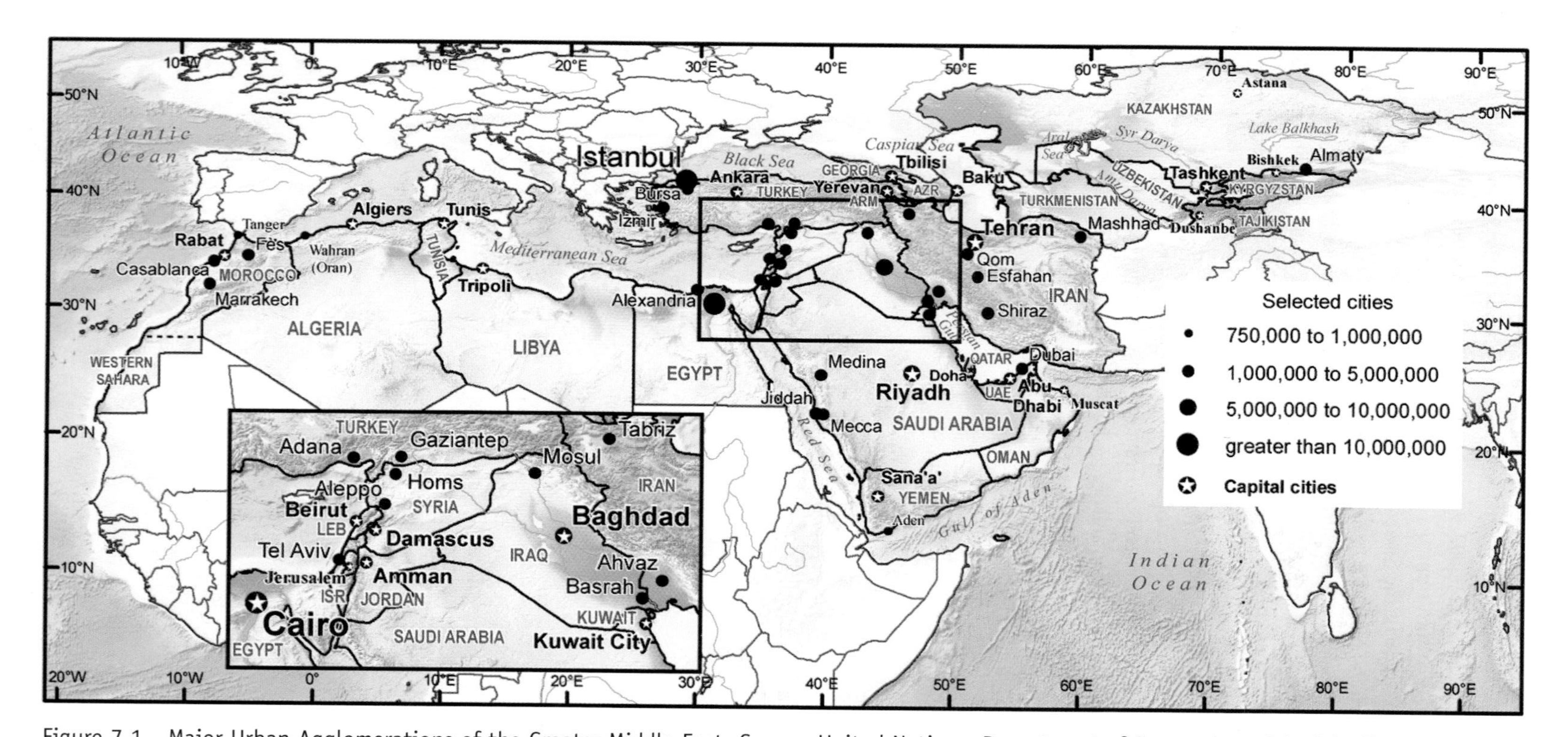

Figure 7.1 Major Urban Agglomerations of the Greater Middle East. *Source:* United Nations, Department of Economic and Social Affairs, Population Division (2014), World Urbanization Prospects: 2014 Revision, http://esa.un.org/unpd/wup/.

7

Cities of the Greater Middle East

ZIA SALIM, DONALD J. ZEIGLER, AND AMAL K. ALI

KEY URBAN FACTS

Total Population	569 million
Percent Urban Population	62%
Total Urban Population	354 million
Most Urbanized Countries	Qatar (99%)
	Kuwait (98%)
	Israel (92%)
Least Urbanized Countries	Tajikistan (27%)
	Yemen (34%)
	Uzbekistan (36%)
Number of Megacities	2 cities
Number of Cities of More Than 1 Million	50 cities
Three Largest Cities	Cairo (18 m), Istanbul (14 m), Teheran (8 m)
Number of World Cities (Highest Ranking)	8 (Dubai, Istanbul, Cairo, Tel Aviv, Beirut)

KEY CHAPTER THEMES

1. Urban landscapes of the Greater Middle East have been shaped by the natural environment and religion (particularly Islam, but also Judaism and Christianity).
2. The world's first cities grew up in the Fertile Crescent, along the Nile, and on the Anatolian Plateau; the locations of all cities have been influenced by the availability of fresh water.
3. Traditional city cores are, or were, walled and dominated by a citadel or kasbah.
4. Urban economic geography has traditionally been shaped by the commerce that coursed across the region, a result of its relative location at a tri-continental junction.
5. Some states have a primate city, some have two or more competing large cities, and a few have fully developed urban hierarchies.
6. The "urban triangle" that defines the region's core has a foothold on all three continents and in three different culture realms (Arab, Turkic, Persian).

7. During the twentieth century, oil and gas revenues have turned some of the least urbanized countries into some of the most urbanized.
8. The urban geography of the oil-rich states has been transformed by petrodollars and millions of "guest workers," particularly from South and Southeast Asia.
9. The domino effects of the revolutions in Egypt and Tunisia stimulated democratic uprising in Libya, Syria, and Yemen, with both positive and negative results.
10. Major urban problems range from assuring supplies of fresh water to coping with rapid population growth and tending to the preservation of heritage resources.

Cities in the Middle East are uniquely positioned in time and space. The foundations of urbanization as we know it can be traced to this region's ancient cities: the oldest continuously inhabited settlements and two of the world's five urban hearths lie in this region. The dense layering of history, seen in other cities around the world, is especially prominent here. In the Middle East, great cities have risen and fallen over the centuries, and they have long served as connection points between the region and the rest of the world. From the cradles of civilization, to dusty cities on the Silk Road, to Roman cities in present-day Turkey, Lebanon, and Jordan, to the megacities of Cairo, Tehran, and Istanbul, to the spectacular cities of the Gulf, the local and the global come together in fascinating ways on the Middle East's urban landscapes (Figure 7.1).

As a vernacular term, the exact geographical delineation of the "Middle East" is difficult. In this chapter, the term *Middle East* refers to the "Greater Middle East," a crescent stretching from Morocco, eastward across North Africa through the lands of southwest Asia, to the steppes of Kazakhstan. Although there is great complexity and diversity within the region, three shared geographic characteristics and history connect this great swath of territory (Figure 7.2). First, the Middle East's physical geography is predominantly arid and semi-arid, though river systems and oases mitigate the region's dryness. Second, the region's cultural geography is marked by the shared history experienced through successive Islamic empires and their interaction with the wider world. Third, in terms of relative location, the Middle East is literally in the "middle"—the middle of the Eastern Hemisphere, a land bridge between Europe, eastern Asia, and Africa.

Although the expanse described in this chapter includes a variety of climates, some characteristics pervade the Middle East's physical geography as a whole: dry and seasonally dry, arid and semi-arid, desert and steppe. Available water comes from winter showers, orographic precipitation, exotic streams, rivers, natural springs, and shallow aquifers. Most of the region suffers a freshwater deficiency; in arid areas, the location of water has historically determined the location and evolution of towns and cities. Further, urban centers had to overcome a set of natural obstacles presented by the dry environment: scorching sun, high daytime temperatures, desert winds, dusty air, and scarce water. The same environment—the desert—offered early cities a set of natural frontiers while serving as a buffer that provided protection from potential invaders. Generally, physical geography (arid climate and rugged topography) has affected the amount of space available for human settlement and habitation, contributing to high population densities in urban areas.

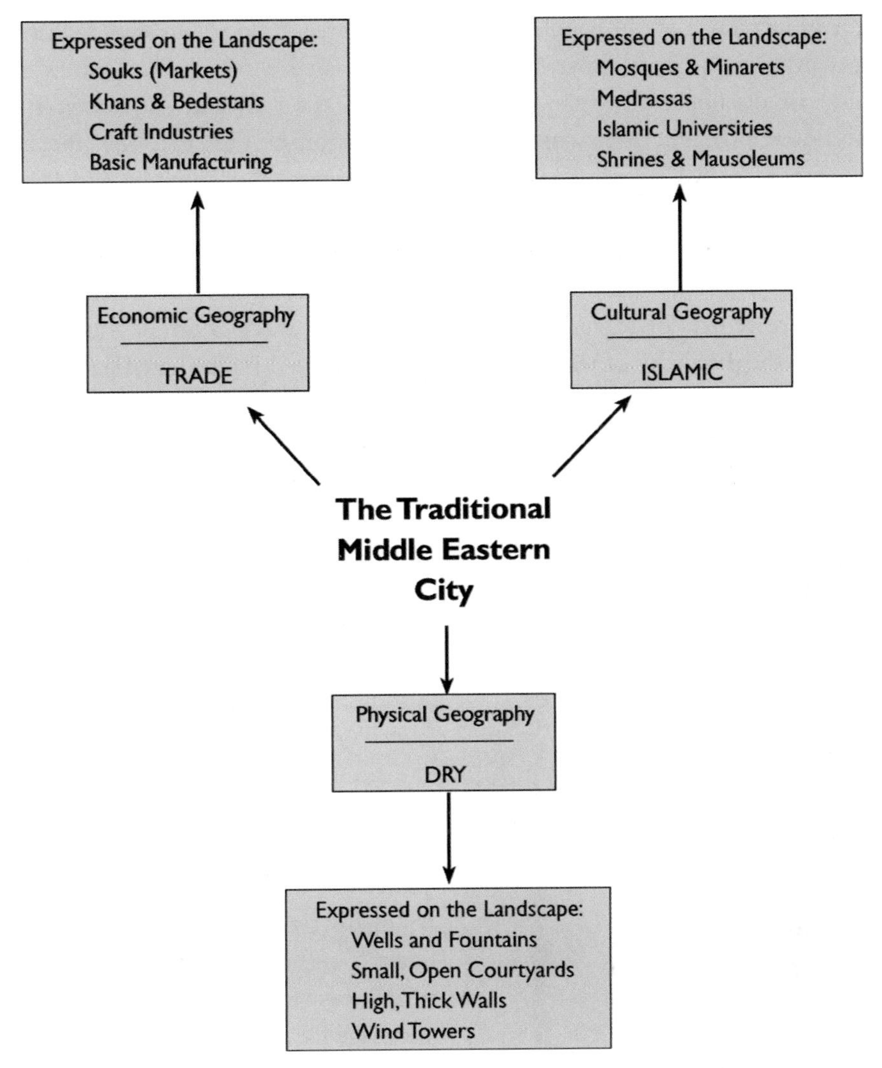

Figure 7.2 The Traditional Middle Eastern City. *Source:* D. J. Zeigler.

Contemporary Middle Eastern cities also share a common element of cultural geography, that of religion. This region is the birthplace of the monotheistic faiths we know as Judaism, Christianity, and Islam. All three have left their enduring stamp on the region; but Islam is most widely associated with the Middle East, having begun there in the early seventh century. Within the region, cultural geography varies from place to place—Arab, Persian, Turk, Kurd—but the cultural matrix is webbed together by past and present impacts of Islam and the religious, social, and political systems that it created. Christians and Jews (and some smaller groups like the Baha'i, Druze, and Zoroastrians) live within the Islamic matrix; but they are the exception, not the rule. Only in a few areas were urban landscapes historically punctuated by anything but the minarets of mosques (Figure 7.3). The Middle Eastern city has always been a center of spiritual and intellectual life and a generator of new ideas about people's relationships with God and each other. Indeed, Islam itself first developed in urban centers (Mecca and Medina); and Damascus, Baghdad, and Cairo have all been centers of religious authority.

The relative location of the Middle East has given its cities a third set of common characteristics: they are centers of trade and commerce. Prior to the discovery of water routes around Africa and around the world in the fifteenth century, trade between the great civilizations had to pass through the dryland crescent separating Europe from eastern Asia and sub-Saharan Africa. Trade among these three regions—by geographical necessity—passed through the Middle East. The configuration of land and water—the interpenetration of the seas (e.g., the Red Sea), land bridges (e.g., Anatolia and Persia), and peninsulas (e.g., Arabia)—provided a multitude of routes through this dry-world barrier. Trade in food and fabrics, gold and copper, spices and perfumes, frankincense and myrrh, helped build the cities of the Middle East. In fact, new ideas about how to create and expand wealth, perhaps even capitalism itself, were born in these cities. Their marketplaces—called *bazaars* in the Persian language, *pazars* in Turkish, *souks* in Arabic, and *shuks* in Hebrew—are among the oldest in the world (Figure 7.4).

Figure 7.3 Rising above every Middle Eastern city are the minarets of mosques. One of the most famous is the Koutoubia, the largest mosque in Marrakech. By tradition, the muezzin issues the call to prayers five times a day from the minaret. *Source:* Photo by D. J. Zeigler.

In other words, the evolution of cities in the Middle East has centered around water, the house of worship, and the marketplace. Until the late twentieth century and the genesis of oil economies, relative city size was, in fact, proportional to the above factors: larger cities evolved in direct proportion to the availability of freshwater, the abundance of "spiritual capital," and the bounty of trade.

It is easy to see how physical geography set the stage for urban development by affording defensibility, facilitating movement, or providing resources. For example, settlements on hills and peninsulas were easily defended, and riverbank settlements afforded safe crossing of rivers. By paying attention to the locations of hills, springs, oases, harbors, and headlands, one can trace patterns of urban location. A peninsula, for instance, set the stage for the evolution of Istanbul, an oasis for Damascus, a hilltop for Aleppo, a spring for Tehran, and a harbor for Beirut.

While site, the local physical characteristics of place, may be responsible for the founding of a city, situation or location relative to wider

Figure 7.4 The traditional markets of Marrakech, Morocco, are some of the most well-known in the world. In Arabic-speaking countries they are known as souks or suqs. *Source:* Photo by D. J. Zeigler.

contexts (such as routes of commerce or seats of power) more often determines whether a settlement prospers or withers and dies. Some cities have the potential to be trading hubs or imperial capitals, while others do not. Furthermore, relative location is dynamic, as a good location in one era may be a bad location in another. The seat of the Islamic caliphate moved over time from Medina to Damascus to Baghdad. The Suez Canal's opening in 1869 drew trade away from the cities of the Fertile Crescent. During Lebanon's civil war (1975–1990), much of Beirut's economic activity, especially insurance and banking, relocated to Manama in Bahrain, and then to Dubai in the United Arab Emirates. All across the Middle East, "dead cities" and archaeological *tells* (hills created as one city was built on top of its predecessors) illustrate how changing relative location influences urban geography.

A final point to make is that the "Middle East" moniker subsumes great cultural diversity. An array of national, ethnic, religious, and linguistic differences marks the social landscape, and these differences often set the stage for contemporary events that hit the news. For example, understanding the size, spatial distribution, and relationships of ethnic groups in Iraq explains why, after the U.S.-led invasion in 2003 and the deposition of Saddam Hussein, the country has been unable to find national unity. Similarly, some aspects of current political conflicts in Yemen, Bahrain, Lebanon, and Syria can be traced to sectarian divides between Sunni and Shiite Muslims. Finally, the Israeli-Palestinian conflict can be understood as an inter-group conflict, although it is more of a conflict over territory and political power than it is a faith-based conflict between adherents of two different religious traditions.

FOUNDATIONS OF THE URBAN SYSTEM

What is past is prologue: Understanding the past helps to explain how present-day urban patterns came to be. Perhaps no other geographical region in the world presents such a long-standing connection with urbanism as the region commonly referred to as the Middle East. Solidly anchored in the tri-continental junction known as the Fertile Crescent are the roots of the Western city. Our word *urban* still carries the name of the world's first truly urban places, Ur and Uruk, in southern Mesopotamia. These early settlements, dating from at least five millennia BCE, offered protection, security, and the ability to trade and control resources. Ideas about urban development and planning originated here and spread as byproducts of commerce and conquest, along with other innovations, such as writing and record-keeping systems. The second oldest urban hearth, the Nile River Valley, also lies in this region.

As small settlements in the Fertile Crescent and the Nile Valley grew into city-states that served production, worship, defense, and trade functions, their growth was stimulated by social stratification and the production, storage, and distribution of agricultural surpluses. Cities were distinguished from the countryside by offering the best that life had to offer, at least for powerful elites and their clients. The oldest cities in the world, though most often classified as proto-urban, were born during the Neolithic period. They are all associated with the beginnings of agriculture. Their locations form a triangle with one vertex in Iraq, one in Palestine, and one in Turkey. In lower Mesopotamia (the "land between the rivers") were the cities of Ur, Uruk, Eridu, Kish, and others. As truly urban places, they emerged in the fourth millennium BCE. They were the largest cities in the world until the rise of Babylon. Only archaeological tells remain, but these forerunners of the modern Middle Eastern city set off a chain reaction in urban innovation that continues to this day. Earlier than this, however, in Palestine, on the other side of the Fertile Crescent, ancient Jericho (now in the West Bank) boasted a wall and watchtower as early as nine millennia BCE. Deep within an arid rift valley, this "city of palms" is located next to a gushing spring, and not far from the *wadi* (riverbed), which brings runoff from the Judean Hills into the sun-drenched Jordan River valley. On the Anatolian plateau, just to the north of the Fertile Crescent, stands the recently discovered prototypical city, Çatal Höyük, which dates to about 6500 BCE. It was located on a small river in the Konya Plain, today a rather desolate area in Turkey. It had an estimated 50,000 inhabitants, and it was probably one of the largest and most sophisticated settlements in the world in its day because of the successful domestication of wheat and other staples. Just as its size proved the viability of largeness, its form proved to be a pacesetter in the development of urban landscapes. Çatal Höyük, Jericho, and Ur all illustrate the principle that civilization and urbanization evolve hand in hand.

The Iranian plateau and the Mediterranean basin gave birth to indigenous empires that were some of the world's earliest. The Sumerians developed advanced irrigation systems and the first forms of writing. The exquisite art and monumental architecture of the Egyptians is unparalleled. The Babylonians codified laws and governed large parts of Mesopotamia from their capital of Babylon, one of the ancient world's great cities. The Assyrians ruled large areas and developed modern banking and accounting systems; their capital was Nineveh. The Phoenicians, with a sea-faring empire focused on the Mediterranean, traded from a series of port cities. The Persians ruled from the Aegean to India and founded the great city of Persepolis. Each conquest opened a new chapter in urban history as the culture of the conqueror transformed the cities of the conquered.

Foreign empires also played a critical part in the development of the Middle East. Greek culture, under Alexander and his successor generals, blended with various "eastern" elements to create a new Hellenistic city. Later, Roman culture transformed Phoenician trading posts and gave rise to a new set of cities in northern Africa and southwest Asia, some built on their Hellenistic predecessors. Empires built by the Persians, Greeks, and Romans succeeded because they were successfully (but sometimes brutally) administered and promoted trade in both goods and ideas. Innovations in urban form and function rolled across the region's urban landscapes, converging around the institutions of government, commerce, and religion. In the Roman city, for instance, forums, basilicas, coliseums, amphitheaters, public baths, and temples provided many elements that linger in the landscapes of the Middle East today. Istanbul, Turkey, for instance, still has its hippodrome, a racetrack for chariots, dating from 330 CE. That same century, the Roman Empire became officially Christian (Figure 7.5). Thereafter, the followers of Jesus transformed the cultural landscapes and social geography of its cities. Later, its successor, the Byzantine Empire, became the guardian of Christianity in the eastern Mediterranean.

Figure 7.5 The Armenians pre-dated the Roman Empire in becoming the world's first officially Christian nation in 301 CE. To commemorate that event's 1700th anniversary, the Republic of Armenia built a new cathedral in Yerevan, here seen on Palm Sunday. *Source:* Photo by D. J. Zeigler.

Nevertheless, a new religion, born on the Arabian Peninsula, was to conquer the Byzantine lands of southwest Asia and North Africa. Between the seventh and tenth centuries, Islam created the unique urban landscape we know today as "the Islamic city." It was a city of mosques, madrassahs (religious schools), and universities, a city built on free thinking and scientific progress, a city of honest trade, tolerance, and justice. Its daily routines, seasonal rhythms, architectural appearance, and governing system were all heavily influenced by Islam. These Islamic cities were characterized by several common elements: mosques, markets, forts, palaces, and city walls. Numerous Islamic empires and dynasties have risen and fallen over the centuries, including the Umayyids, Abbasids, and Fatimids. The most recent one, the Ottoman Empire, continued expanding for about five hundred years; the Ottoman city continued to be an Islamic city.

After the Ottoman Empire was defeated in World War I, much of the Middle East came under the control of European powers. In European style, the colonizers added new sections onto traditional cities. Independence came to the region after World War II, and new governments built skyscrapers in the global style. Today, the "Middle Eastern" city often has a historic core composed of the original Islamic city, and new sectors reflecting European and global architectural influences. And yet, the essence of the distinct Islamic matrix continues to characterize cities throughout the region.

Many countries of the Middle East have an urban tradition that transcends not just centuries but millennia: Iraq, greater Syria, Turkey, Egypt, Iran, and Uzbekistan. Yet, some countries of the region entered the twentieth century without an urban tradition at all. The emirates (principalities) of the Persian Gulf (known as the Arabian Gulf in Arab lands and referred to as the Gulf in the remainder of this chapter for simplicity) knew nothing of city life—until recently. The Gulf was punctuated by small fishing and pearling ports. Arabia's urban population was almost entirely *hajj*-related, with Mecca, Medina, and Jeddah surfacing at the top of the urban pyramid. The ancient cities of the frankincense trade, cities like Ubar, had been reclaimed by the desert, and the few large seaports, Aden and Musqat, served offshore interests more than their hinterlands.

Today, whether a country's cities date back to antiquity or to the recent past, the Middle Eastern population tends to be decidedly urban, a response to the declining viability of nomadism, the rapidly growing numbers of people, the restricted range of arable land, the rise of prosperous fossil fuel economies, increased educational opportunities, accessibility to international economic networks, and the political decisions of powerful elites.

CONTEMPORARY URBAN PATTERNS

According to the United Nations, 177 cities in the region had populations over 300,000 in 2015, and the average percentage of the population that lived in cities had soared from 29 percent in 1950 to 62 percent in 2015. That same year, global urbanization stood at 54 percent. By subregion: In North Africa, the level of urbanization rate is 52 percent; in Southwest Asia it is 70 percent; and in Central Asia it is 41 percent. Urbanization rates vary at the country scale, as well. The most urbanized countries are Israel and the small states of the Gulf, with rates in excess of 90 percent. Lebanon and Jordan are not far behind, at 88 percent and 85 percent urban, respectively (Box 7.1). The most urbanized countries generally have high levels of economic development. Conversely, the least urbanized countries are the least economically developed: less than 3 out of 10 inhabitants in Tajikistan, for example, live in cities. Finally, the dynamism of urbanization varies. Over the past four decades, the ratio between rural and urban dwellers in the Central Asian republics and Egypt has been very stable; in the same period, urbanization rates in countries like Saudi Arabia, Oman, Iran, Turkey, and Lebanon have skyrocketed. Several factors underlie these spatial differences in stability and transformation. Countries with high levels of agricultural productivity (such as Egypt) can support and maintain rural populations; oil wealth, and economic development underlie high urbanization rates in the Gulf; and rural-to-urban migration has driven urbanization rates in Turkey.

The core of the Middle East is defined by a decagon formed by the five seas (Mediterranean, Black, Caspian, Gulf, Red Sea) and five land bridges (Anatolia, Caucasus, Iran, Arabia, Suez). On the perimeter of this core, lie three of the world's 20 most populous metropolises. These large cities are important parts of the global urban system; they anchor an intercontinental, international, and intercultural urban triangle (Figure 7.6):

- Cairo, Egypt, on the continent of Africa, has 18 million people. It is the largest city in the Arab realm.
- Istanbul, Turkey, on the continent of Europe, has 14 million people. It is the largest in the Turkic realm.
- Tehran, Iran, on the continent of Asia, has 8 million people. It is the largest in the Persian realm.

Yet, in 1900, no city in the entire Middle East had more than a million inhabitants. By 1950, only Cairo had grown to exceed 1 million. Today, 50 cities have exceeded the million mark. Except for their historical centers, all are products of the late twentieth century—these are new cities, not old. Yet, the countries that anchor the Middle East's triangular urban core, all have a less developed frontier side, too. Eastern Iran, eastern Turkey, and most of upper (southern) Egypt remind us that urbanization has not entirely transformed even the most urbanized countries in the region.

A variety of urban systems is found in the Middle East. Some states are anchored by a single primate city (e.g., many Gulf States and all of Central Asia), while others have two rival urban cores. Prior to its decline into chaos, Syria provided a clear example of the latter case: Aleppo (the larger) and Damascus (the capital) were quiet rivals. Other countries with dual anchors and intercity rivalries include: Yemen (Sana'a and Aden), mostly as a result of its colonial history, Libya (Tripoli and Benghazi), as it was cobbled together by the Italians, and Israel (Tel Aviv and Jerusalem), split by its secular and religious axes. Iraq, on the other hand, has one clearly dominant city, Baghdad; two smaller anchors, Mosul in the north and Basra in the south, have different cultural geographies that reflect the country's divisions. Finally, some states have complex urban hierarchies. Both Turkey and Iran are punctuated by booming metropolises, regional centers, small towns, and villages. Similarly, Morocco is a country of cities large and small, many of which have assumed highly specialized roles in the Moroccan urban system. Rabat is Morocco's political capital, but three other cities—Marrakesh, Fès, and Meknes—have historically served in that role. In addition, Casablanca is

Box 7.1 Green Space in Beirut

Rana Boukarim, Cofounder, Beirut Green Project

Beirut: A once beautiful green city, now choking on exhaust fumes of cars. A city where free land is quickly converted into a parking lot awaiting its eventual fate as yet another luxury mall. A city with only one large public green space, which is closed off to most citizens but not to foreigners or Lebanese with official permission (which is impossible to get if you are under 30) because our mayor fears that the presence of locals will destroy it. Each of Beirut's citizens enjoys 0.8 m^2 of green space—barely larger than the chair you're sitting on. The World Health Organization maintains that a healthy city should have at least 9 m^2 of green space per person. While cars roam free in ever-expanding spaces, Beirut's second-class citizens (us humans) suffocate on their 0.8 m^2.

Beirut wasn't always like this. It was green. Sadly, it is now gray. The Beirut Green Project (BGP) was born of this frustration. A few like-minded individuals made official-looking signs and placed little 0.8 m^2 patches of grass all over town with ironic signs: "Enjoy your green space!" People were soon noticing BGP installations, agreeing, chuckling, taking pictures. The media and blogosphere quickly joined in. It made quite a stir, so BGP decided to make a bigger splash to attract more like-minded people to join the cause. They took over a busy roundabout for a day, covered it in real grass, and invited everyone they knew. More than 400 people showed up. They brought books, guitars, kites, and kids, and had a lovely day on the grass. Their presence sent a message: We need more public green spaces in Beirut.

Nevertheless, BGP faced resistance not only from politicians but also from the people themselves: "We don't need parks; we can sit on our balconies, or at Starbucks," and "Just go up to the mountains if you want to be in nature," and "We don't have continuous electricity, people are dying of hunger, war, and abuse, our whole system is plagued by corruption . . . And you're worrying about your right to green spaces?!" So, BGP's scope expanded to actively raising awareness of the importance of green spaces through efforts like mapping all Beirut's parks and encouraging people to picnic in a park. Parks are needed in ALL large, stressful cities like Beirut, and are not a superfluous luxury of rich cities. They are necessary for our mental and physical health. Are clean air, a place to exercise, and a place for our children to play, a luxury?

If people haven't lived in a city where green spaces are treasured, if they haven't had politicians who grant them their rights instead of judging if they deserve them, if they haven't been trusted with public goods in any real way, then it is natural that their interest in (and right to) green spaces has been marginalized. Almost everyone is dumbfounded to learn that Beirut has 24 public green spaces; most guess that there are between one to four. BGP's mission for now is to start using the green spaces we do have and eventually demand more.

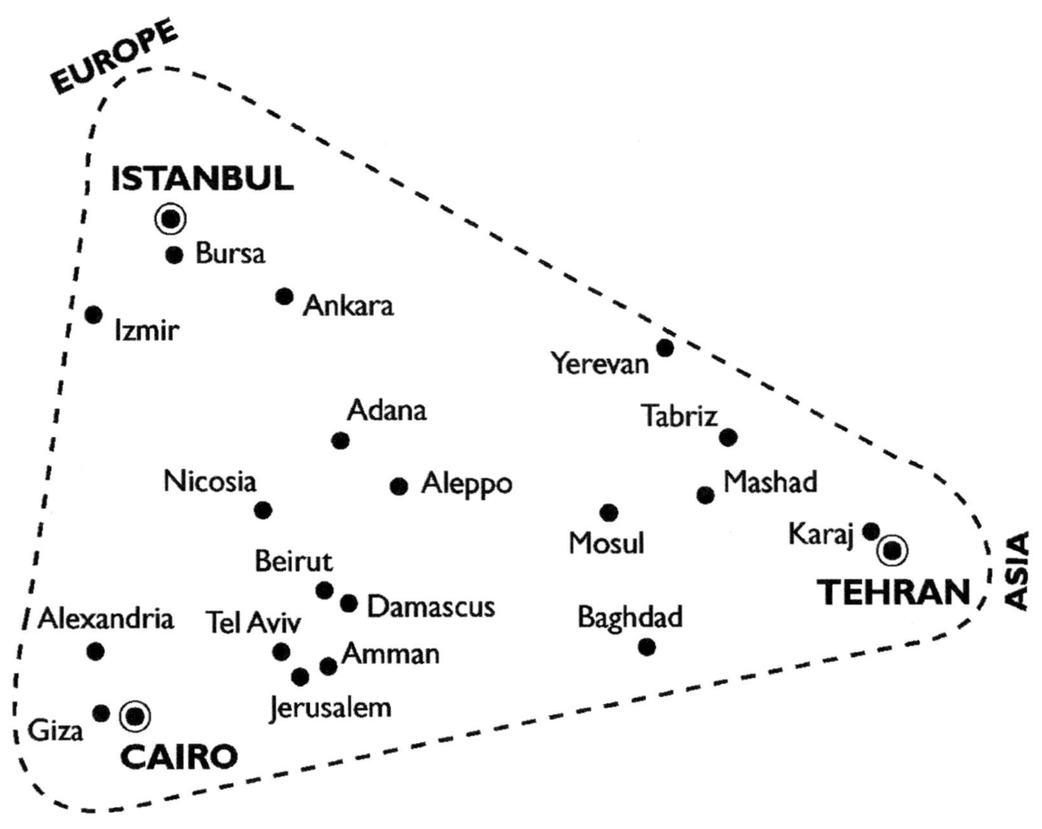

Figure 7.6 The Urban Triangle of the Middle East shows the relative locations of major cities. These cities are in their correct geographical locations, but shown without the base map underneath. *Source:* D. J. Zeigler.

Morocco's unofficial economic capital. Its former diplomatic capital (as far as ambassadors were permitted to venture into the country), Tangier, now serves as a bridgehead to Europe. The city of Ourzazate in southern Morocco has even become one of the movie-making capitals of the world, starring in films such as *Star Wars* and *Lawrence of Arabia*, and providing a location for several seasons of *Game of Thrones*.

Not only have cities in the region increased in population and territorial extent, they have also begun to coalesce in a fashion reminiscent of Jean Gottmann's Megalopolis. In these urban mega-regions, life is at its most intense, a characteristic symbolized by the pace and volume of traffic along connecting thoroughfares. There are perhaps seven megalopolises developing on the Middle Eastern map in the twenty-first century (Table 7.1). The most populous is in Egypt. The only international megalopolis is the one stretching from coastal Saudi Arabia, beginning with the oil-engorged cities of Dhahran (headquarters of Saudi Aramco, the world's largest oil company) and Dammam, and then across the four-lane causeway to Bahrain and its capital, Manama.

An intertwined pair of demographic and economic factors underlies Middle Eastern urbanization patterns. Rates of natural

Table 7.1 Megalopolises of the Greater Middle East

Name	*Range*
Egyptian Megalopolis	Alexandria to Cairo to the Suez Canal
Marmara Megalopolis	Istanbul to Bursa and around the Sea of Marmara
Iranian Megalopolis	Tehran to Karaj to Eslamshahr and Nazarabad
Mid-Mesopotamian Megalopolis	Baghdad to Fallujah
Moroccan Megalopolis	Casablanca to Rabat to Kenitra
Israeli Megalopolis	Haifa and Akko to Tel Aviv to Jerusalem
Arabian Megalopolis	Jubayl to Dammam and Dhahran to Manama

Source: Donald Zeigler

population increase and migration are the main drivers of urbanization in the Middle East. Natural increase is related to several factors: as health and dietary conditions have improved, death rates have fallen, while birth rates have remained high, leading to a net growth in population. Persistently high birth rates can be attributed to a number of factors, including preference for male children and the relatively lower status of women. There are many ways to categorize migration: internal versus international, voluntary versus involuntary. With respect to internal voluntary migration, rural-to-urban migration in some countries and the ending of nomadic ways of life in others, have contributed to urban growth. As economic policies aimed at improving urban conditions have been implemented, increased access to a number of urban amenities (e.g., education, health, employment) has acted as a magnet that draws individuals from the surrounding countryside. International voluntary migration, driven by a demand for labor, has spurred urbanization, most notably in the oil-rich Gulf States, where labor migration has transformed cities into global immigrant destinations. Foreign-born individuals make up approximately 60 percent of the population in Kuwait, 74 percent in Qatar, and 84 percent in the United Arab Emirates. Foreign-born individuals comprise even larger proportions of the labor force in the Gulf States—in Bahrain 54 percent of the population is foreign born, but 75 percent of the labor force is foreign born. Because a portion of international migration in the Middle East is intra-regional (e.g., Egypt and Jordan send migrants to other countries in the region), transnational ties also connect sending and receiving countries. Voluntary international migration also affects other countries. For example, Israel received a million Russian Jews between the mid-1980s and the end of the century; it has also received major immigrant streams from Ethiopia, Argentina, France, the United Kingdom, and the United States, and today more than a quarter of the Israeli population is foreign born.

Conflict and political instability have spurred forced or involuntary migration. Refugees, individuals who have been displaced beyond their country's borders, have significantly impacted urbanization in some cases. For example, Amman, Jordan, grew from a village of 2,000 people in 1950 to a metropolis of 1.1 million people by 2014. Much of this growth is due to waves of refugees: the first wave of refugees moved into Jordan from neighboring Palestine in the wake of the Arab-Israeli conflict; a second wave of refugees fled violence in Iraq, and refugees from the conflict in Syria make up a third wave. Today, Jordan is home to more than 2.5 million refugees. The Syrian conflict has set sizable refugee flows in motion: since the outbreak of the civil war in Syria, more than 4 million Syrians have fled the country to refugee camps in Turkey, Lebanon, Jordan, and Iraq (Figure 7.7). Within Syria itself, 6.5 million people have been displaced as a result of the crisis.

Within the context of protracted conflict, camps and communities for those affected by involuntary migration can acquire semi-permanent status. Although informal economies and social networks provide some social order, the lack of long-term financial support means that the residents of these camps lead precarious lives and face numerous challenges: temporary housing, limited infrastructure, overcrowding, anxiety and depression, and lack of opportunity. Refugees place pressure on existing infrastructure, housing, and services, exacerbating issues of resource scarcity; as a result, relations with host communities may not always be positive.

MODELS OF URBAN STRUCTURE

An analysis of elements such as land use, street layout, and building structure helps to explain the form, and function of the Middle Eastern city (Figure 7.8). It is important to point out that these patterns and models are not static, and that they do not claim to represent the urban diversity of the entire region. However, as they provide useful generalizations, models simplify the task of analyzing and comparing different cities. At the heart of every traditional Islamic city was a fortress: the citadel, *al-qalat*, or (in the Maghreb) the *Kasbah*. It

Figure 7.7 As of 2015, there were 4 million refugees from Syria. Turkey has taken in almost 2 million, with many housed in camps like this one near Karkamish on the border with the self-proclaimed Islamic State, now in control of northern Syria. *Source:* Photo by D. J. Zeigler.

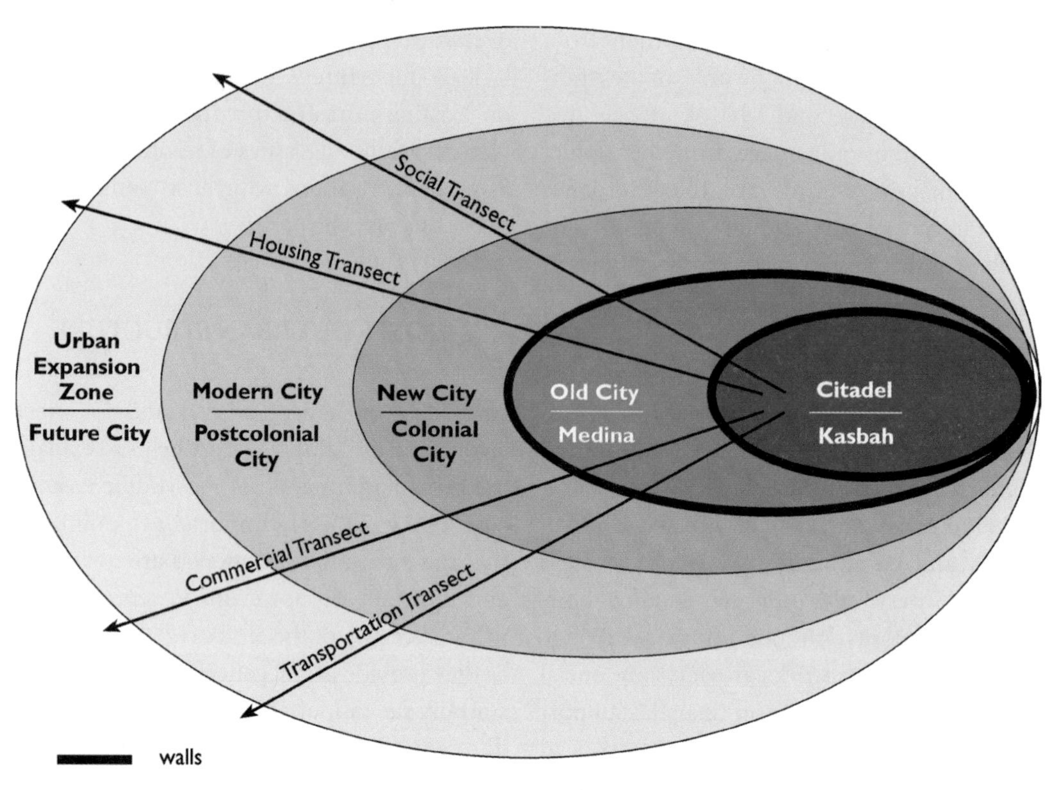

Figure 7.8 Internal Structure of the Middle Eastern Metropolis. *Source:* D. J. Zeigler.

usually covered only a few acres and occupied the most defensible site, often on a hilltop and typically surrounded by a wall. In the past, it would have served as the administrative heart of the city, the site of the palace. Today, the citadel is most likely to be a preserve of history, a valuable visual reminder of the past, important in building national identity, and part of the historic core within the modern city (Figure 7.9).

Surrounding the citadel is the old city itself. In the old city's heyday, the most coveted space was at the center, close to the seats of economic power and social interaction. Palaces and merchants' houses were central elements of the old city's landscape, as were the central souks. Like the fortress, the old city was usually walled, and its often-ornate gates gave access to the world beyond. In the Maghreb, the old Islamic city is called the *medina* (Arabic for "city"). It is the city of antiquity, at least as it has survived conquest, disasters, and well-meaning modernization attempts. The residential population of cities, in the traditional city model, tended to be concentrated in "quarters": sections for Jews, Europeans, different Christian sects, different ethnic groups, and people of different village or regional origin. In the past, many of these quarters had gates of their own. Residential space was highly segregated, yet it also mimicked the security and social cohesiveness of the village by providing a scale of life to which people were accustomed and a set of community institutions which people of like mind could control. Both rich and poor lived within each quarter.

The inclusion of courtyards in homes, mosques, khans, and palaces creates a

Figure 7.9 The citadel, or cale, of Gaziantep, Turkey, occupies a strategically located hilltop that dominates the fertile agricultural region near the Turkish-Syrian border. *Source:* Photo by D. J. Zeigler.

"cellular" pattern. For most people in the world, the landscape of the old city—introverted, compact, congested, cellular, and fortified—provides the stereotype for the region's cities. Walls, along with their watchtowers, differentiated quite sharply between city and country until the twentieth century. Outside the walls were olive groves, grazing lands, cemeteries, quarries, and periodic markets, not to mention potential enemies. Noxious enterprises like tanneries were located at the old city's periphery. The medinas of modern Middle Eastern cities contain, at most, 4 percent of the urban population; residents of modern neighborhoods rarely patronize the old city cores.

In addition to the ravages of age, the old cities (e.g., in Yemen, Egypt, and Morocco) face pressure from rapid urban development. As they have survived, the narrow streets of old residential and commercial quarters tend to be penetrable only by foot traffic and donkey carts. Still, this does not stop the occasional taxi or service vehicle from squeezing through. The cramped feeling is heightened by second and third stories jutting out overhead, sometimes joining and creating aboveground "tunnels." The largest streets lead into the city from gates in the walls, but a large public square dominating the center of town is a rarity (except in Iran). A grid pattern loosely manifests itself in towns with a Roman heritage, and it is common for elements of the pre-Islamic landscape to become parts of the working city, usually unrecognized for their historic value. Roman pavements may lie deep underneath the contemporary street, and what was a wide Byzantine thoroughfare may now be subdivided into three or four narrow, parallel alleys, each serving as a souk of its own. Periodically spaced along the most well-traversed routes, are richly decorated drinking fountains (or their remnants), often given to the public by wealthy merchants.

Beginning with the colonial era (British, French, Italian, or Soviet), a new city typically developed outside the old city walls. In the Maghreb, it was called, in fact, *la nouvelle ville* ("the new city"). It was a city between two worlds, traditional and modern. The traditional elements of urban form, from mosques to bakeries to public baths, were incorporated into the new city, but they were supplemented by modern amenities and architectural styles, including larger stores and hotels, traffic circles and wide boulevards, European-style churches, new government buildings, and corporate offices, all plastered with the language of the colonizers.

The new city was gradually enveloped by the modern, postcolonial city. Courtyard homes all but disappeared from the landscape, replaced by extroverted buildings and apartment blocks, with high-income flats and single-family units becoming increasingly common. The postcolonial city became the zone of international hotels, corporate headquarters, and modern universities. It also became the zone of squatter settlements (which can invade even the inner zones if there is any unoccupied space) where recent in-migrants from the villages find lodging while they work their way up the urban social pyramid.

Beyond the modern city is the urban expansion zone. Here, small villages find themselves undergoing urbanization *in situ*. Here, also, may be new industrial estates, modest new housing tracts, or, in the case of Egypt and Saudi Arabia, new cities, were built from scratch. For the richest countries, those able to afford automobiles and the gasoline to fuel them, it can also be a zone of raging urban sprawl. Most often, the international airport is located here as well.

Urban Transects

The Middle Eastern city today, therefore, can be seen as an interlocking set of concentric zones, patterned in time: citadel, old city (Islamic), new city (European), modern city, and urban expansion zone. As such, the most traditional and often most distinctive part of the city is at the center; when comparing cities, the centers will be the most different and the peripheries will be the most similar. The following transects illustrate the variety of visible changes as one moves from the innermost zone of a city to the outermost.

- A transect along the social axis: The old city is becoming increasingly marginalized by society, particularly as the well-to-do move out. The modern city is attracting the lion's share of new neighborhood investment and the best of social services. Even tourists typically stay in the modern city and depend on air-conditioned buses to drop them off at a city gate for a brief sojourn into the past.
- A transect along the housing axis: The old city (except in Turkey and Yemen) is a zone of traditional two- or three-story courtyard houses. The modern city is a zone of mid-rise and high-rise apartments. In fact, the multistory apartment block is now the most typical component of the Middle Eastern city's residential landscape (and a reminder of how architecture has pulled away from

the physical environment and its indigenous roots).

- A transect along the commercial axis: The old city still displays fully functional commercial districts that house traditional industries and small-scale, family-owned artisanal enterprises. The modern city displays its logo-laden landscape of chain stores, international (and national) franchises, shopping malls, and ever more snippets of English signage (Figure 7.10).
- A transect along the transportation axis: The old city's narrow streets are clogged with pedestrians, taxis, and even donkey carts. The new city is marked by the proliferation of privately owned automobiles, gas stations, and parking spaces.

Arab Cities on the Gulf

The preceding generalizations about urban form and function apply to the old cities of the Middle East. An entirely new set of cities that has grown up on the twentieth century's oil and natural gas fields has followed its own dramatic urban trajectory. These cities are best exemplified by the cities of Kuwait, Saudi Arabia, Bahrain, Qatar, and the UAE (though oil revenues have also transformed cities in Iran, Iraq, Libya, and Algeria). Urban landscapes from Kuwait to Dubai have been built, largely since the 1960s, on revenues from the world's largest oil fields. While lacking the historical depth of a city like Aleppo, they offer a glimpse of what a modern Middle Eastern city can be.

Although the historical cores of Middle Eastern cities are small, they serve as the source of local identity, particularly in layout and architectural styles. Their function is different from the surrounding postindustrial city, with its high-rise office buildings, shopping malls, gardens and golf courses, apartment complexes, sprawling low-rise suburbs, and mosques. The automobile's dominance in the postindustrial city is clearly visible in the wide boulevards, modern highways, bridges, and ring roads, and single-use zoning. It is also visible in the

Figure 7.10 The landscape of Amman, Jordan, shows the signs of global commercialization in the form of this bilingual advertisement for Subway. *Source:* Photo by D. J. Zeigler.

empty sidewalks and underdeveloped public transit options.

It is oil wealth that has built many of the Middle East's postindustrial cities. With the ability to afford the world's most creative architects, cities have been able to blend modern structures with traditional themes. One of the surprising elements of these new urban landscapes is how green they are. Turning itself into a Garden City, in fact, has been one of Dubai's urban planning objectives. Figuratively, oil is turned into water, and water into green space. Oil is also turned into new human geographies. Petro-economies have, by governmental design as well as economic magnetism, virtually eliminated nomadism as a way of life. The rural population has become urbanized. In addition, the faces of the oil-engorged boomtowns have also changed. The economic magnetism of Kuwait City, Dammam, Doha, Abu Dhabi, and others, draws unskilled workers from as far away as India, Pakistan, Sri Lanka, and the Philippines, and skilled workers from Europe and the United States, as well as other parts of the Arab world.

The Gulf's Arab cities are centers of consumption rather than centers of production; the industrial era is missing from the landscapes of Arab cities on the Gulf. Manufacturing is limited to local craft industries and manufacturing-in-transit at the region's free ports, most notably those in the emirate of Dubai. As banking and trading centers of the Middle East, however, some of these cities—Dubai, Abu Dhabi, Manama—have been thriving as increasingly important transactional nodes in the economic systems reshaping the region. The center of Arab world banking has shifted (in part) from Beirut to Manama and Dubai. Doha, Qatar, is now the headquarters of Al Jazeera, the most popular satellite television channel in the Middle East; Qatar is currently slated to host the FIFA World Cup in 2022 (Figure 7.11). Since the 1990s, the port cities of the Persian Gulf have been entrepôts supplying Central Asia with cars, electronics, and other high-end goods. While built on petrodollars, the cities of the Gulf are likely to continue thriving only if they diversify and lay the groundwork for post-petroleum economies.

Figure 7.11 The skyline of Doha seems out of proportion to its role as capital city of a country, Qatar, with only 2 million inhabitants. *Source:* Photo by Zia Salim.

FORM AND FUNCTION ON THE URBAN LANDSCAPE

What factors have shaped the morphology and landscape in the Middle Eastern city? The most prominent influences on the form and shape of the contemporary Middle Eastern city include the physical environment, religion, economic activity, culture, and politics. These factors do not operate independently. Rather, they interact and overlap with each other. The passing down of architectural and design know-how, often over generations, has guided the development of the Middle Eastern city. But, new technologies and international trends have also influenced the form and function of cities.

Albeit principally affected by their location in an arid climate, cities of the Middle East are spread across a variety of climatic zones, including desert, steppe, and Mediterranean. Dry environments influenced the location of cities near oases or wells and along coastlines. Additionally, the traditional city's morphology was profoundly influenced by temperatures. Examples of design elements, at various scales, that promote cooling and minimize heat absorption include the use of open-air central courtyards that interact with the microclimate to facilitate cooling. Air is chilled by wells or fountains located in courtyards, and shade is provided by the use of vines, arbors, and trees. Partially covered, narrow and winding streets also maximize shade and walkability. Other attempts to moderate temperatures include the attachment of multiple homes together (to reduce exposed wall surface and heat absorption), the selection of high-albedo building materials (to increase reflectivity), the minimization of the size and number of outside openings, and the use of wood lattice screens to reduce direct sunlight.

In much of the Middle East, Islam's imprint on urban morphology can be seen in numerous ways. Minarets mark the skyline, and mosques and religious buildings (such as Koranic schools, traditional universities, shrines, and mausoleums) are key features of the landscape. At the heart of the city is a main mosque, known as *Al-Masjid Al-Jami*, where the weekly Friday noon prayers are held. The city is typically divided into quarters, each with a local mosque. Guided by the religious duty of *zakat* (alms), the surrounding businesses and residents help support the mosques and the social services that they provide. An examination of Islamic law and neighborhood building guidelines identifies 12 religious principles and 5 behavioral guidelines that were used to guide urban development. The religious principles range from interdependence, privacy, and respect for the property of others, to more concrete aspects such as the minimum width of streets. Behavioral guidelines include cleanliness, public awareness and responsibility, and trust, respect, and mutual obligation among neighbors. Numerous elements of urban structure, design, and use in the built environment highlight the influence of religion. The prioritization of privacy influences the placement of small windows above eye level, staggered doors on opposite sides of streets (so that doors that are open simultaneously do not allow for others to see inside one's home), walled homes and quarters (in which the entrances to individual homes are often doors in walls and not distinguishable individual units), and the use of wooden latticework to screen windows.

Trade dominated the region long before Muhammad, himself a merchant, revealed a new religion, so it is little wonder that *souks* (markets) are another typical part of the traditional city landscape. Central souks usually evolved around the city's "grand mosque"; smaller souks enveloped the neighborhood mosques. Souks display functional specialization: shoes in one area, copperware in another, produce in another, and traditional fast food available throughout. Merchants dealing in the same product compete with each other, complementary trades are located near each other, and the concept of fixed prices does not exist. Shoppers know the techniques of hard bargaining and expect low prices. In today's city, specialized outdoor souks (e.g., vegetables, fruits, fish, and clothes) are spread across central areas and traditional neighborhoods. However, mass-produced goods are as common as locally produced products. In other realms of the traditional (albeit disappearing) city: bread tends to be purchased daily at the bakery itself; most public baths (*hammams*) have become historical sites since houses are served by water pipes; modern beauty salons are located in almost every neighborhood; cafes and coffee houses take the place of bars for men (while alcohol is not permitted by the Koran, coffee, tea, and the water pipes known as *nargileh* in the Levant and *sheesha* in Egypt and the Gulf are a popular custom); and private life takes place in very private places, like the home (Box 7.2). The informal economy is an important sector, providing employment and livelihoods to large numbers of urban residents.

With the growth of the global internet and the rapid adoption of social media, particularly among the region's large youth populations, all cities in the Middle East are developing into *cyber cities*. The region's universities, manufacturing establishments, and traders are increasingly tied to constant flows of information that arrives by waves, wires, and fiber optic cables. Every computer becomes its own harbor in the informational landscape, and the public is demanding frontage on these "harbors." Today, there are thousands of internet cafes throughout the Middle East, and the percentage of people with mobile internet connections (on phones and other devices) is very high. Home internet connections are still rare. For example, 3.3 percent of Egyptians have fixed internet access at home, but 10 times as many Egyptians have internet access using a cell phone. Similarly, Bahrain has an impressive 110 active mobile-broadband subscriptions for every 100 inhabitants. The implications of increased connectivity are varied. On one hand, cyberactivism is popularly understood as being crucial to the protests that swept cities in the Arab world in 2011, but it is too simplistic to attribute the movement solely to cyberactivism. The long-term outcomes of these protests illustrate some of the limits of internet connectivity. On the other hand, the increasing numbers of tech-savvy young people in the Middle East can lead to new social and economic formations. For example, Iranian e-commerce firm Digikala, the brainchild of two brothers who were frustrated because they could not find camera reviews online in their own language, employs more than 700 people and ships thousands of orders a day in several Iranian cities. In less than a year, Digikala's value has soared to over $300 million.

The Middle Eastern city's landscape has been marked by imperial, colonial, and nationalist, royalist, and autocratic politics. The earliest Middle Eastern cities were controlled by various empires (e.g., Ummayyid, Ottoman), and cities were purposely built as symbolic centers of power and control. Later, Western colonial powers left their mark on the Middle Eastern city. Today, zones of colonial architecture, characterized by modernist planning and layout as well as individual design

Box 7.2 Home Space in Tehran

Farhang Rouhani, University of Mary Washington

Since 1979, the Islamic Republic of Iran's efforts to quell modernization along Western lines have imposed strict policies on Iranian citizens. In urban centers, most notably Tehran, these include the state policing of public spaces such as parks and commercial streets. What is particularly striking, however, is how these politics invade even the most private of spaces: the home.

Home spaces in Tehran were significantly rearranged over the course of the twentieth century. The traditional home was divided into separate male (*birun*, outer, public) and female (*andarun*, inner, private) sections. A lack of street-facing windows accentuated the sense of privacy; indeed, internal courtyards served as the home's central focus. Modernization under the Pahlavi Shahs (1920s–1979) and the demands of rapid population growth ushered in Western high-rise apartment-style living, without the physically gendered division of space, and more public street-facing windows. The importance placed on familial privacy, though, has been maintained in different ways, including the prevalence of walls around the complexes and buzzers to screen visitors.

On top of this urban transformation, the politics of satellite television viewing has brought the middle-class Tehran household to the forefront of Iranian state politics. The clerical government first instituted a ban on the sale, import, and use of satellite dishes in 1994 because of the proclaimed polluting effect of Western media products on Iranian society and particularly the youth.

Under Iran's more liberal-democratic government, the late 1990s witnessed a relaxing of the ban, including the requiring of a warrant for home searches. The current, more conservative-theocratic government, however, has recommenced the policing of public and private morality. While the ban on satellite dishes still exists and periodic raids to confiscate them are made by the morals police, a majority of the population of Tehran has access to the air waves, whether directly in their homes or through family and friends. Ultimately, police regulation and infiltration of people's homes have intensified fears in an already politically insecure society.

The practice of policing the home has led to dynamic conflicts in Iranian state and society over the role of privacy in a democracy, the importance of the home as a space of refuge, and the moral and social effects of the global media. It is within this context that *The Simpsons, American Idol*, and the living room furniture are transformed into a realm of resistance.

elements, can be found outside the older walled cities across the region. The French colonization of most of North Africa, for instance, strongly influenced city cultures and built environments in Morocco, Tunisia, and Algeria. After World War II, the end of colonialism and the rise of Arab nationalism contributed new symbolic elements to Arab cities in the Middle East. A parallel trend is seen in countries ruled by monarchies and autocratic regimes. Monumental squares and public art, statues or other monuments to leaders, and renamed public streets and buildings all reflect nationalist, royalist, and autocratic influences on the Middle Eastern city.

To varying degrees, cities in the Middle East are integrated into international circuits of tourism. Activity in the tourism sector includes heritage and cultural tourism (drawing on a rich inventory of districts, archaeological sites, monuments, and museums) and recreation and nature-based tourism. The first wave of construction of large hotels and other visitor-serving infrastructure was supported by oil revenues in the 1970s; since then, tourism has spread across the region and down the urban hierarchy. In 2014, 50 million international tourists visited the Middle East. In a direct manner, tourism provides significant portions of national GDP in some countries (Egypt, Turkey, Morocco, and Lebanon) and employs tens of thousands of people (more than 1.4 million in Egypt alone). Obviously, the indirect benefits of tourism in terms of GDP contribution and employment are several times greater than the direct benefits. While recent political instability has impacted tourist flows, tourism in the Middle East as a whole is increasing at one of the highest rates in the world. Tourism is an important part of the economic and social landscapes of many cities in the region. A range of cultural events, festivals, and sporting events are used to market cities and attract visitors and capital. Some Middle Eastern cities have dualistic landscapes—tourists see and experience one portion of the city, while residents see and experience another. Tourism also relates to questions of urban preservation and redevelopment. Exactly how architectural heritage and historic buildings should be preserved, within larger socioeconomic and cultural contexts, is often contentious. Further, governments often use the redevelopment of old cities as an economic development strategy, but this can result in islands of gentrification and the creation of romanticized versions of the past.

FROM ARAB SPRING TO ARAB WINTER

On December 10, 2010, a wave of democratic uprisings was born. It began in the small Tunisian town of Sidi Bouzid and quickly spread to the capital, Tunis, then to the cities of other Middle Eastern countries including Egypt, Libya, Syria, and Yemen. These revolts aimed to engender political and economic reform by overthrowing dictatorial leaders who had been in power for decades. They began as affirmations of people's struggles for freedom, democracy, and social justice. Because they took place as the winter of 2010–2011 gave way to spring in the Arab world, the movement was quickly dubbed the Arab Spring. Only time will tell what the long-term democratic and social impacts will be. At present, it is difficult to see anything but an Arab winter. Often called youth revolutions, these movements showed how peaceful protests could oust dictators, as in Egypt and Tunisia, how powerful presidents might lead their countries to civil wars to stay in power, as in Libya and Syria, and how external interventions could complicate the processes of democratization everywhere in the Middle East, with Yemen as just one example.

Tunisians call the overthrow of President Zine El-Abidin Ben Ali the Jasmine Revolution. It began when street vendor Mohamed Bouazizi decided to commit suicide—by setting himself on fire—after he had been harassed and humiliated by local police and municipal officials. Supporters spontaneously took to the streets and the uprising quickly spread to other cities. Less than a month later, Ben Ali, in power for more than 23 years, fled the country. The next year, Tunisia held its first democratic parliamentary elections and set about drafting a new, democratic constitution. In 2014, multiparty elections took place under the newly approved constitution. Islamist parties vied with secular parties for power. Although secular parties won both parliament and the presidency, a variety of violent reverberations ensued. In early 2015, extremists killed 21 tourists at the Bardo National Museum in Tunis. World leaders and tens of thousands of Tunisians marched through the streets of Tunis to condemn terrorism. Later that year, 38 vacationers were killed near Sousse. Islamic extremists were responsible for both attacks. The first revolution of the Arab Spring, however, seems to have pointed Tunisia toward democratization. In other countries where revolutions have taken place, progress has been more tortuous and, in some, the groups which have taken over are worse than the overthrown dictators.

Egypt's January 25 Revolution, in 2011, was sparked by the protests in Tunisia and fueled by youth, and others, demanding "bread, freedom, and social justice." At first, as President Hosni Mubarak resigned and elections followed, it looked as if success was at hand. However, it took a second wave of protests in Cairo and around the country to put Egypt on a path to full democracy. In the first multiparty elections, the Muslim Brotherhood won parliamentary and then presidential control. As they tightened their grip on power through often-debatable means, including the adoption of a controversial constitution, Tahrir ("Liberation") Square once again came alive with mass protests that culminated in the military's ouster of democratically elected president Mohamed Morsi of the Muslim Brotherhood (Figure 7.12). Following a period of military rule and the adoption of yet another constitution, new elections were held, and secular-party candidate Abdel Fattah el-Sisi was elected president (as chief of the Egyptian Armed Forces, el-Sisi deposed Morsi; he then resigned to run for president). The Muslim Brotherhood was outlawed, but as in Tunisia, this led to a violent reaction in the form of extremist terrorism that has hit the government, the Coptic Christian minority, and the Sinai especially hard. Nevertheless, in Egypt, the revolution seems to be making progress.

In other Arab countries where popular uprisings were inspired by Tunisia's Jasmine Revolution, there has been no progress toward democracy and social justice. Libya has become a failed state, a state without a government in control. Although the world welcomed the overthrow of Muammar Gaddafi after his 42 years of absolute repression, there were no institutions in Libya ready to assume the reins of power. In the subsequent power vacuum, old regional and tribal rivalries surfaced, outside powers supplied arms, and the conflict was quickly joined by violent extremists, including the self-proclaimed Islamic State in Iraq and Syria (ISIS). The failure of the Libyan spring has placed a potentially destabilizing failed state in the middle of North Africa.

In Syria, antigovernment protests started in 2011 in the city of Dera'a and quickly spread to Damascus, the capital, and other major cities. Protesters eventually demanded the resignation of President Bashar al-Assad. His response was to use police and military

Figure 7.12 Demonstrations to oust President Mohamed Morsi from power took place in cities around the world as expat Egyptians took to the streets of cities like Amsterdam, shown here on July 7, 2013. Although he was democratically elected, Morsi's abuse of power enraged the public and the Egyptian military. *Source:* Photo by Amal Ali.

forces to suppress the rebellion. Within a few months, Syria's Arab Spring had given birth to a brutal civil war between anti-Assad forces (which were highly fragmented internally) and the Syrian state. The complex conflict soon gained international dimensions: other countries in the Middle East intervened, major powers picked sides, and terrorists from across the region and beyond entered the conflict. Most dangerous to the region has been the ISIS, a violent extremist group that grew out of al-Qaeda in Iraq. ISIS's self-proclaimed caliphate now controls, from its capital at Raqqa, large amounts of territory in Syria and Iraq. The ensuing strife has caused a massive humanitarian crisis, as millions of Syrian civilians have been displaced.

Protests against the rule of Yemen's dictatorial president Saleh also took place in 2011 in the capital of Sana'a and other major cities including Aden and Taiz. Eventually, President Saleh stepped down, a rebel group called the Houthis took over the capital, and a proxy war between Shiite Iran and Sunni Saudi Arabia, backing the Houthis and the government, respectively, ensued. The protests that took place in these four countries, plus others, beginning in 2011, have left some countries a step closer to democracy, while others have become horrifyingly violent killing fields beyond the control of any recognized government.

DISTINCTIVE CITIES

Cairo: The Victorious

Al Qahirah means "victorious," and Cairo has emerged victorious as the most populous city in the Arab World, in the Middle East, and on the continent of Africa. Greater Cairo, with about 19 million people, is well on its way to metacity status. It is known as "the City of 1000 Minarets" since mosques spread across all city neighborhoods. With its movie industry and annual International Film Festival, Cairo is known as the "Hollywood" of the Middle East, as well. The Arabic-language cinema and popular Arab music have made Cairo one of the cultural epicenters of the Arab universe. *Al-Ahram*, a government-owned Cairo daily newspaper with an online English edition, has the largest readership in the Arab world. Several private newspapers such as *Al-Wafed*, *Al Masry Al-Youm*, and *Al-Shorouk* are also published in Cairo to represent critical views of governmental policies. As the capital of Egypt and the headquarters of the League of Arab States, Cairo is also the head city of pan-Arab politics, a role facilitated by both its size and its relative location. It hosts foreign embassies and cultural centers and a major educational complex where Cairo University, Ain Shams University, and other public and private universities are located. Cairo's Al-Azhar University is the world's oldest Islamic university and a major center of Sunni Islamic education; students from around the globe come to Al-Azhar to study Islam and other subjects. Cairo is uniquely positioned between the western Arab world of North Africa and the eastern Arab world of Asia.

Over a thousand years old, Cairo is a multi-layered city; its buildings and neighborhoods reflect the impact of various historical periods. At its dense core lies the "Islamic" city. To the east of the Nile River the city began as a military encampment and grew to include a citadel, mosque, and city walls. Today most walls are gone, torn down as the city's boundaries expanded. However, three of the original gates remain to draw tourists, who mingle with the area's residents. In medieval times, Cairo was an epicenter of world trade; caravans brought luxuries and necessities to the city's famed markets. Today, tourists crowd the Khan al-Khalili market to buy souvenirs, most often trinkets reflecting Egypt's pharaonic history, or to sip tea in traditional coffee shops.

As dams tamed the Nile floods in the nineteenth century, the city expanded onto the river's shores. At the same time, Europe was establishing overseas colonies in Asia and Africa, new neighborhoods in Cairo were being constructed in European design. Houses were built that looked like Italianate villas, parks were constructed to open up the city, and a "new" downtown was created to mimic the design of Paris. Upper-class Egyptians even enjoyed performances in the new opera house, where, in 1871, an opera premiered that has become one of the most popular: *Aida* by Giuseppe Verdi.

Cairo's explosive population growth occurred in the era after World War II, when Egypt became an independent republic. Migrants from the rural areas flooded the capital in search of jobs and opportunities; they created enormous economic challenges for the young government. Massive high-density apartment blocks, with bleak architectural designs and poor quality construction, were built to accommodate the influx. At the same time, the government became concerned about Cairo's massive size, and its military vulnerability.

In an effort to stem the tide of urban expansion onto valuable farmland, the Egyptian government began to redirect growth into the desert. The result was government-built, industry-based cities distant from Cairo. The 10th of Ramadan City, for instance, located on the way to the Suez Canal, was built to have an industrial base anchored by several thousand factories. Since the 1970s, it has offered jobs, housing, and some services. The new towns, however, never met their target population goals and did little to relieve the population pressure on Cairo. Also, new settlements were built along the ring road surrounding Cairo to redistribute the population. Many of these settlements became homes for the middle and upper classes.

Since the 1990s, Cairo's landscape has increasingly reflected the impact of globalization. Chili's, TGIF, Hardee's and other fast-food chains with global reach are ubiquitous. Massive malls, office towers, and new hotels now line the Nile. In addition to the American University in Cairo (AUC), founded in 1919, new international universities such as the German University in Cairo (GUC) have been established to internationalize the city's educational opportunities. The presence of foreign banks (e.g., CitiBank, HSBC, and Scotia Bank) illustrates Cairo's integration into the global economy.

As a megacity, Cairo is really a product of the twentieth century. As it has expanded, however, it has engulfed dozens of predecessor settlements and unique historical landscapes (Figure 7.13). These visual reminders of the past, numbering in the hundreds, make Cairo a vast open-air museum. Large tracts of twentieth-century blandness separate such historical nucleations as the following:

- The great pyramids (and the sphinx) of Giza, on the west bank of the Nile, date back to the Old Kingdom, but they have been encroached upon by an expanding city, deflating some of the excitement of first-time visitors who may be disappointed to find a Pizza Hut practically at the pyramids' base.
- Heliopolis (on the way to the airport) was one of the ancient world's cult centers, but only a single obelisk remains—now in the middle of an urban park.

Figure 7.13 Coptic Cairo, now the city's Christian "quarter," is one of the historical nucleations that has survived from medieval times. Here communal urns provide the neighborhood with water while political posters try to attract attention. *Source:* Photo by D. J. Zeigler.

- Babylon-in-Egypt, now known as Coptic Cairo (because of its Coptic Christian inhabitants), has a history associated with the world's most famous refugee family—Mary, Joseph, and Jesus. It is now engulfed by the modern suburb of Ma'adi.
- Cairo's Citadel was built by Saladin in the twelfth century; it was transformed by the Mamluks and then the Ottomans.

Recently, Tahrir Square has become a major landmark in Cairo. It was the focal point of the Egyptian Revolution that started on January 25, 2011, to demand freedom, social justice, and economic reforms. With the success of the revolution, Tahrir Square has become a place where Egyptians go to demand ever more reforms. It has become the equivalent of London's "Hyde Park Speakers' Corner" in which people openly voice their opinions and debate political issues.

Whether counting people or cars, the growth of Cairo has been meteoric. Today, the city's traffic snarls are of world renown. Since Cairo's metro rail system opened in the 1980s, it has expanded to include a subway line tunneling under the Nile to the west bank. The new metro stations are conveniently sited, brilliantly lighted, and immaculately clean. The metro links some outer suburbs (but not yet the satellite cities) with the center of the metropolis, and at least one car on every train is reserved for women. Trips that at one time took three hours by car, now can take as little as half an hour. As the city's transportation system continues to become more efficient, so will its economy. In early 2015, the Egyptian government announced plans to construct a new capital city on undeveloped land to the east of Cairo. The proposed city would serve administrative and financial functions, and would be planned and developed by the private sector. While the future of this new city remains to be seen, it illustrates the long-standing tensions of managing urban growth while balancing the traditional and the modern.

Jerusalem: City of Three Faiths

Jerusalem occupies neither an attractive site nor a strategic location. It is not central in a geographical sense and lies astride no major trade routes. It is a city that should have been bypassed by time. Instead, Jerusalem has become an epicenter of religious veneration and conflict. Three religions regard it as a holy city: Judaism, Christianity, and Islam. Muslims rank it behind only Mecca and Medina in importance. For them, it is the place from which Mohammed made his "night journey" to heaven to talk personally with God. To mark the place of his ascension, Muslims built the Dome of the Rock in 691 CE (Figure 7.14). To Jews, Jerusalem is the city of David's kingship and Solomon's Temple. The wall of the platform on which the Temple stood is all that remains; it is called the Western Wall and is the focal point of Jewish prayers. To Christians, Jerusalem is the city where Jesus of Nazareth revealed himself to be the Messiah. Since Byzantine times, the Church of the Holy Sepulchre has sheltered the place of the crucifixion, entombment, and resurrection of Jesus.

One always speaks of "going up" to Jerusalem. It began as a hill town at the very southern tip of the western Fertile Crescent. The only feature of the physical environment that commended the site was a spring, now known as the Gihon. The Jebusite village at the spring was destined for prominence, however, largely because its relative location made

Figure 7.14 The Dome of the Rock (venerated by Muslims) and the Western Wall (venerated by Jews) are symbols of a religiously divided Jerusalem. *Source:* Photo by D. J. Zeigler.

a difference 3,000 years ago. The village was conquered by the Hebrew king, David, who needed a centrally located capital between the northern and southern tribes of the Hebrew people. Jerusalem fit the bill. With the decision to move the Arc of the Covenant (containing Moses' tablets of stone) to Jerusalem, the city began to acquire the religious capital needed to sustain its spiritual centrality for three millennia. Jerusalem became the place where the God of Moses and Abraham permanently resided; and, before the first Muslims prayed facing Mecca, they prayed facing Jerusalem. Religion endowed Jerusalem with elements of centrality that geography could not.

The current walls of the old city of Jerusalem date back to the Ottoman period. Within the walls, Jerusalem is divided into four so-called quarters: Muslim, Jewish, Christian, and Armenian (Figure 7.15). The mental map conjured up by such a description, however, belies the reality of the city's cultural geography. In fact, almost the entire old city has an Arab feel to it, save for the Jewish quarter. Furthermore, the boundaries of the quarters no longer (probably never did) define the cultural divisions of the city. Residential patterns and movement into and out of the four quarters challenge the idea that they are homogenous neighborhood groupings of like-minded souls. In the "old city," Muslims and Jews are the primary actors, Christians are diminishing in numbers, and Armenians are doing what they have done best for over 1,500 years,

Figure 7.15 In the Jewish Quarter of Jerusalem, enough archaeological excavation has gone on to bring back the Cardo, or main street, of the ancient Roman city. *Source:* Photo by D. J. Zeigler.

surviving as a culturally distinct Christian minority. Muslim Arabs are expanding into the Christian quarter, the traditional niche of Christian Arabs. Jews are solidly in control of the Jewish quarter; but they are also acquiring property in the other three quarters, which they conspicuously mark with signs, synagogues, and Israeli flags. Jews moving into the old city are more likely to be extremely religious, and those leaving it are more likely to be secular. Armenians, especially seminarians, flow through the Armenian quarter from all over the world, their identity bolstered since 1991 when Armenia reappeared on the map of sovereign states.

The "old city" is only one of two Jerusalems. The other is the sprawling modern metropolis. While the walled city is no larger than a college campus, metropolitan Jerusalem covers at least 50 square miles (129 sq km). Despite being governed as a single municipality, the metropolitan area is bisected by a cultural fault line. West Jerusalem is thoroughly Jewish and provides the site for Israel's parliament, the Knesset. East Jerusalem is primarily Arab (including both Muslim and Christian Arabs), a collection of Arab villages, one of which may someday become the capital of the Palestinian state. "Occupied" East Jerusalem, however, is not as homogeneous as West Jerusalem. In the east, Jewish settlements occupy a dozen hilltop sites, all of them new (post-1967), wealthy, and strategically positioned to maintain control of greater Jerusalem for the Israelis. To the north, south, and east of Jerusalem, where border crossings are patrolled by Israeli soldiers, the Palestinian West Bank begins. The city's relative location between Israel proper and Palestine gives it a frontier feel.

The future of Jerusalem will be determined by the ability of the Israelis and the Palestinians to negotiate a peaceful resolution to their conflicting ambitions. In the meantime, repeated conflict between the Israelis and the Arabs has destroyed the infrastructure of many West Bank cities, such as Ramallah and Jenin. The construction of a separation barrier (also known as the separation wall or fence) by the Israelis further complicates efforts to resolve the conflict. Fifty-six miles (90 km) of the barrier is in Jerusalem, much of it thrusting deep inside the pre-1967 border. The barrier separates Jerusalem from the West Bank. Although it was constructed to increase Israeli security, the barrier also restricts the ability of thousands of Palestinians to reach their jobs, fields, and medical services.

Dubai: Gulf Showplace

A newcomer to the roster of Middle Eastern cities, Dubai's recent and rapid growth now attracts attention on a global scale. Originally a sleepy town on the shores of the Gulf, Dubai was historically focused on pearling and seaborne trade. The town's location was originally dictated by site; it was at the juncture of the Dubai Creek and the Gulf. The Bastakiya neighborhood and the Fahidi fort were located on one side of the creek, and the Deira area, with its markets and wharves, was located on the other. The old city's area is relatively small, giving the rest of Dubai a blank slate upon which to grow. The pace and scale of its growth have been nothing short of staggering. The city's population exploded from 20,000 in 1950 to 2.4 million in 2015. By some measures, its area quadrupled in the space of six years, becoming the Gulf's leading city in terms of business, entertainment, and consumer services.

When the seven emirates that make up the United Arab Emirates joined together to become an independent state in 1971, Dubai was not the largest, the wealthiest, or the most oil-rich. Government strategies have worked to grow Dubai by using funds from its relatively small petroleum deposits to implement diversified development strategies and emphasize non-oil economic activities that put it at the center of the global marketplace. The Jebel Ali port and free-trade zone were established in the 1980s; today, Jebel Ali hosts 6,000 companies and is the eighth busiest container port in the world. Emirates Airlines, owned by the government of Dubai's investment corporation, flies to 140 global destinations, and Dubai's airport has become one of the world's busiest passenger hubs. It ranked 13th in the world in 2010 and has risen to third place today.

Within the city, enclaves have been created to focus on specific activities: the Dubai International Financial Center emphasizes banking and financial services; Media City is a hub for media outlets and internet-based companies; and Knowledge Village houses international educational institutions. This diversification strategy has given Dubai a relatively stable economic base from which to grow. To create an urban identity, a Dubai "brand," the city's entrepreneurial government has actively developed megaprojects, commissioned iconic architecture, and constructed shopping malls, resorts, and other spaces of conspicuous consumption that cater to visitors (Figure 7.16). In addition to the creation of image, these projects help to attract foreign capital and investment.

The highly speculative nature of real estate development in Dubai was made clear by the rapid boom and subsequent bubble economy that was eventually burst by the global recession. A degree of notoriety has accompanied Dubai's success. Some aspects of the city's growth, while impressive, seem more

Figure 7.16 Elements of traditional and modern Arab culture seem to blend harmoniously in the world's largest themed shopping mall, which was named after the medieval Arab geographer Ibn Battuta. It is located in Dubai. *Source:* Photo by Zia Salim.

connected to spectacle than substance: an indoor ski slope is located in a shopping mall; man-made islands in the shape of palm trees jut into the Gulf's blue waters (Figure 7.17) while an archipelago in the shape of a world takes shape offshore; and the tallest building in the world, Burj Khalifa, towers over a city whose skyline already boasted clusters of skyscrapers. Burj Khalifa was finished in 2010 and meant to anchor the new mixed-use downtown. The tower itself has hotel, residential, commercial, and office space, plus an observation deck, all stratified by floor.

Beyond the global and the exceptional, everyday Dubai is a fascinating place. Traditional covered souks specializing in gold and spices anchor older neighborhoods adjoining the Dubai Creek. Traditional wooden ships, called *dhows*, are tied up at the wharves, loaded with all sorts of goods for trips across the Gulf, the Arabian Sea, or the Indian Ocean. Although the skyscrapers are iconic, the largest land use in the city comprises the residential neighborhoods that house Dubai's middle-class population. Dubai has an exceptionally unique social fabric. Estimates suggest that the local Emirati population makes up only about 10 percent of Dubai's population, while South Asian migrants comprise the largest proportion. Although some South Asians are the third and fourth generation to live in Dubai, they cannot gain Emirati citizenship. English is Dubai's lingua franca. Given the fact that upwards of 90 percent of residents are foreign born and that Dubai acts as a global immigration magnet, everyday Dubai is a uniquely diverse place.

A consideration of Dubai's urban future must take into account a number of questions related to economic, environmental, and social sustainability. The city's horizontal sprawl places demands on infrastructure and creates congestion and pollution. Dubai is multinuclear, with several centers that are

Figure 7.17 Palm Jumeirah is one of three palm-tree shaped islands that are being built as a reclamation project in the Gulf. Dubai specializes in landscapes of spectacle that attract the attention of the world. *Source:* Photo by Zia Salim.

separated by multilane highways. Although a new metro rail system has been in place since 2009, public transit and pedestrian movement are both subordinate to the automobile. Water is supplied by desalination, which is currently fueled by abundant and cheap energy. The social dimensions associated with having a large nonnational population, particularly migrant workers living in stark conditions, deserve consideration. Urban space is stratified, as prominent spaces of consumption, including gleaming shopping malls, modern supermarkets, and an annual shopping festival, segregate those who can afford to consume them from those who cannot.

All in all, Dubai abounds in contrasts. Its urbanization exemplifies, albeit at a magnified scale, other cities in the Gulf. Although the relative wealth, urban entrepreneurialism, and unique demographics of Dubai and other Gulf cities (Doha, Abu Dhabi, Manama) make them outliers in the overall roster of Middle Eastern cities, they are a significant group in and of themselves.

Mecca: City of the Hajj

Mecca is the city at the heart of the Islamic world. It has singular religious significance in the lives of the world's billion-plus Muslims—devout Muslims face Mecca five times a day in prayer, and a pilgrimage (*hajj*) to Mecca once in a lifetime is a pillar of Islam. Mecca is located in the Hijaz mountain range along the Arabian Peninsula's western extent. Site controls the city's form: mountainous topography channels the city's growth and pushes development away from the core through valleys into the surrounding desert plains.

Besides its innate religious importance, numerous locations in and around Mecca are associated with Islam's beginnings: the place where Mohammad was born, the mountain where he received his first revelation of the

Koran, and the hill from which he made his final sermon. However, while Mecca has spiritual importance, it was never Islam's political capital. Within a few dozen years after the death of Mohammad, the political center successively shifted from Medina to Damascus and later Baghdad. As a succession of caliphs assumed both spiritual and secular roles, Mecca never acted independently of the new centers of power. But urbanism and Mecca's religious and historical importance are intertwined. The Great Mosque and the *hajj* impacted the city's morphology and functions; at a regional scale, pilgrim traffic to Mecca was one of the reasons for the relatively high degree of urbanization in this part of the Middle East during the late Arab and Ottoman eras.

Mecca's current population is about 1.7 million. While the city is a year-round religious destination, it experiences extreme seasonal swells as the pilgrimage brings an annual influx of the devout (both internal and foreign). In 2014, 2 million pilgrims visited Mecca. The city's non-pilgrim population is ethnically diverse, as there is a history of people from many nationalities completing the pilgrimage and remaining in Mecca. This migration has created a cosmopolitan city, and the Mecca's non-pilgrim immigrant population is large enough to rank it as one of the largest in the world. Finally, a politics of exclusion operates in Mecca, as non-Muslims are not allowed to enter the city. While the creation of sacred space through exclusion occurs in other religious traditions, its application at the scale of an entire city is a singular example of how exclusion is used to create space and community.

Mecca's core is dominated by the mosque (instead of a fortress, as in the Middle Eastern city model). The monumental mosque that is Mecca's spiritual and physical heart has grown as several cycles of development have expanded the structure and provided more modern amenities. Most recently, the Saudi government has made impressive efforts to serve visitors to Mecca. An ongoing expansion project will, at the expense of surrounding neighborhoods, increase the area of the mosque and associated external courtyards. At this project's completion, the mosque complex alone will be able to accommodate a staggering 1,500,000 visitors at a time.

The mosque is similar to a Central Business District (CBD)—it can be thought of as a Central Religious District. In the core-frame model, the CBD continually grows in different directions, alternately absorbing and discarding the surrounding building stock. In Mecca, there is no zone of discard, only a zone of accumulation. Commercial properties near the mosque are almost exclusively geared toward religious visitors; consequently, land values at the core are extremely high, regardless of the age, size, and condition of the property. The mosque's immediate environs are increasingly dominated by large capitalistic developments. International hotel chains such as Sheraton, Intercontinental, and Hilton are located in mega-developments that integrate shopping and restaurants to create high-end consumer spaces. Given the guaranteed draw of religious mobility, Mecca has been relatively insulated from the booms and busts of the global economic slowdown. In fact, the pace of development has recently increased. In the past 30 years, local and international construction firms, such as Dubai's Emaar, have worked in tandem with the state to develop increasingly ambitious projects in Mecca, some of which seem out of place. Other cities with structures of special significance have enacted height restrictions to prevent significant buildings from being eclipsed, but Mecca has been less fortunate. The new towers mentioned above loom figuratively and literally over the mosque. These mega-developments are framed as visitor-serving *nahda umranniah* or "urban progress," but the question about whether this is really progress (and who it serves) is pertinent.

The older parts of Mecca extend away from the mosque, nestled in the valleys and hanging onto the less accessible lower portions of the mountain slopes, safe (for now) from the bulldozers and cranes. Still visible in parts of the core, older parts of the city are tightly organized into residential zones or quarters with small, densely packed houses, narrow, winding lanes, and an organic plan. However, the creative destruction associated with the mosque's expansion and associated development has eroded the old city's imprint in successive waves of demolition. The area north of the mosque is a massive construction site, with no traces remaining of the dense urban neighborhood that existed there only five years ago; in other places, older neighborhoods have had wide streets cut through them. In general, neighborhoods in the older part of the city have filtered down and are of lower quality. The rest of Mecca has grown to dwarf the remaining old city area.

The historic preservation seen in other world regions is lacking in Mecca. The city's old gate is a remnant exception and a reminder of the time when the city was enclosed by walls. In addition, some buildings in older neighborhoods are still utilized as they were in the past. However, in a city with great historic significance, relatively few historically significant structures remain.

Moving outward from the core, the modern city, the everyday Mecca that nobody hears about, is where the city's 1.7 million residents live, work, and play. The modern city sprawls into the valleys away from the core in all directions. It has single-family homes, apartment blocks, small parks, schools, offices buildings, and suburban shopping malls, on a planned street pattern. Residential structures here are a combination of single-family, two- to three-floor, walled villas and mid-rise apartment buildings. The Saudi government's Real Estate Development Fund provides long-term, interest-free loans to private-sector builders, and most residential structures are individually financed and built by construction firms and contractors. To varying degrees, the urban impacts of Islam can be seen throughout Mecca; to accommodate the annual influx of hundreds of thousands of pilgrims, hundreds of hotels have been built in the modern city, typically along major transportation corridors. However, because of the seasonality of demand, these blocks of contemporary towers are occupied for only two months a year at most. Mecca's outskirts contain large areas of new government-built housing tracts, in master-planned developments with regular street layout and integrated park and recreational space.

Istanbul: Transcontinental Hinge

Istanbul has existed for almost 27 centuries (since 657 BCE), for 16 of them as an imperial capital. Its original name was Byzantium, but it was rechristened the New Rome in 330 CE. Almost immediately the people began calling it Constantinopolis, Emperor Constantine's city. Today, it appears on the map as Istanbul. It has been the dominant city of the eastern Mediterranean realm for more than a millennium, having surpassed a million inhabitants by 1000 CE. Today, the population of the urban agglomeration is 14 million and growing (Box 7.3).

Istanbul's location makes it a hinge between continents. From its situation on the European side of the Bosporus, it is positioned to control

Box 7.3 Istanbul's Double-edged Crisis of Urban Ecology and Democracy

A critical perspective on what is happening in cities requires us to understand that urban development is seen differently by different parts of the community. For every narrative that helps us to understand a city like Istanbul, there are many counter-narratives. Captured from the blogosphere, the assessment below paints the unbridled destruction and reconstruction of Istanbul as a threat rather than an achievement. The Reclaim Istanbul blog is written by Yaşar Adnan Adanalı; his original post had the provocative title "Blood Architecture" to highlight the construction worker lives lost in turning Istanbul into a world city. Adanali sees Istanbul as:

(1) An economy dominated by the construction sector: The present government came to power in 2002 in the aftermath of one of Turkey's worst financial crises. Since then, the government has initiated and supported urban and rural interventions at a grand scale to resolve the country's capital surplus absorption problem. Today, after more than a decade, economic growth in Turkey is heavily dependent upon the construction sector.

(2) Massive-scale urban transformation: Turkey is currently experiencing an urban transformation at a massive scale. The expected number of housing units in Turkey to be demolished and redeveloped is around 7 million, a substantial part of which is located in Istanbul.

(3) Rapid and unlimited access to urban and rural land: Rapid and unlimited access to urban and rural land is at the center of this economic model. In 2013, 60 percent of all decisions made by the Council of Ministers were related to real estate development and construction. The dispossession of the urban poor, the loss of public spaces, and the threat to urban ecology are unavoidable repercussions of the search for further urban land to be developed.

(4) Istanbul becoming global: Being at the center of this economic policy, Istanbul, in the last ten years, has rapidly become a mega construction site where around 30 percent of the national GDP is produced. It is a Global City in the making. The fact that Istanbul was ranked first for real estate investment and development in Europe in 2012 underscores this assessment.

(5) Uneven social development: The construction boom that the city has been undergoing is accompanied by a highly uneven social development. On the one hand, Istanbul is now number five on the list of world cities with the highest number of dollar billionaires, yet on the other hand, Turkey is ranked last among the 31 OECD countries in terms of social justice.

The construction frenzy not only dispossesses the urban poor, encloses public spaces, or endangers fragile urban ecology, but also claims the lives of the workers. In 2013, at least 1,235 workers lost their lives in Turkey, with most of these deaths taking place in the construction sector.

Sources: Reclaim Istanbul (http://reclaimistanbul.com); Mutlu Kent (http://mutlukent.wordpress.com)

overland trade between Europe and Asia, and the shipping lanes between the Mediterranean and Black Seas. The huge empires that Istanbul commanded—Roman, Byzantine, Ottoman—also gave it the ability to control overland access to Arabia, the Indian Ocean, and eastern Asia. Until the sea route around Africa was fully opened in the sixteenth century, Istanbul was able to control trade between north and south, east and west. As the imperial capital of the Ottoman realm since 1453, it reached its peak in the 1500s when the emperor, Suleyman, commanded so much wealth that he was known to the world as "The Magnificent." His city was at that time larger in population than London, Paris, Vienna, or Cairo. Later that century, however, the power of Constantinople began to wane. No longer did Ottoman subjects hold a monopoly on the ancient silk routes across Asia or on the Fertile Crescent caravan trade from the eastern Mediterranean to the Persian Gulf. As technology enabled mastery of the sea, caravels replaced camels as the most reliable and economical modes of transport. Well before the end of the nineteenth century, the Ottoman Empire was the "sick man of Europe," and "Stamboul" was a city in decline. Modern Turkey was born out of the Ottoman Empire, thanks chiefly to the secular nationalism inspired by Mustafa Kemal Atatürk (Figure 7.18).

The core of Istanbul, the historical city, occupies a peninsular site with deep water on three sides. Crowning the peninsula, and visually dominant from the sea, are seven hills, just like Rome. The peninsula is bordered on the south by the Sea of Marmara, on the east by the Bosporus, and on the north by the Golden Horn, the large, sheltered harbor that enabled the city to dominate the shipping trade. The Bosporus and its companion strait, the Dardanelles, enabled oceangoing vessels to penetrate central Eurasia. On the Black Sea's northern shore, the ancient Greeks implanted colonies. The fertile

Figure 7.18 Ataturk, the revered father of modern Turkey, continues to be memorialized on the urban landscape. In this case, his visage is positioned to welcome those approaching Izmir from the airport. *Source:* Photo by D. J. Zeigler.

hinterlands of these colonies became a breadbasket, producing wheat for the Aegean core of the Hellenic world, wheat that went to market via the Bosporus. The first "world-class" city to dominate these straits goes back to the Bronze Age. Its name was Troy, and it was located at the southern end of the Dardanelles. Troy was the Istanbul of its day.

In the twentieth century, the Bosporus provided one of the Soviet Union's few outlets to the world ocean and was consequently a point of strategic significance during the Cold War. By controlling the strait at Istanbul, NATO could deprive Moscow of dominating one of the world's most strategic locations. Even now, the Bosporus continues to be important to Ukraine and the Russian Federation. It has also taken on a new strategic significance in ensuring the steady flow of crude oil from the landlocked Caspian Basin fields, much of which transits the Bosporus, already one of the world's busiest straits.

Oil is not the only commodity important to Istanbul's economy, however. During most of the twentieth century, Istanbul's European hinterland was all but severed by the Iron Curtain and the animosity of neighboring Greece. Now, however, Eastern European nations have opened their borders. The routes of commerce between Europe and Asia are once again funneling traffic across the Bosporus. The growth of trade is nowhere more powerfully symbolized than by the growing volume of truck traffic navigating the transcontinental Bosporus bridges, completed in 1974 and 1988. Now, a new rail line designed to carry passengers and freight has begun operating through an immersed tube under the Bosporus. The Eurasia Tunnel is under construction and will be yet another intercontinental crossing, this one designed to link Europe and Asia by a highway tunnel under the strait. Istanbul also aspires to be a more important gateway to Central Asia, where the Turkish people originated and where most of the languages spoken are Turkic in origin.

URBAN PROBLEMS AND PROSPECTS

Urban issues in the Middle East (and elsewhere, of course) are interconnected. For example, residents in Cairo's informal settlements have limited access to freshwater and sanitation systems and their living conditions often cause environmental degradation, which can, in turn, lead to negative social and political outcomes. Similarly, high land costs and the lack of affordable housing can be a flashpoint for other grievances that ultimately result in political instability. In this section, we profile some of the more acute issues faced by cities in the Middle East. The issues and processes discussed here operate at a variety of scales, from the region to the nation to the city to the neighborhood. Further, these issues are related to larger structural issues. For example, social inequality is a structural factor that exacerbates the impacts of some of the individual issues discussed below.

Water

Urbanization in the Middle East is critically affected by questions of water security, which also affects growing populations, agricultural demand, and climate change. Urban expansion depends on the development of water resources for new homes, businesses, and industries. Also, given that upwards of 80 percent of water use is in the agricultural sector, questions of food security for burgeoning urban populations have to be balanced against water security; policies aimed at achieving an assured source of water have significant implications in terms of food self-sufficiency and employment in the agricultural sector. All in all, the balance between supply of water and demand on it is extremely tenuous.

Stress on freshwater supplies is especially critical in arid environments where water resources are scarce and new sources are expensive. Thirteen of the world's most water-scarce countries are in the Middle East. Groundwater and surface water resources are being utilized to the maximum throughout most of the region. Sana'a in Yemen threatens to become the world's first capital city to go dry, as groundwater extraction, primarily for agricultural purposes, has seriously depleted the basin's aquifer. Other aquifers, in Syria and Jordan, are also being used beyond their rates of recharge. Renewable aquifers, while easy to use, are far too easy to damage if a balanced, regulated approach is not used. The Gulf States and Israel are pursuing desalination to provide freshwater supplies (Figure 7.19), but desalination has significant energy costs. Further, effluent from desalination discharges hot, high-salinity brine and trace metals, which can harm marine life and biodiversity in coastal zones. Jordan, with one of the world's lowest levels of water-resource availability per capita, is making significant investments in costly infrastructure in an attempt to provide freshwater.

On the other hand, issues relating to water are exacerbated by poor management and infrastructure; these are further magnified by larger structural factors such as urban growth and socioeconomic inequality. One key point is that most cities do not use their water resources efficiently. High subsidies and

Figure 7.19 The Sorek seawater desalination plant, one of the largest in the world and one of five in Israel, became operational in 2013. Israel is a world leader in the field despite the drawbacks: the immense amount of energy needed for desalination and the environmental costs of disposing of the brine. *Source:* Photo by Ben Sales/JTA.

low water tariffs have created a fiscally unsustainable water supply network. What water exists could go further if leaky pipes were repaired, if demand was managed more efficiently, if water-conserving technologies were used, and if irrigation systems made do with less. Plus, water problems are not simply a matter of quantity; they are also problems of quality. Virtually every city must concentrate on upgrading its water treatment operations so that tap water is safe to drink. Although some countries have shifted from focusing on water infrastructure to improving management of water resources, results are mixed due to the overall complexity of the water sector. The implementation of demand management (rationing) and increased tariffs are important parts of a balanced solution to water scarcity, but they are politically complicated. Finally, as with the question of the environmental disamenities discussed below, the poor lack access to efficient water services.

Environmental Degradation

Environmental degradation in Middle Eastern cities is uneven, reflecting the varying levels of infrastructure quality and socioeconomic status. Environmental degradation spans a gamut of issues, including waste disposal, air pollution, water purity, pests, noise levels, and chemical pollution. Environmental degradation and urbanization are connected in a circular relationship: environmental degradation can detrimentally affect urban residents, and urbanization can cause or exacerbate environmental challenges. Further, these factors can be compounded by population concentration, density, and growth. Poverty and environmental justice are also factors, as the poorest residents tend to live in the most polluted or hazardous areas.

Urban pollution is a significant challenge in the Middle East. Air quality has been impacted by increases in the number of cars, underinvestment in public transportation and infrastructure, and the region's increasingly sprawling cities. When increasing numbers of cars enter the medinas on ancient narrow streets that were never designed to accommodate them, they cause congestion and pollution. Water pollution occurs when unregulated industrial activities, especially small-scale enterprises, such as tanning, and agriculture's chemical inputs add contaminants to water sources. Inadequate or nonexistent urban sanitation infrastructure causes solid and liquid waste to be disposed of improperly: Waste may be dumped in water channels that are used for household uses, solid waste may be dumped in neighborhood dumps, or, in the case of some municipal dumps, simply incinerated.

The Middle East has particular vulnerabilities to climate change. The IPCC's 4th Assessment Report indicates that the region is projected to gradually become hotter and drier. Results will include a greater risk of drought, increased water stress, desertification, reduced agricultural productivity, loss of hydropower, and the unsustainability of some existing crops. Sea-level rise is particularly worrisome for cities in low-lying areas. More than $30 million has already been spent to install sea walls along the Egyptian coast. Similarly, sea-level rise will increase salt water intrusion into the coastal aquifers that cities in the eastern Mediterranean and North Africa rely on. Climate change also carries attendant political, economic, and social impacts. For example, the UN has estimated that in Egypt alone, a 0.5m rise in sea levels would affect nearly 4 million people and cause $35 billion in losses. Other socioeconomic impacts include resource-based migration, increased tension between the countries that share water resources, and heightened stress on natural resources, all of which can cause or exacerbate political instability.

Environmental degradation has a variety of other impacts. Car exhaust and noxious industrial effluents pose public health hazards; the prevalence of respiratory illnesses due to air pollution is a troubling phenomenon; and uncollected solid waste and informal dumps shelter a wide range of disease vectors. Significant economic costs are associated with environmental degradation, including use of high levels of energy in the transport sector as a result of gridlocked traffic, loss of productive agricultural land as a result of urban sprawl, and higher mortality and health-care costs as a result of low air quality. For example, in Cairo and Alexandria, the damage caused by urban air pollution has been estimated at 2 percent of GDP, and 20,000 people a year die due to air pollution–related causes.

The urban poor often bear the brunt of environmental disamenities (e.g., industrial air and water pollution) and hazards (e.g., risks from natural disasters, such as rockslides and landslides). Similarly, the public health impacts of environmental change disproportionately impact the poor because of their lack of access to resources and public health infrastructure.

Housing

Given that residential space typically forms the largest single land use within a city, housing is a dimension of urbanization that directly affects the daily life of urban residents. The growth of informal housing, primarily on urban peripheries, is a modern trend that is driven by rural-to-urban migration, growing populations, and economic conditions at a range of scales. For example, *gecekondus*, "houses built (without permission) overnight," have been estimated to house half of Istanbul's population. Informal housing in Turkey started appearing in the 1950s as economic opportunities associated with urban industrialization attracted migrants from Turkish villages. This, coupled with real estate speculation that drove the cost of formal housing in the cities out of reach of the average middle-income household, started the first wave of *gecekondu* construction. Similarly, about 25–30 percent of Cairo's population (perhaps 4 million people) live in *ashwaiyyat* or "informal zones" in and around the city. Conditions in these zones can be very poor, as inadequate infrastructure for water, electricity, and sewage, combined with crowded conditions and reduced access to education, health care, and other important governmental services lead to negative social and environmental consequences.

Another issue related to housing is that of socioeconomic polarization and the suburbanization of some types of housing. Urban residents are faced with limited property availability, limited access to credit, and lack of affordability; further, government policies have encouraged planned residential developments on urban peripheries. The resultant suburbanization of metropolitan populations has implications in terms of sprawl and infrastructure provision. Suburbanization also intersects with the question of socioeconomic polarization: The contemporary flight of the middle and upper classes from cities has new implications in terms of urban inclusiveness as exclusive housing developments and gated communities have sprouted up in cities in Egypt, Turkey, Lebanon, and Saudi Arabia, among others. As higher-income groups move away from density and crowding into gated communities, additional polarization is added to the city's social geography. In Cairo, the large suburban gated communities have names like "Beverly Hills" and "Dreamland."

Planning responses to housing-related issues have been uneven. In some countries, such as Egypt, squatter settlements have remained poor, underserved by urban infrastructure, and socially marginalized. In other countries, such as Turkey, squatter settlements were quickly legitimized by the government, and upgrading was implemented. Some affordable housing and social housing programs have been developed across the Middle East and, as a legacy of socialist planning, are prominent in Central Asia. However, limited public-sector budgets and rapid urban growth have meant that housing demand outweighs supply. For example, in Egypt, the average home price is 18 times the typical annual salary and in Israel the average home price is 16 times the typical annual salary (the comparable figure for the United States is 3.3 times). In many countries, planning authorities or government officials are increasingly looking to the corporate real estate sector as a model for urban planning and development, as evinced by reliance on modern makeover plans and megaprojects. However, questions remain about the long-term economic sustainability of this model and the types of outcomes that it produces.

CONCLUSION

Urban life in the Middle East is not utopian, but it is not dystopian, either. Cities in this region have their problems, some seemingly intractable, just as in other world regions. Authoritarian regimes and political violence at a variety of scales are problems that have real impacts on cities. Geopolitical tensions simmer. The ecological sustainability of cities is impacted by environmental degradation, climate change, and water security. However, to characterize the entire region as troubled, unwelcoming, or uniformly "violent" would be a broad generalization and a gross mischaracterization. Middle Eastern cities do many things well. First, they reflect the hospitality of their inhabitants, people who easily talk to visitors, who are eager to communicate despite linguistic barriers, and who have (and take) time to spend in casual conversation on the street. Arabs, Turks, and Iranians are among the friendliest people in the world, and their cities make you feel at home.

Second, Middle Eastern cities, with few exceptions, are safe day and night. Social networks are strong. There are "eyes upon the street" all the time, whether you can see them or not; and family networks, undergirded by strict codes of conduct, hold family members accountable. Third, the generations mix freely. Neither the old nor the young are warehoused; parents are seen with children; teenagers use the same streets as the elderly; and young apprentices are common in the city's businesses. Households are often multigenerational, and the extended family is more prominent, socially, than the nuclear family. Fourth, homelessness, although it does exist, is less common than in Western cities. It is taken for granted that some people will not be able to live self-sufficient lives, so families compensate for personal inadequacies, and many social needs are taken care of by the Islamic emphasis on required almsgiving and charity. Fifth, food is central, and sharing a meal with friends and family is a leisurely and enjoyable event. Almost every city takes pride in its food, whether served in sit-down restaurants or on the street (Figure 7.20). Middle Eastern cuisine reflects and helps to define both national cultures and urban life. It often is healthy food, not overprocessed, and rarely fried. Sixth,

Figure 7.20 When you have a business that is mobile, you can move with the market, which is exactly what this street vendor of qanafeh (a sweet pastry always made in round pans) does in Amman. *Source:* Photo by D. J. Zeigler.

cities are generally well served by a variety of transportation. Cars are not required; and in the old cities they may be a hindrance. City buses, taxis, service taxis (often 12-passenger vans), and fixed-rail lines (in a few cities) always make it possible to get around at very low cost. The cores of cities are often compactly organized, so walking is a possibility.

Over the centuries, Middle Eastern cities have produced many of humanity's most enduring achievements and legacies. These flourishing cities anchored empires and gave rise to religions. Today, cities in the Middle East are complex, bustling, and vibrant places. A strong sense of community characterizes life in Middle Eastern cities. The human element of place, seen in food, art, music, culture, and literature, adds the important element of culture to studies of cities across the region. Grassroots advocacy on real issues, from politics to the lack of greenspace, illustrates the productive power of the city (Box 7.4). Combining these bottom-up approaches with top-down approaches such as proactive urban planning can make cities in the Middle East even more inclusive and sustainable places.

Box 7.4 A Hopeful Vignette: Cairo's Al-Azhar Park

While the Middle East's problems may seem intractable, there are many reasons to remain optimistic about cities in the region. One example comes from Cairo. Over the span of 500 years, generations of Cairenes had dumped their debris and household garbage on a 30-hectare site originally located just outside the Fatimid city walls. In some places, the debris and trash were 130 feet (40 m) "deep." Today, several dense neighborhoods surround the site. Both Al-Azhar Mosque and Al-Azhar University are minutes away, and the ancient Citadel is directly to the south.

After a 1984 conference on urban growth in Cairo, Aga Khan decided to finance and create a park as a gift to Cairo's residents. Cairo was severely park-poor: one analysis indicated that every resident of Cairo had 54 in^2 (350 cm^2) of green space, approximately the area of two adult footprints. Given the difficulty of locating open space in such a densely populated city, the dump was suggested as a potential park location. After the site's selection, years of extensive geotechnical surveys, excavation (80,000 truckloads of material, equivalent to more than half the Great Pyramid's volume, were removed), soil remediation, landscaping (experiments were conducted for five years to determine the most suitable plants), and construction followed. Finally, Al-Azhar Park opened to widespread acclaim in 2004. The *New York Times'* architecture critic argued that it reversed "a trend in which unchecked development has virtually eradicated the city's once-famous parks." The park's design is inspired by Islamic gardens from the Persian and Mughal Empires, its landscape architecture optimizes irrigation and includes plants adapted to arid climates, and it includes one of Cairo's few public children's playgrounds. But Al-Azhar Park's story does not end with the vital green space or the impressive vistas it has provided.

During construction, a completely buried section of Cairo's Ayyubid-era defensive wall was uncovered, complete with gates, towers, passageways, and galleries. This 5,000-foot section (1,500 m) has been preserved and now serves as a connection between the park and the Ayyubid city. Following the Islamic endowment system, income from the park's tickets, parking, and restaurants are used for park maintenance. Al-Azhar Park's development purposely called for urban upgrading in the surrounding neighborhood. The local community prioritized a list of neighborhood rehabilitation efforts, and income from the park provided funds for a community center, training programs, micro-loans for small business owners, and the renovation of mosques, schools, and homes. The park's construction and operation created jobs. Local artisans were employed to do stonework, in a revival of ancient craft techniques that had been in danger of dying out. Today, residents of the adjacent neighborhood are given preferential hiring for jobs at the park. Al-Azhar Park also provides infrastructure: three giant reservoirs that had been slated to be sited in the dump to provide drinking water for Cairo have been placed below ground in the park and landscaped to minimize their visual impact. Cairo may have turned the corner in restoring some open space to one of the world's largest cities.

SUGGESTED READINGS

Abu Lughod, J. L. 1971. *Cairo: 1001 Years of the City Victorious*. Princeton, NJ: Princeton University Press. A chronicle of Cairo from 969 to 1970, and a glimpse of how to make sense of any urban landscape.

Benvenisti, M. 1996. *City of Stone: The Hidden History of Jerusalem*. Berkeley: University of California Press. Offers a balanced view of Jerusalem's urban landscapes, boundaries, and demographics.

Dumper, M. 2014. *Jerusalem Unbound: Geography, History, and the Future of the Holy City*. New York: Columbia University Press. Presents Jerusalem as a city of enclaves that undermine Israeli control of the city.

Elsheshtawy, Y. 2004. *Planning Middle Eastern Cities: An Urban Kaleidoscope*. London and New York: Routledge. Presents urban planning in the context of globalization, with separate chapters on Cairo, Dubai, and Algiers.

Hitti, P. K. 1973. *Capital Cities of Arab Islam*. Minneapolis: University of Minnesota Press.

Thoughtful profiles of historical capitals: Mecca, Medina, Damascus, Baghdad, Cairo, and Cordova.

Hourani, A. H., and S. M. Stern. 1970. *The Islamic City*. Philadelphia: University of Pennsylvania Press. Delves into the question of whether there is an "Islamic" city, with specific reference to Damascus, Samarra, and Baghdad.

Kheirabadi, M. 2001. *Iranian Cities: Form and Development*. Syracuse, NY: Syracuse University Press. A thorough treatment of the spatial structure and physical form of Iranian cities.

Salamandra, C. 2004. *A New Old Damascus*. Bloomington, IN: Indiana University Press. Presents a portrait of Damascus' cultural anthropology, with considerable attention focused on the problems of historical preservation.

Saliba, R. 2015. *Urban Design in the Arab World: Reconceptualizaing Boundaries*. Farnham, UK: Ashgate. Draws on case studies to articulate a regional geography of urban design, which is conceptualized as discourse, discipline, research, and practice.

Serageldim, I., and S. El-Sadek, eds. 1982. *The Arab City: Its Character and Islamic Cultural Heritage*. Riyadh, Saudi Arabia: Arab Urban Development Institute. Photographs, drawings, and readable text on city form and urban planning.

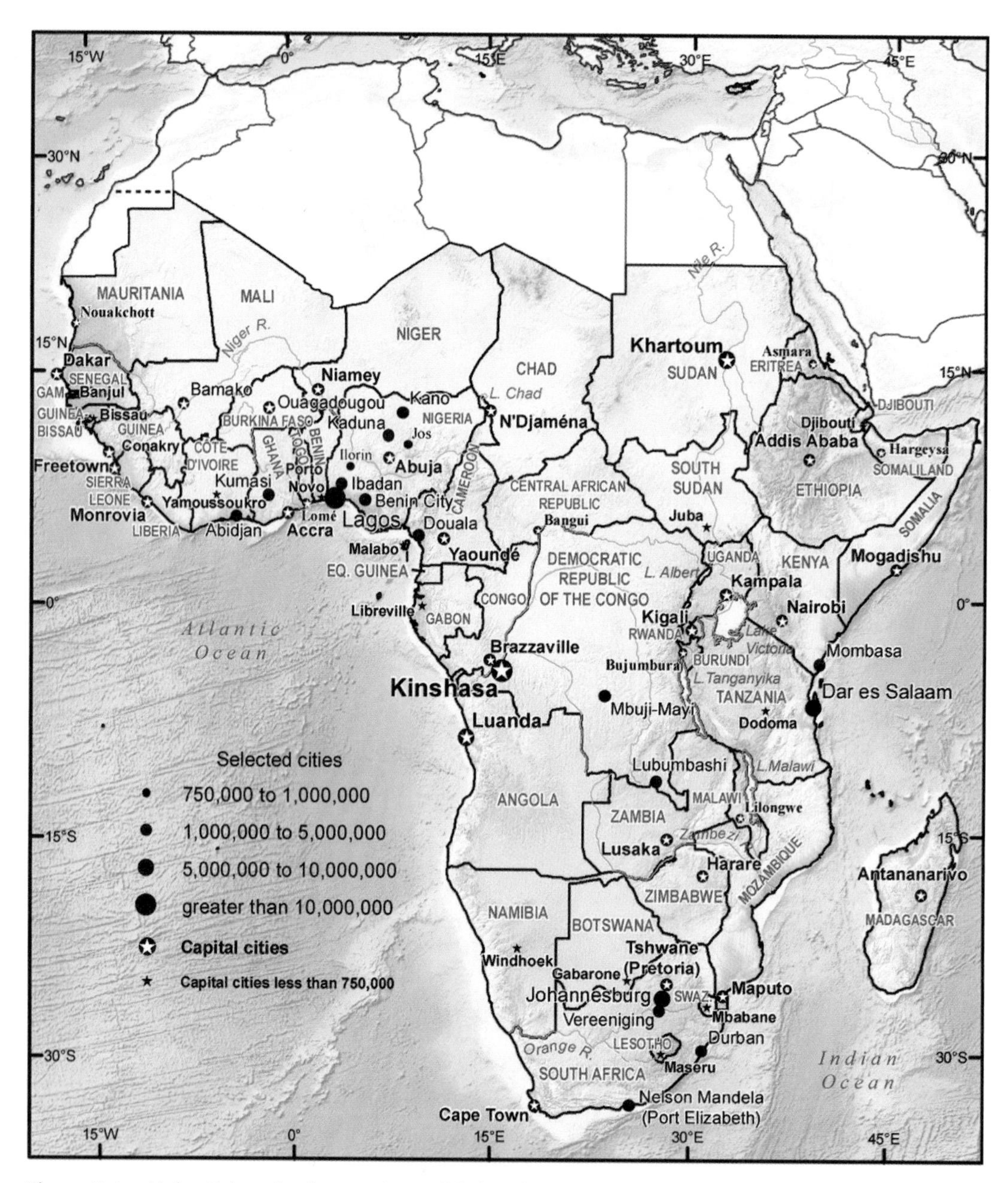

Figure 8.1 Major Urban Agglomerations of Sub-Saharan Africa. *Source:* United Nations, Department of Economic and Social Affairs, Population Division (2014), World Urbanization Prospects: 2014 Revision, http://esa.un.org/unpd/wup/.

8

Cities of Sub-Saharan Africa

GARTH MYERS, FRANCIS OWUSU, AND ANGELA GRAY SUBULWA

KEY URBAN FACTS

Total Population	963 million
Percent Urban Population	37%
Total Urban Population	359 million
Most Urbanized Countries	Gabon (87%)
	Djibouti (77%)
	South Africa (64%)
Least Urbanized Countries	Burundi (12%)
	Uganda (16%)
	Malawi (16%)
Number of Megacities	2 cities
Number of Cities of More Than 1 Million	43 cities
Three Largest Cities	Lagos (13 m), Kinshasa (11 m), Luanda (5 m)
Number of World Cities (Highest Ranking)	2 (Johannesburg, Cape Town)
Emerging World Cities	Nairobi, Lagos, Durban

KEY CHAPTER THEMES

1. Sub-Saharan Africa (SSA) is among the least urbanized of the world's regions, but it has some of the world's most rapidly urbanizing countries.
2. A rich urban tradition preceded the arrival of colonialism in several parts of Sub-Saharan Africa.
3. Colonialism had profound impacts on urban development, particularly in the creation of what would become primate cities along the coast.
4. Rates of urban primacy are generally high across the region, with a few exceptions, and economic production and political power are concentrated in the primate cities.
5. Many, though not all, primate cities are also capital cities.

6. Many SSA cities have experienced major impacts from cultural globalization, as in changing patterns of consumption and personal security, but minimum impacts from economic globalization, in terms of production and investment.
7. Most SSA urban land-use patterns and urban economies develop outside of formal regulation, but with significant overlap of the "formal" and the "informal" urban structures and economies, both of which are highly gendered spaces.
8. Many Sub-Saharan African cities are characterized by spatial, socioeconomic, and gender inequalities and high rates of urban poverty.
9. Water is a major concern in Sub-Saharan cities, as urban growth and climate change impact water availability and quality.
10. Great cultural diversity and creativity help shape very dynamic urban life experiences for residents of the region's cities.

SSA's interlocking urban environmental problems are magnified by shortcomings in management and oversight by both governments and the private sector. Patrick lives in Dar es Salaam, Tanzania, where he works as a chef at a Chinese restaurant. Patrick is a mixed-race South African, born in Cape Town. He worked for many years as a cook on oil tankers, where many of the crew members were Bangladeshi, Filipino, or Tanzanian, the latter often from the Zanzibar islands or from Dar es Salaam. After a stint cooking for an offshore oil rig in Cabinda, Angola, he took up an offer from a Tanzanian friend to come start a new high-end restaurant in the rapidly gentrifying inner city Kariokoo neighborhood in Dar es Salaam. Patrick has found it exciting to learn KiSwahili, which will be his sixth language once he conquers it, but he posts messages on Facebook and tweets for his South African friends around the world in Afrikaans or English. He loves the mix of foods and cuisines available in Dar es Salaam, but favors Chinese food, which gained a foothold in Tanzania and much of Africa along with Chinese investments in the region's cities in the early twenty-first century. He hopes that the restaurant will take off, with an eclectic mix of Tanzanian African, Asian, and European customers.

Meanwhile, across the world in Houston, Jamila, a Nigerian-born software engineer, receives a text message from her father in Calabar asking her to call home. She knows what this means, but hesitates because she wants to have good news for him before she calls. She sends an email to the secretary of the local hometown association for southeastern Nigerians in Texas, and asks for an update on her plea for help in raising funds—because her father will be telling her, she knows, to come home for her mother's funeral. Jamila has the money for her own plane ticket, but she knows the family will expect her to pay all of the funeral costs and to bring her twin daughters with her. She is conflicted, since she knows that the whole extended family feels that they have invested in her education and emigration with the expectation that she will provide support through remunerations. She has succeeded for many years in sending enough money home to build her parents the nicest house in their neighborhood in Calabar; but her husband's recent death has put a major strain on her household financially, to say nothing of her sadness. The hometown association secretary tells her the news she has been waiting for: in just two days, the large southeastern Nigerian community in Texas has raised more than $10,000 on her behalf. She does not know how she will ever thank these people, many of whom she does not know and only a handful of whom would ever have met her mother. She would do the same for them, she tells herself because "all of us in what we call The Remote Lands have to stick together to help our Motherland." She calls her father in Calabar, Skype-to-Skype since it is free, with the good news, and on her laptop video camera box, through the Skype software, she sees tears on her father's face, for the first time in her life.

AFRICAN URBANIZATION

SSA has long been among the least urbanized world regions. But many Sub-Saharan countries have been urbanizing rapidly since the 1960s (Figure 8.1). This rapid urban growth has come with limited opportunities for employment in the formal economy or for effective governance. African cities also suffer from a lack of decent and affordable housing, failing infrastructure and basic urban services, alongside increasing inequalities (Figure 8.2). But negative views of contemporary cities are overly simplistic and pervasive. African cities are also creative engines of cultural change and dynamic centers of political and associational life (Figure 8.3). Many accounts of cities in SSA miss the resourcefulness, inventiveness, and determination of millions of ordinary people who manage to negotiate the perils of everyday life, to make something out of nothing (Box 8.1).

Cities in Sub-Saharan Africa are diverse and heterogeneous. Scholarly efforts to construct an ideal model of a generic "African

Figure 8.2 Chronic flooding necessitates near-constant, major efforts to drain residential areas of Pikine, an informal city on the outskirts of Dakar, Senegal. Many of SSA's informal settlements are flood-prone, yet their residents often experience the deprivation of limited access to clean drinking water. *Source:* Photo by Garth Myers.

Figure 8.3 Bustling markets, such as this one in Monrovia, Liberia, are common features of Sub-Saharan cities. *Source:* Photo by Robert Zeigler.

city" in terms of urban structure have failed to find a single profile that fits all cases. Anthony O'Connor, for example, tried to fashion such a general scheme more than 25 years ago, but his effort led him toward not one but six possible types. O'Connor identified and diagrammed city morphologies that he classified as the indigenous city, the Islamic city, the European city, the colonial city, and the dual city with examples across the continent. His sixth category—what he termed the hybrid city—actually functions as a kind of catch-all for cities with multiple morphological characteristics. Over time, more cities in Africa seem to have become hybrid cities.

The paths to such African hybrid cities are complex and sometimes contradictory. While many cities came into existence as overseas extensions of European colonial powers seeking to establish beachheads on the African continent, their subsequent growth and development did not conform to one pattern. City-building processes that took place under the dominance of European colonialism often left an indelible imprint on the original spatial layout, built environment, and architectural styles of cities in Africa. Yet with time, these features have sometimes been modified beyond recognition. Thus cities that were built specifically for Europeans, such as Cape Town or Nairobi, still have clear European influences, but these have been overwhelmed by African urbanism. The colonial urbanisms, too, have been dramatically transformed by what amounts to 50 years of independence for most cities. Likewise, indigenous, Islamic, and dual cities have in nearly all cases witnessed the steady overlay and erasure of their original forms through colonial and postcolonial impacts (Figure 8.4).

In SSA cities, previous patterns of government dominance in urban centers have been replaced by more reliance on the private sector and/or nongovernmental institutions.

Box 8.1 Water, Water, Everywhere

Water is one of the most complicated aspects of urban Africa. Half of all large SSA cities are within 50 miles of the coast, and many of them are on river mouths, estuaries, or deltas. Many others (including some of the region's largest urban areas, such as Kinshasa, Khartoum, and Brazzaville) are located in low-lying riverine settings. This means that a great many cities in SSA face significant flood risks, which are often most severe in poor settlements. Even in cities at relatively high average elevations, poorer areas and informal settlements are typically at lower elevations in zones subject to seasonal flooding. Khartoum, Dar es Salaam, Dakar, and many other major urban areas have experienced severe flooding in the last few years alone. Moreover, the Climate Change Vulnerability Index points toward high or severe risk from rising sea levels along Africa's urban coastline. Urban flooding has also increased water-borne disease threats, including the spread of malaria into East African highland cities, such as Addis Ababa and Kigali, that were previously malaria-free.

In a bitter irony, many SSA cities face potential water shortages caused by climate change and rapid growth. The most severe threats of freshwater shortages, unsurprisingly, appear to be in cities in arid and semi-arid areas such as the Sahel. But, even cities with plentiful precipitation have failed to keep pace with the rise of consumer water demand under conditions of rapid urbanization. Most everywhere, potable water supply shortages are severe and increasing.

Water may present certain risks in SSA cites, but it also presents abundant opportunities. After all, so many cities in Africa are on coasts or navigable rivers because of trade opportunities that continue to increase. African cities are among the world's leaders in urban agriculture. Parks, preserves, forested areas, and natural open spaces are also widespread in SSA cities, despite stereotypes of African cities as "cities of slums." Centers of higher learning across the continent tend to be urban, and the curricula in many countries feature environmental education lessons from primary levels onward that are improving popular knowledge of healthy and efficient water usage. Environmental awareness and activism are on the rise across the continent, especially in urban areas, and water issues are often central to this activism. In the peri-urban slum of Pikine outside Dakar, hip-hop artists, other musicians, graffiti artists, and Senegalese professional wrestlers have played major roles in raising awareness of flood relief and prevention. Water provisioning can also be an important entrepreneurial arena, particularly in informal settlements (see Figure 8.2). A Nigerian-American geography research team has highlighted the policy implications of the ignorance of officials about the struggles of poor Niamey residents to obtain water, and the crucial role of entrepreneurial water vendors in providing this most basic need. Water and sanitation activists in Nairobi also built a highly successful business, Ikotoilet, from environmentally efficient toilets. Innovation and creativity abound in water and sanitation services, proving that one must be cautious about seeing water only as a source of problems in urban Africa.

Sources: A. Bontianti et al., Fluid experiences, *Habitat International* 43 (2014): 283–92; R. Fredericks, The old man is dead, *Antipode* 46, no. 1 (2014): 130–48.

Figure 8.4 The Victoria and Albert Waterfront is a major shopping destination, center of tourist activity, and gathering place for Cape Town's diverse population. *Source:* Photo by Garth Myers.

Municipal authorities have not kept up with the demand for infrastructure, social services, or access to resources. Many urban residents have looked outside the formal economy and conventional administrative channels to gain access to income, shelter, land, or social services (Box 8.2).

The wide range of seemingly unsolvable problems has led some to conclude that cities in Africa just "don't work." Others, like the urban scholar AbdouMaliq Simone, prefer to see them as "works in progress," driven forward by inventive ordinary people. In city after city, urban residents rely on their own ingenuity to stitch together their daily lives. SSA cities are often distressed places in need of good governance, management, or infrastructure, greater popular participation in decision making, sustainable livelihoods, and expanded socioeconomic opportunities. Yet they are much more than some form of failed urbanism. To see SSA cities more complexly, we must appreciate the historical specificity and heterogeneous cultural vibrancy of different cities in Africa.

HISTORICAL GEOGRAPHY OF URBAN DEVELOPMENT

Simply because SSA is often considered among the least urbanized world regions, outsiders assume that its cities must be recent. Because European colonialism was such a pervasive regional experience, it is also assumed that the urbanization of Africa ought to be attributed to colonialism. In fact, many SSA urban settlements are much older than the colonial era, and the relationships between formal colonialism and the urbanization process in Africa

Box 8.2 Multiple Livelihoods Strategies

The economic crisis that spread across Africa in the 1970s and 1980s and the structural adjustment programs that were introduced have caused major upheavals in the livelihood strategies of millions of people in African cities, including formal-sector employees.

Since the early 1990s, an increasing number of studies have documented how people of various socioeconomic backgrounds seek additional income by engaging in multiple economic activities. Many formal-sector employees supplement their incomes with part-time informal-sector jobs, such as cab driving or petty trading. Other members of their households may also supplement the family income by engaging in similar activities. For instance, many civil servants in Kampala engage in urban agriculture and poultry keeping, own taxis or operate small kiosks, and about two-thirds of households in Accra are engaged in at least two income-generating activities. Such multiple livelihood strategies have become "the way of doing things" in many African cities. As a result, the traditional distinction between formal sector and informal sector has become more blurry and complex.

The proliferation of multiple livelihood strategies has significant implications for urban planning in the region. First, it signifies the need to revise African city models to include urban cultivation, a ubiquitous feature of the urban landscape, as a legitimate urban activity. This requires documenting the benefits and disadvantages of urban agriculture and finding ways of creatively integrating the practice into the urban fabric. Second, the house or dwelling as a mono-functional (residential) unit is increasingly out of sync with the reality in many African cities. Many urban residents of different socioeconomic backgrounds have economic enterprises that are located in their homes. Urban planners need to introduce relevant changes in zoning regulations and housing design standards to accommodate home-based enterprises. Third, multiple livelihood strategies also challenge the conventional definition of households and the distinction between urban and rural residence. Historically, households, especially in southern Africa, have used migration as a strategy to overcome limitations of particular local economies. But, involvement in multiple livelihood strategies requires different and more creative living arrangements that allow members to participate in multiple urban and/or rural economies. The final issue relates to the increased involvement of public-sector employees in multiple economic activities and the implications for public-sector efficiency. While participation in multiple economic activities by public-sector employees benefits those directly involved in the practice, the overall impact on society is often negative. As the involvement of civil servants in multiple income-generating activities becomes widespread, the moral authority of supervisors to reprimand moonlighting staff is compromised, especially when the officials themselves are guilty of the same.

Sources: Francis Owusu, "Conceptualizing Livelihood Strategies in African Cities: Planning and Development Implications of Multiple Livelihood Strategies," *Journal of Planning Education and Research* 26, no. 4 (2007): 450–63.

are more complicated than they first appear. Roughly speaking, we may divide contemporary African cities into categories, including urban areas with origins in the: (1) ancient or medieval precolonial period; (2) period of the trans-Atlantic slave trade or European trade and exploration; (3) period of formal colonial rule; and (4) postcolonial period. However, it rapidly becomes difficult to differentiate cities by these categories. For instance, take the case of Zanzibar, Tanzania, where an indigenous urban center with origins in the 1100s was refashioned under the domination of outsiders from Portugal in the 1500s and Oman in the 1690s; the city then became caught up in the slave trade and trade with Europe and the Americas in the 1700s and 1800s, then became a British colonial capital, and then the symbolic heart of a postcolonial socialist revolution. Like so many hybrid cities of contemporary Africa, Zanzibar has elements of its fabric that belong to all four of the categories above. Rather than making sharp breaks between city types based on their origins, it is more helpful to simply lay out some of these different types of origin stories and to appreciate that most contemporary African cities are woven together from threads of each origin.

Ancient and Medieval Precolonial Urban Centers

Many urban centers that were prominent before 1500 CE—and in some cases, prominent before the Common Era even began—are ruins now. Other prominent centers of ancient and medieval times were bypassed by the new economic geographies that arose in Africa's relationships with Europe and the New World after 1500, which developed strong associations with coastal urbanisms.

There were at least five major centers of urbanism before 1500, with the oldest being the ancient Upper Nile/Ethiopian centers of Meroë, Axum, and Adulis (Figure 8.5). The medieval Sahelian (or Western Sudan) cities of West Africa's great trading empires, such as Kumbi Saleh, Timbuktu, Gao, and Jenne, arose as middle-agent ports of a network of caravan routes that crisscrossed the Sahara. They achieved significance in the medieval world as nodes of empires, trading entrepots, or centers of learning. Timbuktu, Gao, and Jenne were widely regarded for their scholarship in medieval times, but they disappeared or stagnated after the fifteenth century. Timbuktu has about the same population as it had seven hundred years ago, while Gao and Jenne no longer exist. Other early Western Sudan urbanisms survived the new circumstances of the post-1500 world and developed into important contemporary settlements, for instance the Hausa cities of today's northern Nigeria and southern Niger, particularly Kano, because their nineteenth-century rulers derived great strength from Islamic religious *jihad* movements.

This adaptation and growth after 1500 was even more common for many ancient and medieval cities of Nigeria, like Oyo, Ibadan, and Benin in the Benin-Yoruba area of early urbanism. The Yoruba cities of southwestern Nigeria had developed metalwork artistry and skill unsurpassed in the first millennium world. Benin-Yoruba cities, and neighboring urban areas further to the west, were well positioned to capitalize on the new trade with Europeans after 1500, as is seen below.

Some of the trading city-states along the Swahili coast and the East African coast more broadly, including Mogadishu and Mombasa, also grew after 1500; but many coastal settlements, like the settlements further to

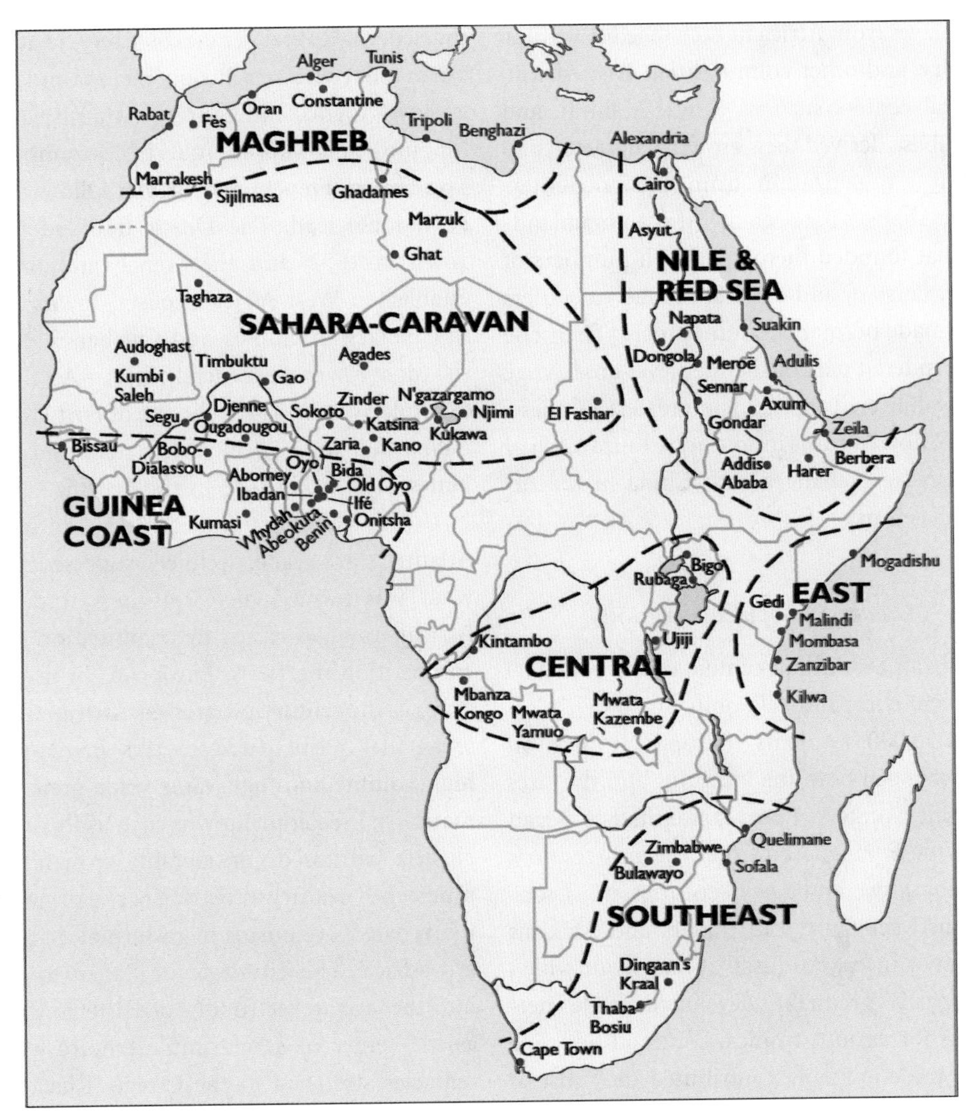

Figure 8.5 Historical Centers of Urbanization in Africa. *Source:* Assefa Mehretu.

the southern interior (the Zimbabwean zone of urbanism, in particular) with whom they traded, largely disappeared. The ruins of the Great Zimbabwe in today's Zimbabwe still demonstrate the remarkable organizational and architectural features of the medieval empire whose central city was located there. The southern interior cities were connected by trade to those along the coast for many centuries before 1500.

Coastal trading centers on the Red Sea and Indian Ocean arose in ancient times, and an extensive trade linking the African interior from Zimbabwe north to Lake Victoria with the Arab and Persian peoples of Asia flourished for more than a thousand years. Beginning in the ninth century, the significance of the Swahili coast ratcheted upward with increased trade with the Arabian peninsula and the Gulf area, based around the export of gold, ivory, and slaves from Africa in exchange for textiles, jewelry, and other commodities. East African coastal centers such as Kilwa, Malindi, and Mombasa derived their growth, character, and political organization from the encounters and exchanges between the African mainlanders that founded them and small numbers of Arab, Persian, and even South Asian settlers who made permanent homes there. Their rise is considered part of the medieval golden age of Swahili civilization. The greatest of these, Kilwa, now in ruins in southern Tanzania, had diplomatic exchanges with China in the fifteenth century.

Urban Development after 1500

Nearly all SSA urban centers of the pre-1500 era were comparatively quite small, with less than 50,000 residents. Europe's impact on SSA changed both the locations and the sizes of major centers. European influence began with the Portuguese in the fifteenth century. For about two and a half centuries, most contact between European traders and Africans occurred in coastal installations, from which Europeans gradually developed trade networks for various tropical commodities. The slave trade arguably contributed the most to the development of many coastal trade centers between about 1500 and 1870, but this impact was not an unambiguously positive one. During those years, more than 20 million Africans were forcibly relocated to the Americas or died en route; roughly an equal number died or were displaced within Africa. Nonetheless, it is remarkable how many of the major and secondary cities of coastal West and Central Africa in particular grew up in the midst of the trans-Atlantic slave trade.

The Portuguese established the first of these towns, St. Louis, in the 1440s at the mouth of the Senegal River, later creating centers at Bissau in today's Guinea-Bissau; Luanda and Benguela in Angola; and Lourenço Marques (now Maputo) and Mozambique in Mozambique. The Dutch, French, and British followed the Portuguese lead. The Dutch founded Cape Town in 1652, and the French and British established West African coastal towns such as Conakry in Guinea and Calabar in Nigeria. Most towns were merely forts—Accra, for example, was originally the site of Fort Ussher, established in 1650 by the Dutch, and Fort James, founded in 1673 by the British.

During the nineteenth century, the trans-Atlantic slave trade declined, superseded by what was termed the "legitimate trade" in African raw materials. In combination with competition between European firms and states as the century progressed, African urban areas that participated in this increasingly high-volume and high-value trade grew dramatically. Precolonial towns such as Ibadan in Nigeria witnessed considerable growth. The nineteenth century also saw the rise of new or rejuvenated urbanisms in eastern and southern Africa. The city-state of Zanzibar grew into the "island metropolis of Eastern Africa" as the center of a mercantile empire whose tentacles stretched to the Congo. Khartoum emerged in the Sudan. A number of South Africa's major cities, including Port Elizabeth, Durban, Bloemfontein, East London, and Pretoria, were founded via European settlement.

During the period of European contact before formal colonialism in the 1880s, SSA's urban geography began to take form, but under constraints. First, most European contributions in settlement development were coastal with minimal impacts in the interior. Second, many coastal settlements were intended as transshipment points for trade and lacked regular urban facilities, except those structures that served as European housing or as port and defense establishments. Third, there was a lack of diffusion of European technology and culture to the interior's indigenous urban centers.

African Urbanization in the Era of Formal Colonial Rule

The European Scramble for Africa lasted from the 1880s through the 1914 outbreak of the First World War. By that point, virtually the entire continent had fallen under European domination. Ethiopia and Liberia remained independent states, and South Africa became an independent, white-minority-ruled state in 1910; but the British, French, German, Italian, Portuguese, Belgian, and Spanish colonial powers controlled the rest of SSA. Urbanization followed suit, since social and physical aspects of urban development followed the social and political objectives of these European powers. Colonial regimes moved aggressively into the interior of their colonies; and urban settlements sprang up or expanded from existing towns along infrastructure lines (roads or railroads), near mines or large-scale plantation areas, or in regions requiring administrative centers. Virtually all coastal ports and railheads from Dakar to Luanda became the capitals and/or primate cities in their respective countries, with external trade as their major function. In East Africa, where the resource hinterlands are far in the interior, towns such as Kampala, Nairobi, and Salisbury (Harare) were linked by railways to ports in or near each country, such as Mombasa in Kenya, and Beira in Mozambique. Other East African centers, such as Dar es Salaam (Figure 8.6) and Maputo, became important ports.

Figure 8.6 The historic African CBD of Dar es Salaam, Kariakoo, has undergone rapid gentrification in the twenty-first century, where the pace of new construction has outrun the ability of the government to provide basic services. *Source:* Photo by Garth Myers.

In South Africa, the pattern was somewhat different. Major European settlement in the interior pre-dated the formal colonial era of most of SSA (i.e., the 1880s to the 1960s), and major mining and agro-industrial towns were well established by 1900. In South Africa, as a result, there are now numerous urban centers in the interior served by a number of ports all around the southern tip of the continent. The railway pattern is much more intensive, with a high degree of connectivity between urban centers in the plateau hinterland as well as between the interior settlements and the port cities.

In most of SSA, under European colonialism, little real industrialization occurred.

Colonial regimes prioritized the export of minerals, metals, or primary goods to Europe, so that industrial development was most intensive in places like the Zambian Copperbelt (in cities such as Ndola or Kitwe) and the neighboring mining province of Shaba (Katanga) in Congo. In many colonies, and in white-ruled South Africa, severe limits were placed in African residency in urban areas. To support the scale of trade that flowed between Africa and Europe and to control the colonies, larger administrations emerged, leading to an outsized service sector for the comparatively shrunken state of secondary sector activities. Urban services were also generally quite warped by race and class.

As a result of the limited economic opportunities and restrictions on movement, many SSA urban centers remained relatively small until after World War II. The so-called second colonial occupation of the postwar era, when colonial regimes invested in African development largely in an effort to shape decolonization movements away from the influences of the Soviet Union, led to the growth of investments in many urban areas. Relaxation of migration and residency regulations with independence brought massive rural-to-urban migration in SSA.

There are differences between the respective colonial powers (particularly Britain, France, and Portugal) in terms of their legacies in urban areas, but there are also facets of their legacies held in common. British colonies with substantial white settlement developed more highly segregated urban settlement patterns regulated by more rigid building rules and land laws than would be the case for colonial cities in the interior areas of many French West African colonies, for instance. Cities with significant white populations in the colonial era tended to have larger investments from colonial states for infrastructure and from the private sector for industrial development. Yet exceptions to these differentiations existed, and the distinctions between different colonial powers' strategies in urban areas are often overridden by the commonalities. One can still see some distinctly British features of eastern and southern African cities in architecture (many colonial government buildings that are still in use were designed by the British architect Herbert Baker and an army of his protégés) or urbanism more generally (the many small urban parks just adjacent to Central Business District (CBDs) with the same strict use rules on signs at their entrance that one sees in London or Hong Kong). Many of the neighborhoods formerly segregated by race are now just as segregated, but by class, as illustrated by the dramatic air photo from the late 1990s in Lusaka, Zambia, of what was until 1964 the whites-only and separately governed township of Roma and the informal settlement of Ng'ombe on its eastern edge (Figure 8.7). Today, Roma is populated predominantly by the African professional class and the political elite of Lusaka, and their maids and gardeners—also African—still live in Ng'ombe, although it is beginning to gentrify (Figure 8.8). Distinctively French architectural or planning legacies are also in evidence up until today in former French colonies. But over time, cities all across the region are becoming more and more alike in their hybrid form and function as the postcolonial era brings unprecedented urban growth to SSA (Box 8.3).

Postcolonial Urbanization

From the 1960s through the 1980s, SSA contained the world's most rapidly urbanizing countries. Eastern and southern African

Figure 8.7 A dramatic air photo of Lusaka, Zambia, today shows the formerly all-white township of Roma. *Source:* Photo by Garth Myers.

countries have led the world in urbanization rates for nearly half a century. Even during the 1990s and 2000s, when many observers noted a slowdown in African urbanization, several countries had estimated urban growth rates near or above 5 percent. In less than 50 years, some eastern and southern African countries have gone from being largely rural societies to being places where almost half the people live in or around cities.

The rapid growth of many cities and the path of urbanization in most countries have been somewhat distinct from what has been seen in other regions, particularly in wealthy European or North American settings. With some exceptions, the extraordinary story of urbanization in Sub-Saharan Africa has not accompanied a substantial economic transformation of society toward industry and manufacturing. In some countries,

Figure 8.8 A billboard advertising a new, high-security elite housing enclave, Silverest Gardens, on the outskirts of Lusaka, built by the Henan-Guoji Development Company. It is one of nine such neighborhoods built by this Chinese company in SSA cities since 2010. *Source:* Photo by Garth Myers.

notably South Africa, industrial development occurred with the urban-ward trends. But in much of SSA, the ever-expanding numbers of urban residents have become increasingly dependent on what are termed informal activities—small-scale, low-technology manufacturing, petty wholesale trading, and informal service provision—for basic needs and daily life (Figure 8.9). Many of these informal economic activities are highly gendered activities that are defined by—and often challenge—traditional understandings of gendered divisions of labor (Table 8.1). Especially in the case of southern Africa, the informal sector is sometimes associated with another problem, HIV/AIDS.

Current Urbanization Trends

Compared to the other regions, SSA still has one of the lowest levels of urbanization. According to UN estimates, only 41 percent of the region's population will live in urban areas by 2020. There are, however, significant differences in the levels of urbanization within the region. Coastal western Africa and southern Africa have the most developed urban hierarchies. Eastern Africa is the least urbanized (Table 8.2).

Unlike the other regions of the world that play dominant roles in the globalization process, SSA lacks many major "world cities" given its marginality to the world economic system. Although Johannesburg plays a dominant regional role and Lagos is growing in economic importance regionally, most large SSA cities are centers of national economies. Urbanization in the region has continued, but at a slower pace over the last decade in many countries. The proportion of its population in urban areas was 15 percent in 1950; it then jumped to 25 percent in 1970; and it is

Box 8.3 BRICS, Urban Investment, and the Middle Class

The last ten years have brought a profound shift in the sources of foreign investment in SSA, with China becoming the region's biggest trading partner and donor. Trade between China and Africa is now valued at more than $200 billion. India, Russia, and Brazil are also increasing in significance as trading partners, investors, and donors in the region, and South African investment in the rest of Africa has increased. Of the new players in the development and investment game in Africa, the group of five states which calls itself the BRICS (standing for Brazil, Russia, India, China, and South Africa) are the most significant, especially since they formed a development bank to serve as an alternative to the World Bank, in part with African development in mind.

Traditionally, the interest of BRICS in the region has been tied to natural resource extraction, often impacting urban areas only indirectly. However, a surprising trend has been the extent to which these countries have entered the urban housing and real estate markets. Some of the most dramatic examples have involved the construction of gated communities, high-security suburbs, or even entirely new satellite cities, often built by private firms from the BRICS countries. For example, a Russian firm registered in Cyprus, Renaissance Capital, developed and financed plans for satellite cities in South Africa, Nigeria, and Kenya, while China's Henan-Guoji Development Company created nine high-security suburbs in eight different African countries (see Figure 8.8).

This trend is often treated with near-derision, either because of media frenzies over glitzy new ghost cities or because of the galling inequity of luxurious developments in some of the world's poorest cities. To wit, Kinshasa's Cité du Fleuve, built by a Chinese-Zambian engineering team on two artificial islands in the Congo River, opened in 2015 with a plan for more than 10,000 luxury apartments with asking prices beginning at $175,000 and a minimum down payment of $50,000—on the edge of a megacity with per capita income estimated at $280 per year. On the other hand, there is certainly a degree to which the new satellite cities and secure suburbs meet the growing demand for high-end housing from SSA's expanding urban middle class. The most infamous of the BRICS-built "ghost cities," Luanda's Chinese-built satellite city of Kilamba, quietly gained 40,000 Angolan residents between 2012 and 2014 by lowering the asking price for its apartments; even though the new low prices were well out of reach of the average Luanda citizen, other Angolans clearly had the cash.

There is a healthy dose of skepticism surrounding the much-ballyhooed rising tide of SSA's economies, since so much of the growth is in the familiar oil-and-minerals sector, with its notoriously limited benefits for ordinary urban Africans. Yet, across the region, there is no mistaking the expansion of high-end consumerism, as mall after mall emerges in Dakar, Accra, Nairobi, and many others urban settings. A great many cities are witnessing a housing and construction boom to accommodate the new growth of this urban middle class, and the BRICS donors, investors, and engineers are a big part of the story. It remains to be seen how durable SSA's growing middle class will be, and the same may be said for the BRICS' interest in them.

Figure 8.9 Along Great East Road in Lusaka, Zambia, the informal economy punctuates the streets as vendors sharpen the pitches that they need to clinch each sale. *Source:* Photo by Angela Gray Subulwa.

now projected to exceed 50 percent by 2040. Between 1950 and 1995, SSA's urban population increased by an average of 5 percent per annum—this represents about twice the average population growth rate of the region. Since then, however, the growth rate has slowed and it is projected to be slightly above 3 percent by 2030. There are also significant variations in the urban population growth rates of the countries in the region.

Although the overall growth of urban population in SSA has slowed in recent years, most of the major cities continue to increase their populations. For instance, the 2015 population of Lagos, Nigeria's metropolitan area is estimated to be over 13 million; Kinshasa-Brazzaville is 11 million; greater Johannesburg 9.9 million, and Abidjan 4.9 million. Between 1990 and 2010, Luanda tripled its population from 1.6 million to 4.8 million, and Conakry doubled its population to 1.7 million. Also, some of the secondary urban centers have experienced some growth since the 1960s due to deliberate government policies to slow the growth of the capital cities.

Table 8.1 Female and Male, age 15–24, in Informal Employment (in percent)

	Female	*Male*	*Year*
Benin	79.8	69.8	2006
Burkina Faso	82.8	20.7	2003
Cameroon	69.9	62	1998
CAR	96.8	75.2	1994
Chad	83.6	60.2	2004
Comoros	93.4	54.5	1996
Congo	92.5	52	2005
Cote d'Ivoire	77.6	35.3	1998
Ethiopia	69.9	16.8	2005
Gabon	75.6	71	2000
Ghana	85.2	30.1	2003
Guinea	98.6	–	2005
Kenya	63.8	5.3	2003
Madagascar	77.7	–	1997
Malawi	72.6	–	2000
Mali	91.2	53.3	2001
Mozambique	70.9	8.5	2003
Namibia	38	–	2000
Niger	92.1	57.2	1998
Nigeria	59	16.8	2003
Rwanda	60	23.2	2005
Senegal	84	23.9	2005
South Africa	39.3	–	1998
Togo	94.3	60.1	1998
Uganda	74.4	14.9	2001
Tanzania	70.6	4.7	2004
Zambia	68.7	11.4	2002
Zimbabwe	53.6	–	1999

Source: Global Urban Indicators Database 2010.

Table 8.2 Urban Population as Percentage of Total Population

Regions	*1990*	*2020 Projected*
SUB-SAHARAN AFRICA	28.2	40.7
Eastern Africa	17.7	27.3
Middle Africa	32.4	46.1
Southern Africa	48.8	62.7
Western Africa	33.2	49.9

Source: UN Habitat (2014): 266–67.

Most of the largest cities in SSA are the national capitals of their country. In most cases, the other urban centers are much smaller, except when such cities house major economic activity like a mine or a port. Development efforts have focused on such national cities while ignoring the many smaller urban centers. Most administrative, transport, communications, commercial, educational, and industrial functions are concentrated in the major national cities.

Many SSA cities derive their importance from the role that they played during the colonial and/or postcolonial eras. An important postcolonial SSA urban development phenomenon is the creation of newly planned cities. The first of such cities was the port of Tema, Ghana, built in the early 1960s in anticipation of the country's industrial development. Several countries have since established new cities as their capitals, including Dodoma in Tanzania, Lilongwe in Malawi, Yamoussoukro in Côte d'Ivoire, and Abuja in Nigeria. These new capitals were meant to give the nation a "fresh start" and to direct growth away from existing cities (Figure 8.10). However, considering that none of the new capitals, other than Abuja, have grown to much more than about half a million inhabitants, one can say that these new cities have not had significant influence on the growth of the already established cities. And in Abuja's case, the staggering rates of growth that the city has experienced (its population jumped from 800,000 to 2 million between 2000 and 2010 alone) have far outstripped Nigeria's planning capacity for coping with it.

Another characteristic of SSA urban centers is the importance of port cities. Apart from the landlocked countries in the region and those that have created new capitals, many of the rest of the countries have port

Figure 8.10 A downtown shopping street in Dodoma, Tanzania. Tanzania's socialist government relocated the national capital from the colonial port of Dar es Salaam to the deliberately non-monumental new capital of Dodoma, beginning in the 1970s, as an attempt to overturn the colonial legacy. *Source:* Photo by Garth Myers.

cities as their capital city (see Figure 8.1). This is often a carryover from the colonial period when the main function of the capital city was to provide access to the metropolitan country. In addition, SSA's role during the colonial era as the producer of natural resources led to urban development that was based on resource exploitation. Zambia provides a good example of urban centers that grew out of the copper mining centers. For instance, Chingola grew up around the Nchanga copper mine; Kitwe is at the site of the Nkana mine and Luanshya stems from the Roan Antelope copper mine. Another important feature of SSA's urban evolution is the increasing importance of tourism cities. Mombasa, the second-largest city in Kenya and the center of the coastal tourism industry, continues to attract immigrants from the interior of Kenya because of the employment opportunities in the tourist industry. Gorée Island, located just off the Dakar Peninsula also attracts many tourists annually because of its slave history. Similarly, Cape Coast and Elmina in Ghana attract many tourists, who are interested in the experiences of the trans-Atlantic slave trade, to the castles from where the slaves were shipped to the New World.

DISTINCTIVE CITIES

Kinshasa: The Invisible City

About 40 percent of the Democratic Republic of the Congo's (DRC) population lives in cities, with that percentage expected to top 50 percent by 2040. Kinshasa's population was conservatively estimated to be 11 million in 2015. This capital city has more than 15 percent of the DRC's population. Instability and warfare (especially from 1996 to 2002) have hindered Kinshasa's economic development, even while enhancing incentives for Congolese people to migrate to it. Its rapid growth in the last half-century has outstripped the government's political and economic capacity to provide for its needs.

For Sub-Saharan Africa's second-largest city, Kinshasa has a relatively brief and notably turbulent history. The British-American explorer-agent for the Belgians, Henry Morton Stanley, built a new city just adjacent to a set of preexisting settlements in 1881, naming it Leopoldville to honor Leopold II, the Belgian king. A railway connection with the coastal port of Matadi soon made Leopoldville a key town for linking the vast interior of the basin with the world economy, and in 1923 Leopoldville became the Belgian Congo's capital. Eventually, the Belgians extended the city's boundaries; at independence, this expanded entity became Kinshasa.

Kinshasa had only about 30,000 residents in the 1880s, but this had risen to 400,000 by independence in 1960. As was the case in many cities in SSA, colonialism held the population down by enforcing restrictions on urban residence. At independence, the new government ended these controls, opening the doors for massive rural-to-urban migration. For much of the last half-century, the annual growth rate of the city's population has been above 5 percent; it is estimated to have slowed to 3.8 percent for 2010–2020.

Until 1945, most of Leopoldville's Africans lived not in the city itself, but in adjacent riverine settlements. After World War II, new neighborhoods arose, some planned for African workers by the colonial regime. These planned neighborhoods were nearly the only serious investments in African areas of the city made during the Belgian era; Leopold II's Congo Free State and the Belgian Congo that replaced it in 1908 are considered by many scholars to demonstrate the worst case for colonialism's negative impacts, with extremely limited investments in human welfare or security in the colony's capital city. Thus, Kinshasa's infrastructure woes are not entirely the result of warped postcolonial era's politics—the colonial regime failed to provide urban services to African areas even when investing heavily in European areas of the city.

The governments that have ruled Kinshasa since 1960 and the private-sector entities that have invested in the Congo's vast resources have not improved matters. Both government and the formal private sector failed to keep pace with Kinshasa's housing, infrastructure, or employment demands. Despite this, migration to Kinshasa continues to rise. Warfare, violence, hunger, insecurity, and the departure of industries from rural areas drive people to Kinshasa. Their perceptions do not match with realities—some 60 percent of Kinshasa's workforce is estimated to be unemployed; housing and sanitation conditions remain poor; and environmental health problems are rampant.

Population-wise, the largest zones in the city are at the far eastern and far western edges of its urban expanse. Growth in these and other areas is mostly unregulated and uncontrolled. Postindependence efforts to provide public housing, credit facilities, or transport have been marred by gross corruption, mismanagement, and negligence, particularly under the notorious dictatorship of Mobutu Sese Seko, who ruled the DRC (which he renamed Zaire) with brutal inefficiency from 1965 to 1997. As a result, Kinshasa's residents do as much as they can informally, outside of the state's purview or the formal private sector. So much of what comprises Kinshasa in both physical and economic terms is undocumented, giving rise to discussion of it as an "invisible city." Geographers Guillaume Iyenda and David Simon estimate that three-fourths of Kinshasa's houses are self-built by their owners, often in such close proximity to one another as to prohibit sufficient road construction. Roads, railroads, airports, port facilities, river transport, bridges, and public vehicles in Kinshasa have deteriorated steadily.

Industry in Kinshasa has been declining for 30 years or more. Rioting, looting, and urban violence in the 1990s and 2000s reduced the city's industrial capacity still further. Kinshasa's manufacturing sector still produces many lower-order goods, but in declining volume. The service sector dominates Kinshasa's economy, accounting for three-fourths of all urban activities. Yet, the extraordinary degree of urban primacy that Kinshasa still maintains means that it continues to dominate the DRC's economy, accounting for between 19 percent and 33 percent of all firms or establishments.

Despite the negativity that surrounds most descriptions of and scholarship about Kinshasa, this megacity is a thriving center of the arts, particularly for popular music. Kinshasa's musicians have produced chart-toppers and dance-hall favorites across SSA and Europe for decades, inventing new styles of music and dance and pushing on through every new twist in the city's political and economic malaise. The 2006 democratic elections in the DRC marked a turning point toward peace and stability, and Kinshasa is beginning to benefit from the DRC's tremendous base of natural resources. DRC was once among the top five producers of industrial diamonds in the world, and these are still estimated to account for more than half of the country's export earnings, alongside extensive copper, cobalt, coffee, palm oil, and rubber exports. Such riches have attracted many foreign investors to the DRC, some of whom are investing in real estate and housing ventures in Kinshasa, as the DRC's government begins to invest in long-overdue infrastructure for this megacity. Widespread street protests in 2014 against efforts to change the constitution to allow the president a third term in office, though, were reminders that Kinshasa still faces many challenges on the road to more equitable and democratic government (Box 8.4).

Accra: African Neoliberal City?

Ghana's level of urbanization in 2014 was 53 percent, and it is projected to increase to over 63 percent by 2030. The two most important cities in the country are Accra, along the coast, and Kumasi, in the interior. Accra, however, with an estimated 2015 population of 2.3 million, is the undisputed primate city of Ghana. Accra's dominance manifests itself in the political, administrative, economic, and cultural spheres. The country's open economic policies and its relative political stability in a region characterized by instability have also elevated Accra's influence internationally. Accra has experienced a surge in business and industry, becoming a destination for many foreign visitors to West Africa. At the same time, a significant proportion of the city's residents have not benefited from the good economic fortunes.

Accra began as a coastal fishing settlement of the Ga-Adanbge people in the late sixteenth century. Although there were other trade and political centers in the inland of the country at the time, there is no evidence that they were connected in any way to Accra. During the seventeenth century, a number of forts were established in the area by the

Box 8.4 Kinshasa's Imaginative and Generative Side

Kinshasa is often seen as one of the worst examples of what has gone wrong in SSA's cities. It has grown very rapidly without corresponding industrial manufacturing growth, enduring decades of mismanagement amidst severe governance crises in the DRC. Its vast sprawl and poor infrastructure are part of why it is often portrayed as an example of relocalization: where a city becomes a set of villages distinct and cut off from each other. And yet at the same time, Kinshasa has endured as a major engine of creativity in music and the arts, and its people display tremendous ingenuity in manufacturing the means to survive. In recent years, its residents have farmed its cemeteries and reclaimed stretches of the Malebo Pool in the Congo River as arable land. Filip de Boeck has estimated that Kinois (the people of Kinshasa) have now empoldered more than 800 hectares of the Malebo Pool. More than 80 farmers' associations govern this vast urban agricultural garden belt essentially outside of government control. What de Boeck calls an "organic approach to the production of the city" certainly does not occur without conflicts, but its freedoms and innovations need to be recognized in any future attempts to come to grips with this megacity. Unfortunately, the creativity of Kinois residents is more frequently subjected to harsh and capricious crackdowns, street-sweeps, or programs of demolition.

The new democratically elected regime of Joseph Kabila has, since the 2006 election, invested heavily in remaking Kinshasa's downtown, with Chinese, Indian, Pakistani, UAE, or Zambian engineers, contractors, or investors. Ubiquitous billboards advertise the global ambitions for Kinshasa among the DRC's new elites, including a proposed gated condominium community, *La Cite du Fleuve*, to be built on two artificial islands in the Congo River. Kinshasa's artists, such as Bodys Isek Kingelez, have reimagined Kinshasa as well. Kingelez's most striking piece, *Projet pour le Kinshasa du troisième millénaire*, is a multimedia model imaginarium for the DRC's megalopolis in the future, the "Third Millennium." In all likelihood, the future of Kinshasa belongs to the farmers reclaiming Malebo Pool for their gardens more than it does to the dreamland of billboards or dioramas of its glory. But Kinshasa, far from an "invisible" city, actually makes itself a visible symbol of all that is wrong, but also all that is marvelous, about the DRC.

Source: Filip de Boeck. 2010. "Spectral Kinshasa: Building the City through an Architecture of Words," paper presented to the workshop, "Beyond Dysfunctionality: ProSocial Writing on Africa's Cities," Nordic Africa Institute, Uppsala, Sweden.

Europeans. The rise of Accra as an urban center began in 1877 when it replaced Cape Coast as the capital of the British Gold Coast colony. Unlike many SSA capital cities such as Dakar that were selected because of preexisting economic advantages, the choice of Accra was influenced by the colonialists' desire to find a newer area that would protect Europeans from native-borne diseases. Accra's new status as the capital made it an attractive location for many merchants and investors, and by 1899 the city had been transformed into the busiest port on the Gold Coast with the largest number of warehouses. The colonial administration used legislation to limit the development of manufacturing in the city, so at independence in 1957 Accra had developed a reputation not as a factory city but a warehouse city. As the first city in Africa to become the capital of a new nation after World War II, Accra also became an important political center for the struggle for independence in Ghana and in Africa.

Postindependence governments in Ghana continued to promote the city's development by concentrating governmental functions and economic opportunities in the city and ignoring the other important cities in the country, such as Kumasi. As a result, Accra expanded as the administrative functions for the entire country expanded. In addition, the development of Tema port led to the abandonment of Accra harbor as a commercial port. However, like many cities in Ghana, Accra's growth began to slow down significantly in the 1970s and 1980s because of the economic crisis that engulfed the country.

The Ghanaian government accepted a World Bank-supported economic reform package in 1983 and agreed to pursue neoliberal economic policies, including the privatization of state-owned enterprises, deregulation of currency markets, promotion of the private sector and foreign direct investment, reduction in the public sector, and trade liberalization. These free-market policies are essential for understanding the contemporary urban economy. The policies helped to transform the state-controlled business environment in the country and encouraged the development of the private sector. It became easier to import many commodities, including building materials, leading to rapid residential development in the city and an expansion in the number of motor vehicles that clutter Accra's roads (Figure 8.11). In addition, the infrastructure-building program created visible signs of development in the city such as major road

Figure 8.11 A long line of drivers wait for gas at a station in Accra. One of the great ironies in many SSA cities appears in situations where Africans experience shortages of a major export commodity of their own country. Here, the irony is that Ghana is an exporter of petroleum, yet has not been able to keep up with demand in its own capital city. *Source:* Photo by Francis Owusu.

construction (new thoroughfares and flyovers), upgrading of the international airport in Accra and the Tema port, and the creation of export-processing zones to attract foreign investors. As the hub of Ghana's economic activities, Accra has also become the host to a number of the national, regional, and multinational financial and business institutions. These economic activities have exceeded the ability of the old central business district to house them, and as a result many of the headquarters are located around the outskirts of the city. The proposed "Hope City" near Accra is one such effort and is expected to transform Accra into an important IT hub and increase its global competiveness.

Yet, not all residents in Accra have benefited from free-market policies—negative effects of the policies are also visible on the urban landscape. Income levels of most residents have not kept up with the rising cost of living. Lack of employment opportunities for the majority of the residents has widened the gap between the rising, yet small, middle class and the poor majority. The unequal distribution of wealth can be seen from the proliferation of new housing developments on the outskirts of the city, including development of gated communities, luxury apartment buildings, expensive urban shopping malls, and the increased securitization of architecture presumably for protection against crime. Although the crime rate in Accra is low (compared to other SSA cities such as Lagos), the growth of the private security industry in the city reflects the feeling of insecurity among the residents and emphasizes the need to address this emerging problem. The poverty in the city can also be seen from the increasing number of street traders on busy intersections and other hot spots. Many of the poor, including young children, make a living by hawking anything that they can find, especially the ubiquitous water in plastic bags. The effect of the liberalization policy has also been the flooding of the city with vehicles which, combined with lack of comprehensive transportation planning, has created insurmountable traffic problems in the city.

Ghana celebrated its 50 years of independence in 2007, with the country booming with many activities of varying proportions; and, as the nation's capital, Accra played an important role. In the same year, Ghana also discovered oil offshore and began pumping in 2010. It is hoped that the nationalistic overtones of the 50th independence anniversary and the country's oil resources will help address the challenges facing the majority of the city's residents who so far have not benefited from the liberalization of the city but are bearing the brunt of the government's policies.

The optimism over the future of Accra and Ghana is being tampered by the recent performance of the Ghanaian economy. Following years of growth, Ghana's economy slowed down to 5.5 percent growth in 2013. This was accompanied by large budget deficiencies, currency devaluation, high interest rates, and rising inflation. To compensate, the government raised electricity and water tariffs. Lack of infrastructure has already been cited as one of the major barriers to growth and this economic decline caused further cuts to public expenditure. Specifically, deficiencies in electrical and transportation systems impact both commerce and residential areas throughout Accra.

Lagos: Largest Megacity of SSA

Nigeria's population is more than 47 percent urbanized. Lagos, with over 13 million inhabitants in 2015 and over 18 million projected for 2025, qualifies as SSA's most populous city and one of the world's megacities. The development of the petroleum industry in Nigeria has given a boost to urban development, including that of Lagos. It is often said that Lagos owes its growth and dynamism to European influence. Yet, it is also true that Lagos, in its development dynamic, owes much to early African urban development. Lagos was established in the seventeenth century, when a group of Awari decided to cross over the lagoons and settle in a more secure setting on the island of Iddo. They later crossed over to Lagos Island in search of more farmland. In this manner, the three important parts of the city of Lagos were founded as fishing and farming villages by the indigenous population well before the impact of major external influences in the eighteenth century were felt.

Another important historical factor in the development of Lagos is its significance in the slave trade between 1786 and 1851, in which Africans, especially the Yoruba, played an important facilitating role. Lagos was not a slave market until 1760, but it soon became one of the most important West African ports in the slave trade. Lagos Island became an important center where slaves were barricaded as they awaited their export along with primary commodities, particularly foodstuffs and Yoruba cloth, which reached markets as distant as Brazil. Although in 1807 the British passed an act to abolish the slave trade, Lagos, because of its locational advantage, continued the trade until it was halted by the British invasion of the city in 1851, which also caused a temporary decline in the city's population. With the cession of Lagos to Britain as a colony in 1861, the colonial era for Lagos had begun.

People continued to move into the "free colony," leaving behind slavery, war, and instability in the interior. Freed slaves also returned from Brazil as well as Sierra Leone and made their homes in Lagos. Toward the end of the nineteenth century, Britain intervened to stop internal hostilities and established a protectorate over the whole of Nigeria. A railway from Lagos, begun in 1895, reached Kano in 1912. As its effective hinterland now expanded to the interior of Nigeria, Lagos became even more important as a trade and administrative center. By 1901, the city had a population of more than 40,000, and by this time the future prominence of the "modern metropolis" was pretty much established.

Lagos has experienced spectacular population growth and spatial expansion in the past four decades. Many of city's current problems are rooted in its rapid growth. It has been called the "biggest disaster area that ever passed for a city." That may be overstating it, but Lagos has acute, sometimes incomprehensible, problems of congested traffic, inadequate sanitation, housing and social services, and urban decay.

Lagos is a primate city. The disparity in socioeconomic status between the elite and the mass of urbanites is very wide. That also means the city reflects two contradictory modes of living: one that is an extension of European style brought about by those who can afford the luxuries of a high level of technology and another that is an extension of the traditional mode of living, which has been distorted to fit an urban milieu. This curious amalgam, as it reflects itself in an African urban setting, loses the beauty, charm, and convenience of both of its parts and becomes a nuisance, as exemplified by the traffic congestion and slum dwellings of Lagos.

As a megacity, Lagos has the typical problems of rapid population growth and insufficient employment opportunities. The net effect of these problems is enormous. It depresses urban wages to almost marginal subsistence levels and adds to the pressure on urban amenities and housing, as well as to numerous other social problems, especially in the slums of Lagos and peripheral residential communities. In fact, an estimated 64 percent of the residents of Lagos reside in slum areas. Lagos is a good example of an African primate city whose growth rates and attendant problems in distorted consumption patterns have created a stultifying effect that a weak and often disorganized city government is incapable of handling.

There are, however, some positive developments underway. Abuja, in the vicinity of the confluence between the Niger and Benue rivers, was designated as the new capital city of the country and all government functions have steadily moved to that more central location. This has meant a major step toward decentralization, and it has reduced the concentration of functions in Lagos. Another major development in Lagos that hopes to reshape Lagos' relations with the global political economy and transform it into a world, or perhaps even a global, city in time is Eko-Atlantic. This new development is expected to include 3,000 new buildings zoned in 10 separate districts on reclaimed land with waterfront areas, tree-lined streets, efficient transport systems, and mixed-use plots that combine residential areas with leisure facilities, offices, and shops. It is projected to house many businesses, 250,000 residents, serve as the workplace for 150,000 people, and support an additional 190,000 commuters. It is expected to help to reverse coastal erosion and relieve some of the pressure on land and resources in Lagos as well as privately administer and supply the city with electricity, water, mass transit, sewage, and security. Despite these promises, there is a risk in abandoning traditional African cities through the promotion of detached new cities that are explicitly geared toward serving elites and international capital. Further, Eko-Atlantic does not address the problems of Lagos that are discussed above and may indeed exacerbate them. For instance, it is unclear how many jobs would be created beyond those that will be available during the construction phase, and the exclusivity of the new city could promote existing inequality.

Lagos, even in the colonial period, had less than 5,000 expatriates. Hence, compared with Dakar, Nairobi, and Kinshasa, the character of the city and its spatial organization were considerably less a function of the impact of the Europeans. The process that Lagos is undergoing, if there is any recognizable process at all, may throw light on the problems of indigenization of African primate cities that have been, and in most cases still are, enclaves of European economic systems and often are as alien to their people as cities in Europe. Lagos is a bona fide African city and, as disorganized as it is, it may offer a lesson on the transition from a colonial to an indigenous urban environment.

Nairobi: Urban Legacies of Colonialism

East Africa has commonly been taken to be SSA's least urbanized region and, until quite recently, it was estimated that only 20 percent of Kenya's population lived in cities. That percentage had risen to 24 percent by 2010, and it is expected to hit the 40 percent mark by 2040. Nairobi, the capital and primate city of Kenya, is currently estimated to have 4 million residents. It is a transport hub for much of East Africa, as well as a key site for international diplomacy, and it serves as the headquarters or conference space for many international organizations. Despite its short history, Nairobi has grown into the major industrial urban center in its region, and the economic engine of Kenya. Nonetheless, it is a city with substantial rates of poverty, great disparities between rich and poor, and faltering urban services.

Nairobi is something of an accident of geography. Sparsely inhabited forest and swampland in the 1890s, Nairobi was by 1906 the site of the new capital of British East Africa (renamed Kenya Colony in 1920). Nairobi became the headquarters of the BEA's Uganda Railway, conveniently positioned near the Nairobi River and the mid-point of the rail line. The site's drainage and health problems did not prevent the colonial rulers from seeing the new railway headquarters as an ideally situated forward capital for the colony.

By 1906, the new city contained more than 13,000 people. By 1931, this population had grown to 45,000, nearly 60 percent of whom were Africans. Nairobi became the most important colonial capital in the region, as the seat of Britain's High Commission for East Africa (including the colonies of Kenya, Uganda, Tanganyika, and Zanzibar). By 1948, the city had more than 100,000 people, and its growth continued steadily through to independence in 1963.

Nairobi's colonial legacies continue to haunt it. The first element of this legacy is its physical location. It might have been convenient as a site for railway administration and management, and its geographical centrality might have assisted the efficiency of colonial rule; but physically, colonial Nairobi was, as the early colonial administrator Eric Dutton once put it, "a slatternly creature, unfit to queen it over so lovely a country." Sanitation and urban services more generally lagged—and continue to do so—in part because much of the city lies in or near wetlands.

The second legacy scars Nairobi more heavily, and that is the legacy of colonial segregation. Nairobi was built to be the capital of what its small population of European settlers claimed as a "white man's country"; and, though Africans were a majority in the city by 1922, most were not legally given rights of residency under colonial rule. When the colonial regime did begin to formally plan for the African areas in Nairobi in the 1920s, these were consistently laid out in the lowest-lying, least-desirable eastern areas of the urban zone. Whites invariably were situated in the higher elevation areas west of Nairobi's downtown. Since the colonial regime at times also encouraged the immigration of Indians and Pakistanis to East Africa, Nairobi quickly developed a substantial Asian population that took up residence in the middle, literally and figuratively.

Under colonial rule, the Europeans controlled the government, the resources, and the finances of Kenya despite the paltry percentage of the population that they represented. More than half of the urbanized land of Nairobi in the 1960s still remained in white hands, when whites comprised less than 5 percent of the city population. The Asians of the colonial era were mostly shopkeepers, merchants or skilled artisans. Eventually, much of the land of the eastern, African-dominated areas of the city came to be Asian owned. From the beginning, in Nairobi, Africans occupied the lowest ground and the lowest rungs of the economy. Most Africans lived in rented housing built by the city or their employers. The African residential zone of Eastlands, for example, was characterized by high turnover rates, high unemployment, and poor environmental conditions.

Although the tripartite racial geography of the city has faded somewhat in the 48 years since independence, Nairobi remains a heavily divided city, in class terms now as much or more so than in racial terms. The western and northwestern suburbs remain low density elite—albeit increasingly multiracial—areas. Some African elites have moved into traditionally European and Asian neighborhoods, but this is a minority in the upper echelons of political society. Upper Nairobi and the "Nairobi Hill" residential areas continue to be dominated by European single-unit and fashionable homes complete with servants' quarters. Well-to-do Asians inhabit Parklands, adjacent to the historically European sector. Poorer Asians live in Eastleigh. Some of the Asian population has moved to a second Asian quarter in Nairobi South. The CBD and middle-class or working-class zones predominate in Nairobi's geographical center, and most of the city east of downtown is dominated by informal squatter settlements. The unregulated growth of the latter has been the major story of the postindependence landscape of Nairobi, and occasionally these interrupt the general geographical pattern. Extensive efforts have gone toward their formal upgrading, but the more significant process of transformation has been toward their densification, as private investors replace shacks with mid-rise tenements.

Although both its growth rate and its economic health have declined in the past 25 years, Nairobi has become a major African metropolis. Its primate city role for Kenya is augmented by its role as an international center for East Africa and even SSA more generally. Nairobi is a prime African example of a "splintering urbanism," where one portion of the city is highly integrated with the world economy while another larger portion is disintegrated, literally and figuratively. Compared to many SSA cities, Nairobi has had a good record in industry and has an important financial services sector in the CBD. The average European, African elite, or Asian in the city lives in a comfortable home in Upper Nairobi or Parklands and works in the CBD or some other similar enclave. Many African residents are not integrated into the core functions of the city and find themselves locked out of the Nairobi economy that most Western visitors see—as in the development of gated communities for elites in the city. The informal economy provides the overwhelming majority of job opportunities and residences in Nairobi now. Although Nairobi shows evidence of a growing middle class and middle-class housing estates, over 40 percent of the population still resides in informal settlements.

The CBD of Nairobi represents one of the busiest spots in the continent. Its most prominent functions are commerce, retailing, tourism, banking, government, international institutions, and education. It gained unwanted international notoriety with the 1998 bombing of the U.S. embassy and an adjacent office building that caused 284 deaths, all but ten being Kenyans. It returned to the news again with a wave of postelection violence in 2007–2008 that left thousands of Nairobi residents displaced for months afterward. In 2013, al-Shabaab Islamic militants claimed responsibility for a horrific assault on the Westgate Mall just north of the CBD, resulting in more than one hundred deaths and the mall's total demolition. Despite these tragedies, the Nairobi CBD still provides SSA with arguably its most picturesque and captivating skyline of multistory buildings, and the CBD is ringed by several, even larger, elite shopping malls besides the now-destroyed Westgate.

Box 8.5 Crisis Mapping from Kenya to the Globe

As the postelection violence unfolded in the streets of Nairobi in 2007, a group of bloggers, citizen journalists, and software developers created a website to map reports of political violence in Kenya. The original developers, led by Kenyan activist Ory Okollah, were all current or former residents of Kenya committed to providing real-time, open-source, participatory spatial information and maps of the violence.

The initial website was named Ushahidi—Swahili for "testimony"—and, by most accounts, it outperformed mainstream Kenyan and international media in covering the grounded realities of postelection violence. The Ushahidi site utilized text messages and Google Maps and allowed crowd-sourced information to be visually displayed on an online, interactive map that built-up "hot spots" of activity. Due to the highly accessible platform, the original Ushahidi site was able to accurately capture the multifaceted dimensions of political violence—mapping both fatal and nonfatal events —with input from over 45,000 users.

Since its inception in 2008, the Ushahidi idea has evolved and expanded into a company with a platform that extends beyond a single website focused on mapping postelection violence in Kenya into a platform with numerous products, developments, and prominent global deployments. The main Ushahidi platform has undergone significant updates since its first release (1.0 Mogadishu, 2.0 Luanda, 2.3 Juba), with release 3.0 currently under development. In addition to the original site, the Ushahidi Platform also provides a number of other products, such as CrowdMap, CrisisNET, Ping, and SMSsync, all designed to provide tools for crowd-sourced mapping and information sharing. The Platform was used extensively in 2008 to map and track anti-immigrant violence in South Africa. The Ushahidi Platform has been deployed in numerous crisis situations outside of the continent as well, most visibly during the 2010 Haitian earthquake. In addition to extensive use of CrisisNET data and CrowdMap to target humanitarian assistance in post-earthquake Haiti, the Platform has also been used to organize global Occupy Movement events, track pharmaceutical supplies in Kenya, Malawi, Uganda, and Zambia, monitor elections in Mexico and India, and build a pollution map in post-Deepwater Horizon oil spill in Louisiana. Ping is a two-way, multichannel alert and group check-in system to ask, "Are you ok?" during an emergency. Like the original Ushahidi site, Ping developed directly from events unfolding in Nairobi, specifically the need for real-time group check-in during the 2013 Westgate Mall attacks. Most recently, the Ushahidi Platform was utilized to map, monitor, and track the Ebola outbreak in Guinea, Liberia, and Sierra Leone.

According to the Ushahidi team, the core platform is "built on the premise that gathering crisis information from the general public provides new insights into events happening in near real-time." The team is committed to covering events and geographic regions that typically exist at the margins of mainstream media, with particular attention to sub-Saharan Africa. In addition to these commitments, the Ushahidi Platform is deliberate in showcasing that a cutting-edge software company can arise and thrive from the continent. These commitments were recognized with a MacArthur award in 2013.

Source: Ory Okolloh, "How I Became an Activist," TEDGlobal 2007.

Nairobi's massive congestion, gigantic billboards, neon signs, glitzy hotels and casinos, and elite residents from all over the world make it the ultramodern heartbeat of Kenya and, for many, of SSA. The reconciliation and constitutional changes that have followed the cessation of postelection violence in 2008 brought hope to many Nairobi residents that renewed government attention and reinvigorated foreign investments would reverse their city's steady decline. The peaceful and fair 2013 elections conducted under Kenya's popularly endorsed 2010 constitution further solidified this hope. Despite numerous setbacks for the post-2013 government, including trials for the president and vice president in the International Criminal Court for their alleged culpability in the 2007–2008 postelection violence, Nairobi continues to catapult forward toward its ambitions to be a world-class city-region by 2030. While no real reversal of Nairobi's declining fortunes is in evidence, the city remains a lively and creative cultural center for Kenya and the region around it (Box 8.5).

Dakar: Senegal's City of Contradictions

Senegal is about 43 percent urbanized. With 3.5 million inhabitants, Dakar is a principal primate city in West Africa. The city is known for its beauty, modernity, charm, and style, as well as its agreeable climate, excellent location, and urban morphology. But, of course, this image applies to only part of Dakar. As with Nairobi, Dakar is a city of phenomenal contradictions (Figure 8.12).

The city was founded in 1444, when Portuguese sailors made a small settlement on the tiny island of Gorée, located just off the Dakar Peninsula. In 1588, the Dutch also made the

Figure 8.12 Fishing boats at Soumbedione fish market in Dakar. *Source:* Photo by Garth Myers.

island of Gorée a resting point. Although the French came to the site in 1675, they did not move onto the mainland until 1857; they used Dakar as a refueling and coal bunkering point. A number of developments expanded Dakar's functions, leading it to be, in a short time, the most important colonial port on the west coast of Africa. In 1885, Dakar was linked to St. Louis, the old Portuguese port, by rail; and this gave it an added importance as a trading center. Because of its situation and site advantages, Dakar soon became a focus for French colonial functions in the region. In 1898, Dakar became a naval base and in 1904 it became the capital of the Federation of French West Africa. Dakar's location on the westernmost part of the continent made it the most strategic point for ships moving between Europe and southern Africa and from Africa to the New World. As capital of French West Africa until 1956, it served a hinterland stretching from Senegal in the west to the easternmost part of Francophone West Africa, which included Mali, Burkina Faso, and Niger (Figure 8.13).

After the French moved from the island of Gorée to the peninsula, there was some uneasiness about living in quarters surrounded by African villages. Although a policy of racial segregation was not officially pursued, the French settlers had always wanted to keep the two communities separate. However, only because of a natural calamity that befell the Africans could the French finally accelerate the establishment of their exclusive holdings. Progressive displacement of African dwellings was underway before the outbreak of a yellow fever epidemic in 1900, but the Europeans, invoking sanitation requirements, displaced the Africans at a greater rate afterwards, pushing them northward. Between 1900 and 1902, numerous African homesteads were burned

Figure 8.13 The influence of Dakar extends well inland to the landlocked states of Mali, Burkina Faso, and Niger via the Trans-Sahel Highway. These residents of Mali's capital, Bamako, share a language with the residents of Dakar: French. *Source:* Photo by Jared Boone.

down as a "sanitary measure," and the occupants were relocated after receiving compensation. Another epidemic in 1914 again brought destruction of African homesteads in the south and more relocation of Africans to the north. On the eve of World War II, the French succeeded in almost completely dominating downtown Dakar, often called Le Plateau or Dakar Ville, concentrating Africans in what became known as the African Medina, in the north-central part of the peninsula. The problem of "cohabitation," as the French called it, was at the root of the whole displacement campaign. Although the colonial authorities would never admit a policy of official segregation, many recommendations were made to openly enforce a system based on race. A commission charged with the study of Dakar in 1889 put forth a recommendation for separate residential quarters for European and African populations. In 1901, another report proposed relocating the Africans outside the confines of the city. A new plan, implemented in 1950–1951, gave further excuse to the colonial administrators to displace more Africans.

The present internal morphology of Dakar reflects this historical background. The city is composed of four main divisions. Although rigid, exclusive ethnic domains are no longer in evidence, Le Plateau still contains one of the most westernized sectors in Africa. It compares easily with any European city—with high-rise buildings, expensive shops, exclusive restaurants, business offices, and many European residents. Characterized by its white-painted, tree-lined boulevards, Le Plateau is the most modern sector of the city, and contains upper-class residential quarters, commercial and retail functions, and government offices and institutions. The African Medina, by contrast, reflects its background as a concentration of Africans into high-density housing projects and *bidonvilles* (squatter settlements). It is the popular area of the city, is still densely populated, and houses many popular markets and clubs. Its functions are primarily residential, but it also contains shops, markets, and cultural features. It contains the industrial laborers and those employed in the informal sector, both outside and inside the Medina. The population of the Medina, and the adjacent *bidonvilles* of Ouakem and Grand Yof, is uniformly poor and lives in poorly serviced parts of the city. Recent expansions of the city have also resulted in the development of a sector called Grand Dakar, which contains a variety of neighborhoods ranging from the well-to-do through middle-income and poor sectors and includes a mixture of modern residential quarters, industries, and *bidonvilles*. There is also the Dakar industrial sector, which houses the bulk of the city's industrial activities.

Another important part of Dakar that deserves attention is Gorée Island. This island served for many centuries as one of the principal factories in the triangular trade between Africa, Europe, and the Americas. The popular *Maison des Esclaves* (Slave House) built by the Dutch in 1776 serves as a poignant reminder of Gorée's role as the center of the West African slave trade. The Slave House with its famous "Door of No Return" served as a place where Africans were brought to be loaded onto ships bound for the New World. The Slave House has been preserved in its original state and attracts thousands of tourists each year.

As with many African primate cities, Dakar faces the problem of rapid population growth. In 1914, the city had a total population of 18,000; by 1945 it had 132,000; and by 2015 it had 3.5 million. Clearly most of the growth is attributable to rural-urban migration, which is characteristic of all Sub-Saharan primate cities. The rate of natural increase of the city's population has also been much higher than the national average on account of better sanitation and medical services.

There is no doubt that Dakar is still an important center whose functions reach far beyond its national boundaries. Its ideal location still makes it a center of maritime as well as airline traffic. Many international organizations are located in Dakar because of its geographic situation and agreeable urban environment. Many international conferences and meetings are held there. Above all, it is one of the most favored vacation spots in West Africa for European tourists, especially those from the Mediterranean, who find a familiar climatic comfort in exotic surroundings. Dakar's position on the coast, however, also leaves the area vulnerable to the projected increase in sea level and coastal erosion and the threats of more flooding and storms throughout the urban region. As many of the tourism locations are situated along the coast, this degradation could also bring a decrease in this important revenue source for the region.

The future of Dakar, nevertheless, depends largely on confronting two challenges. One is to stem the tide of rural-to-urban migration by a sound policy of regional and rural development, including development of satellite cities and subsequent decentralization. Senegal is attempting to stem rural-to-urban migration through the development of "ecovillages." This strategy focuses on equipping existing communities with the means to utilize solar energy, sustainable water storage, and waste management techniques such as composting. As Dakar continues to grow, developing strategies to sustain rural livelihoods such as ecovillage development could curb the intense urban migration and hold growth at a manageable level for the city. The other challenge, as in other African primate cities, is to bridge the gap between the city's ultramodern sectors and the *bidonvilles* and make Dakar a true African city.

Johannesburg: A Multicentered City of Gold

South Africa has long been among the most urbanized countries in SSA. Some 64 percent of its population now lives in cities. With more than 4.4 million inhabitants, Johannesburg is the largest city in South Africa. Gauging its population is, however, both easier and more complicated at the same time. On the one hand, the relative wealth of the Republic of South Africa affords it the possibility of keeping more regular and reliable census figures than most SSA countries, and the South African Cities Network is one of the continent's best repositories of urban data. On the other hand, the reorganization of municipal and local government in South Africa and the presence of Johannesburg in the geographical center of Africa's greatest example of a polynodal conurbation (many-centered urban area) make it more complicated to decide where Johannesburg begins and ends. The major cities of Ekurhuleni (formerly East Rand, 3.2 million), Tshwane (formerly Pretoria, 2.9 million) and Vereeniging (1.1 million residents) are within the metropolitan area of Johannesburg, and other big towns adjoin it as well. Thus the metropolitan conurbation is said to contain more than 11.6 million residents.

South Africa also has the most deeply developed urban hierarchy in SSA. The common issues surrounding primacy in African urban hierarchies are moot here. Johannesburg's population of over 4.4 million is nearly equaled by Cape Town (3.7 million); eThekwini (formerly Durban) has 3 million, and both Nelson Mandela Bay (formerly Port

Elizabeth) and Vereeniging are over 1 million, giving South Africa seven municipalities with more than 1 million residents. Five more cities have more than a half a million. This significantly dilutes any primacy Johannesburg might claim, although, when one considers the immediate proximity of Tshwane and Ekurhuleni, it is still possible to recognize the greater Johannesburg area and Gauteng Province as the core of the South African economy.

Johannesburg is frequently the only SSA city to be considered a world city (though the global connectivity of Cape Town, eThekwini, and Tshwane is increasingly significant, along with a few other SSA cities). It holds the largest mining and industrial center on the African continent. It is home to Africa's largest stock exchange, busiest airport, most diverse manufacturing sector, and the ugliest urban racial history in Africa.

Johannesburg owes its establishment and phenomenal growth to the discovery of gold in 1886. The rush of settlers from the south to share in the riches of the land caused the town's population to increase to 10,000 within a year of its birth. By 1895, hardly ten years after its establishment, the city had about 100,000 people, half of whom were European. Johannesburg was the creation of the mining companies, which until recently probably had more to do in determining the spatial organization of the city than did the civil authorities.

The City of Gold has the unfortunate distinction of having been at the heart of a notorious experiment in social engineering. This experiment was built around the notion of separate development for settlers and the indigenous African population. Johannesburg's separate development started with the assertion in 1886 that no native tribes could live within 70 miles (112 km) of the site of the new town. When the "native" problem first arose in 1903 and when it surfaced again in 1932, with the creation of the Native Economic Commission, the European settlers argued that Johannesburg had been built by the Europeans, for the Europeans, and belonged to them alone. They maintained that the "natives" were needed for unskilled labor and came to the city to work but not to live in it, mainly because of their inability to handle European civilization. Africans were barred from living in the city and denied permanence of dwelling while they worked in the city; they were restricted to guarded compounds or distinct townships during the tenure of their urban employment. The "pass law" (requiring all Africans to carry passes, or internal passports) begun in 1890 and the "compounding system" (restricting Africans to certain residential areas) contributed to severe urban structural problems with which Johannesburg still has to cope.

Johannesburg became Africa's largest manufacturing center and a principal center of culture and education. The prosperity of the city was derived from the labors of all the races, but was appropriated by the European minority who enjoyed perhaps one of the highest living standards in the world. Today, the city is clean, with well-planned streets, skyscrapers, and plush residential quarters. The downtown area is similar to that of any industrial city in Europe and North America, with high-rise development to house the offices of the numerous companies, trading firms, and government institutions. In the suburbs of Johannesburg, such as Sandton (which is increasingly an alternative white-oriented CBD), are residential homes for the well-to-do Europeans, whose architecture and amenities match those of their European and North American counterparts.

Separate development became a formal national policy under the name *apartheid*, after the 1948 election of the white racist National Party in whites-only polls. The Nationalists' apartheid built on decades of gradual evolution and enforced unequal separation in an extremely geographical manner. Apartheid was a socially engineered, hegemonic tool to maintain a privileged status for the European settler population so that its small number could appropriate the vast amount of wealth that was being generated in the country. The impact of this policy was perhaps felt more in places like Johannesburg than anywhere else in the country. Apartheid's application was severely tested in the dynamic environment of Johannesburg, which was attracting a great number of Africans to supply the labor requirements of a rapidly growing industrial conurbation. The authors of apartheid followed a myopic vision by engineering an unsustainable institution of separate development for Africans in their native homelands. They were later forced to tolerate settlements such as Soweto, growing by leaps and bounds in the shadows of Johannesburg. Apartheid was doomed to succumb to the social disorder that its authors never anticipated. The 1976 race riot in Soweto was a watershed in the development of non-racist South Africa. The impatience of the world community with the brutal regime brought moral outrage from outside and increased violence from within. Under the weight of these two dynamics, and aided by a visionary leader Nelson Mandela, South Africa emerged from its nightmare by the early 1990s when apartheid came to a formal end, symbolized by Mandela's election as president in 1994.

The future of Johannesburg lies in how the root causes of urban instability created by apartheid are dismantled while maintaining the city's ability to continue as South Africa's most important industrial and business center. The challenges that Johannesburg faces are evident from what has been happening in attempts to resolve the severe socioeconomic disparities. With the enforcement mechanisms of the apartheid influx control laws gone, an orderly transition from a divided city into an integrated city has been a daunting task for policy makers and city planners. Johannesburg residents long suffered from high rates of violent crime and continuing insecurity. Johannesburg is becoming a megalopolis, as a series of mining towns and industrial areas merge together. The previously marginalized townships such as Soweto and Alexandra are now firmly integrated into metropolitan life. Though still high, income inequality within Johannesburg has declined, as has the violent crime rate. Johannesburg (home to two of the stadiums used to host the final games) led the rest of South Africa in celebrating the success of the 2010 FIFA World Cup in soccer—the first time the world's biggest sporting event had ever been held in Africa. While controversies persisted over the massive investment in stadiums and infrastructure that hosting the World Cup necessitated, Johannesburg's new Gautrain rapid transit railway, remodeled international airport, fully upgraded highway system, and many other manifestations of FIFA 2010 remain on the landscape, alongside a deep pride among urban residents across the country in South Africa's ability to produce a first-rate World Cup without the mishaps and fiascos many outsiders (almost gleefully) seemed to expect.

With a nonracial central government at the helm and with ample human and physical resources for economic progress, all signs continue to point to a more progressive trajectory for cities like Johannesburg. Success depends on whether people of all ethnic groups living in the city can deal responsibly with the history of their relations and choose to build a diverse society in which everyone has a stake in the new South Africa's development.

URBAN CHALLENGES

Urban Environmental Issues

Given the diversity of environments across the continent, the impacts of global climate change in SSA cities are extremely varied. Places in northern and southern Africa are projected to become much hotter and drier in the summers, with increased risks for drought, while drier subtropical regions will become warmer than the wetter tropics. Many of SSA's largest cities are located along the continent's coast and are particularly vulnerable to the threat of rising sea levels (Figure 8.14). Because most SSA cities have experienced more limited industrialization processes than similarly sized cities of Europe, Asia, or the Americas, specifically *urban* environmental problems are often assumed to be of a smaller magnitude. Yet problems of solid-waste management, air and water pollution, toxic-waste disposal, and environmental health are profound issues in much of urban SSA. In part, due to the generally smaller formal industrial sector and smaller manufacturing value-added base in urban Africa, revenues that accrue to urban local government are typically not significant enough to support the broad array of urban services expected of city governments. This array includes environmental management services such as solid-waste management, water and sanitation supply, as well as any form of environmental monitoring or oversight. As a consequence, formal and regulated supply of these services in SSA cities is often

Figure 8.14 African cities located in low-elevation coastal zones, such as Monrovia, Liberia, are vulnerable to severe flooding from sea-level rise. *Source:* Photo by Robert Zeigler.

in very short supply. Even in cities with some service provision, access to basic services is typically skewed toward high-income groups and neighborhoods.

Solid-waste services illustrate one crucial example of the interlocking environmental problems facing SSA cities. Many SSA cities with more than 1 million inhabitants report that the proportion of the residential solid waste produced that actually makes it to a landfill ranges from 3 percent to 45 percent, meaning that the majority of solid waste remains in urban neighborhoods. In an earlier era of smaller settlements, the burial or burning of such waste was not a significant problem because its content was overwhelmingly organic and biodegradable. The increasing use of plastics and other inorganic materials, along with ordinary source items for toxic waste (such as batteries and household insecticides in aerosol canisters), and the staggering growth of settlements mean that the lack of proper solid-waste management is a severe crisis for SSA's urban environments. Buried in congested neighborhoods, solid wastes can and do pollute the water supplies of untold millions of urban Africans. Pollutants sourced to uncontrolled landfills have been shown to enter into the fruits and vegetables urban farmers pluck from downstream gardens in Dar es Salaam, Lusaka, and elsewhere. For example, just along the Roma-Ng'ombe border (see Figure 8.8), urban women gardeners utilize a "rich" section of sewage-infested wetland for growing tomatoes, vegetables, and sugar cane for sale in downtown Lusaka, as well as in the surrounding compounds such as Ng'ombe. Burned on the surface, the waste causes serious damage to the air quality of such neighborhoods. Left in ditches, the waste can inhibit proper drainage, leading to flooding or the increased presence of standing water that becomes a breeding space for malarial mosquitoes. Mounds of waste left for months on the surfaces of African urban neighborhoods provide habitat for vermin that carry serious public health risks.

The interconnected water, sanitation, waste, air quality, and environmental health problems of African cities may in fact be as bad or worse, proportionally, as the infamous environmental crises of Southeast Asian or Latin American megacities. This is because the problems have great potential to magnify one another in the absence of regulation or amelioration. Even where significant industrial development is associated with African urbanization, such as in Nigeria's oil-rich Niger Delta, Zambia's Copperbelt, or South Africa's Gauteng Province (i.e., greater Johannesburg), colonialism, transnational capitalism, or repressive governments (or, in the latter case under apartheid, all three at once) make for a heady combination of roadblocks to environmental control. The levels of heavy metals pollution downstream from the Copperbelt's largest copper smelter, for example, are mind-boggling, and yet even as the technology exists to prevent or significantly reduce the smelter's air and water pollution, successive colonial and postcolonial governments for many decades now have shown reluctance to force environmental controls onto an industry that provides more than 90 percent of all of Zambia's export earnings.

Despite such limited or politically circumscribed capacity for urban environmental management, many African cities are witnessing substantial efforts to bring environmental crisis points under control. In line with the prevailing development models of the day, many cities are experimenting with private-sector urban-service provision, including in environment-related sectors. Dar es Salaam, as the pilot city for the United Nations Sustainable Cities Program, privatized solid-waste management services and produced an increased rate of deposition from under 10 percent of residential waste to more than 40 percent in less than a decade. Other cities have privatized water supply or even sanitation services. Still more have attempted public-private partnerships where private-sector companies have joined forces with governments to provide services.

Not all of the new innovations have been driven by the private sector, nor have they been automatically friendly to the environment. For instance, South Africa's post-apartheid regime has carried out a policy of free basic water provision that did increase the supply of clean water for the poor; but many critics point to the limitations in that system that are causing poor urbanites to seek unclean water alternatives, exacerbating a cholera epidemic. In other cities, the driving forces for change in urban environmental management belong with grassroots community groups, such as Nairobi's Mathare Sports Club, whose local environmental planning and consciousness raising earned them global attention at the World Summit on the Environment in Rio de Janeiro in 1992 that, in turn, improved the club's soccer match gate revenues. Regardless of the paths taken, though, it is clear that much more needs to change for African cities to gain control over the daunting array of environmental problems confronting them.

Primate Cities

Urban primacy continues to dominate the African scene (Figure 8.15). Indeed, one of the more significant factors in urban transformation in Africa in the post–World War II and postindependence period has been the dramatic growth of the primate cities. Primate cities contain more than 25 percent of the total urban population in SSA. In places such as Lesotho, the Seychelles, and Djibouti, primate cities contain 100 percent of the urban population. Primate cities in SSA are also not limited to small countries such as Burkina Faso or Guinea-Bissau; they can also be found in such large countries as Angola or Mozambique. Generally, in those countries where urbanization has had a relatively long history, the ratios are lower. But the degree of primacy will continue to be a significant pattern in Africa's urban development for quite a long time (see Table 8.2).

In the 1960s, most primate cities in Africa accounted for about 10 percent of the urban population. By the year 2000, many cities, such as Kinshasa, Lusaka, Accra, Nairobi, Addis Ababa, Luanda, Dakar, and Harare, increased their share of their respective nation's urban population to over 20 percent. Currently, many primate cities account for over 30 percent of the urban populations in their respective countries. It is also important to note that, since the second half of the 1980s, some encouraging signs of deconcentration around primate cities have been observed and the ratios of urban populations that reside in primate cities seem to be stabilizing. For instance, between 1990 and 2005, the share of the urban population in the primate city in several countries has declined. The decreases in the percent of urban population in the largest city were highest in Angola, Burkina Faso, and Guinea.

While the dominance of primate cities in SSA has historical roots traceable to colonial administration policies, postcolonial governments have perpetuated this pattern by making them the centers of modern development

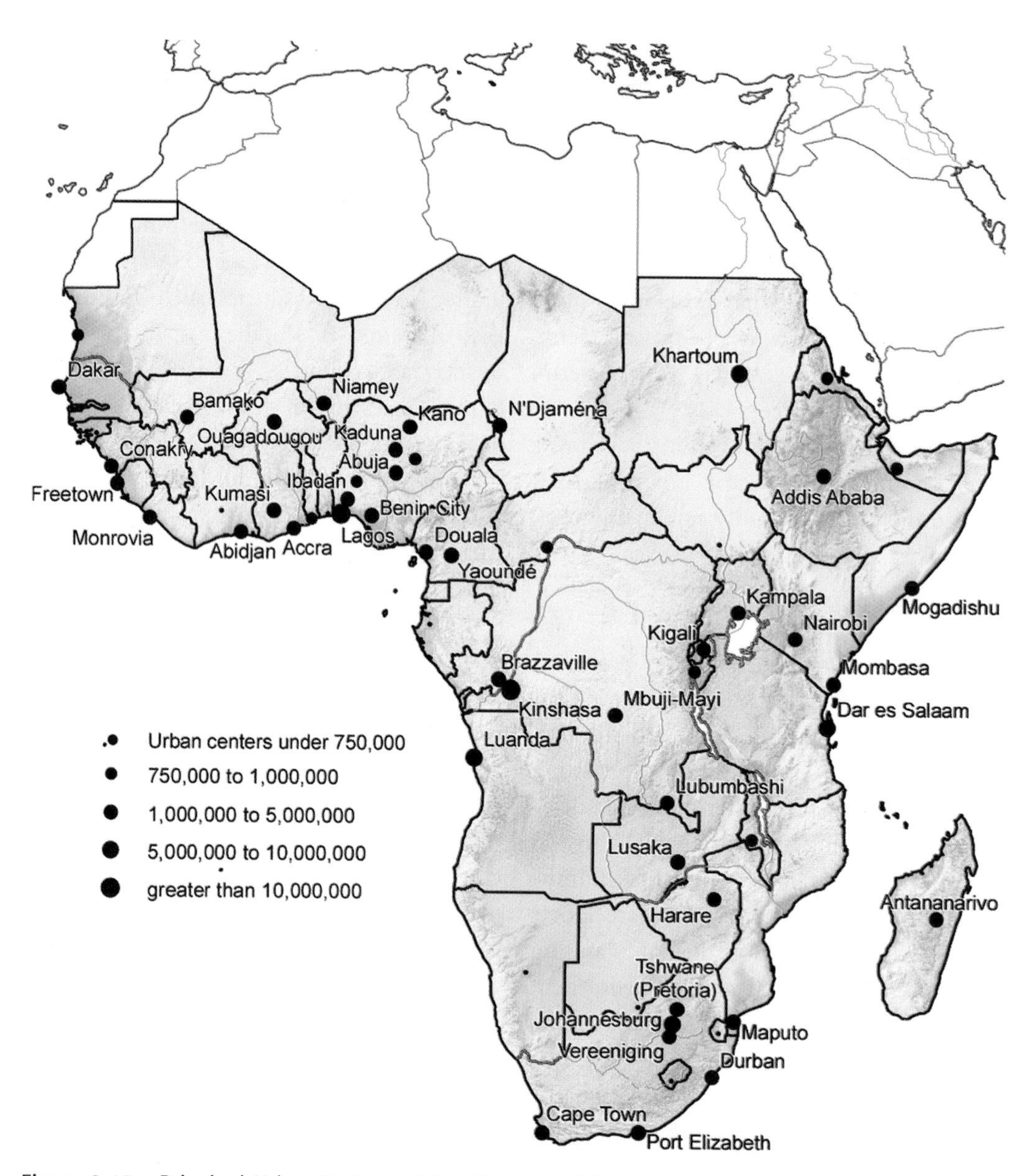

Figure 8.15 Principal Urban Centers of Sub-Saharan Africa, many of which are primate capital cities. *Source:* United Nations, Department of Economic and Social Affairs, Population Division (2014), World Urbanization Prospects: 2014 Revision, http://esa.un.org/unpd/wup/.

and governance. SSA's primate cities dominate the political processes, often reinforcing the status quo but sometimes creating avenues for change (Figure 8.16). For example, urban political processes in some SSA cities have created new spaces of engagement, particularly for professional women. For example, women comprise over 30 percent of parliament in Burundi, Mozambique, Rwanda, Tanzania, Angola, Uganda, and South Africa (in comparison, the female representation in the U.S. Congress is less than 20 percent) and over

Figure 8.16 By using billboards to help change human behavior, Lusaka, Zambia, tries to create a greener capital city as a role model for the nation. *Source:* Photo by Angela Gray Subulwa.

25 percent of ministerial positions in Niger, Burundi, Mozambique, Gambia, Uganda, Lesotho, South Africa, and Botswana. Formal and informal solidarity organizations and movements of women, operating in SSA's primate cities, have also emerged as significant agents of change. The Liberian peace movement, for example, was heavily influenced by such coalitions of women and ultimately ushered in SSA's first female head-of-state, Ellen Johnson Sirleaf.

Primate cities in SSA not only tend to be the capital city of the country; they often also have more disproportionate influence than is warranted by the magnitude of the populations living in them. They dominate the political, economic, infrastructural, and cultural scene of their countries. The strong influence that they exercise also enables them to preempt a good portion of the national social and industrial investments. Concentration of power in these cities has produced large disparities in standards of living between those who live in them and the population in the rest of the country. Primate cities also have some of the most serious urban problems in the region. These include mounting unemployment and the resultant increased crime and youth unemployment; severe housing problems, reflected in overcrowding and the spread of slums and squatter settlements (over 80 percent of the urban populations reside in slums in Mozambique, Niger, Angola, Chad, Central African Republic, and Sierra Leone); and immense pressures on urban infrastructure and services such as water, sewage, and transportation.

Rural-to-Urban Migration

Between 50 and 60 percent of the urban growth in SSA comes from rural-to-urban migration of young adults seeking jobs and other livelihood opportunities in urban areas. This migration is one of the most crucial problems facing African governments at present. A massive flow of people into the few urban centers that became the locus of power and investment, particularly in the period following decolonization of the continent, has strained the carrying capacity of most urban centers. Concern about the rate of SSA urbanization would hardly seem justifiable, considering the very high magnitude of its rural population. However, the urban growth pattern, often dominated by the primate city and the high rate of growth of such centers, is far beyond the capabilities of the urban socioeconomic system to generate the needed employment opportunities, housing, and social services. As a consequence, the primate city has become, in most instances, a liability to the overall development process.

The paradoxical fact that rural-urban migration continues to grow in spite of rising unemployment in urban centers of the Global South has given rise to a migration theory based on rural-urban income differentials. This theory assumes that migration is "primarily an economic phenomenon" and that the potential migrant makes a calculated move in order to realize much higher "expected" earnings with varying probabilities. According to this theory, the wide differences between urban and rural wages, coupled with the fact that the long-run probability that a migrant could secure wage employment in the urban area, explains the motives behind the increases in rural-urban migration. The structural adjustment programs implemented across the region were meant to bridge this gap by increasing the prices of agricultural exports and instituting payment for previously free or subsidized services consumed mostly by urban dwellers.

Rural-to-urban migration in SSA is also explained with reference to "push factors" that operate in rural areas and "pull factors" that attract migrants to urban areas. The push factors include the deteriorating socioeconomic conditions in rural areas, including lack of access to agricultural land, which literally forces people to leave rural areas. The pull factors emphasize the attractions and socioeconomic opportunities available in urban areas. The economic opportunities in urban areas, as well as social, cultural, and psychological factors including escaping social controls in rural areas, also attract people to urban areas.

Whether rural-to-urban migration in SSA is caused primarily by the difference in expected wage from migration (urban wage) versus an agricultural wage or the balance between pull and push factors might be open to debate. It is important to note that more recent evidence seems to suggest that rural out-migration has not only abated in SSA, but that its counterstream (i.e., urban out-migration) has progressed and even, in some cases, surpassed the rural-to-urban flow of people. For instance, a number of major towns and cities in Ghana and Zambia experienced a negative migratory balance between the 1970s and 1980s. Zambian census data also indicated that the population of some urban areas decreased between 1990 and 2000, due in large part to growth in some secondary cities, but also due to opening up large farm blocks and resettlement schemes designed to attract some urban dwellers and retirees. Similar patterns have been observed in the Francophone West African countries of Burkina Faso, Guinea, Côte d'Ivoire, Mali, Mauritania, Niger, and Senegal, where many secondary towns registered a negative net loss of migrants between 1988 and 1992. In some of these countries, the net migration rates of rural areas suggest that rural-to-urban migration may not be as important as expected or that the reverse movement has increased.

While the apparent slowdown of rural-urban migration is important, the fact remains that the SSA primate city is a major factor in the contradictions between urban and rural life. Because of its monopoly of economic opportunities and political power, the city has created a perception on the part of rural-urban migrants of certain and immediate opportunities for socioeconomic improvement, a perception that is rarely realized. The phenomenal growth of shantytowns and squatter settlements and the proliferation of informal employment in the cities of SSA are the result of this miscalculation (Figure 8.17).

A HOPEFUL VIGNETTE

Namushi comes from a family of seven—four boys and two girls living with their mother—and was born in Mabumbu village in western Zambia. Throughout her childhood, Namushi and her family survived on what they managed to grow from their fields, together with any small temporary jobs, to access limited cash. Often times, Namushi's family struggled to provide more than one meal a day. At the age of 20, Namushi, along with four boys from

Figure 8.17 Getting hair cut and styled is one of the basic services provided by every culture. Around Kaunda Square in Lusaka, entrepreneurs earn a bit more by adding telephone services to their business model. *Source:* Photo by Angela Gray Subulwa.

her village, made the (mis)calculation to set out for the capital city of Lusaka in search of a better life. According to Namushi, she set out with an unquestioned belief in the socioeconomic possibilities that life in Lusaka would offer.

Upon arrival, Namushi quickly realized that her beliefs and perceptions about life in Lusaka were indeed miscalculations. Her childhood in rural, western Zambia did not equip her with the tools (linguistic or otherwise) to easily navigate and negotiate the complexities of life in Lusaka. Although she struggled to learn Nyanja and English, she remained committed to the idea that succeeding here would translate to success for her entire family back home in Mabumbu. For the first few days (before they could locate anyone who lived in the city from Mabumbu's nearby villages), Namushi and her four travel companions slept at the bus station and set out each morning in search of employment. Luckily, Namushi was able to secure employment as a housemaid in the suburbs of Lusaka by her third week in the city.

For the next year, Namushi worked as a housemaid, while sending a significant portion of her earnings back home to her village. It was during this time that Namushi was able to improve her English, which later helped secure better employment at a newly opened gas station. While working at the gas station, Namushi realized that she would continue to struggle and fail to realize her goals if she continued to send all of her earnings directly home to Mabumbu. She decided to reduce her remittances to the village and direct some of her income into savings, with the goal of opening a small business in the nearby compound, Kaunda Square. Over the course of a year, Namushi saved enough cash to construct a small, mobile store (*katemba*) from which she sold vegetables, candy, candles, salt, and the daily essentials of life in the compound. As her business stabilized, Namushi continued to save cash, hoping one day to open a larger, permanent-structure grocery in the compound market area. Within three years of saving and navigating the intricacies of grocery marketing in the compounds, Namushi finally succeeded as she opened her grocery in Kaunda Square in 2005.

When asked, Namushi characterizes her successful navigation of life in Lusaka as a story of constant negotiations and calculations of risk, coupled with the determination to survive that she learned from her grandmothers back in Mabumbu. Namushi reflected on the overlapping challenges that she faced—coming from a rural area, navigating the informal urban economy, lacking any support system or network, and most critically doing all of this as a woman. As a woman, Namushi felt compelled to carry herself a "bit rough" in order to guard against those who saw an opportunity to take advantage of a young, single, rural woman in the city. Another problem she faced as a woman entrepreneur in the informal economy was the issue of transportation (from the wholesalers to the compound). Namushi found it difficult to secure transport without the fear of being taken advantage of by the overwhelmingly male drivers—fears ranging from simply being overcharged to fears of thieves and even fears of being raped along the way. Another obstacle Namushi faced came from the wholesalers themselves (in Lusaka, the wholesale market is dominated by men of Indian descent). Namushi often found herself served last even if she arrived first. And while her male counterparts were able to negotiate for small credits and loans on their wholesale purchases, Namushi was unable to negotiate

Figure 8.18 Namushi and her grocery shop on Kaunda Square in Kinshasa. *Source:* Photo by Angela Gray Subulwa.

similar deals. In the face of all of these obstacles, Namushi did succeed in opening her small, permanent grocery shop on a busy corner near the Kaunda Square vegetable market in 2005 (Figure 8.18).

By 2010, she had expanded her original grocery shop and had opened another six shops in Kaunda Square (three additional groceries and three cosmetics/pharmacies). Namushi returned to Mabumbu, collected her four brothers, and returned with them to Lusaka. The boys assisted Namushi in maintaining and expanding her shops, often mediating some of the gender constraints that remain in Lusaka's informal economies. While her brothers helped her with the shops, Namushi sent her nieces to school, in the hopes that they will do greater things than she had accomplished herself.

SUGGESTED READINGS

Bekker, S., and L. Fourchard, eds. 2013. *Governing Cities in Africa: Politics and Policies*. Cape Town: HSRC Press.

Charton-Bigot, H., and D. Rodrigues-Torres, eds. 2010. *Nairobi Today: The Paradox of a Fragmented City*. Dar es Salaam: Mkuki na Nyota Publishers.

Locatelli, F., and P. Nugent, eds. 2009. *African Cities: Competing Claims on Urban Spaces*. Leiden: Brill.

Murray, M. 2011. *City of Extremes: the Spatial Politics of Johannesburg*. Durham, NC: Duke University Press.

Myers, G. 2011. *African Cities: Alternative Visions of Urban Theory and Practice*. London: Zed Books.

Obrist, B., V. Arlt, and E. Macamo, eds. 2013. *Living the City in Africa: Processes of Invention and Intervention*. Berlin: Lit Verlag.

Parnell, S., and E. Pieterse, eds. 2014. *Africa's Urban Revolution*. London: Zed Books.

Pieterse, E., and A. Simone, eds. 2013. *Rogue Urbanism: Emergent African Cities*. Cape Town: Jacana Media & African Centre for Cities.

Quayson, A. 2014. *Oxford Street, Accra: City Life and the Itineraries of Transnationalism*. Durham, NC and London: Duke University Press.

United Nations Habitat. 2014. *State of African Cities 2014: Re-imagining Sustainable Urban Transitions*. Nairobi: UN Habitat.

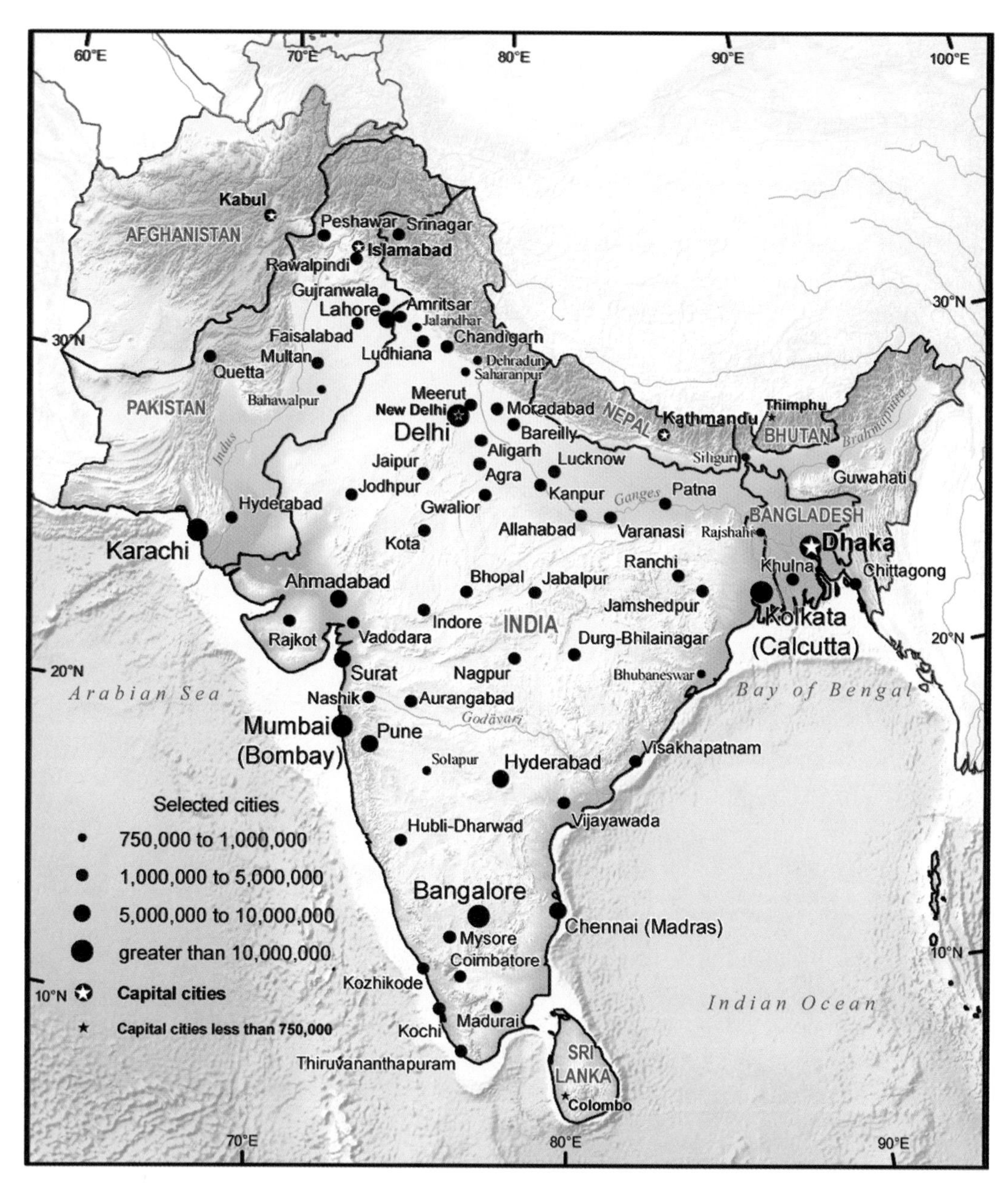

Figure 9.1 Major Urban Agglomerations of South Asia. *Source:* United Nations, Department of Economic and Social Affairs, Population Division (2014), World Urbanization Prospects: 2014 Revision, http://esa.un.org/unpd/wup/.

9

Cities of South Asia

ASHOK DUTT, GEORGE POMEROY,
ISHRAT ISLAM, AND IPSITA CHATTERJEE

KEY URBAN FACTS

Total Population	1.7 billion
Percent Urban Population	33%
Total Urban Population	552 million
Most Urbanized Countries	Maldives (45%)
	Pakistan (38%)
Least Urbanized Countries	Nepal (18%)
	Sri Lanka (18%)
Number of Megacities	6
Number of Cities of More Than 1 Million	72
Three Largest Cities (*Metacities*)	*Delhi* (25 m), *Mumbai* (21 m), Dhaka (17 m)
Number of World Cities (Highest Ranking)	5 (Mumbai, New Delhi, Bangalore, Chennai, Karachi)

KEY CHAPTER THEMES

1. The duality of prosperity alongside poverty in the midst of a vibrant mosaic of language, ethnicity, and faiths make cities in South Asia unique.
2. There are three basic types of South Asian cities: bazaar based, colonial, and planned.
3. There have been five major influences on the development of South Asian cities: the Indus Valley civilization, the Aryan Hindus, the Dravidians, the Muslims, and the Europeans.
4. The current urban system most distinctly reflects the dominance of Presidency towns during the colonial era.
5. The urban form of South Asian cities is reflected in two basic models: the colonial-based city model and the bazaar-based city model, with permutations of both.
6. India has a relatively well-balanced urban hierarchy; Pakistan has a dominant southern city and dominant northern one; all other South Asian countries are characterized by urban primacy.

7. Massive rural-to-urban migration has led to exploding urban populations, the growth of squatter settlements, and emerging inabilities to supply urban residents with clean water and other urban services.
8. Civil wars and political instability have been major contributing factors to the destabilization of urban areas over the decades, most recently in Afghanistan and Sri Lanka.
9. Planned cities and new towns have played an important but subsidiary role in the region for a long time, with Islamabad (Pakistan) and Chandigarh (India) as notable recent examples.
10. Globalization, which began with the adoption of economic reforms, has resulted in growing affluence among a rising middle class; but it has also increased urban poverty, spatial exclusion of the poor, urban violence, and perhaps worsening environmental quality.

In the cities of South Asia, a vibrant optimism and a newfound confidence abounds (Figure 9.1). Led by the 300 million-strong Indian middle class, a rampant consumerism illustrates a giddy self-assurance and sense of hope. One merely needs to step into the bright, flashy, and glamorous automobile showrooms, where eager upper-middle-income buyers may be seen purchasing not just a car, but in some cases a *fifth* family car and at prices over $23,000—*more than 46 times* the average per capita-income in India. Glossy malls, housing dominant retail giants like Gucci and Prada, boutiques selling designer clothes, discotheques packed with hip youngsters, McDonalds and Pizza Huts filled to the brim with kids wanting the "American experience," amusement parks and movie theaters run by the latest digital technologies, and beauty parlors and spas define the landscape of fast-globalizing South Asian cities. Globalization is here to stay, and the growing South Asian middle class wants more of it. Computer literacy, software proficiency, and business management skills define the youth component of the middle class, who are increasingly acquiring lucrative jobs in the local branch offices of global corporate giants. The grim determination of the postindependence era (post-1947), to produce scientists, engineers, and doctors who could build the nation, is slowly being replaced by a global dream to produce CEOs, accountants, software professionals, who could afford the consumptive lifestyle of the American middle class. Mumbai, the dominant financial, commercial, and movie hotspot in India, is considered a world city of the first order because it sends and receives massive financial, commercial, and cultural flows. Delhi, Bengalūru (Bangalore), Hyderabad, and Kolkata (Calcutta) in India, Dhaka in Bangladesh, Colombo in Sri Lanka, and Karachi and Lahore in Pakistan are also world cities because they are economically and culturally integrated with global flows of goods, investments, images, and people. Beginning in the late 1980s, the countries in South Asia dissolved protectionist economic systems that shielded their domestic markets through tariff walls, licensing, and quotas. The "License Raj" (as it was known in India) was abolished, and through rounds of structural adjustments, a New Economic Policy of liberalization or free-market globalization was adopted. The adoption of this policy opened up South Asian markets and also its people to global corporations and their investments. The economic reforms have produced tremendous urban impacts, many of which have been contradictory and controversial (Box 9.1). While the

Box 9.1 Call Centers, SEZs, and Sweatshops

Have you done any of the following of late? Called to reserve a rental car? Telephoned technical support for help with your new computer? Spoken to someone via phone to straighten out a credit card issue?

If you have, then chances are good that the person on the other end of the line was in India. If, indeed, this was the case, then it was your personal encounter with "outsourcing," a phenomenon that is transforming the way business is done across the globe. With the adoption of global free-market policies, corporations now have the freedom to take their production activities outside their home country and situate it anywhere they find more advantageous. The result is a dismantling of factory-based manufacturing and the beginning of a more flexible-style of production— different parts of the production process can be geographically dispersed or *outsourced*. Outsourcing first gained attention as U.S. automakers began to subcontract the manufacturing of certain auto components to other firms in the United States. For example, Ford would contract with a smaller, independent firm for wheel assemblies.

Offshore outsourcing is when a firm takes activities and moves these overseas. Giant corporations in the United States, Western Europe, and Japan prefer outsourcing, because it allows them access to cheap labor, tax rebates, and relaxed environmental norms, hence higher profits. The rise of *business process outsourcing firms (BPOs)* shows that even service-sector employment can be outsourced. Business process outsourcing involves taking accounting functions, customer services, computer programming, and other activities outside and usually offshore.

India has several advantages, which make it an attractive destination for BPO opportunities. First, it has a well-developed system of universities and technical colleges that produce a large supply of technically qualified and well-educated personnel. Second, English proficiency, a legacy of British colonial rule, has provided postsecondary graduates with the language skills needed to work in "call centers." Third, the cost differential between hiring U.S. workers and hiring those in India may be as high as 10 to 1. This represents a potential savings—hard for any firm to ignore. Finally, with the rise of modern information technologies (telephone, internet), distance has "collapsed" and the cost of doing business over great distances has in many ways vanished. Bengalūru, Mumbai, Pune, Hyderabad, and Chennai represent important BPO destinations; Accenture, Citibank, Dell, IBM, Infosys, Microsoft, Office Tiger, Verizon, and Wipro are some of the corporations setting up shop there. The call center workers function as customer service representatives answering 1-800 calls; they undergo accent training, are briefed about American sports and weather so that they can politely chat with a customer. They are often given more "relatable" names like Dave or Nancy so that customers are comfortable.

Outsourcing has become the site for heated debates because of its controversial impacts on labor and the environment. A large portion of the outsourced jobs include flexible-style manufacturing jobs—a shirt can be stitched in China, the label sown in Guatemala, and the buttons stitched on in Mexico, before it comes back to the American consumer. On the one hand, this represents a job loss from the outsourcing nation and has therefore become the site of political debates in outsourcing nations. On the other hand, corporations outsourcing jobs to cheaper locations are accused of exploiting labor and environment in the outsourced locations. The result is a world of sweat shops and Special Economic Zones (SEZs). Since India's economic integration, SEZs have expanded at a rapid rate. These SEZs require huge land areas, often a minimum of 1,000 hectares and usually include appropriation of agricultural land adjoining major cities to be used for processing export goods. The SEZs are known to employ labor at exploitative rates—wages are 34 percent lower than non-SEZ jobs and the workers are forced to work longer. Women and children bear the brunt of this exploitation, because SEZs prefer women and children as they are seen as "nimble" and "compliant."

City municipalities often encourage corporations like Nike, Reebok, and Adidas by providing them land, giving them tax rebates, and relaxing environmental laws, because they look upon outsourced ventures as contributing to the export earnings. Manufacturing hand-stitched soccer balls in South Asia has become a controversial case—5–14 year olds in Pakistani cities are employed by global corporations like Nike to work for 12 hours a day in near-slavery conditions. United Students Against Sweatshops (USASS) is an international grassroot organization of students trying to pressure their respective universities to ensure that university clothing is not produced in sweatshop conditions.

middle class has become a global labor source, manning call centers, foreign banks, and corporate offices, the majority of urban poor have been relegated to sweatshop-like conditions in Special Economic Zones (SEZs). Many others have lost formal sector jobs as manufacturing units closed down, unable to keep up with global competition in this free-market regime. Still others have suffered evictions under city greening, beautification, and slum demolition policies.

While annual rates of urban growth have slowed in the region, massive migrations of the rural poor to the cities have put tremendous pressure on urban infrastructure. Increasing urban poverty has been coupled with an increase in low-paid informal jobs resulting in visible landscapes of poverty. Mumbai, the city which best demonstrates the affluence and consumption noted above, also presents urban poverty at its most daunting. Within the city proper, 9 million slum dwellers comprise about 62 percent of the population. Across the Mumbai urban agglomeration these numbers swell further. Dharavi, with over 800,000 people (and perhaps as many as 1 million!) crowded into an area of just under a full square mile (2.6 sq km), is Mumbai's largest slum and perhaps the best known (though, contrary to rumor, not the world's or even South Asia's largest). Estimates of the number of "pavement dwellers," those who have no shelter at all and sleep on sidewalks, doorsteps, and the like, range wildly from as low as 250,000 to over 1 million. Across each of the region's megacities, the story is the same—tremendous numbers of people living in conditions of poverty with inadequate

Figure 9.2 As cities fill up with people, streets become more congested with not only cars, but bicycles and camels as well. *Source:* Photo by George Pomeroy.

shelter, lack of clean water, and filthy living conditions. Already racked with unemployment and underemployment, each swells with new migrants each and every day (Figure 9.2).

This duality of prosperity alongside poverty in the midst of the vibrant mosaic of language, ethnicity, and faiths of South Asia is what makes cities in this region unique. Throughout South Asia, cities remain receptacles of hope and serve as powerful engines of social and economic change.

URBAN PATTERNS AT THE REGIONAL SCALE

South Asia has 552 million urban dwellers, and this number will nearly double to over 1.13 billion by 2050. This means that South Asian cities will expand by more than the populations of the United States, Canada, and Mexico combined! Most of the urban population is in India, Pakistan, and Bangladesh, which currently rank as the world's second, sixth, and eighth most populous countries (Table 9.1). Together, the three comprise 97 percent of

Table 9.1 South Asia's Twelve Largest Urban Agglomerations

City—Country	*Population (Est. 2014)*
Delhi—India	25.70
Mumbai—India	21.04
Dhaka—Bangladesh	17.60
Karachi—Pakistan	16.62
Kolkata—India	14.87
Bangalore—India	10.09
Chennai—India	9.89
Hyderabad—India	8.94
Lahore—Pakistan	8.74
Ahmadabad—India	7.34
Pune—India	5.73
Surat—India	5.65

Source: United Nations. *World Urbanization Prospects: 2014 Rev.*

South Asia's total population; three out of four live in India alone.

With its immense population, India, at 33 percent urban, largely determines the overall regional average, which is also about 33 percent. Pakistan and Bangladesh have just over one in three living in urban areas. The smaller countries of Afghanistan, Nepal, and Sri Lanka have less of their populations in cities. Bhutan, although also small, has demonstrated a burgeoning urban expansion in the recent decades; its cities are growing at an annual rate of 3.7 percent. While most of South Asia has seen steady urbanization, Afghanistan, Sri Lanka, Nepal have suffered civil conflict, revolt, or insurgency, making city growth and development more difficult and erratic.

While cities of all sizes in South Asia have been growing, the six megacities have grown the most. Delhi in the last two decades has asserted itself as India's leading city, eclipsing Mumbai by growing at twice the pace of its rival. Indeed, the growth has been remarkable to peg Delhi as the second-largest city in the world with a 2015 population of nearly 26 million. Although just second in the region, Mumbai ranks fifth in size globally, followed by Dhaka (11th), Karachi (12th), and, despite decades of slower growth, Kolkata (14th). Astonishingly, South Asia now has five of the fourteen largest cities in the world! Even more incredible: by 2030, five of the world's ten largest cities will be South Asian.

Including cities further down the urban hierarchy, there are an additional six cities with between 5 and 10 million people. Altogether, 73 urban centers in South Asia top a million residents. 58 of those "millionaire cities"—a term made obsolete by the astounding magnitude of urban populations—are located in India, ten in Pakistan, three in Bangladesh, and one each in Nepal and Sri Lanka. If one takes into consideration all cities of 300,000 or more in population, the number is 207.

Within India, six megacities dominate the urban hierarchy. Mumbai, Delhi, and Kolkata are the anchors of a northern urban triangle. Chennai, Bengalūru, and Hyderabad form a southern urban triangle. Together, these six cities make up over one-fifth of India's urban population. Cities of the two triangles have been tied together by a great highway-building project known as the Golden Quadrilateral (Figure 9.3). Completed in 2012, the Golden Quadrilateral is a critical part of India's version of the U.S. Interstate Highway system. This project is a component of the country's most ambitious plan to improve transportation infrastructure since independence. The Quadrilateral runs through 13 Indian states and connects the nation's 4 largest cities with 3,600 miles (5,794 km) of four- and six-lane highways. The overall scheme is to widen and pave an additional 40,000 miles (64,374 km) of highways over a 15-year period.

Pakistan's urban hierarchy is dominated by a pair of cities with a combined population of over 45 million: Karachi dominates southern Pakistan, and Lahore (the smaller of the two) dominates the north. Urban primacy characterizes the remaining countries of the region. In Bangladesh, Dhaka is nearly four times the size of second-ranked Chittagong. It is a clear primate city, with all attendant traffic problems (Box 9.2). The same holds true for Kathmandu and Pokhara in Nepal. Both Kabul, Afghanistan, and the extended metropolitan region of Sri Lanka's capital of Colombo are over ten times the size of the next largest city in each country.

South Asian cities may be divided into three basic types: *traditional cities, colonial cities*, and *planned cities*. Traditional cities are those which were part of a thriving urban system

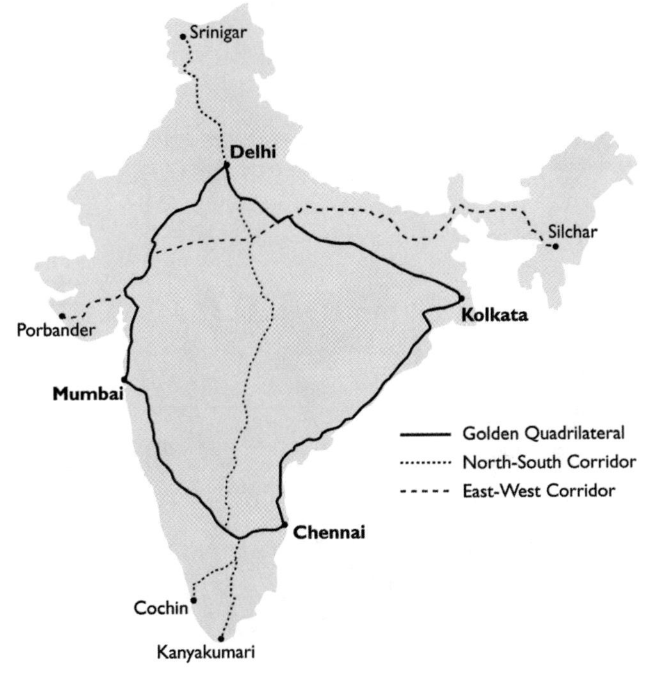

Figure 9.3 The Golden Quadrilateral of express highways links the anchor cities of India's urban hierarchy: Delhi, Mumbai, Kolkata, and Chennai. *Source:* Compiled from various sources.

prior to Western colonialism, whether as centers of trade and commerce (such as Surat on India's west coast), centers of administration, or centers of pilgrimage. Varanasi, on the River Ganges, for example, remains the leading pilgrimage destination in South Asia for Hindus and Jains; it attracts millions of pilgrims each year. Several of the region's largest cities, most notably Mumbai, Kolkata, Chennai, were the products of British colonialism; but, since 1947, they have evolved further to serve India's expanding economy.

South Asia's third basic urban type is the new planned city, of which there are two kinds: (1) political and administrative centers such as Islamabad, Pakistan, and Chandigarh, India; (2) industrial centers for steel and other related heavy industrial activities, such as Durgapur, in the state of West Bengal and Jamshedpur in Bihar, both in India. Islamabad was built in the 1960s to take the place of Karachi as Pakistan's capital city. Chandigarh was also built after independence as the new state capital of India's Punjab. When a new state was carved out of Punjab in the 1960s, Chandigarh found itself on the border and has since served as capital of both Punjab and the then newly formed Haryana.

A recent and remarkable change in the urban fabric of South Asia is in India. With prosperity and growth has come the development of "urban corridors" broadly akin to Jean Gottmann's Megalopolis in the United States. Spurred by the construction of new express highways, with high-speed rail in the planning stages, city-regions are emerging

Box 9.2 The Humble Rickshaw

Figure 9.4 On a Delhi roadside, the driver of a cycle rickshaw takes time for a mid-day nap. *Source:* Photo by D. J. Zeigler.

"Overwhelming" may best characterize the assault on the senses when an outsider first encounters traffic in South Asian cities. The congestion, apparent chaos, and dizzying array of transportation modes, accentuated by dust and the incessant beeping of horns, can be intimidating. What probably strikes outsiders most are the modes of transportation rarely seen in the developed world: bullock carts, large numbers of motorcycles and scooters, and, of course, rickshaws in their motorized and non-motorized forms (Figure 9.4). Auto rickshaws are sometimes referred to as *tuk-tuks*, *samosas*, *bajaj*, and other names across Asia and beyond. They are typically small three-wheeled, gas- powered vehicles employed as taxis. Battery-powered versions are beginning to appear, and some auto rickshaw drivers are even using apps similar to Uber to secure fares!

Auto rickshaws are the motorized versions of the traditional ones which were pulled or bicycle powered. They are rapidly disappearing in many parts of the world, including South Asia, but they still remain plentiful across selected bits of northern India, Nepal, and Bangladesh. A common perception is that this type of work is inhumane, degrading, and exploitative, with drivers being treated as human beasts of burden. However, the reality of bicycle and hand-pulled rickshaws is more complex and nuanced than one initially thinks. Many argue that although the work is arduous and poorly paid, it is often the only opportunity for those with few skills to earn a living. So when the West Bengal state government attempted to curtail and even ban human-powered rickshaws, the biggest protesters were the drivers themselves.

Another common perception is that because they are slow moving, these vehicles contribute to road congestion. Advocates argue that rickshaws take less road space, less space for parking and storage, and make fewer demands on infrastructure. Finally, rickshaw advocates claim that with no emissions they are environmentally friendly, provide affordable transportation for those without other options, and are actually safer and often more convenient.

Most bicycle pullers rent their rickshaws. For the most part, they are newer arrivals to the city with no skills and no investment capital. Other barriers to ownership are the cost of storage and maintenance, the bureaucratic red tape involved with licensing, and the fact that sometimes the work is seasonal. The typical puller will rent his (very few women pull rickshaws) from a local entrepreneur. Rentals are generally done by referral and application, without collateral or a deposit, since that would make the cost prohibitive. The puller then wades into the sea of competition to secure fares. Fares vary, but a typical one may be the equivalent of 25 cents to a dollar for a distance of perhaps several city blocks. Common tasks include shuttling children to school or office workers to work. Competition for fares may be fierce. At the end of the day, the puller returns his rental to the garage, pays the owner a hiring fee of 50–60 cents, and keeps everything leftover for average daily earnings of $2–$3.

With an estimated 400,000 bicycle rickshaws, Dhaka is the undisputed "Rickshaw Capital of the World." Rickshaws are colorfully decorated, especially those belonging to owner operators. Despite being sometimes harassed by police and car owners, they transport nearly half of the city's passengers. To the narrow lanes and slower speeds of Dhaka's crowded old town they are much better suited. When new transportation projects are being planned, however, it is often with a "bigger is better" mentality that focuses on elevated highways and automobiles. However, there are signs that some transportation planners are taking a new look at this long- underappreciated mode of transportation.

along transportation corridors. An example is the urban corridor that begins in Pune, sweeps northward through Mumbai, on to Surat, and thence to Ahmedabad, thus accounting for four of India's ten most populous cities.

HISTORICAL PERSPECTIVES ON URBAN DEVELOPMENTS

The cultural diversity (Figure 9.5) and urban fabric of South Asia are derived from five distinctive influences. Chronologically, these are (1) the Indus Valley civilization, 3000 to 1500 BCE; (2) the Aryan Hindus since 1500 BCE; (3) the Dravidians since about 200 BCE; (4) the Muslims since the eighth century; and (5) the Europeans since the fifteenth century.

Indus Valley Era

The Indus Valley is one of the "cradles of civilization" and among the world's oldest urban hearths. The preeminent cities of the Indus Valley civilization, Mohenjo Daro and Harappa, located in what is now Pakistan, were established as planned communities as early as 3000 BCE. They flourished for about 1,500 years.

Figure 9.5 The Sikhs, neither Hindu nor Muslim, are a major part of India's cultural diversity, seen here in their main *gurdwara*, the place where they worship. *Source:* Photo by D. J. Zeigler.

As the largest urban centers of the region, they anchored an extensive settlement system that included at least three other large urban centers and perhaps over 900 smaller ones. First excavated in 1921, the ruins of Mohenjo Daro reveal a carefully constructed city reflective of a highly organized and complex society. No other city outside the Indus civilization possessed such an elaborate system of drainage and sanitation, signifying a generally high standard of living. This urban civilization came to an end around 1500 BCE, when the newly arrived Aryans—less civilized but more adept in warfare than the Indus people—overpowered the Indus civilization and turned the northern part of the South Asian realm to a mixture of pastoralism and sedentary agriculture.

Aryan Hindu Impact

Eventually the demands of trade, commerce, administration, and fortification gave rise to the establishment of sizable urban centers, particularly in the middle Ganges Plains. Originating from a modest fifth-century fort, Pataliputra developed into the capital of a notable Indian empire, the Maurya (321–181 BCE). Its location coincides with that of present-day Patna. Pataliputra was organized to conform to the functional requirements of the capital of a Hindu kingdom: residential patterns based on the function-based, four-caste social system along with the requisite royal administrative features.

While caste is not as important in determining contemporary residential patterns in cities as it was in Pataliputra, it remains socially significant today. Caste, the designation of social status by birth through the caste of one's parents, places each person into one of four broad groups—*Brahmins*, *Kshatriyas*, *Vaisyas*, and *Sudras*. Within the *Sudra* designation are those of very "low" caste and even those without caste or status. The people "without caste" are today more popularly referred to as *Dalits*. In the past, Dalits were often referred to as "untouchables" because the touch or even a shadow cast by one was thought to "pollute" someone of higher caste. Traditionally, each caste and its subcastes had certain designated occupations; for example, "washer-men" or *dhobi-wallahs*" (Figure 9.6). *Brahmins* (priest caste) are at the top, followed by *Kshatriyas* (warrior caste), *Vaisyas* (commercial and agricultural caste), and *Sudras* (manual labor caste). Caste still plays a major role in society. Marriages are generally within caste and there remain broad connections between the caste status on one hand and income, quality of life, and social connections on the other. This remains true even after years of legal reforms such as a "reservation system" (similar to affirmative action in the United States) and broader social campaigns led by people such as Mahatma Gandhi, spiritual leader of India's independence movement. Even today, urban land-use patterns and socioeconomic structures somewhat reflect the caste system.

In ancient Pataliputra, one could clearly see the spatial distribution of castes. Near the center and a little toward the east were the temple and residences of high-ranking *Brahmins* and ministers of the royal cabinet. Farther toward the east were the *Kshatriyas*, rich merchants, and expert artisans. To the south were government superintendents, prostitutes, musicians, and some other members of the *Vaisya*. To the west were the *Sudras*, including untouchables, along with ordinary artisans and low-grade *Vaisyas*. Finally, to the north were artisans, *Brahmins*, and temples maintained for the titular deity of the city. A well-organized city government, hierarchical street network, and an elaborate drainage system accompanied this functional distribution of population in Pataliputra.

After the eclipse of the Maurya Empire in the second century BCE, rulers of the Gupta

Figure 9.6 The *dhobi-wallahs*, or "washer-men" make their living washing (and drying) clothes. *Source:* Photo by D. J. Zeigler.

Empire (320–467 CE) made it their capital, but the city lost its importance thereafter and was eventually buried under the sediments of the Ganges and Son Rivers. Only small parts of the old city have recently been excavated. Other Hindu capitals that developed in both north and south India changed and modified the many urban forms used in Pataliputra.

Dravidian Temple Cities

In contrast to the subcontinent's north, Hindu kingdoms in India's south gave rise to distinctly Hindu forms of city development. The rulers of south India constructed temples and water tanks as nuclei of habitation. Around the temples grew commercial bazaars and settlements of Brahmin priests and scholars. The ruler often built a palace near the temple, turning the temple-city into the capital of his kingdom; Madurai and Kancheepuram are examples of such lofty and grand temple cities. Such city forms were also exported to Southeast Asia; Angkor Wat in Cambodia is one example.

Mentioned by Ptolemy, Madurai, the second capital of the south Indian Hindu kingdom of Pandyas, dates to about the beginning of the Christian era. Though Madurai is now several times the size of the old walled city, with imprints from a brief Islamic period and a longer British dominance, its religious importance is nearly comparable to Varanasi in north India, which is the foremost Hindu pilgrimage center.

Muslim Impact

The first permanent Muslim occupation that significantly influenced the subcontinent began in the eleventh century CE and resulted in adding many Middle Eastern and central

Figure 9.7 The Taj Mahal has become the single most recognized icon of India. It was built in Agra as a tomb for Shah Jahan's wife and is now a UNESCO World Heritage Site. *Source:* Photo by George Pomeroy.

Asian Islamic qualities to the urban landscapes of South Asia. Shahjahanabad is a particularly good example of Muslim impact. The Moghul emperor Shah Jahan, who planned the Taj Mahal as a tomb for his wife in Agra (Figure 9.7), moved his capital from Agra to Delhi (about 125 miles or 200 km) and started the construction of a new city, Shahjahanabad, on the right bank of the Yamuna River. The city was built near the sites of several previous capital cities and took nearly a decade to complete (1638–1648). In its architecture was a fusion of Islamic and Hindu influences. Though the royal palace and mosques, with their arched vaults and domes, adhered to Muslim styles, Hindu styles were found in combination. The Muslim rulers were most concerned with the magnificence of their royal residences and courts, and massiveness of their fortresses. These features are represented most vividly in Shahjahanabad. Surrounded by brick walls without a moat, the city was completely fortified.

Situated at the east end of the city was the Red Fort (Figure 9.8). Planned as a parallelogram with massive red sandstone walls and ditches on all sides except by the river, the fortress had an almost foolproof defense. Inside were a magnificent court, king's private palace, gardens, and a music pavilion. All were built of either red sandstone or white marble. The Red Fort remains a central feature of Delhi today, serving often as a platform for political proclamations and as a well-known tourist destination. A main thoroughfare, Chandni Chowk ("silver market") ran straight westward from the Red Fort toward the Lahore Gate of the city. The Chandni Chowk was one of the great bazaars of what was then called "the Orient."

Shahjahanabad ceased to be the capital of India, more precisely north India, when the British rule started in the eighteenth century,

Figure 9.8 The Red Fort, in Old Delhi, remains a potent feature of Indian nationalism. *Source:* Photo by D. J. Zeigler.

Figure 9.9 To the left is a Muslim neighborhood and to the right a Hindu one in Old Delhi.
Source: Photo by John Benhart, Sr.

but it continued to be a functional city. Today, the area is known as "Old Delhi" and is part of the Delhi metropolis. Most of the city walls are gone, though all the basic structures of the Red Fort and Jama mosque remain intact. Chandni Chowk continues to be a busy traditional bazaar. It is very densely populated with a mixture of Hindus and Muslims (Figure 9.9). Large parts of Old Delhi are gradually being transformed into commercial and small workshop uses.

Colonial Period

After Vasco de Gama discovered the oceanic route via Africa's Cape of Good Hope and landed on the southwestern coast of India in 1498, the European powers of Portugal, Holland, France, and Britain became greatly interested in developing a firm trade connection with South Asia. Though initially all four powers obtained some kind of footing in India, the sagacious diplomacy and "divide-and-rule" policy of the British succeeded in ousting the other Europeans from most Indian soil. Eventually, the British established three significant centers of operation—Bombay, Madras, and Calcutta, seaports all, for the convenience of trading and receiving military reinforcements from Britain. Hence, these cities were designed as the headquarters for the three Presidencies into which the British divided South Asia for administrative purposes. Consequently, the cities are referred to as the "Presidency towns." In the 1990s, they were renamed Mumbai, Chennai, and Kolkata, respectively, to reflect indigenous cultures and further downplay India's colonial heritage.

The Presidency Towns

When Mumbai, Chennai, Kolkata, and Colombo were established as Presidency towns, their nuclei were forts. Outside these forts were the cities. Inside the cities, two different standards of living were set for two different classes of residents: Europeans and "natives," each with their own parts of the city. The rich were composed of absentee landowners from rural areas, moneylenders, businesspeople, and the newly English-educated elite and clerks. The poor comprised servants, manual laborers, street cleaners, and porters (Figure 9.10). The rich needed the service of the poor, and therefore the houses of the native rich in many instances stood by the houses of their poor, native service providers.

As local industries grew in the nineteenth century, a new working class developed. In Kolkata, the industrial workers worked mainly for jute mills and local engineering factories; in Mumbai, for the expanding cotton and textile-related industries; and in Chennai, for tanning and cotton textiles. As these trades grew, the Presidency towns turned from water to railways and roads for inland transportation. Train services were started in South Asia in 1852. The Presidency towns also developed huge hinterlands that catered to the needs of the colonial economy. The hinterlands supplied the raw materials to the three seaports for export to the United Kingdom; in return, the British sent consumer-type manufactured goods through the same ports. Thus, the Presidency towns became the main focus of the colonial mercantile system, and their architecture exhibited the colonial influence of Western Gothic and Victorian styles.

Figure 9.10 Labor is cheap in India, so porters are often called upon to transport bulk goods from one part of the city (in this case, Mumbai) to another. *Source:* Photo by D. J. Zeigler.

Table 9.2 Topological Characteristics of South Asian Cities

	Land Value	*Population Density Gradient*	*Physical Aspect*	*Land Use Composition of the City Center*	*Historical Roots*	*Mixture of Three Forms*
Bazaar City	Highest at the center; declines as one moves to the periphery	Highest at the center and declining inversely as one moves to the periphery	Narrow streets, commercial establishments at the center occupying the road; front, back, and second-/third-floor residential; generally congested and dirty	Retail and wholesale business mainly, with limited recreation and combined with high-density residential	May have origins from ancient, medieval, or recent times and accordingly may have imprints from Dravidian, Hindu, Muslim or Western forms at the periphery	When a bazaar city was implanted with colonial aspects, garden-like, semi-planned "civil lines" were added; similar addition of planned neighborhoods may also be at the periphery after independence
Colonial City	Highest at the center; generally declines as one moves to the periphery, but relatively higher in the European town compared to the "native" town	Center with minimum density, with highest densities around the CBD, creating a "carter effect" at the center; thereafter declines as one moves to the periphery	Wide streets at the center and the European town, with garden-like, affluent appearance; the "native" town characterized by narrow, sinuous, streets and generally shabby condition	Offices, banks, main post office, transport headquarters, government buildings with large open space, hotels, retail and recreational activities, and residential use	Origins no earlier than sixteenth century; Victorian, neo-Gothic, and other Western forms widely prevalent; native forms also implanted	Parts of colonial city—particularly adjacent to the CBD and some specific locations in the "native" town—evolving characteristics of the bazaar center; planned neighborhoods added, particularly adjacent to the periphery of the European town after independence
Planned City	May vary according to predetermined values for different locations	May vary at different locations of the city, but the initial plan is for low density	Organized and generally pleasing appearance	Combination of retail, office, and recreation in a systematic fashion	May have origins in any historical period, but over time bazaar aspects will begin to dominate the center if restrictions are not strictly adhered to	Planned towns with addition of "civil lines" at the periphery and bazaar central forms evolving at one or many locations

Source: A. K. Dutt and R. Amin, "Towards a Typology of South Asian Cities," *National Geographic Journal of India* 32 (1995): 30–39.

MODELS OF URBAN STRUCTURE

Early on in the study of South Asian cities, some theorists would mechanically try to apply Western models, such as Burgess's Concentric Zones, irrespective of their applicability. No comprehensive model explaining the growth patterns of indigenous cities' structures has been offered, though three basic models have been proposed to explain the distinctive form of South Asian cities: the *bazaar-based city model*, the *colonial-based city model*, and the *planned-city model*. The basic characteristics of the three types of cities have been summarized in a matrix (Table 9.2). Whichever model is used, however, it is important to note that two principal influential forces—colonial and traditional—have combined to create the existing forms of South Asian cities.

The Colonial-Based City Model

The need to perform colonial functions demanded a particular form of city growth that produced the following characteristics (Figure 9.11).

1. The need for trade and military reinforcements required a waterfront because the colonial power operated from Europe. A minimal port facility was the starting point of the city.
2. A walled fort was constructed adjacent to the port with white soldiers' and officers' barracks, a small church, and educational institutions. Sometimes inside the fort, factories processed agricultural raw materials to be shipped to the mother country. Thus, the fort became not only a military outpost but also the nucleus of the colonial exchange.
3. Beyond the fort and the open area, a "native town" or town for the native peoples eventually developed, characterized by overcrowding, unsanitary conditions, and unplanned settlements. It serviced the fort and the colonial administration.
4. A Western-style Central Business District (CBD) grew adjacent to the fort and native town, with a high concentration of mercantile office functions, retail trade, and low-density residential areas. The administrative quarters consisted of the governor's (or viceroy's) house, the main government office, the high court, and the general post office. In the CBD, there also were Western-style hotels, churches, banks and museums, as well as occasional statues of British royals and dignitaries.
5. The European town grew in a different direction from the native town. It had spacious bungalows, elegant apartment houses, and planned streets with trees on both sides; clubs for afternoon and evening get-togethers and with European indoor and outdoor recreation facilities; churches of different denominations; and garden-like graveyards.
6. Between the fort and the European town (or at some appropriate nearby location), an extensive open space (*maidan*) was reserved for military parades and Western recreation facilities such as race and golf courses, soccer, and cricket. On Saturdays, for instance, whites and a few moneyed native people frequented the horse races to gamble.
7. When domestic water supply, electric connections, and sewage links became available, the European town residents utilized them fully; whereas, their use was quite limited in the native town.

8. At an intermediate location between black towns and white towns developed the colonies of Anglo-Indians. They were the offspring of mixed marriages, and they were Christians. Never were they fully accepted by either the native or the European community.
9. Starting from the late nineteenth century, the colonial city became so large that new living space was necessary, especially for the native elite and rich people. Extensions to the city were made by reclaiming the lowland or developing in a semi-planned manner the existing nonurban areas.
10. From the very inception of such colonial cities, population density was very low at the center, which housed Europeans, while a much larger group of "natives" lived outside the colonial center. When the European center was gradually replaced by a Western-style CBD during the second half of the nineteenth century, there was a further decline in the population of the center, giving rise to a density gradient with a "crater effect" at the center.

As the colonial system became deeply entrenched in the Indian subcontinent and an extensive railway network was made operational, waterfronts accessible to oceangoing ships were no longer a prerequisite for a colonial headquarters. Kolkata, Mumbai, Chennai, and Colombo were not the only suitable locations on the subcontinent for high levels of administration.

Cantonments, railway colonies, and hill stations were three other lesser (but numerous)

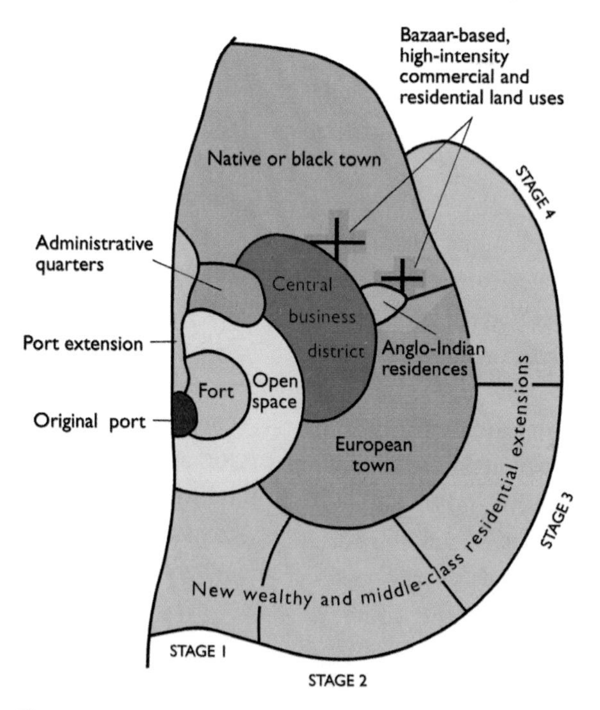

Figure 9.11 A Model of the Colonial-based city in South Asia. *Source:* Ashok Dutt.

colonial urban forms that were introduced to the subcontinent to serve very specific purposes.

Cantonments (from the French word *canton*, meaning "district") were military encampments, some 114 in all by the mid-nineteenth century, which housed a quarter of a million soldiers (both European and natives). Strict segregation by class and ethnicity was practiced in these camps.

Railway colonies surrounded a railroad station or a regional headquarters for railway operation and administration, also with strict segregation in their design. Often situated near urban centers, they eventually formed part of the greater urban area.

Hill stations, at altitudes between 3,500 and 8,000 feet (1,067–2,440 m), served as resort towns for Europeans to escape hot summers on the plains and spend time in the midst of a more exclusive European community. By the time of independence, there were 80 such stations, including Simla and Darjeeling.

The Bazaar-Based City Model

The traditional bazaar city is widespread in South Asia and has certain features that date to precolonial times. Ordinarily, the city grows with a trade function originating from agricultural exchange, temple location, transport node, or various administrative activities (Figure 9.12). Usually, at the main crossroads a business concentration occurs where commodity sales dominate. In north India such an intersection is known as *Chowk*, around which cluster houses of the rich.

The bazaar, or the city center, consists of an amalgam of land uses that cater to the central-place functions of the city. The land use that dominates the center accommodates both retail and wholesale activities. Perishable goods, such as vegetables, meat, and fish—which are bought fresh daily because many homes lack refrigeration facilities—are sold in specific areas of the bazaar. These areas often lack enclosing walls and instead have a common roof. In the process of bazaar evolution, functional separation of retail business occurs: textile shops stay together, attracting tailors; grain shops cluster with each other near the perishable goods market; and pawnshops are adjacent to jewelry shops. Sidewalk vendors are present almost everywhere in the bazaar.

Wholesale business establishments also form part of the bazaar landscape (Figure 9.13). Situated near an accessible location, they tend to agglomerate according to the commodities they deal in, with separate areas for vegetables, grains, and cloth, depending on the size of the city. Traditionally, public or nonprofit inns provide modest overnight accommodation in the bazaar for a nominal fee. However, as a result of Western impact, some hotel accommodations are now available in the medium-sized and larger cities. Prostitutes or dancing girls, once a source of evening entertainment in the bazaar, have been supplanted by cinemas that in turn have declined due to the prevalence of television, VCRs, and DVD players. Traditionally, shops selling country-made liquor were never located in the bazaars, probably because drinking alcohol in public places was considered ill-mannered by both Hindu and Muslim societies. Only in recent years have Western bars and liquor stores started to appear in city centers. Barbers, who used to work outdoors, now have regular shops like those in Western countries, but many still operate on the sidewalks. Long-distance private telephone centers along with internet-access enterprises have become commonplace in city centers.

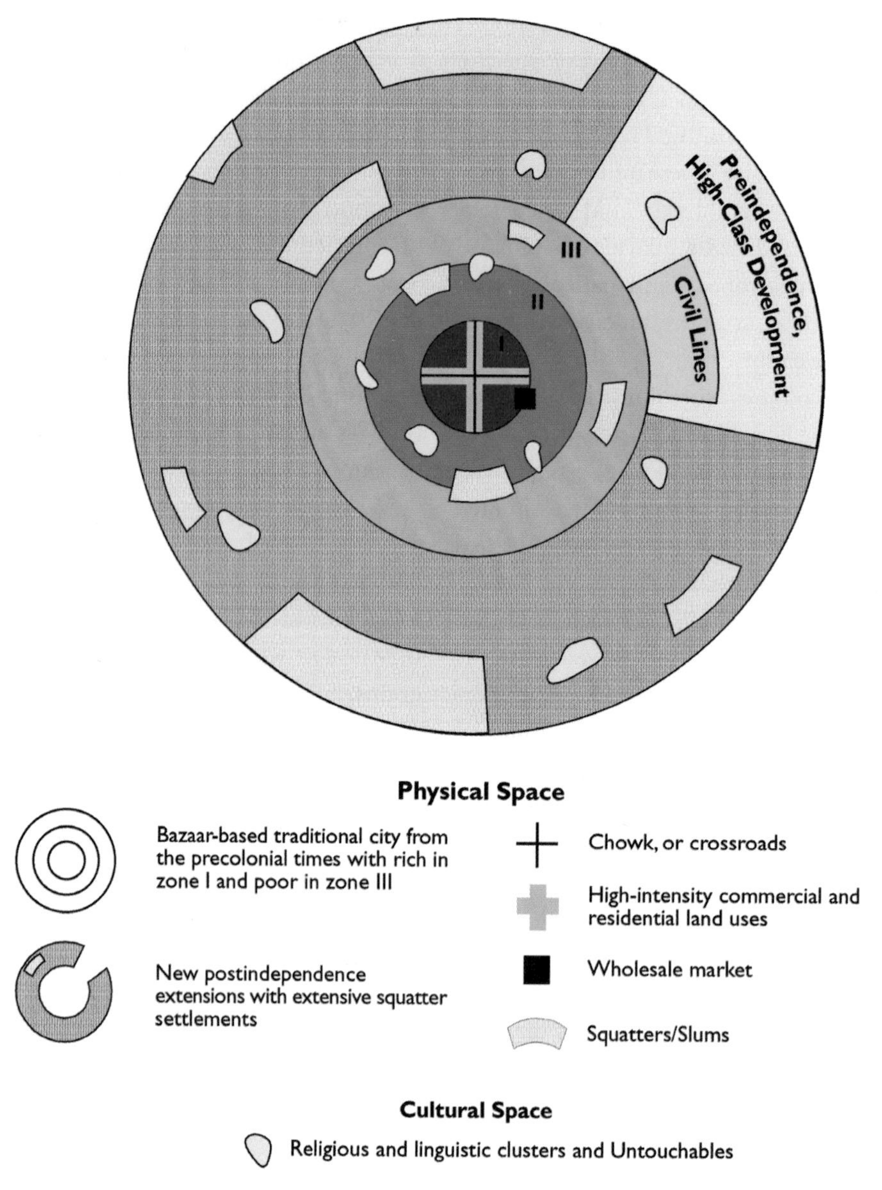

Figure 9.12 A Model of the Bazaar-based City in South Asia. *Source:* Ashok Dutt.

Beyond this inner core, in a second zone, rich people live in conjunction with poorer servants, but not in the same structure. The rich need the poor as domestic servants, cleaners, shop assistants, and porters. The residences of the poor surround this second zone in a third area, where the demand for land is less and its price low. Beyond the third zone, Civil Lines were established during British colonial rule. Here, particularly after independence, the native rich and middle class settled in neighborhoods and squatter settlements developed alongside.

As bazaar cites grew, ethnic, religious, linguistic, and caste neighborhoods were formed in specific areas in accord with the time of

Figure 9.13 A produce vendor in Chennai typifies the bazaar-based city. *Source:* Photo by George Pomeroy.

settlement and availability of developable land. The "untouchables" always occupied the periphery of the city, although sometimes other housing developed later beyond their neighborhoods. In Hindu-dominated areas of India, Muslims always formed separate neighborhoods. Similarly, Hindu minorities in Muslim-majority Srinagar and Dhaka lived in enclaves of the old cities. Often migrants from other linguistic areas formed specific neighborhoods of their own.

Planned Cities

Although there were several planned historic cities in the subcontinent (Mohenjo Daro and Pataliputra, for instance), they did not survive. There were, however, others that were planned during precolonial, colonial, and independence periods that not only survived but also formed nuclei for major urban agglomerations. Jaipur is an example from precolonial times, and Jamshedpur from the British colonial period. Jaipur, India's tenth largest city with an estimated 3 million people, was founded as a planned city in 1727. A hierarchy of streets divided it into sectors and neighborhoods. Though the planned city covered only 3 square miles (5 sq km) and is now surrounded by a built-up area of about 22 square miles (35 sq km), an urban morphology provided by eighteenth-century planning has endured.

Similar is the case with Jamshedpur, planned as a company town for the first steel mill in the subcontinent by the Mumbai-based industrial family of the Tatas. Jamshedpur is about 150 miles (90 km) southeast of Kolkata. The raw materials for the steel making—iron ore, coal, and limestone—are found nearby. The city underwent four different plans during colonial times and one after independence. As the city approaches its 100th anniversary, the metropolitan population approaches 1.5 million residents and much of the industry is still part of the Tata conglomerate. True to the characteristics of the planned-city model, Jamshedpur (like Jaipur) has a planned central core that has undergone modifications over time and is surrounded by unplanned traditional developments and semi-planned postindependence extensions.

Mixtures of Colonial and Bazaar Models

The functional demands created by activities in colonial, bazaar-type, and planned cities generated interaction among city types. British administrative requirements in the traditional cities resulted in the establishment of *Civil Lines*, generally on the urban periphery. The Civil Lines were composed of residential quarters for high administrative and judicial officials, a courthouse, treasury, jail, hospital, public library, police facilities, and club houses. The streets for the Civil Lines were well planned, paved, and had trees planted on both sides.

During the twentieth century, when the local rich needed to build houses of their own, many traditional cities developed planned extensions on their peripheries. For the most part, however, colonial cities never grew as conceived by their colonial masters. Traditional factors played an unavoidable role in altering colonial forms. The traditional bazaar, inherent to the indigenous cityscape, always interacted with other city forms. The bazaar thrived side by side with the CBD. As a result, to correctly model classic colonial cities, such as Kolkata, Mumbai, Chennai, and Colombo, it is essential to consider the impact on them by the traditional bazaars. All colonial cities bear imprints from bazaar forms, just as the bazaar cities were impacted by colonial functional demands. Planned cities, too, as they expanded, often added colonial and bazaar city forms, and sometimes the two latter city types created new planned entities as their expanded appendages.

DISTINCTIVE CITIES

Mumbai: India's Cultural and Economic Capital

Although the UN recognizes Delhi as South Asia's largest urban agglomeration, Mumbai remains the region's largest and most cosmopolitan city. Two dimensions have been critical to its rise to preeminence, one commercial and one cultural. The skyscrapers in the Nariman Point area are the heart of the city's—and nation's—corporate and financial sectors and signify the city's role as the single most important command-and-control point in the national economy. The two largest stock exchanges of Mumbai handle an overwhelming majority of India's stock transactions and figure among the world's largest in volume and value. In addition, 40 percent of the country's foreign trade is conducted through the city. Mumbai has also emerged as the nation's cultural capital through its prolific film industry, which is among the world's largest: Films produced in "Bollywood" (over 1,500 each year) are eagerly consumed not only by the viewing public in India, but also in Bangladesh, the Middle East, and Africa (Figure 9.14).

In 1672, Mumbai became the capital of all British possessions on the west coast of India. The seventeenth-century British possession of the seven islands, which now form the oldest part of this city, initiated the construction of the fort. Beyond the fort grew a "native town," where sanitary conditions were miserably poor and drainage was a serious problem. The "European town" grew around the fortress on higher ground, and protective walls were erected around it. The native or black town

Figure 9.14 "Bollywood" films are popular all across the Indian subcontinent and beyond, including here in Calcutta. *Source:* Photo by Jared Boone.

was separated from the European town by an Esplanade, which was kept free of permanent houses. A main spur to Mumbai's development occurred when Britain's supply of raw cotton temporarily diminished in the 1860s during the U.S. Civil War. India became an important supplier of cotton, most of which moved through Mumbai. This resulted in an amassing of huge reserves of capital by Mumbai-based businessmen, and the city became the main cotton textile center of the realm. In 1853, the opening of railways eventually connected Mumbai with a hinterland covering almost all of west India. Mumbai further prospered with the opening of the Suez Canal in 1869, which enhanced the city's trade advantage by further cutting the distance to Europe. This closer proximity helped earn the city its nickname: "Gateway to India."

When the fort area developed into a Western-style CBD, the British, followed by rich Indians, moved to Malabar Hill, Cumballah Hill, and Mahalakshmi on the southwestern portion of the island. These remain exclusive neighborhoods today. In the 1940s, the rich settled in another attractive area of the island, Marine Drive (now renamed) which lay along the Back Bay (Figure 9.15). Because of the ever-increasing demand for commercial and residential land in the fort area and the lack of land on the narrow peninsula and the island city, dozens of skyscrapers have been erected since the 1970s at Nariman Point, generating a skyline resembling a miniature Manhattan. The most recent phenomenon of Mumbai's commercial land-use change is the partial shifting of office- and financial-related activities to the newly built high-rise buildings of Nariman Point, though the old Fort area is still considered as the main core of the CBD. Mumbai is an expensive city to conduct business in and ranks 6th in the world with respect to office occupancy costs. The poor, the middle class, and a few native

Figure 9.15 Marine Drive, with Nariman Point in the background, serves as the setting for the annual Mumbai Marathon. *Source:* Photo by D. J. Zeigler.

businessmen settled mostly at the center and the north of the island. At present, the colonial influence created by European settlements can be observed in the southern part of the city proper. The more traditional influences have remained observable in the north. Unsanitary slums, built of flimsy materials and serving as poverty-ridden habitat, mushroom all over the city.

As a state capital and the largest metropolis in South Asia, Mumbai has also become the largest port of the entire subcontinent; not only does it handle the largest share of foreign trade, it collects 60 percent of India's duty revenues. Though employment in cotton textiles manufacturing remains important, other sectors including general engineering, silk, chemicals, dyeing and bleaching, and information technology (IT) are now emerging as important employment sources. Total industrial employment has declined and the service sector is increasingly prominent. Still, the Mumbai Metropolitan Region accounts for a disproportionately large share of India's industrial employment and fixed capital, and the Mumbai-Pune corridor is India's second most important center of employment for IT.

Mumbai attracts an enormous number of migrants from the western and central parts of India, thus giving it a religious and linguistic diversity that surpasses all other cities in South Asia. Yet, it also shares some religious characteristics with other South Asian cities. For example, the decline in the Muslim population resulted from the partitioning of British India into India and Pakistan in 1947, prompting the mass exodus of Hindus and Sikhs from West Pakistan (now Pakistan) and Muslims from India. The partition also led to the flight of many Hindus from East Pakistan (now Bangladesh) to India. Today, Mumbai's population is 67 percent Hindu, 19 percent Muslim; 5 percent Buddhist; 4 percent each of Jain and Christian; and smaller populations of others. Two other minority religious groups play a socioeconomic role far beyond their numbers. First, the Zoroastrians, or Parsis as they are called in Mumbai, are a very significant minority group. Even though their numbers in Mumbai are small and declining, more Parsis live here than anywhere else across the globe. Also significant are the Jains, mainly migrant businessmen from nearby Gujarat State, who were drawn by Mumbai's increasing commercial attraction. In terms of linguistic characteristics, no other metropolis of the subcontinent has Mumbai's uniqueness. The regional language, Marathi, is spoken by less than half the population.

Bengalūru and Hyderabad: India's Economic Frontier

When asked to identify economic success stories in South Asia, two cities immediately leap to mind: Bengalūru (Bangalore until 2006) and Hyderabad. Globalization is the vehicle that both cities have ridden to prosperity, as each has become a center for IT development and business process outsourcing. The success of each is built upon the country's supply of capable, technically skilled, and English-proficient (but underemployed) college graduates, combined with the forces of technology and globalization that have reduced distances and hence costs. Other elements distinctive to these two cities are the presence of an entrepreneurial spirit, government flexibility, and critical investments in infrastructure. Both cities also serve as state capitals.

Bengalūru's association with IT dates to the arrival of Texas Instruments in the mid-1980s. Even before that, however, the city had become a center of India's aerospace and defense manufacturing industries. By the late 1990s, so many multinational firms had established operations here that the city had been christened "India's Silicon Valley," an appropriate nickname because it accounts for over one-third of the nation's software exports. The concomitant wealth and affluence has given the city a rather cosmopolitan and trendy reputation.

Hyderabad's emergence as an IT center came in part through the visionary efforts of the state's chief minister during the late 1990s. He pulled out all the stops in providing incentives and infrastructure for high-technology-related development. Today, the city prides itself as being referred to as "Cyberabad." It hosts a Genome Valley and a Nanotechnology park, outgrowths of the city's leading role in the nation's pharmaceutical industry. Microsoft's largest development center outside Redmond, Washington, is located here, as are many other multinational firms.

Delhi: Who Controls Delhi Controls India

Delhi, the seat of India's capital, combines a deep-rooted historical heritage with colonial and modern forms (Figure 9.16). The attraction for Delhi as the capital site was rooted in South Asia's physiography, locations of advanced civilization centers, and migration/invasion routes. Delhi occupies a relatively flat drainage divide between the two most productive agricultural areas of the realm, the Indus and Ganges plains, where the most notable centers of civilization and power developed in the past. The control of Delhi was so vital to the rule of north India that a popular saying arose: "Who controls Delhi, controls India."

Old Delhi (the former Shahjahanabad) is a traditional bazaar-type Indian city, with Chandni Chowk as the main commercial center. Here sanitation used to be one of the main problems, but after independence the area has been fully provided with underground sewers and piped water. Rich merchants and ordinary working people live close to each other (Figure 9.17). West of the Yamuna,

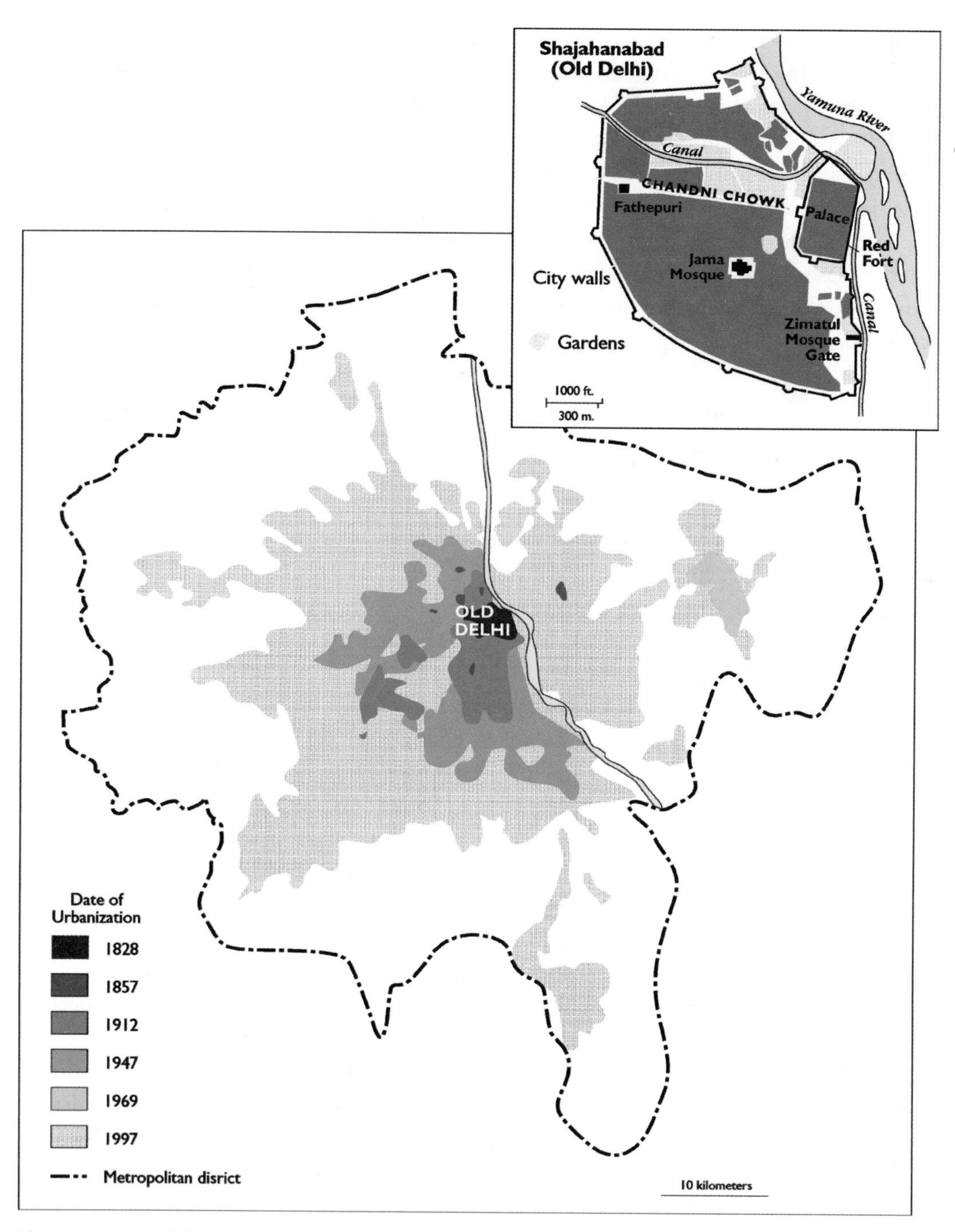

Figure 9.16 Delhi and Shajahanabad (Old Delhi). *Source:* Ashok Dutt and George Pomeroy.

connected by bridges with the main city, is a postindependence semi-planned area as well as a squatter development, where one-third of Delhi's population live—some in unhealthy slums with very little or no basic facilities of water, sewer, electricity, or paved roads.

Figure 9.17 Any service you can think of is available on the streets of India's cities. Here in the Karol Bagh neighborhood of Delhi, for a few rupees, you can get your pants pressed. *Source:* Photo by D. J. Zeigler.

New Delhi, situated south of Old Delhi, is a majestic colonial creation that emerged as a new city after the capital of British India was moved from Kolkata. The capital was temporarily moved to Delhi's Civil Lines and Cantonment in 1911 before being installed in New Delhi (1931). New Delhi was planned by a British architect, Edwin Lutyens, in a geometric form that combined hexagons, circles, triangles, rectangles, and straight lines. Spacious roads, a magnificent viceroy's residence (now the President's Palace), a circular council chamber (which is now Parliament House), imposing secretariat buildings, a Western-style shopping center (Connaught Place) with a large open space in the middle, officers' residences in huge compounds, and a garden-like atmosphere formed the main elements of New Delhi. The new capital was separated from the congested, unsanitary, and generally poor conditions of Old Delhi by an open space.

The main economic base of New Delhi is government services. After India's independence, an increasing demand for housing was created by new government employees, which led to large-scale public-housing developments around earlier settlements of New Delhi. The most noticeable feature of such developments was the segregation of larger neighborhoods according to the rank of the government employees and foreign residents. Class rather than caste determined the new neighborhood composition. Delhi has expanded in a planned manner as the Delhi Development Authority (DDA) has worked to coordinate the land development process with an innovative revolving funding scheme. The dark side of this strategy is that the lower and middle classes are left without affordable housing options. That one in five Delhi residents live in a slum blurs any gains registered.

Delhi's problems are so intense it is numbered among the five worst cities in the world with respect to air pollution, generated by motor vehicles (the leading source) and industrial activities (Box 9.3). Steps taken to ameliorate the problems over the last decade include banning leaded gasoline, conversion to compressed natural gas (CNG) buses, mandated use of low-sulfur diesel fuel, and other tighter emission restrictions. With these steps, pollution levels have stabilized.

Kolkata: Premier Presidency Town

Kolkata, though time, has inspired a number of nicknames, including geographer Rhoads Murphey's "City in a Swamp," Dominique

Box 9.3 Two Billion Life Years Lost

To the international observer, this headline may sound familiar: "US Embassy Monitor Measures Air Quality at 'Very Healthy Levels'." It refers to India's capital, New Delhi, a city whose air quality is, for the most part, worse than China's capital of Beijing. Poor air quality isn't just a problem for New Delhi, either. According to the World Health Organization, thirteen of the world's twenty most polluted cities are in India; three others are in Pakistan; and two more in Bangladesh. All of these cities are on the Gangetic Plain, making the region notorious for urban air quality. Altogether, the human cost of poor air quality in India alone is a reduced life expectancy of 3.2 years, on average, for the 660 million Indians who live in areas with high particulate matter. *This grimly translates into two billion life years lost!*

Just how bad is the air quality in these cities? Let's consider one common pollutant, particulate matter (PM), which is a term for the dust, dirt, soot, and smoke in the air. Of greatest concern is PM 2.5, which refers to particulates less than 2.5 micrometers in diameter, about 1/30 the width of a human hair. These materials are small enough to penetrate deeply into the lungs. Effects include asthma, respiratory disease, stroke, heart failure, and premature death. India, thanks to poor indoor and outdoor air quality, has the world's highest death rate from chronic respiratory disease.

For Delhi, PM 2.5 levels averaged an outrageous 226 micrograms per cubic meter during December 2014 and January 2015, and over 40 percent higher than the same period the year before. For comparison, this is more than double the "bad" air quality in Beijing, nearly 6.5 times the U.S. regulatory standard, and 22 times the World Health Organization (WHO) standard. Incredibly, during the first three weeks of 2015, one monitor's *average* daily peak reading was 473, with several days of readings above 500! Things are not much better in a string of other large cities across northern India.

Sadly, air quality may get much worse before it gets better. Traditional fuels of firewood, agricultural waste, and dried cow dung, which are not burned cleanly by less efficient traditional stoves, are still widely used in both urban and rural settings. In addition, cities are now choked with emissions-generating traffic. Delhi alone has around eight million vehicles, and India has plans to double the consumption of coal in the next five years to fuel economic growth. The country has plenty of coal, but little natural gas or oil, and India's coal is of a poorer quality. And. in Dhaka, Bangladesh, brick factories proliferate, belching smoke everywhere.

In the past, few seemed especially concerned about air quality. However, with alarming rates of illness and more awareness about the hazards of air pollution, attitudes are beginning to shift. Steps are being taken to increase air quality monitoring, to better understand where and how pollutants are being generated, and to consider what remedial steps, regulatory and otherwise, might be taken. In the meantime, how many more millions, or even billions, of life years will truly be lost before air quality improves?

LaPierre's "City of Joy," Rudyard Kipling's "Cholera Capital of the World," and "City of Pavement Dwellers." As administrative capital for much of the colonial period and as an important commercial center since before then, Kolkata set the tone for urban imagery in South Asia.

"City in a Swamp" appropriately describes the city even today. Sited on the levees sloping east and west from the riverbank, the 40-mile (64 km) long metropolitan district is for the most part less than 22 feet (7 m) above sea level. This flood-prone elevation is further aggravated by the monsoon, which brings most of the city's annual rainfall of 64 inches (1,600 mm) between June and September, coinciding with the river's highest level. Waterlogged soils and extensive flooding substantially impacts a majority of slum dwellers. Despite these physical disadvantages, the city's location on the River Hugli (a distributary of the Ganges) provided advantages for industrial growth. Its location 60 miles (97 km) upriver from the Bay of Bengal allowed access to nineteenth-century ocean vessels that provided a populous hinterland with rich mineral and agricultural resources. The city's seaport facilitated the import of wholesale machinery for the jute industry, which became most significant industrial activity of the metropolis. Finally, the establishment of the trading post and military garrison by the British in 1756 provided the mechanisms through which trade was conducted.

Spatial growth of the city, and the business district in particular, centered on the original fort and continued from that point even after its relocation several years later. A European component grew from the south end and a native town in the north. The "native town" included a wealthier and middle-class component that resided at the northern edge of the CBD. This area, the Barabazar, is reminiscent of traditional bazaars (Figure 9.18). Immediately to the south is the Western-style CBD, mainly an office, administrative, and commercial district with a low density of residential population. As the city grew, its northern part reflected more traditional characteristics, while the southern section presented more of a European look. New areas were later reclaimed in the southern and eastern portions of the city to be inhabited mostly by wealthy Bengalis, the native inhabitants of the state.

Postindependence manufacturing activity was initially hurt by the partition of the subcontinent, which severed the jute mills of the city from their supply areas, contributing to

Figure 9.18 Fishmongers are widespread in Kolkata. Not only does the city have a huge consuming population, but it is also along the coast. *Source:* Photo by Ipsita Chatterjee.

the city's relative industrial decline. Today it is the engineering industry that is an important component of the local economy. Other important industries are paper, pharmaceuticals, and synthetic fabrics. Information-technology firms and related employment are growing, but Bengalūru and the Mumbai-Pune corridor remain ahead of Kolkata. Commercially, the city serves as the headquarters of native business firms, banks, and international corporations.

Before the partition, Kolkata attracted migrants from many different parts of northeastern India and East Pakistan (now Bangladesh). The city, therefore, demonstrates a multilingual demography. Two-thirds of the population speaks Bengali, another one-fifth speaks Hindi, and one in ten speaks Urdu. Hindus constitute 83 percent of the population; most of the remainder is Muslim.

Karachi: Port and Former Capital

Situated by the western edge of the Indus River delta and with approximately 16 million people, Karachi is Pakistan's largest city and former capital. It is highly industrialized and relies upon cotton textiles, steel, and engineering for its economic base. It remains by far the nation's leading port, and has become an important center for educational and medical services. Despite its leading role as a vibrant and cosmopolitan hub of finance and trade, Karachi has gained a notorious international reputation for its lawlessness.

Karachi's job opportunities have long provided an urban "pull" not only for Pakistanis, but also for large numbers of Muslim refugees leaving India, especially in the period immediately following partition. In the 1960s only 16 percent of Karachi's citizens were native born, while 18 percent were in-migrants from different parts of Pakistan, and 66 percent were Indian Muslim immigrants from the late 1940s.

While the city has become more religiously homogenous as a result of the in-migration and the effects of the 1947 partition, it has also become more linguistically and ethnically diverse. The partition led to the departure of all but a few thousand Sindhi Hindus, while Urdu-speaking Muslim immigrants came in large numbers from the north and central parts of India, and there was a preexistent base of Gujarati Muslim migrants from India since the pre-partition times. Thus, apart from English, three languages are prevalent in the city: Urdu, Sindhi, and Gujarati. A considerable influx of Pushtu-speaking Pathans from the Northwest Frontier Province (NWFP) and Afghanistan in the 1980s often clashed with the Urdu-speaking migrants from India. The combination of linguistic, religious, and ethnic differences, exacerbated by ineffective and corrupt law enforcement and a bureaucratic judicial system and availability of arms supplied by the United States during the Afghan War in the 1980s have contributed to an alarming level of violence. This violence has escalated and continues today, and, along with declining employment opportunities and a crumbling infrastructure, has led to urban discontent and near anarchy.

The center of Karachi still conforms to a true bazaar model: high population density, high intensity of commercial and small-scale industrial activity, and a relatively higher concentration of rich people. Toward the east from the center of the city were the planned cantonment quarters that originated during the British occupation dating back to 1839. After independence, new suburban residential developments occurred surrounding the eastern two-thirds of the colonial city, while planned industrial estates were built mainly toward the northern and western fringes.

Dhaka: Capital, Port, and Primate City

Dhaka possesses a colorful history of 400 years since it was founded on the left bank of River Buriganga, a distributary of the Ganges. Dhaka's history is routed in two major factors: one is political power and the other commerce and industry. From the first decade of the seventeenth century, for a hundred years, Dhaka served as the capital of the Subah Bangla of the Mughal Empire; and, it served as the secondary capital till the end of the eighteenth century. Under the British, in 1905, Dhaka regained its status as the capital of Eastern Bengal and Assam, only to have that status annulled in 1911. At the end of British rule in 1947, Dhaka became the capital of East Pakistan. And, in 1971, Dhaka became the capital of Bangladesh after it separated from West Pakistan and took on a new name. Although its political history has followed an undulating path, the city has always maintained a connection with the commercial lifeline of the country. The geographic location of the city places it in an advantageous position for trading, particularly by means of water.

Dhaka, a megacity of 17 million, dwarfs all others in the country. It accommodates one out over every ten Bangladeshis and one-third of the country's urban population. Over the last several decades, the city has had some of the highest growth rates among Asian cities, with many rural poor migrants being added to the population. About 95 percent of Dhaka's population is Muslim; Hinduism is the second-largest religion and compromises 4 percent of the population. People speak Bangla, and the second language is English, spoken only by the educated. The literacy rate of population of Dhaka was about 73 percent in 2010 which was significantly higher than the national average of 57 percent.

Dhaka is the commercial, administrative, and educational hub of Bangladesh. Pre-Mughal Dhaka consisted of 52 bazaars. During the Mughal reign (1606–1764), old Dhaka continued to flourish. The *chawk* (square) was the main market place of Mughal Dhaka. Though somewhat transformed, it remains a vibrant major wholesale area of the city.

Major challenges of poverty, pollution, crime, congestion, and often political violence face Dhaka today . The city's population comprises primarily middle- and low-income people. About one-third reside in slum areas. They live in very unhygienic conditions without proper water and sanitation facilities. Dhaka always attracts a large number of migrants from all over the country.

Tejgaon and Hazaribagh are the major industrial areas of Dhaka. The Export Processing Zone in Dhaka was set up to encourage the export of garments, textiles, and other goods. Exports from the garments sector in Dhaka amounted to over 19 billion dollars in 2013. In the last two decades, Dhaka has experienced booming construction. Over time, the skyline of this megacity has changed, and it has begun to take on the some of the flavor of a world city. But, at the same time, attributes like pollution, poverty, unplanned growth, and environmental degradation became characteristic features of Dhaka. The farm land and low-lying lands in the fringe areas of Dhaka are converting very fast to urban uses. Unplanned development has become a major concern from the environmental and social perspective. Industrial and domestic wastes are polluting rivers, canals, and water bodies of Dhaka. The alarming conversion of waterbodies is a major threat to the environment.

Transportation in Dhaka is always identified as a major problem. Traffic congestion is diminishing the quality of life of residents. Both motorized and nonmotorized vehicles occupy the city's roads. Absence of an organized mass transport system is the key reason for the problem. Even though a large number of people walk, pedestrian facilities are very limited. In recent years, investment in road infrastructure has been significant.

Despite its problems, life in Dhaka is quite vibrant. The city's growing economy indicates the high motivation of its residents. During the national festivals and occasions, the people of Dhaka enjoy public life by observing cultural and social events (Box 9.4).

Kathmandu, Colombo, and Kabul: Cities on the Edge

Colombo, Kabul, and Kathmandu are the premier cities and national capitals of Sri Lanka, Afghanistan, and Nepal, respectively. Officially, however, the capital of Sri Lanka has moved to Sri Jayawadenepura, located within the Colombo metropolitan area. As the former capital, the city of Colombo itself continues to be the seat of administration and decision making. These cities on the edge of South Asia differ greatly in size: Kabul with 4.6 million; Colombo with 700,000; and Kathmandu with 1.2 million.

Kabul is the oldest of the three cities, even being referred to in the *Rig Veda*, a 3,500-year-old Hindu scripture, and by Ptolemy in the second century CE. Kabul's strategic position by the side of the Kabul River and at the western entrance to the famous Khyber Pass gave the city great political and trading significance. Sited at an elevation of 5,900 feet (1,800 m), it has served as a regional or national capital for numerous regimes over the centuries, most notably for the Moghul Empire (1504–1526) and, after 1776, for an independent Afghanistan. Later attempts by both the Russians and the British to subjugate the country and its capital failed. After 1880, modern buildings and gardens were constructed but these have not diminished the identity of the old bazaar city.

The city has suffered greatly since the 1970s due to international conflicts being played out on its soil. The overthrow of the king in 1974 initiated a series of events that included the establishment of a Soviet-backed government and subsequently, the U.S. intervention to arm mujahideen (Afghan insurgents) against the pro-Soviet government (1970s), the establishment of a weak puppet state (1989), a period of warlord-dominated chaos (1992), and establishment of the Taliban regime (1996). Allegations that Afghanistan's difficult physical geography and a sympathetic Taliban regime provided havens for extremist elements involved in the World Trade Center attacks of September 11, 2001, invited retaliation by the United States leading to intense and destructive bombing of Kabul, as well as Kandahar and other cities. Later that year, the Taliban regime was toppled, but the post-Taliban government seems to be only nominally in control of the country beyond the vicinity of Kabul.

Kathmandu, in a valley of the same name, lies in the mid-mountain region of the Himalayas, at an elevation of about 4,400 feet (1,350 m). The city occupies a central position for most of Nepal and is the most important commercial, business, and administrative center of the country. The Gurkha ethnic group, after conquering Nepal, made Kathmandu its capital in 1768. Reconstruction efforts after the devastating 1934 earthquake and post–World War I developments added

Box 9.4 Festivals in City Life

Even days before a festival begins in Dhaka, residents can feel the lively pulse of the city beginning to change. New life comes to public spaces; crowds throng the shopping areas, and colorful dresses worn by young women appear on the streets. They signal a break in mechanical city life. Some will celebrate the best of Bangladeshi culture and others will mark religious holidays.

The most widely celebrated cultural festival is *Pohela Baisakhi*, Bengali New Year's Day (April 14). People from all walks of life, irrespective of religion, income, gender, or ethnicity come together to celebrate: Women wear sarees, adorn their hair with flowers, and wear local jewelry; men wear *paizama*. Everyone begins the celebration with songs written to welcome the new year, and musical troupes perform. In Dhaka, Ramna Park is the center of festivities, though programs are arranged in open spaces everywhere. The *Boishakhi* Parade, with its bird and animal figurines and many masks is one of the day's main events, as are *Boishakhi* fairs, where small craftsman and food vendors earn a little more than usual. City streets are ornamented with beautiful *alpona* (painted pavements). The age-old tradition of this day is to observe *Haalkhata*, the ritual of closing the old Ledger and opening a new one by traders involved in gold, clothing, and food businesses. They invite their customers in their shop and entertain them with sweets.

Perhaps Bangladesh is the only country where every religion's festival days are celebrated as government holidays. Eid-ul-Fitr is the largest religious festival for Muslims, who constitute about 90 percent of the population of Dhaka. It is celebrated after the fasting month of Ramadan. Large prayers are performed in every mosque, people join the special Eid prayer, and everyone exchanges greetings with others. Good food and new clothing are the major priority of Eid. In recent years, residents of old Dhaka have been trying to revive the old custom of the Eid parade. Many of the city's Muslims, however, leave for their ancestral homes to enjoy the festival with their extended families and old friends.

Hinduism is the second religion of Bangladesh, and the *Durga Puja* is the major Hindu festival. It is widely celebrated in Dhaka. Special *puja mandaps* (shrines) are established in different locations in addition to the temple sites. The *mandaps* are nicely decorated and attract Muslims also to share the joy. In addition, Christian and Buddhist rituals are performed in churches and pagodas, respectively, during Christmas and Buddha *Purmina*. Some national programs are also arranged to observe these religious occasions.

Although the people of Dhaka love these celebrations, their enjoyment is often disrupted when they get stuck in horrible traffic jams on days close to the occasion. In fact, people often have a hard time arranging rail, bus, or ferry tickets to reach their friends and families safely. In most of the long holidays, many residents leave the crowded city to enjoy a relaxed vacation elsewhere in the country.

more buildings. Though the overwhelming majority of Kathmandu residents are Hindus, there are also some Buddhists. Autocratic rule by the king and sporadic violence by Maoist guerillas made situations unstable; consequently, in 2008, the monarchy was abolished and a succession of coalition governments has been democratically elected. The city remains a major center of South Asian tourism and is a launching ground for treks in the Himalaya Mountains. In 2015, large areas of Kathmandu, including the historic center with its bazaar characteristics, were leveled by two major earthquakes.

Situated on the west coast of the pearl-shaped island nation of Sri Lanka (formerly known as Ceylon to the West), Colombo has functioned as an important port city since at least the fifth century. Nonetheless, it was Western contact, beginning with Portuguese settlement in 1517, which really started the growth of Colombo as the key port for the island. The Dutch occupied the port in 1656, but the British replaced them in 1796 and turned the city of Colombo into their main administrative, military, and trading place in Sri Lanka. After independence in 1948, Colombo became the country's capital, but it has been challenged to build a single nationality in a country where the majority Buddhist Sinhalese and minority Hindu Tamils have been at odds for most of Sri Lanka's history and engulfed the country in a civil war from 1983 to 2009.

Colombo continues to be Sri Lanka's most important city in terms of business, administration, education, and culture. It is a colonial-based city. The CBD-like center, with high-level government offices situated within the former colonial fort area, lies by the side of the old section of the city, Pettah ("the town outside the fort" in the Tamil language). Pettah represents the characteristics of the Bazaar city enclave. Cinnamon Gardens, the former cinnamon-growing area of the Dutch period, has been turned into a high-class, low-density residential quarter. The city has expanded significantly since independence and has industries that mainly process the raw materials that are exported through the port of Colombo. In order to diversify the national economy through industrialization, an export-oriented Free Trade Zone has been established near the port of Colombo.

GLOBALIZATION, CITY MARKETING, AND URBAN VIOLENCE

The contemporary geography of South Asian cities is defined by globalization and its various economic, cultural, and political impacts. Urban landscapes of South Asia represent a complex mixture of global influences and local particularities. Globalization has been defined as the intensification of interaction between previously faraway places and people so that local happenings are now shaped by distant events. Economic liberalization, moves toward free-market reforms, and the information and technology revolution are considered to be the main forces propelling increased spatial interaction. Telemarketing, electronic banking, plastic money, and high-speed internet have made national borders porous. Economic liberalization has encouraged the erosion of tariff barriers and the free flow of investments across national borders. As a result, the cities in South Asia have increased abilities to tap global business in the form of corporate offices, export-processing endeavors, tourism (spiritual and medical), and retail industries. Corporations headquartered in advanced nations search for emerging markets among the growing middle class in

South Asia. Global corporations also look for cheap pools of unorganized labor in populous South Asian cities. This give-and-take allows for "glocalization" as cities engage with global flows and ground them in locally specific ways so that both the cities and global flows are altered. For example, McDonald's, a global fast food chain, is localized in Indian cities when it sells the "no-beef" Chicken Maharaja Mac because Hindus do not eat beef.

South Asian cities now manifest a hybridization that symbolizes this tension between local imperatives and global impetuses. City municipalities are increasingly urged to "go global" in their search for funds. This results in a "new urban politics" of city-versus-city competition to acquire foreign business and investments. In the early twenty-first century, for instance, India's, central government launched the Jawaharlal Nehru Urban Renewal Mission (JNNURM). Its purpose was to select 63 cities, which would be given funds to "go global" by becoming entrepreneurial, profit seeking, and "world class," like New York, Tokyo, and London. The JNNURM urged city governments to shed their pro-poor social agendas (e.g., providing cheap infrastructure and slum upgrading), and, instead, focus on marketability. Redistributive measures like rent control, which were put in place to control the concentration of wealth, were abolished to allow entrepreneurialism and private investments to flourish. This push toward urban entrepreneurialism has been answered differently by different cities, but the dominant strategy has been "place marketing"—repackaging urban areas to make them attractive to global capital (Box 9.5). One place-marketing strategy includes green entrepreneurialism, which involves greening, cleaning, and developing urban gardens and parks, introducing manicured traffic islands, and switching to CNG instead of gasoline. The idea is to present a global image of an environmentally friendly, sustainable, and smart city that will be attractive to foreign business and tourists. This green entrepreneurialism has become controversial, because greening is often accomplished by evicting the poor, and forcing them to adopt the more expensive CNG, while the urban rich are allowed to own multiple numbers of gasoline-operated vehicles. Greening also concretizes uneven geographies within the city where affluent neighborhoods gain parks and open spaces, while poor neighborhoods continue to suffer from stagnation. These exclusions brought about by greening strategies have often been touted as "bourgeois environmentalism" or "elitist environmentalism." For example, in Ahmedabad, India, greening under the Green Partnership Program has benefited the more affluent west Ahmedabad, while the poorer parts of east Ahmedabad lack open spaces. The exclusionary politics of enforcing CNG in Delhi has also been well documented.

Another place-marketing strategy includes city beautification through urban renewal. Beautification involves giving the city a facelift so that it can project the image of an efficient growth engine—a spruced-up city can potentially outcompete other cities in attracting investment. City governments therefore, go out of their way to sell communal or public land to private construction companies who are then supposed to "upscale" the city with promenades, boulevards, water parks, state-of-the-art offices, malls, parking lots, and high-speed transit corridors. The cultural impact is often a homogenization of once unique landscapes. Mumbai, Delhi, Colombo, and Karachi demonstrate this growing loss of cultural diversity as small businesses, local food cultures, indigenous handicrafts, and local embroidery disappear to make way for the world's McDonalds and the Benettons. Delhi, for instance, has seen a growing grassroots movement of hawkers challenging the government's policy of evicting street vendors (Figure 9.20). A more problematic dimension of urban renewal is that it is often achieved by "liberating" spaces through demolition of

Box 9.5 Devastation in the Kathmandu Valley

Keshav Bhattarai, University of Central Missouri

Friction between the Eurasian and Tibetan tectonic plates put Nepal in one of the world's most vulnerable seismic zones. Over the past 100 years, earthquakes originating in Nepal or nearby areas have had disastrous consequences (Table 9.3). Two temblors, with 400 aftershocks, hit the country in 2015. The highly urbanized Kathmandu Valley was especially hard hit, with the Barpak epicenter in the district of Gorkha. Estimated losses included 8,623 deaths, 16,808 injuries, the displacement of 2.8 million people, and $10 billion worth of infrastructure damage (Figure 9.19). Almost half a million buildings fell or were rendered uninhabitable by the earthquakes. Thousands of people were forced to live under the open sky in adverse weather conditions, and half a million residents from Kathmandu and its vicinity fled the city for the countryside fearing more earthquakes. Despite the history of tectonic activity in Nepal, the country's government demonstrated lackluster performance in disaster management and relief. Adding to the tension, coordination among the state agencies in relief distribution and rehabilitation works in the quake-affected areas led to donor fatigue, growing political instability, and anti-government protests.

Table 9.3 Earthquake Occurrences in and near Nepal

Date	*Place*	*Fatalities*	*Magnitude*
June 7, 1255	Kathmandu	30% population	N/A
August 26, 1833	Kathmandu/Bihar	N/A	8.0 Ms
July 7, 1869	Kathmandu	N/A	6.5 Ms
August 28, 1916	Nepal/Tibet	N/A	7.7 Ms
January 15, 1934	Nepal/Tibet	10,600	8.0 Mw
June 27, 1966	Nepal /India border	80	6.3 Ms
July 29, 1980	Nepal/Pithouragarh, India	200	6.5 Ms
August 20, 1988	Kathmandu/Bihar, India	1,091	6.6 Ms
September 18, 2011	Sikkim	N/A	6.8 Ms
April 25, 2015	Kathmandu/Tibet	8,623	7.8 Mw
May 12, 2015	Chinalkha, Dolakha	157	7.3

Ms, Surface-Wave Magnitude; Mw, Moment Magnitude

Figure 9.19 Infrastructure damage resulting from the Kathmandu earthquakes amounted to 10 billion US dollars. *Source:* Photo by Damodar Sharma.

Post-earthquake assessments revealed that excepting a few commercial complexes and apartments, many buildings did not follow the recommended building codes. It was revealed that in urban areas people got government permission to construct two-storied houses and later added more flats not intended for load bearing. In addition, government regulations were ignored while raising multistoried buildings, resulting in tussles between insurance companies and property owners. None of the buildings constructed of brick and mud were spared. Detailed inspections revealed that only 40 percent of the houses in Kathmandu Valley were safe to live in, while 60 percent either needed repair or complete rebuilding. Over 100 historical and cultural sites were destroyed; the UNESCO World Heritage sites in Kathmandu—temples, shrines, and monasteries—were fully damaged. Tourism, both cultural and mountain-climbing, which contributed around 10 percent of GDP, was paralyzed in the immediate aftermath of the quakes, and full recovery is uncertain.

In view of the repeated earthquakes in Kathmandu, consideration is being given to shifting the federal capital to Chitwan, and building international airports in the nearby districts of Bara (Nizgarh) and Rupandehi (Bhairahwa). Chitwan sits at a lower elevation, is more centrally located, and is already noteworthy as Nepal's medical capital and for nature tourism. Other countries have relocated their capital cities to avoid disasters, so there would be ample precedent for the move. Nevertheless, all Nepalis are anxious for Kathmandu to be rebuilt and are calling for a Disaster Management National Council to prepare a long-term action plan to mitigate risks and to respond to disasters like earthquakes. Furthermore, disaster risk and response information should be incorporated into school and university curricula.

Figure 9.20 Three generations of women position themselves on the curb to sell what produce they can to passersby in Mumbai. *Source:* Photo by D. J. Zeigler.

slums with little compensation for the slum dwellers. Mumbai, Delhi, and Ahmedabad have become dominant sites for massive demolitions. In the post-liberalization era, all of these cities have become "world class," and in a desperate effort to outcompete each other, have produced various forms of public-private partnerships that have engineered violent evictions. The municipalities in these cities represent the public entity that maintains a rhetoric of "liberalization with a human face," while an entourage of private construction companies are given the go-ahead to do the "needful." In Ahmedabad, a gigantic project called the Sabarmati River Front Project was launched to develop the riverfront into a fast track "world-class" corridor. NGOs claim that over 6,000 families will be evicted by this project.

City marketing also manifests as gated communities replete with gyms, sports facilities, swimming pools, and shopping complexes, mimicking the "good life" of the American middle class. In Bengalūru, where a burgeoning group of software professionals have quickly become rich in India's own Silicon Valley, the gated communities represent spaces of social mobility. Professional elites with sizable disposable incomes are increasingly seduced by the globalization of home-garage-pool lifestyles. These spaces of affluence have been touted as spaces of exception—they geographically materialize the growing gap between the urban rich and urban poor. The gated communities also cripple informal economies of vendors and hawkers who are no longer allowed to enter the "sanitized spaces" of the rich.

The global-local tensions of glocalization are manifested not only in the uneven geographies of affluence and deprivation, but also through violence. Cities in South Asia, therefore, not only internalize the daily violence of exclusion, but are often also the sites of inter-community riots and global terrorism. In India, Hindus are the majority religious group, and Muslims account for the largest minority community (13 percent approximately). Hindus and Muslims share a contentious history because of the colonial policy of "divide and rule," the partition of India and the creation of Pakistan, and the horrific violence that resulted from it. In the postindependence context, Hindu-Muslim violence has been concentrated in cities. The Mumbai riots of 1992–1993 and the Ahmedabad riots of 2002 were the deadliest in post-partition India. The Mumbai riots claimed 2,000 lives—a regional far-right political party is said to have engineered the riots and was allegedly responsible for horrific atrocities against Muslims. Many Muslims were killed along with Hindus, and many others were displaced. The 2002 riots in Ahmedabad lasted for two-and-a-half months and involved systematic destruction of Muslim homes, property, and businesses. Mobs led by cultural and political affiliates of another far-right party engineered the riots. Victims claim that the rioters came with lists of addresses of Muslim homes—2,000 Muslims were killed, 100,000 displaced, mosques were demolished and replaced with temples and roads; the displaced now live in all-Muslim ghettos outside the city. The urban riot machinery in India is fueled by a political ideology that asserts that Indianness equals Hinduness, and therefore other religious minorities are considered foreigners; hence, their patriotism is always suspect. In the post-September 11 context, the global narratives of "war on terror" and "terrorism and Islamophobia" are also adopted and localized. This "golden age of Hindu India" is not envisioned as isolated from global economic integration and global discourse of terrorism. A local politics is creatively juxtaposed with economic reforms, where non-Hindu foreign capital and foreign corporations are welcome. Local "Islamophobia" is also juxtaposed with global narratives of terrorism—the global-local tensions of glocalization are creatively imprinted on urban space.

In the Sri Lankan context on the other hand, the majority Sinhala Buddhist community, benefiting from small-scale self-enterprises under a preeconomic liberalization regime, faced increased hardship due to the crowding out of indigenous firms in the post-economic-liberalization context. Increased urban hardship led to increased ethnic polarization in the 1980s when the minority Hindu business community was targeted by the majority Buddhist community in the 1983 ethnic violence in Colombo and elsewhere. Apart from intranational conflicts, South Asian cities have also become a hotspot for international terror. The Taj Mahal Hotel shootings in Mumbai in 2008 were touted by the media as "India's September 11."

URBAN CHALLENGES

The euphoria of globalization and economic growth poorly masks the enormity of the challenges facing South Asian cities. Most countries inherited a colonial legacy of poverty and an extreme rural-urban dichotomy. In the contemporary context, urban realties depict a simultaneous juxtaposition of affluence and poverty. An expanding middle class, proliferation of malls, extreme poverty, inadequate housing, lack of public services, unemployment, and environmental degradation inscribe the South Asian urban landscape.

In the postcolonial period, most South Asian nations pushed for food self-sufficiency and basic industrial development subsidized by the government. The idea was to develop the villages, with towns and cities acting as complementary industrial hubs. However, rural land distribution was extremely unequal because of a feudal land tenure system that was never rectified in the colonial era. In the postindependence period, inability to launch cohesive land reforms exaggerated rural poverty. In that context, the primate cities, which had already experienced infrastructural development in the colonial period, continued to attract masses of rural poor in the postindependence period. Although manufacturing received a major boost in most countries through government-initiated import-substitution industrialization, the rate of growth of manufacturing jobs could not keep pace with rural-to-urban migration. Moreover, most rural migrants were unskilled and hence incapable of employment in modern industries. The result was a swelling informal sector consisting of low-paid jobs that offered no security (e.g., porters, rickshaw pullers, domestic servants, construction workers and other manual laborers). These informal workers often had to live on pavements, in railway and bus stations, and in other interstitial spaces. Others were "lucky" enough to find a one-room home in already overflowing slums.

Estimates of the number of pavement dwellers vary widely. For Mumbai alone, estimates vary from 250,000 to 2 million. In most places, they are concentrated in the central areas of the city and irregularly employed in low-skill occupations that pay the least. For Kolkata, about one-half of the pavement dwellers are employed in transport. They are predominantly males aged 18–57 years, though nearly one-third are children, most of whom supplement the family income by working as child laborers, or by begging and scavenging.

Slums have developed in almost all the major cities of South Asia. The name *bustee* is used in Kolkata and Dhaka, *jhuggi* is used in Delhi; *chawl* in Mumbai. The *bustee* has been defined by the Indian government's Slum Areas Act of 1954 as a predominantly residential area where dwellings (by reason of dilapidation, overcrowding, faulty arrangement, and lack of ventilation, light, or sanitary facilities—or any combination of these factors) are detrimental to safety, health, and morals. Moreover, the slums mainly consist of temporary or semi-permanent huts with minimal sanitary and water supply facilities and are usually located in unhealthy waterlogged areas. Although they can be found throughout any metropolitan area, there is always a greater concentration of large slums away from the CBD. They begin with temporary settlements, sometimes started by landlords, but oftentimes by illegal squatting on public lands, sides of railroad lines or canals, unclaimed swamp-like lands, public parks, and vacant lots. In Mumbai, it is estimated that nearly 60 percent of the population lives in slums; this proportion is representative of most large cities in South Asia.

South Asian nations were forced to embrace structural adjustment programs at the behest of the World Bank and the International Monetary Fund. These programs forced open their economies under the policies of free-market liberalization. Governments were supposed to roll back their control over the economy and to stop subsidizing industries and other sectors like health and education: The task of development was to be left in the hands of the market. Opening up markets brought foreign corporations, their Toyotas and Macs, their call center jobs, and dreams of a consumptive lifestyle. The English-educated, computer-literate middle class took advantage of the economic reforms; many acquired jobs with salaries equal to their First-World counterparts. These inflated salaries in poor countries afforded conspicuous consumption and life in bungalows and gated enclaves. More consumption drew more global business. Construction companies, encouraged by the boom in foreign investment and domestic consumption, pushed for renewal of the cities. The old, ugly, and poor gave way to gloss and glitter. City municipalities were often incentivized by the central government to decentralize and become market oriented. Place marketing and urban renewal were adopted to achieve world-class urban status. This trend has been described as "Manhattanization" or "Shanghaization"; others have called it "bourgeois urbanism." Place marketing calls for greening, cleaning, and beautifying the city to create affluent spaces so that the rising middle class and global business and service industries can find their niche. Culturally, this means that cities lose their personality and uniqueness, homogenized through the impact of "McDonaldization." Politically, this means that South Asian cities are increasingly acquiring symbolic capital and are more integrated into the global geopolitics of violence, often becoming prominent targets for extremist groups. Socially and economically it means an increased gap between the rich and poor and the places that they occupy. The poor find themselves increasingly pushed out through urban renewal, greening, and beautification schemes that demolish their already flimsy *bustees*, *jhuggis*, and *chawls* without any promise of relief or rehabilitation. Right-to-the-city struggles of the poor are rising in many cities of South Asia. These struggles aim to reclaim the city, alter the vision of urban development, and push for a more inclusive urbanism. South Asian cities therefore embody the tensions between local imperatives and global push.

SUGGESTED READINGS

Ahmed, Waquar, Amitabh Kundu, and Richard Peet. 2010. *India's New Economic Policy: A Critical Analysis.* New York and London: Routledge. A critique of the economic forces that are reshaping India's cities.

Boo, Katherine. 2014. *Behind the Beautiful Forevers: Life, Death, and Hope in a Mumbai Undercity.* New York: Random House. Embedded journalistic account of life and survival in a Mumbai slum based on the author's three years among the residents.

Chapman, Graham P., Ashok K. Dutt, and Robert W. Bradnock. 1999. *Urban Growth and Development in Asia: Making the Cities.* 2 vols. Aldershot, England: Ashgate. Includes chapters devoted to cities, urbanization, development, and planning.

Gayer, Laurent. 2014. *Karachi: Ordered Disorder and the Struggle for the City.* Oxford and New York: Oxford University Press. Chronicles the criminality and violence of Karachi.

Hossain, Shahadat. 2010. *Urban Poverty in Bangladesh: Slum Communities, Migration, and Social Integration.* London and New York: I.B. Tauris. Provides an examination of the understudied slums of Dhaka.

King, Anthony D. 1976. *Colonial Urban Development: Culture, Social Power, and Environment.* London: Routledge and Kegan Paul. A comprehensive analysis of colonial urban forms of New Delhi, the hill station of Simla, and cantonment towns.

Nair, Janaki. 2005. *The Promise of the Metropolis: Bangalore's Twentieth Century.* New York: Oxford University Press. A timely, well-written, informed, and empirically rich case study of the South Asian city most closely associated with globalization.

Noble, Allen G., and Ashok K. Dutt, eds. 1977. *Indian Urbanization and Planning: Vehicles of Modernization.* New Delhi: Tata McGraw-Hill. A classic work containing more than 20 chapters contributed by leading geographers and planners.

Ramachandran, R. 1989. *Urbanization and Urban Systems in India.* New Delhi: Oxford University Press, 1989. Straight-forward and comprehensive treatment of urban India.

Turner, Roy, ed. 1962. *India's Urban Future.* Berkeley and Los Angeles: University of California Press. Of particular interest in this classic collection of articles is the contribution by John E. Brush, "The Morphology of Indian Cities."

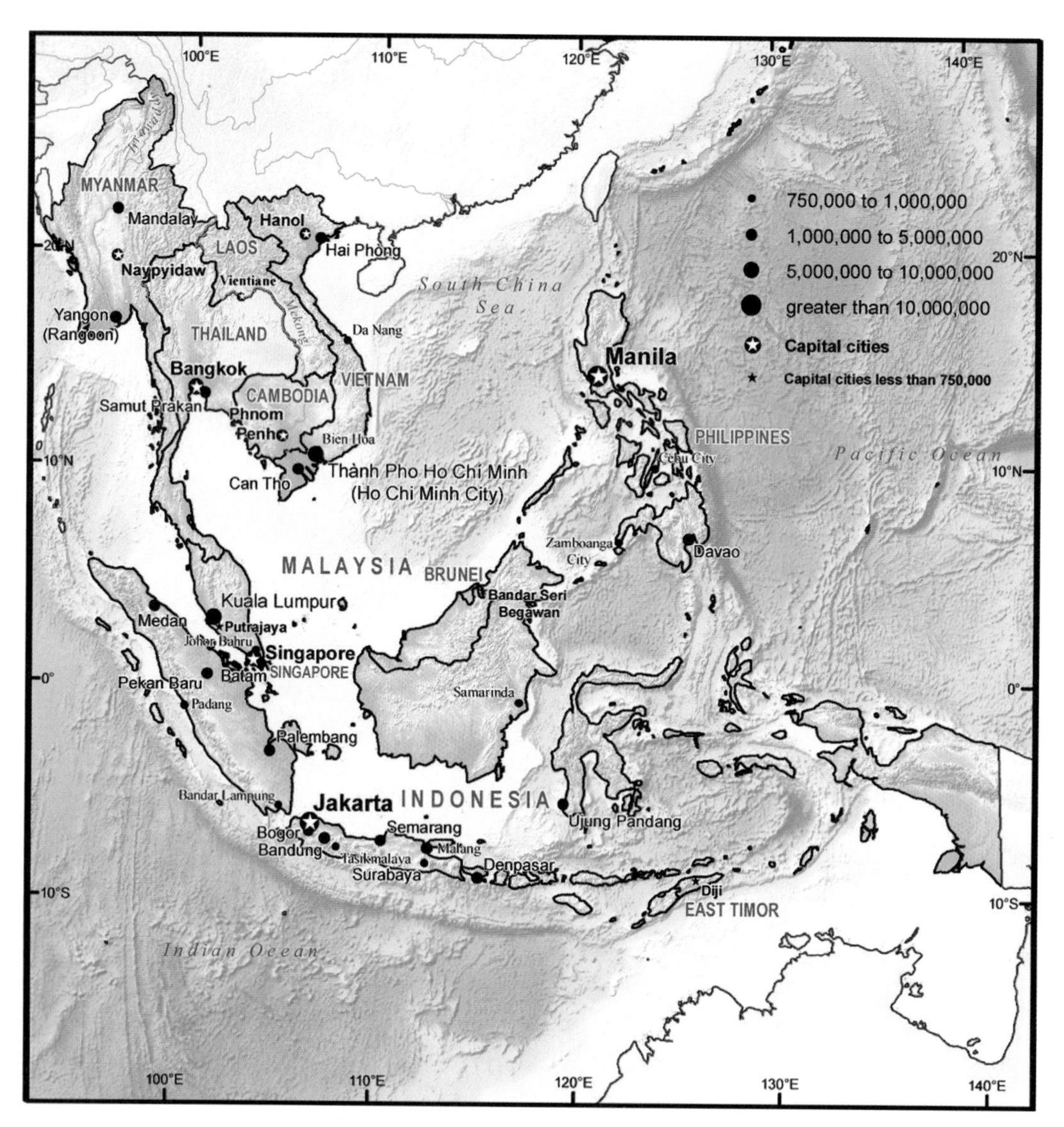

Figure 10.1 Major Urban Agglomerations of Southeast Asia. *Source:* United Nations, Department of Economic and Social Affairs, Population Division (2014), World Urbanization Prospects: 2014 Revision, http://esa.un.org/unpd/wup/

10

Cities of Southeast Asia

JAMES TYNER AND ARNISSON ANDRE ORTEGA

KEY URBAN FACTS

Total Population	626 million
Percent Urban Population	47%
Total Urban Population	294 million
Most Urbanized Countries	Singapore (100%) Brunei (77%) Malaysia (74%)
Least Urbanized Countries	Cambodia (21%) Timor-Leste (32%) Myanmar (34%)
Number of Megacities	2
Number of Cities of More Than 1 Million	25 cities
Three Largest Cities	Manila (13 m), Jakarta (10 m), Bangkok (9 m)
Number of World Cities (Highest Ranking)	6 (Singapore, Kuala Lumpur, Jakarta, Bangkok, Manila, Ho Chi Minh City)

KEY CHAPTER THEMES

1. Urban landscapes of Southeast Asia have been shaped by Chinese, Indian, Malay, and international influences, especially colonialism and more recently globalization.
2. All of the world's major religions are represented in the landscapes of Southeast Asia's cities.
3. All of the major cities of the region have experienced rapid population growth and rising environmental challenges, including problems of water quality and quantity, since independence.
4. Primate cities (notably Manila, Jakarta, and Bangkok) dominate the region, but the key urban center of Southeast Asia is the city-state of Singapore.
5. Foreign influences, especially through foreign direct investment, play a critical role in Southeast Asian cities today.

6. Land reclamation is increasingly used in port areas to provide space for urban expansion.
7. Land-use patterns in the cities are very similar throughout the region.
8. Many cities are restructuring their economies to become IT ("information technology") cities.
9. Some of the world's largest cargo ports—notably Singapore—are located in this region.
10. Transnational cities, which reach across international boundaries in their influence, are becoming more important.

Towering glass-encased skyscrapers, flashing Coca-Cola signs, McDonald's restaurants—the increasingly universal symbols of central cities around the world—are very evident in the cities of Southeast Asia, especially the larger ones, giving the cities a deceiving sense of familiarity. Closer examination, however, reveals many subtle, and sometimes not-so-subtle, differences. Southeast Asia as a whole is a cornucopia of cultures, with hundreds of different languages and many distinct religions. Nestled between two dominant cultural hearths, China and India, and exhibiting a storied colonial past, Southeast Asia is a blend of indigenous and foreign elements. This diversity, not surprisingly, has been and continues to be inscribed on the region's urban landscape, from the lotus-blossom-shaped stupas of Buddhist temples in Bangkok, to the brightly colored Hindu temples in Singapore; and from the golden-domed Muslim mosques of Kuala Lumpur to the Roman Catholic cathedrals of Manila and Ho Chi Minh City.

Yet for many travelers, the extent of Southeast Asia's urban regions comes as a surprise. The typical image of the region is agrarian: thatched huts perched atop stilts and brilliant green rice paddies with water buffalo. The reality is very different. Flying over Manila, Bangkok, or Ho Chi Minh City is like flying over Los Angeles, New York, or Tokyo. The landscape reveals not a dense green three-tiered canopied jungle but instead a dense concrete jungle of apartment complexes, shopping malls, financial districts, and amusement parks.

Southeast Asia's major cities are focal points of political and cultural activity and centers of commercial circulation and exchange (Figure 10.1). Some, such as the "post-socialist" cities of Ho Chi Minh City, Hanoi, and Phnom Penh are undergoing phenomenal political and economic changes; others, such as Rangoon (Yangon) remain aloof from broader global trends. The cities of Southeast Asia are also sites of vast inequalities between the rich and the poor as well as the healthy and the malnourished. In Bangkok, Manila, and Jakarta, Toyota Land Cruisers and Louis Vuitton designer stores are as much a part of the urban landscape as are shantytowns and raw sewage.

Beyond the limits of Southeast Asia's primate cities are a host of medium, or intermediate, cities. Many, such as the Philippines' Cebu and Thailand's Chiang Mai and Chiang Rai, are fast becoming important regional urban centers in their own right. And still other, predominantly rural areas, such as the Central Highlands of Vietnam, are sites of contestation and conflict resulting from indigenous land-use practices and national urban policies.

The urban and urbanizing areas of Southeast Asia are more than just containers of people and commodities. They are agents in their own right and, in the coming years, will continue to influence, and be influenced by, local, national, and global affairs.

URBAN PATTERNS AT THE REGIONAL SCALE

Downtown Phnom Penh, the capital of Cambodia, is dominated by the mustard-colored Central Market (Figure 10.2). Built in the Art Deco style of the 1930s, the market is cruciform in design, with four halls radiating out from a central, cavernous dome. Inside, hundreds of venders ply their wares. Care to buy handwoven silks or a traditional Khmer scarf (known as *krama*)? Perhaps you're in the mood for some fresh vegetables or pork? No matter your taste, whatever you seek can probably be found at Phnom Penh's Central Market. And if not, it is but a short journey by motorbike to visit the city's Russian Market. On the surface, the two markets are very much alike, with many of the same fruits, vegetables, and souvenirs found in each. However, the sweeping arches and vaulted ceilings of the Central Market give way to the Russian Market's dimly lit and claustrophobic feel. The Russian Market is a rabbits-den of activity, as shoppers jostle elbow-to-elbow with merchants and tourists. Inside, the air is stifling, a sweltering mix of too many people and too many cooking pots bubbling stews of fish and vegetables.

Figure 10.2 The Central Market in downtown Phnom Penh was built in 1937 in art deco style. It is the soul of the city, a place where you can purchase just about anything. *Source:* Photo by James Tyner.

Stepping inside any of Southeast Asia's historic (and even new) markets is like Alice stepping into Wonderland. Whether you find yourself wandering the stuffy aisles of Binh Tay Market in Ho Chi Minh City, window shopping in Bangkok's upscale River City shopping complex, or sipping an iced coffee while shopping along Orchard Road in Singapore, you are guaranteed to be dazzled with new sights, sounds, and smells. The intoxicating aroma of sandalwood incense combines with the smells of fresh fruits, vegetables, and spices to provide an aromatic bouquet that is found nowhere else. And unlike the sedate, antiseptic shopping malls of North America (which are increasingly popping up in Southeast Asia), the labyrinthine markets of Cambodia, Thailand, Vietnam, and elsewhere seem to embody much of the region's urban geography.

As a whole, Southeast Asia remains one of the least urbanized regions of the world. Only four countries—Singapore, Brunei, Malaysia, and the Philippines—are more than 50 percent urban. Other states are considerably more rural in character: Cambodia, Laos, Burma, and Vietnam, for example, are all less than 30 percent urbanized. However, recent years have seen many countries registering startling urban population growth rates in excess of 3 percent per year; and Cambodia and Laos stand at more than 6 percent.

Such urban growth is not new to Southeast Asia (Figure 10.3). Before the Portuguese ever arrived in Malacca, or the Spanish landed in

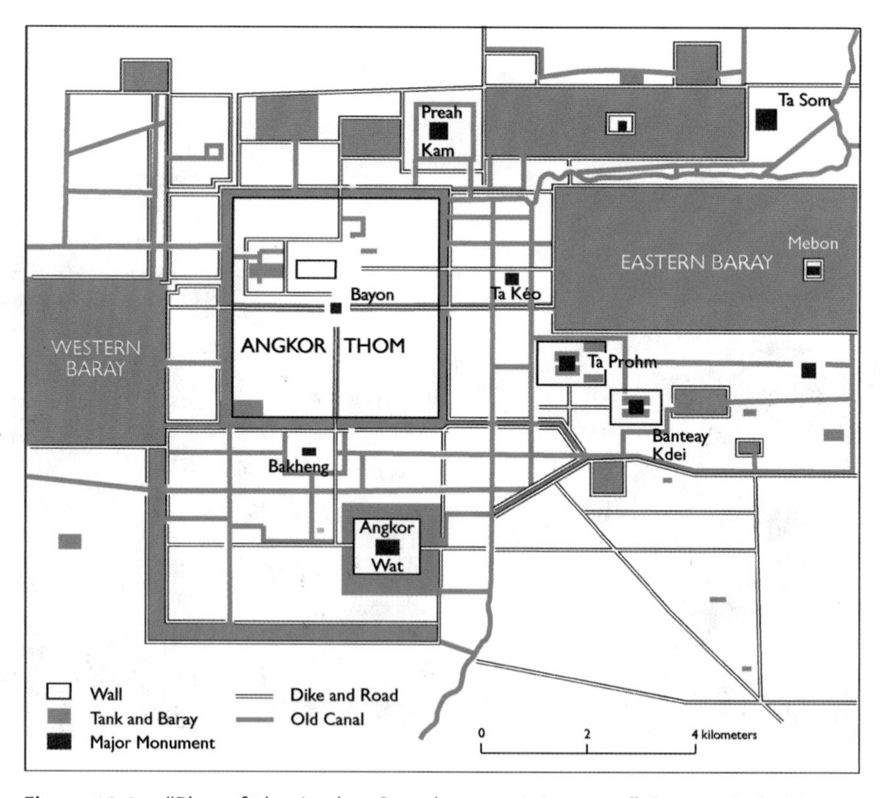

Figure 10.3 "Plan of the Angkor Complex, ca. A.D. 1200." *Source:* T. G. McGee, *The Southeast Asian City* (New York: Praeger, 1967), 38.

the Philippines, Southeast Asia was home to some of the world's most impressive cities: Angkor in Cambodia, Ayutthaya in Siam (now Thailand), and Luang Prabang in Laos. Their names continue to evoke rich histories of commerce and conquest. For it was through Southeast Asia that the fabled spice trade coursed. And it was through Southeast Asia that ships laden with goods from China, India, and beyond sailed. Then and now, the cities of Southeast Asia were centers of economic, religious, and cultural exchange.

Historically, one or two urban areas would dominate the region. The Kingdom of Angkor, for example, exerted its influence between the ninth and fourteenth centuries over much of present-day Cambodia, Laos, and Thailand (Figure 10.4). Centered on the city-state of Malacca, the Srivijayan empire ruled much of insular Southeast Asia from the late fourteenth century to the early sixteenth century. And today, most countries in Southeast Asia continue to exhibit phenomenally high levels of urban primacy. In Thailand, the capital city of Bangkok stands as second to none. Not to be outdone, both Jakarta and Manila exert their dominance over Indonesia and the Philippines, respectively.

But Southeast Asia's urban patterns of the twenty-first century reveal many remarkable differences from previous eras. Whereas many earlier cities were densely populated and compact, the cities of Southeast Asia today are densely populated and sprawling. Rapid urban growth is occurring on the peripheries of Manila, Bangkok, Phnom Penh, Jakarta, and Ho Chi Minh City. This growth, some planned, some not, has led to conflicts over land use; it has threatened once-prime agricultural lands; and it has spurred attendant

Figure 10.4 Angkor Wat, built between 1113 and 1150 by Suryavarman II, is one of but hundreds of wats spread throughout Cambodia. Because it symbolizes Cambodia's golden age, its image can also be found on the nation's flag. *Source:* Photo by James Tyner.

economic problems of land speculation and landlessness (Figure 10.5).

In the light of persistent problems of over-urbanization—traffic congestion, pollution, unemployment—local government officials throughout Southeast Asia have initiated regional economic development projects. Often these projects are multipurpose in scope: promoting economic growth and development in more peripheral regions (i.e., northeast Thailand) and lessening the burdens of primate cities. Still other governments have relocated entire cities for unknown reasons. The always-secretive leaders of Burma, for example, relocated in 2005 its capital from Rangoon (Yangon) to the interior, semi-rural district of Pyinmana. The few western journalists who have been fortunate enough to visit the new Burmese capital—named Naypyidaw ("Seat of Kings")—write of expansive residential areas, reminiscent of any suburban development found in North America, and

Figure 10.5 New residential, leisure, and commercial developments rise on the outskirts of Manila, taking the place of former sugar cane plantations. *Source:* Photo by Arnisson Andre Ortega.

Figure 10.6 In Pleiku, Vietnam, a woman makes a living by selling fresh fruits and vegetables—proudly displayed as in an American supermarket—to shoppers in the early morning hours. *Source:* Photo by James Tyner.

impressive office buildings. But hardly any one lives there! Indeed, most Burmese are denied access to the new city, thus lending credence to the new capital's unofficial designation as a "ghost city."

Similar to Phnom Penh, the cities of Southeast Asia reflect the past and the future. They are centers of intense commercial activity and cultural exchange (Figure 10.6), and, while sharing many commonalities, they—like Phnom Penh's Central and Russian Markets—exhibit remarkable differences.

HISTORICAL GEOGRAPHY OF URBAN DEVELOPMENT

Precolonial Patterns of Urbanization

Southeast Asia is characterized by more coastline than perhaps any other major world region, and much of this coast is accessible to sea traffic. It is understandable, therefore, that maritime influences have contributed significantly to the Southeast Asian urbanization process.

The region, but most especially mainland Southeast Asia, also contains many fertile river valleys, which gave rise to densely populated settlements. These include the Chao Praya, Irrawaddy, Mekong, and Red Rivers, along with their tributaries. Bangkok, Phnom Penh, Hanoi, and Ho Chi Minh City all continue to reflect the importance of these highways of water.

Southeast Asia's physical geography, its complex environment of river systems and coastlines, contributed to the region's importance as a crucial crossroads of commerce between China, India, and beyond. And it was this factor that precipitated the urbanization process of Southeast Asia. Although it is commonplace to speak of the global economy as beginning in the sixteenth century, it is important to recognize that international trade existed long before European states like England and Spain began colonizing the Americas, Africa, and Asia. Indeed, long-distance trade existed between China and India, and linked these areas with places as far afield as Africa, as far back as the early centuries of the first millennium CE. The importance of long-distance trade in the eastern Indian Ocean and South China Sea regions, in fact, led to the appearance of a series of cities and towns along the coast of the Malay Peninsula and on the islands of Sumatra and Java. In time, Southeast Asia would be home to some of the largest urban centers in the world. Indeed, prior to the era of European colonialism (from the early sixteenth to the mid-twentieth century), Southeast Asia was one of the world's most urbanized regions. As late as the fifteenth century, for example, the population of Angkor (in present-day Cambodia) had a population in excess of 180,000; Paris, in contrast, had a population of only 125,000.

Southeast Asia's geographical location made it a natural crossroads and meeting point for world trade, migration, and cultural exchange. A century before the Christian era began, seafarers, merchants, and priests traversed the region, contributing to the urbanization process. In turn, the nascent towns and cities of Southeast Asia became centers of learning through the diffusion of new religious, cultural, political, and economic ideas. Most Southeast Asian societies (with the exception of Vietnam and the Philippines) were influenced primarily by India, and this is most pronounced in the religious and administrative systems of the region. The process of "Indianization," however, was not marked by a mass influx of population like the movement of Europeans into North America. Neither was this a process of replacing indigenous Southeast Asian culture with Indian elements. Rather, the influence of India on Southeast Asia represented a more gradual and uneven process of exposure and adaptation. China provided the other major cultural impetus, although this impact was greatest in Vietnam and through tributary arrangements with various maritime Southeast Asian kingdoms bordering the South China Sea.

Two principal urban forms emerged in precolonial Southeast Asia: the *sacred city* and the *market city*. Although both types of cities performed religious as well as economic functions, the two exhibited many differences. First, sacred cities were often more populous; wealth was gained from appropriating agricultural surpluses and labor from the rural hinterlands. Market cities, in contrast, were supported through the conduct of long-distance maritime trade. Through the market cities passed the riches of Asia, including pearls, silks, tin, porcelain, and spices. Second, sacred cities were sprawling administrative, military, and cultural centers, whereas market cities were mostly centers of economic activity. In physical layout, sacred cities were planned and developed to mirror symbolic links between human societies on earth and the forces of heaven. Monumental stone or brick temples commonly occupied the city center. Market cities, in contrast, tended to occupy more restricted coastal locations, and thus had more limited hinterlands. These cities were more compact in their spatial layout, with much activity associated with the port areas. Lastly, compared to sacred cities, market cities were ethnically more diverse, populated by traders, merchants, and other travelers from all parts of the earth.

The earliest city to emerge in Southeast Asia was apparently Oc Eo, located along the lower reaches of the Mekong Delta, in present-day Vietnam. Flourishing between the first and fifth centuries ad, Oc Eo was an important center for the exchange of cargo, ideas, and innovations. It served as an important city for both Chinese and Indian traders, as well as other seafarers from as far away as Africa, the Mediterranean, and the Middle East.

After the decline of Oc Eo, Srivijaya emerged as an important maritime empire, flourishing between the seventh and fourteenth centuries. It depended on international maritime trade and China's sponsorship through a tributary system, which meant paying tribute (goods and money) to the Chinese emperor in exchange for independence. Located on the straits of Malacca on the island of Sumatra, Srivijaya controlled many important sea lanes, including the Sunda Strait. Evidence suggests that the Srivijayan Kingdom had numerous capitals, one of which was Palembang, located on the southern end of Sumatra. Palembang provided an excellent, sheltered harbor and served as an important Buddhist pilgrimage site. To this day, Palembang remains an important port city and marketplace in Indonesia.

Another example of a market city is Malacca (Figure 10.7). Founded around 1400 on the western side of the Malay Peninsula, Malacca was a counterpart to Palembang and emerged as an important entrepôt and a key node in the spice trade. Although Malacca never had a permanent population of more than a few thousand, it was an extremely vibrant city, inhabited by many foreigners as well as indigenous Malays. In recognition of its multicultural heritage and its role in blending cultures from east and west, Malacca was recently named a UNESCO World Heritage Site. Other important market cities located throughout Southeast Asia included Ternate, Makasar, Bantam, and Aceh.

Figure 10.7 For 130 years, Malacca was a Portuguese colony. Today, a miniature version of the fort has been rebuilt, primarily to enhance Malacca's status as a World Heritage City. *Source:* Photo by D. J. Zeigler.

Sacred cities often occupied more inland locations. One of the earliest was Borobudur, situated on the island of Java. It is at Borobudur that the world's largest Buddhist temple is located. Built between 778 and 856 CE, the ten-level Borobudur temple corresponds to the divisions within the Mahayana Buddhist universe and is one of the great cultural treasures of Southeast Asia. A UN-sponsored program rebuilt the complex several decades ago to preserve the site for future generations.

Arguably, the best known and most famous of all inland sacred cities is Angkor. Centered at the northern end of the Tonle Sap basin, the Angkorian Empire, at its peak, included present-day Cambodia and parts of Laos, Thailand, and Vietnam. The Angkor Kingdom was founded in 802 CE and by the twelfth century contained a population of several hundred thousand. According to some historians, it may even have exceeded 1 million. The temples at Angkor—there are more than 70 recognized sites—were designed to mirror the complex Hindu and, later, Buddhist cosmologies.

By the sixteenth century, many of the once-prosperous inland sacred cities were in decline. In part, internal factions, economic collapse, and foreign intervention hastened the collapse of Angkor and other empires. Coastal market cities, however, continued to thrive on maritime trade. For the region as a whole, the coming years of European colonial dominance would irrevocably alter the course of urbanization in Southeast Asia.

Urbanization in Colonial Southeast Asia

Nutmeg and cloves, cinnamon and sandalwood—these were the prized commodities that drove the world's economy for hundreds of years. And these were the goods that spurred European colonial activity in Southeast Asia.

Five hundred years of colonial and postcolonial influence dramatically affected cities in Southeast Asia. Compared to other world regions, Southeast Asia was relatively urbanized by the time of European colonialism. By the sixteenth century, there were at least six trade-dependent cities that had populations of more than 100,000: Malacca, Thang-long (Vietnam), Ayutthaya (Siam, present-day Thailand), Aceh (Sumatra), and Bantam and Mataram (both on Java). Another half-dozen cities had at least 50,000 inhabitants. Only the Philippines, because of its more peripheral location vis-à-vis the major sea lanes, lacked an urban tradition. But even there, by the early sixteenth century, the seeds of urbanization had been planted. The sultanate of Brunei had extended his authority into the Philippine archipelago and with this came the spread of Islam.

In 1511, however, a Portuguese fleet captured the port city of Malacca, thus ushering in nearly 500 years of European colonialism. The Portuguese came primarily to gain access to and control of the lucrative spice trade. They were soon followed by the Spanish (1521), the British (1579), the Dutch (1595), and the French (mid-seventeenth century). Other colonial activities, such as religious conversion, were present but less important at this time.

The early years of European colonialism in Southeast Asia were similar to the colonial practices found in Africa and the Americas. Europeans captured or built garrisons in coastal cities, established treaties with local rulers, and thus brought about a transformation of the urbanization process in Southeast Asia. Many former empires and kingdoms and their cities suffered tremendous population declines. Malacca, for example, once the premier entrepôt on the strait that shares its name, declined in size and importance after its capture by the Portuguese. From a peak of over 100,000 inhabitants, its size dwindled to 30,000 inhabitants in a very short time.

During the first three centuries of colonialism, European influence was most pronounced in two regions: in Manila (the Philippines) under the Spanish and in Jakarta (Indonesia) under the Dutch. The first permanent Spanish settlement, Santisimo Nombre de Jesus (Holy Name of Jesus), was established in 1565 on the Philippine island of Cebu. Five years later, the Spanish occupied a site on the northern island of Luzon, situated on the Pasig River and proximate to Manila Bay (Figure 10.8). Two existing fishing villages known as Maynilad (from which the present city takes its name) and Tondo were occupied and expanded. Apart from accessibility, defense was often an important consideration in early city planning. In the Philippines, for example, the Spanish had to contend with European rivals, namely the Dutch and Portuguese, as well as Chinese pirates. Consequently, after 1576, construction began on a fortified structure known as the *Intramuros* (walled city). In time, Manila would become the commercial hub of the Philippines and a key node in the Spanish galleon trade that stretched from India to Mexico.

The early European presence was also pronounced on the island of Java in Indonesia. The Dutch East India Company during the seventeenth century established a few permanent settlements, one of which, Batavia, would become the largest city in the region. It is now known as Jakarta. From its inception, Batavia exhibited numerous situational advantages. Geographically, it was located near both the Sunda Strait and the Strait of Malacca, thus allowing easy access to maritime trade. The first Dutch building, a combination

Figure 10.8 A statue in Manila honors Raja Solayman, the city's Muslim prince, who defended the town against the Spaniards in the 1500s. *Source:* Photo by Arnisson Andre Ortega.

warehouse and residence, was built in 1611, and by 1619 a plan was laid out for the city. Much of early Batavia was modeled after the cities of Holland; canals were dug; and the narrow, multistoried Dutch residences were also copied. However, the architecture found in Europe was not functional in hot, humid locations such as Java, so building styles were altered to better fit the tropical environment. Batavia would emerge as the preeminent city of Java and serve as the key node of the Netherland's Southeast Asian empire.

Many of the great cities of Southeast Asia today trace their roots to European colonialism (Figure 10.9). Singapore (from the Malay words *singa*, lion, and *pura*, city), for example, began as a small trading post located on an island south of the Malay Peninsula at the southern entrance to the Strait of Malacca. The town was known as Temasek (Sea Town) before 1819, when Sir Stamford Raffles of the British East India Company signed an agreement with the Sultan of Johor allowing the British to establish a trading post at the site. Benefiting from its strategic location and deep natural harbor, the incipient Singapore began to attract a large number of immigrants, merchants, and traders. As a British colony, Singapore emerged as one of the paramount trading cities of the world—a role it has maintained to this day.

Saigon (now Ho Chi Minh City), which became the capital of French Indochina, likewise began as a small settlement, one that included a citadel and fortress, surrounded by a Vietnamese village (Figure 10.10). Saigon's location in the Mekong delta region—one of the world's great rice granaries—provided the city with an important function as an agricultural collection, processing, and distribution

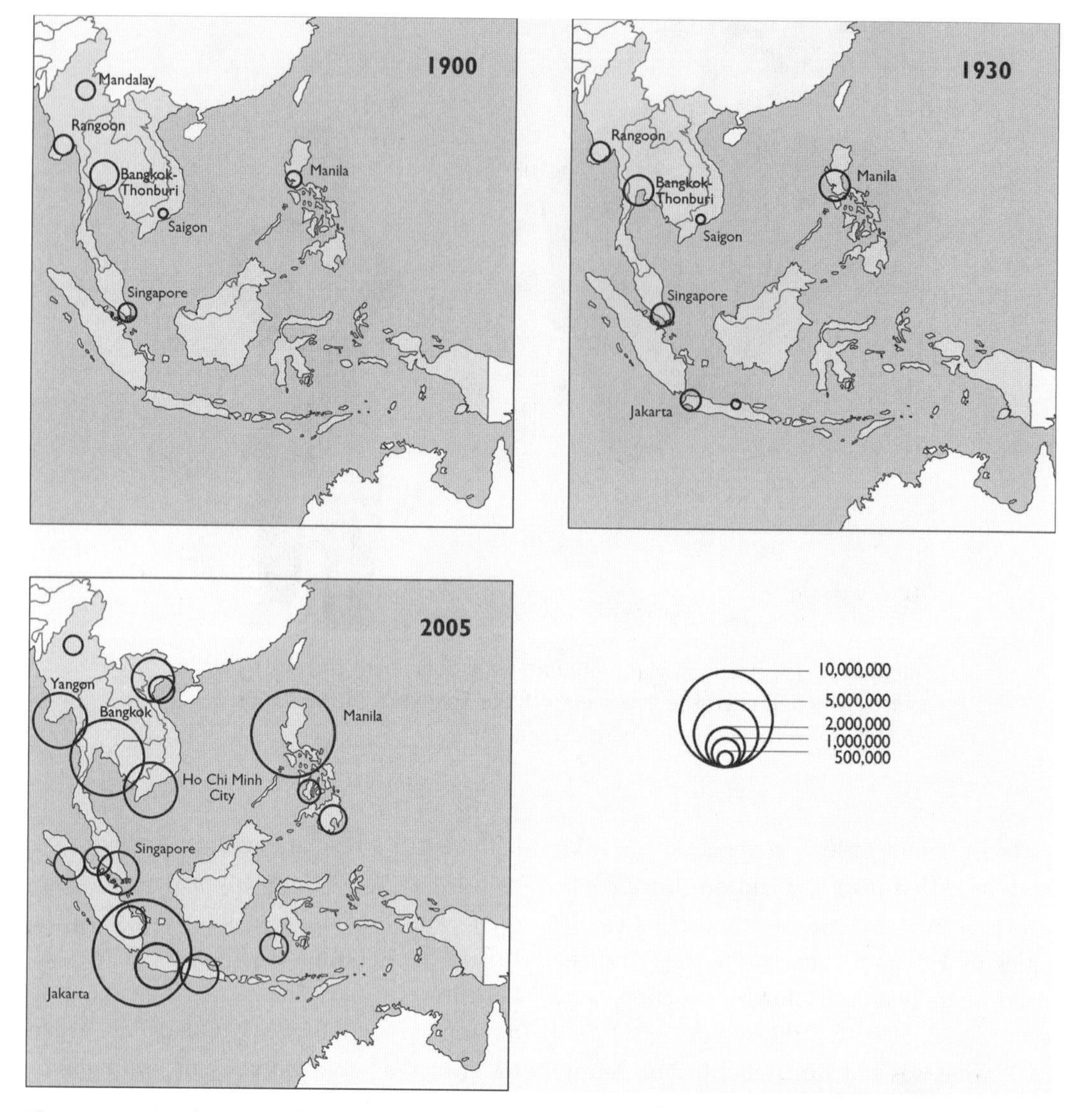

Figure 10.9 Urban Growth in Southeast Asia, 1900–2005. *Source:* Compiled by authors from various sources.

center. The French, also, were intent on refashioning their colonial capital as a microcosm of French civilization. In 1880, the French erected Notre Dame Cathedral in the heart of Saigon. Stones were imported. Ten years later saw the completion of Saigon's French-built Central Post Office. This latter structure, complete with an iron and glass ceiling, was designed by Gustave Eiffel.

Bangkok, the current capital of Thailand, was never colonized by the European powers, yet it still reflects considerable Western influence. Bangkok proper is a relatively new city; it was not founded until 1782. Prior to this date the capital of Siam, as Thailand was then known, was located at Thon Buri along the Chao Phraya River. Beginning in the 1780s, however, construction began on an easily defensible but

Figure 10.10 Fast food—or "good food fast"—is widely available on the streets of Southeast Asian cities. Here, early morning breakfast is served in Ho Chi Minh City (Saigon). *Source:* Photo by James Tyner.

swampy site located opposite Thon Buri, on the east bank of the Chao Phraya. Later, major public works were initiated, often with Western advice and assistance. This is particularly evident in the expansion of rail and road networks, port facilities, and telegraph services.

The most significant impact of colonialism was the establishment of urban nodes, such as Bangkok, Manila, Batavia, Saigon, and Rangoon (Yangon), that would grow into primate cities. Geographically, these cities were located at sites that provided access to seas or rivers. They afforded the European colonizers easy access, so that their ships could export primary products from the region and import secondary products from Europe and elsewhere. Dependence on maritime trade, and the subsequent concentration of political and economic functions in these selected cities, contributed to the decline of other, inland cities. In this manner, the urban system of Southeast Asia was turned inside out, as development was encouraged on the coast and suppressed inland.

The primate cities also served multiple functions. Thus, political, commercial, financial, and even religious activities were concentrated in these urban areas. Saigon, for example, was an administrative and manufacturing center, as well as the dominant trading port of Indochina. Primate cities also became exceptionally large. Stemming from the increased concentration of economic and political functions, these cities served as magnets for both internal and international

Figure 10.11 Bricktown is one of the historic, and now gentrified, neighborhoods of Kuala Lumpur. It was settled by Indians, mostly Tamils, brought in by the British to make bricks. *Source:* Photo by D. J. Zeigler.

migration. Moreover, their populations were characteristically diverse. In many colonies, the colonizers encouraged contract labor. The British, for example, actively encouraged the importation of labor from China and the Indian subcontinent (Figure 10.11). By the nineteenth century, immigration facilitated the establishment of Chinese communities in cities such as Singapore, Manila, and Bangkok, and an Indian community in Singapore. As a result, segregated foreign quarters and Chinatowns emerged, for example, Cholon in Saigon and Binondo in Manila. It was not uncommon for these segregated areas to arise as a result of European force and prejudice. In Manila, for example, the Spanish government issued a series of decrees that required, with few exceptions, all Chinese, Japanese, and even Filipinos to leave the *Intramuros* section of Manila before the closing of the city gates at nightfall. Additionally, Spanish authorities enacted regulations enforcing ethnic segregation and commercial activity.

Apart from the establishment of primate cities, a second impact that European colonialism had on Southeast Asian cities was the establishment or transformation of smaller cities. This included the establishment of mining towns, such as Ipoh in Malaysia, or regional administrative centers such as Medan on the island of Sumatra and Georgetown on the Malay Peninsula. Also notable was the emergence of upland resort centers or hill stations. To escape the oppressive heat and humidity of the lowlands, colonial powers would erect cities high in the mountainous regions. Bandung, for example, a small city located in a deep mountain valley on the island of Java, was established by the Dutch. The cool climate of Bandung served as a welcome relief from the tropical climate and also facilitated the cultivation of coffee (still known as "Java" worldwide), cinchona (for quinine), and tea. Other examples of hill stations include Dalat, a French-built city located 4,800 feet (1,463 m) above sea level in the Central Highlands of Vietnam; Baguio City, a mountain resort atop a 4,900-foot (1,493 m) plateau in the Philippines, developed by the Americans in the early twentieth century; and the Cameron Highlands of peninsular Malaysia.

Lastly, European colonialism significantly affected the development of regional transportation and urban systems within Southeast Asia. Along the Malay Peninsula, for example, the British-built railways ran from the Perak tin-mining areas to the coast and were later expanded along a north-south axis to provide access to the ports and tin-smelting facilities in Penang and Singapore. Consequently, the major urban areas along the western coast of the Malay Peninsula—Kuala Lumpur, Ipoh, Seremban, and Singapore—formed an interconnected urban system that remains evident today. Similar patterns are also visible in Indonesia, where road and rail networks reflect access between sites of resource extraction (plantations and mines) and ports. To a lesser extent, a similar colonial-derived infrastructure remains in the Philippines and the former French Indochina. Colonial powers also exploited river systems (such as the Irrawaddy in Burma, now Myanmar) for internal transportation systems, to exert political and economic control, to link administration functions, and to provide access to sites of resource extraction.

RECENT URBANIZATION TRENDS

In 1940, no city in Southeast Asia registered a population of more than a million; by 1950, just two cities had surpassed this mark. By the twenty-first century, however, 13 cities exceeded the 1 million threshold. Three megacities—Jakarta, Manila, and Bangkok—have also emerged. These massive urban agglomerations contain a disproportionate share of the region's urban population and stand as exemplary primate cities. Bangkok, for example, contains more than 54 percent of Thailand's urban population, while Manila accounts for nearly a third of the Philippines' urban population. Such growth is indicative of the pattern of urbanization that has characterized Southeast Asia throughout the past century.

Recent urban growth in Southeast Asia is the result of three basic demographic processes. First, urban areas in Southeast Asia have increased in population size resulting from the excess of births over deaths. In general, this natural increase accounts for about one-half of urban population growth in Southeast Asian countries. It is important to remember, however, that while overall natural increase may contribute to *urban growth*, it may not contribute significantly to *urbanization*, that is, the increasing proportion of people living in urban areas relative to rural areas. Such urban growth has dramatically altered countless lives throughout the region (Box 10.1)

Second, cities in Southeast Asia have increased through a net redistribution of people from rural-to-urban areas through migration. Studies reflect that, overall, rural-to-urban migration has resulted from larger regional and global economic transformation, and subsequently has played a major role in both the rapid urbanization and urban growth in Southeast Asia over the past two decades. Internal migration is extremely important to urban growth in Thailand, for example. In 1990, the Thai census recorded more than 1.5 million rural-to-urban migrants, although

Box 10.1 A Geography of Everyday Life

Mong Bora is a 12-year-old boy. He lives in a stilt-house with his mother, father, and four sisters. The village in which he lives is approximately one hour north of Phnom Penh, Cambodia. During the rainy season much of the area is inundated with water, hence the necessity to live in houses perched on stilts. In the surrounding vicinity of Bora's village are acres of rich fields and fish ponds. His diet is typical of many Khmers: a staple of rice and fish, coupled with fresh fruits and vegetables. Bora is particularly fond of ripe mangoes, watermelon, and papayas.

Increasingly, Bora's village is being encroached upon by urban sprawl from Phnom Penh. This is considered both a blessing and a curse. On the one hand, villagers are concerned about maintaining their way of life. On the other hand, they recognize that urban growth may translate into better economic opportunities. Many residents of Phnom Penh, for example, visit the area for Sunday picnics, seeking a respite from the chaotic hustle-and-bustle of the capital. And many villagers are able to supplement their incomes from these weekly picnickers. Villagers, like Bora's mother, sell lotus-blossom seeds as snacks, or bottled water. Others rent "cabanas" or tent awnings, under which the visitors can escape the intense heat.

Bora's house sits near the base of two hills, the larger of which is called Phnom Reach Throp, or "hill of the royal treasury." In part, it is because of this hill that people, both locals and international travelers, come to the area. Between 1618 and 1866, Cambodia's capital was located at this site. Known as Udong (meaning "victorious"), the former capital once dominated the landscape. Today, however, little remains of Udong's former glory. The passage of time, but mostly the effects of armed conflict, has devastated much of Udong's former architectural greatness. Several *stupas* remain at the site, as does a colossal Buddha figure. Many of these are currently in the process of being restored, as the site is widely acknowledged as a key piece of Cambodia's cultural heritage.

On any given morning Bora walks 2 km (1.2 miles) to the other side of Phnom Reach Throp to attend school. Among his favorite classes is English. He also enjoys learning about Cambodia's ancient history and geography. It is Bora's dream that these subjects will help him in his chosen career. You see, when Bora is not attending school, or playing soccer with his friends, he is informally working as a tour guide for the many visitors who travel to Udong. Walking step-by-step with tourists, Bora happily details the specifics of the former capital: how many stairs from one *stupa* to the next, the height of the Buddha, the dates of former kings. In this way Bora gets to practice English and earn some extra money to help his family.

Perhaps, if your future travels include the ancient site of Udong, you may be approached by a young man by the name of Bora who will offer to guide your tour. Be sure to accept, for Udong is Bora's home.

there were more than twice as many rural-to-rural migrants. Currently, approximately one out of every seven urban residents in Thailand is classified as a recent migrant. In Malaysia, with economic growth concentrated in Kuala Lumpur and Johor Bahru, the city closest to Singapore, these cities have attracted sizable numbers of migrants in recent years. The Central Highlands of Vietnam have also experienced rapid urban growth through both government-sponsored and spontaneous migration. At the beginning of the twentieth century, for example, the four provinces that comprise the Central Highlands (Kontum, Gia Lai, Dak Lak, and Lam Dong) had a population of approximately 240,000; today, the region's population exceeds 4 million residents. The majority of these inhabitants, approximately 75 percent of the total, are lowland migrants, their children, or refugees. Unfortunately, such remarkable urban growth has resulted in an escalation of land conflicts and political violence.

Third, immigration has contributed to the growth of urban areas in Southeast Asia. This occurs because, in general, overseas migrants tend to move to urban areas as opposed to rural areas. Singapore and, to a lesser extent, Kuala Lumpur, are two of the major immigrant receivers in Southeast Asia. That said, many governments (including Singapore's) often try to prevent immigrants from settling permanently. Overall, however, urban growth in Southeast Asia through international migration remains a relatively insignificant component. Emigration also is inscribed on the urban landscape.

A final component of urban *change* that must be introduced is "reclassification." As a result of bureaucratic decisions, urban populations may change simply through administrative acts. In Malaysia, for example, the number of people needed for an area to be classified as urban changed from 1,000 to 10,000 in 1970. Such statistical changes reflect the changing perception of urbanization by people and their governments.

Within Southeast Asia, over the past two decades, the combined components of migration and reclassification account for the largest portion of urban growth (Table 10.1). There was also considerable variation among countries. Urban growth in Burma, the Philippines, and Vietnam, for example, has occurred primarily through natural increase; whereas, natural increase has assumed a lesser role in Cambodia, Thailand, Malaysia, and Indonesia. The importance of internal migration to the urbanization process of Cambodia, of course, is a consequence of the forced displacement of urban-based people during the murderous Khmer Rouge regime. At 100 percent urban already, the Republic of Singapore's population growth consists mostly of natural increase.

Aggregate numbers such as these mask significant social and economic changes that are occurring. Internal migration in Southeast Asia, for example, is increasingly dominated by female migrants, a process that mirrors that occurring elsewhere in the world, such as China. In Thailand, for example, the share of female migrants increased to more than 62 percent of all Bangkok-bound migrants in the 1980s. The increased feminization of internal migration in Thailand is related to structural changes occurring in both rural and urban areas. The majority of these women—most of whom are in their early twenties—originate in northeastern Thailand, one of the most impoverished regions in the country. Faced with minimal prospects in the rural areas, these women are increasingly moving to Bangkok to obtain employment in factories, the service

Table 10.1 Components of Urban Growth in Southeast Asia (percentage of urban growth)

	1980–1985		*1990–1995*		*2000–2005*	
	Natural Increase	*Migration and Reclassification*	*Natural Increase*	*Migration and Reclassification*	*Natural Increase*	*Migration and Reclassification*
Southeast Asia	49.1	80.9	44.9	55.1	41.7	58.3
Cambodia	70.9	29.1	49.5	50.5	30.6	69.4
Indonesia	35.2	64.8	37..0	63.0	36.7	63.3
Laos	43.8	56.2	44.7	55.3	43.8	56.2
Malaysia	22.0	78.0	38.0	62.0	40.0	60.0
Myanmar	110.0	−10.0	63.2	36.8	44.5	55.5
Philippines	66.0	34.0	62.4	37.6	57.0	43.0
Singapore	100.1	−0.1	100.1	−0.1	98.9	1.1
Thailand	39.6	60.4	31.4	68.6	31.2	68.8
Vietnam	71.7	28.3	50.5	49.5	38.1	61.9

Source: Graeme Hugo, "Demographic and Social Patterns," in *Southeast Asia: Diversity and Development*, edited by Thomas R. Leinbach and Richard Ulack, 74–109 (Upper Saddle River, NJ: Prentice Hall, 2000), table 4.17.

sector, or the informal sector. Also, a certain number of these migrants find employment in the sex trade and end up working in brothels, massage parlors, or strip clubs. Likewise, internal migration to Jakarta, Manila, and Phnom Penh has also become more feminized in response to the structural transformations occurring in these cities.

Not all internal migration is permanent. Indeed, many cities in Southeast Asia, but especially Bangkok and Ho Chi Minh City, are impacted by daily or seasonal circular migration. Three factors are readily identifiable. First, circulation is highly compatible with work participation in the urban informal sector. Migrant laborers are able to circulate between rural villages and urban sites depending on the season. Circulation thus offers a flexible solution to the seasonality of labor demands; laborers are able to work on the nearby farms during the peak agricultural period, while during downtimes these same workers are able to participate in the informal economy in the city. Second, circular migration diversifies families' income-generating activities. Depending on the relative economic strength of urban and rural areas, workers may alternate their activities accordingly. Third, circular migration has, with advances in transportation systems, become a more viable option. Improvements in mass transportation systems, such as paved roads and mass-transit bus lines, have permitted people to move with greater ease, thus contributing to the growth of suburban residential areas. The spectacular urban growth to the south and west of Ho Chi Minh City is indicative of this process.

Globalization, Urbanization, and the Middle Class

The most significant development in the world economy during the past few decades has been the increased globalization of economic activities. The transnational operations of multinational firms have given rise to a new international division of labor, one that has witnessed a shifting of manufacturing sector enterprises from developed to developing economies, and the emergence of new corporate headquarter activities, producer services, and research/development sites. Final assembly and testing of audio-video equipment are located in Singapore and Penang, Malaysia; the assembly and packaging, low-skilled and labor-intensive, in Bangkok, Jakarta, and Manila; and marketing and sales functions, mid- to high-end manufacturing, in Singapore. In Southeast Asia, these far-reaching changes are evidenced by the spectacular growth of assembly plants in Phnom Penh as well as the emergence of Cyberjaya, Malaysia's high-technology city (the "Silicon Valley of the East") that forms the hub of that country's Multimedia Super Corridor.

Shifts in the structure of Southeast Asia's economies have led to remarkable societal and occupational changes. Declines in agricultural workers are matched by increases in the number of workers employed in the service and manufacturing centers. Consequently, the increasing portion of clerical, sales, and service workers, in particular, has translated into redefined social categories. One change that is especially salient is the emergence of a new middle class. And given that many of the economic transformations have occurred disproportionately within the urban areas of Southeast Asia, it should come as no surprise that the emergent middle class in Southeast Asia is likewise urban based.

The rise of Southeast Asia's new urban middle class has drastically altered the urban landscape. Demographically, Southeast Asia's middle class tends to have smaller families; economically, it tends to have high and rising levels of consumption and to spend money on nonessential items, such as luxury cars. Many members of this emergent class express "Western" middle-class fantasies to materially project their newfound social status. In terms of housing, Southeast Asia's middle class demands more space and more privacy (hence leading to demand for Western-style housing in the form of detached and semi-detached single-family dwellings). Having the ability to purchase an automobile, the middle class is able and willing to commute longer distances to work, thus fueling the sprawl of cities into traditional agricultural hinterlands. Others prefer to live closer to the traditional downtown districts, fueling the proliferation of condominiums and apartment complexes. The emergence of the middle class, and its growing spending power, is likewise reflected in the mushrooming of shopping malls and country clubs and in the proliferation of leisure activities and nightclubs. It is seen in the growth of gourmet restaurants, coffee bars (Starbucks are becoming all too pervasive), theaters, galleries, and boutiques.

Cities in Southeast Asia have historically been segregated. During the colonial years, British, French, Spanish, and America authorities often restricted residential and commercial activities by ethnic classification in their respective colonies. The Spanish, for example, disallowed Filipinos from living in *Intramuros* Manila; the French restricted Vietnamese settlements in Saigon. Today, segregated areas remain, but these often reflect class differences as much as anything. This segregation is epitomized by the rise of gated communities.

The desire among the new middle class for gated communities results from demands for privacy, security, and prestige. Many of these new housing developments are equipped with strictly controlled gates that are fully secured by armed private guards and monitored by CCTV. In exclusive villages, entrance is permitted only to residents with proper photo

ID or to their friends and acquaintances. Also, many of these villages carry Western-themed names and architecture and are fully equipped with top-notch amenities such as tennis courts, club houses, golf courses, swimming pools, and spacious houses; some even provide heliports for their residents. Everyday life in these communities is heavily controlled by home-owner association rules from the kinds of designs permitted for houses to curfew hours.

The growth of Southeast Asia's middle class is a major contributor to the sprawl of its cities. A key prerequisite for the construction of gated communities, for example, is land. In Jakarta, Ho Chi Minh City, and Manila, local governments have allowed the spread of new middle-class enclaves on their peripheries and the conversion of old land uses for middle-class condominiums in the cities. Many lower-cost housing units in Phnom Penh, for example, have been razed in order to erect higher cost apartments and condominiums for the middle class. In Manila, informal settlement communities have been demolished or were dubiously destroyed by fire to make way for new mixed-use commercial business districts. Such changes have resulted in conflict.

Often, these new middle-class enclaves sit side by side with ever-rising numbers of the urban poor, who continue to migrate toward cities in search of jobs. Displaced by land scarcity and the mechanization of agriculture, these rural-to-urban migrants must compete with the wealthy, and now with the middle class, for limited resources in the cities. The poor provide their own housing, usually makeshift structures of corrugated tin, cardboard, or plywood. Often, these structures sit in sharp contract next to the golf courses and gated, elitist communities of the better-off. The urban landscape of Southeast Asia thus reflects the social and economic transformations of a region enmeshed in much broader changes; the growing disparity between the "haves" and the "have-nots" is all too apparent.

MODELS OF URBAN STRUCTURE

For more than three decades, thinking about Southeast Asian urbanization has developed out of T. G. McGee's model of city structure (Figure 10.12). Building on a long tradition of urban modeling in North America, McGee's model presumed, first, that no clear zoning characterizes land use of the large cities of Southeast Asia. Instead, he proposed that only two zones of land use remained relatively constant, these being the port district and, on the periphery of the city, a zone of intensive market gardening. In between were areas of mixed economic activity and land use with other areas of dominant land use, such as spines of high-class residential areas and clusters of squatter settlements.

As cities in Southeast Asia have experienced rapid urban growth, in part associated with more intense integration into the global economy, our understanding of these places has changed considerably. While remnants of McGee's initial model are still visible, for example, a central port zone and mixed land-use patterns, other aspects of Southeast Asian cities have been radically transformed. Urban scholars now speak of *extended metropolitan regions* (EMRs). This distinctive form of Asian urbanization has its origins in the way in which Asian cities (not limited to Southeast Asia) have been incorporated into the global economy. The colonial-influenced primate cities are increasingly penetrating their surrounding hinterlands, urbanizing the countryside, and drawing rural populations deeper and deeper

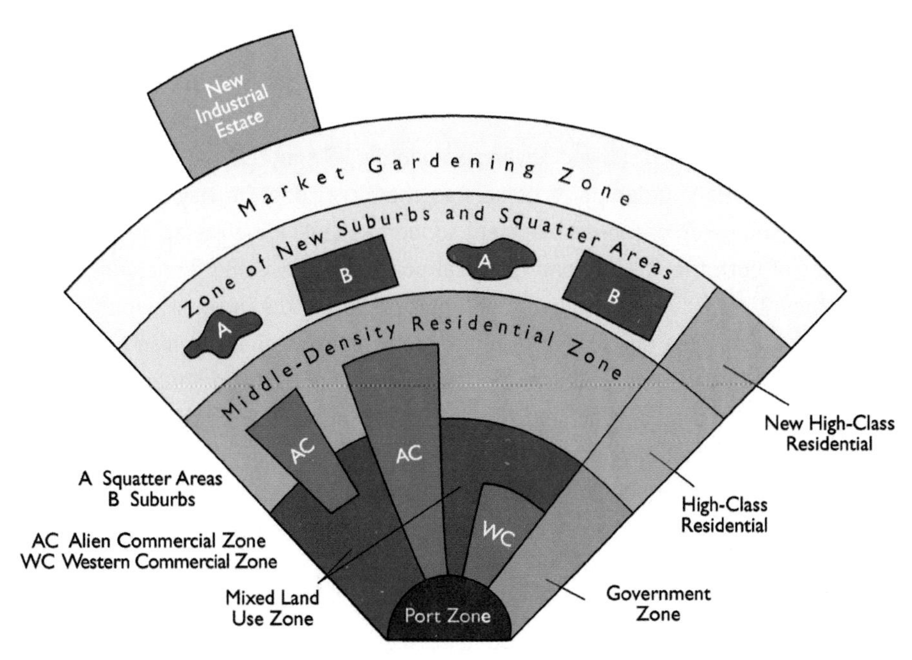

Figure 10.12 A Generalized Model of Major Land Use in the Large Southeast Asian City. *Source:* T. G. McGee, *The Southeast Asian City* (New York: Praeger, 1967), 128.

into the urban economy. In certain respects, EMRs are similar to metropolitan regions in the United States. However, EMRs in Southeast Asia differ from their North American counterparts in that the former exhibit a greater population density in both the urban cores and surrounding rural periphery.

EMRs may be differentiated into three basic forms. The first is the *expanding city-state.* Singapore provides the only example. In recent decades, Singapore has extended its political and economic influence into the territory of its neighbors, Indonesia and Malaysia. A second type is the *low-density* EMR, exemplified by Kuala Lumpur. These EMRs have been able to maintain relatively low population densities through the successful development of satellite cities that form a fringe around the dominant urban area. In other words, low-density EMRs reflect controlled and managed growth, as opposed to the more rapid and unplanned growth of other cities. Ho Chi Minh City likewise reflects aspects of a low-density EMR. The third and most prevalent type is the *high-density* EMR. This form, epitomized by the massive cities of Jakarta, Manila, and Bangkok, exhibits a chaotic spillover of urban economic functions into the rural hinterland, with an accompanying conversion of agricultural land to residential and industrial development.

Related to extended metropolitan regions is a new urban form, identified by McGee as a *desakota.* The term itself is derived from the Indonesian words for village (*desa*) and town (*kota*) and is meant to capture the process whereby urbanization overtakes its surrounding hinterland. A number of elements have been associated with *desakotas.* First, these cities exhibit considerable diversity in their land use. Characteristically, a *desakota* region encompasses cities with mixed residential and industrial land uses, as well as a densely

Box 10.2 From Hacienda to Mixed-Use Suburbia

"You don't have the right!" ("*Wala kayong karapatan!*"), shouted the lead architect of a development firm to a large group of farmers on the picket line. On May 21, 2010, violence erupted between a group of pro-development supporters (including the architect, surveyors, and members of both the military and the local police) and resisting farmers in Buntog, an upland community in Canlubang, Laguna, Philippines. During the incident, around a hundred farmers and activists were hurt. Eleven people, including a pregnant woman and a 70-year-old grandmother, were illegally jailed, two other people were badly beaten by the police, and an elderly man suffered a heart attack. This is just one of the many cases of violence faced by farmers resisting urban development in the past decade.

Buntog is part of the former Canlubang Sugar Estate, or Hacienda Yulo. The 7,200-hectare (17,800-acre) estate is owned by one of the Philippines' most powerful elite families, but the farmers of Buntog have been living in the area since the early 1900s when the region was still an unoccupied forest. The farmers cleared the forest and planted coconut trees and other crops. In the ensuing years, the vast tract of land near Buntog was purchased by an American conglomerate, Ehrman-Switzer, who established the Calamba Sugar Estate. Nationalist calls for independence and World War II provided an effective context that transferred the estate to the hands of Filipino elite families. Thus it was that Jose Yulo purchased the estate using a war reparation loan and renamed the hacienda "Canlubang Sugar Estate." Local farmers, deprived of access to lands that they historically farmed, were forced to find employment as sugarcane plantation workers. Later, many of these farmers filed a formal petition to acquire the title of their land; but, surprisingly, found out that Yulo's sugar estate had extended into their community.

In 1988, the Philippine government enacted the Comprehensive Agrarian Reform Program that sought to redistribute large landholdings to tenant farmers. But the Yulos of Canlubang had different plans. In fact, they had been converting parcels of the whole estate into industrial sites and golf courses a decade prior to land reform. They successfully reclassified the area as industrial and therefore received an exemption to redistribute their lands to farmers in the area. In 1996, the sugar estate closed and farming was prohibited in large sections of the estate. Through their many corporations, the Yulo family entered into joint-venture agreements with top real estate firms to develop mixed-use urban projects that combined gated communities, golf courses, and commercial districts. The most ambitious of these development projects is Nuvali, a 1,700-hectare (4,200-acre) joint-venture master-planned township project with Ayala Land Corporation. This "city of the future" is promoted as the Philippines' first "eco-community" and boasts of blend of "environmentally sustainable" residential, commercial, and recreational developments.

Not all is what is seems, however. Numerous cases of military intimidation and illegal cutting of coconut trees have been reported. The entire estate has been littered with security posts that serve to control the movement of residents in the area. Famers have been prohibited from taking their produce to nearby markets and from bringing in much-needed materials for house repair. In addition, a water tank facility project in the community was halted by the landlords, who argued that the whole area is their "private" property.

The contestation over land in Buntog is not isolated. For many former haciendas in the Philippines, land-use conversion and urbanization have become effective means by landlords to circumvent land reform and to squeeze land for profit. Around Manila and its hinterlands, many new mixed-used urban developments are built on lands with painful histories of dispossession and violence.

populated wet-rice agricultural area. Within a *desakota*, moreover, there is significant interaction between village and town. This is made possible by an integrated transportation system that permits high levels of population mobility. Indeed, the increased daily commuting patterns seen in Bangkok testify to this increased circulation within that *desakota*. Second, these regions also are strongly integrated into the global economy. Foreign investment is generally important in these areas, as multinational corporations tap into large and readily available labor surpluses. Lastly, it is the *process* of formation of the *desakota* that is perhaps more important than the resulting pattern itself, because surrounding rural areas become urbanized without the transfer of population that occurs in, say, rural-to-urban migration. On many occasions, urban transformation of rural areas involves dispossession and displacement of farmers and other rural residents. For example, there has been a large-scale conversion of agricultural lands in provinces adjacent to Metro Manila into industrial parks, ecotourism developments, and gated communities. These developments have effectively displaced many farming communities and often involve violence between resisting farmers, the military, and landlords' security forces (Box 10.2).

DISTINCTIVE CITIES

Singapore: World City of Southeast Asia

There is truly no other city in the world quite like Singapore (Figure 10.13). Its striking modernity, orderliness, and Disneyland-like cleanliness seem unreal, especially if one has just arrived from the sprawling, chaotic, and noisome cities of Bangkok, Jakarta, or Manila. To some, Singapore is a model of efficiency, a mosaic of well-manicured lawns, efficient transportation, and planned development. To others, Singapore represents a draconian police state masquerading as utopia. The reality, of course, lies somewhere in between, and depends on one's personal tastes and values.

Singapore is unique within Southeast Asia, and in the world, in that it is both a city as well as a sovereign state. And it is small. At just 246 sq miles (640 sq km), it is one-fifth the area of Rhode Island. But despite its size, Singapore is also very prosperous. In per-capita income, the city-state is second only to Japan in all of Asia. In part, Singapore's economic success has been tied to its geography. Both historically and now, the city benefits from its strategic location and superb natural harbor. Furthermore, effective government policies and dynamic leadership have propelled Singapore into its role as Southeast Asia's

Figure 10.13 The Singapore River was at the very heart of commercial life in Singapore. A hundred years ago, it would have been packed with junks, with wharves and warehouses along both sides. *Source:* Photo by D. J. Zeigler.

leading port and industrial nexus, as well as a leading banking and commercial center. Indeed, Singapore ranks along with Tokyo and Hong Kong as one of Asia's three key urban centers in today's global economic system.

Singapore's economy is based on a strong manufacturing sector (initially processing raw materials such as rubber, but more recently electronics and electrical products), oil refining, financial and business services, and tourism. The economic success of Singapore has translated into a very favorable quality of life. Singapore registers the region's lowest infant mortality rate and the lowest rate of population increase, in addition to the highest per-capita income. This quality of life is facilitated through subsidized medical care and compulsory retirement programs.

Singapore is a truly cosmopolitan world city. The affluence of Singapore is vividly seen along the city-state's major shopping and tourist corridor, Orchard Road. In the 1830s, this area was home to fruit orchards, nutmeg plantations, and pepper farms. Now, the mangoes, nutmeg, and peppers are gone, replaced by jeweled necklaces, designer clothes, and perfumes. Orchard Road currently extends 1.5 miles (2.4 km) and is lined with major shopping centers, upscale boutiques, luxury hotels, and entertainment centers such as the Raffles Village (built around the carefully restored classic Raffles Hotel, one of the great hotels of the past).

True to its place at the center of Southeast Asia, the urban landscape of Singapore reflects its rich ethnic heritage. Approximately 75 percent of Singapore's population is ethnic Chinese; Malays and Indians constitute 15 percent and 7 percent of the population, respectively. Consequently, Singapore has four official languages: English, Mandarin Chinese, Malay, and Tamil, a language of southern

India and Sri Lanka. The immigrant history of Singapore, moreover, is well preserved in the city-state's architecture (Figure 10.14). The Chinatown area of Singapore, for example, located at the mouth of the Singapore River, began in 1821, when the first junk load of Chinese immigrants arrived from Xiamen, in Fujian province. In 1842, these Chinese immigrants completed the Thian Hock Keng Temple and dedicated it to Ma-Chu-Po (Matsu), the goddess of the sea. Nearby is the Nagore Durgha Shrine, built by Muslim immigrants from south India, and further down the road is the Al-Abrar Mosque, also known as Indian Mosque, which was built between 1850 and 1855. Because of land reclamation projects, the ultramodern skyscrapers of Singapore's financial district now eclipse Chinatown's oceanfront view.

Figure 10.14 This colorful and finely detailed Indian temple in Singapore is one of the best-known cultural landmarks of the city. *Source:* Photo by James Tyner.

Given the large number of historical temples and monuments, urban preservation is important in Singapore, although this has not always been the case. During the 1960s and 1970s, for example, many older buildings were demolished to make room for a more modern infrastructure. However, a movement to preserve Singapore's urban history was initiated, both for reasons of national prestige and tourism. Today, many areas, such as the waterfronts along Collyer Quay and Boat Quay, have been renovated.

Singapore is also instructive for its public-housing programs. Beginning in the 1960s, the Singaporean government, primarily through the efforts of the Housing Development Board (HDB), moved to ensure adequate housing for its population. The result was the establishment of numerous new towns and housing estates. The Queenstown housing estate, for example, located in the central region of Singapore, was one of the earliest estates developed by the HDB. Tao Payoh New Town and Ang Mo Kio New Town, both located within the Northeastern Region, were initiated in 1965 and 1973, respectively. A more recent development is Woodlands New Town, located on the northern coast of Singapore. These new towns are designed to be self-sufficient communities of many thousands of residents. Moreover, in recognition of Singapore's ethnic diversity, the government has mandated ethnic-based occupancy rates to offset the emergence of hyper-segregated enclaves. By 2000, nearly 90 percent of Singaporeans lived in one of the high-density housing estates built by the HDB.

Despite these successes, urbanization in Singapore remains, and will continue to remain, hindered by two physical obstacles. First, Singapore cannot readily—or cheaply—expand

Box 10.3 A Thirsty Singapore

Water is arguably the single most critical site-need for a city to survive and prosper. Singapore provides a surprising illustration of this need. Given that Singapore occupies an island located in the tropics and receives an average of 240 cm (95 in) of rainfall per year, it is ironic that water constitutes one of the greatest obstacles to the city-state's future. Residential and industrial water consumption amounts to approximately 1.3 million cubic meters per day.

Water is a serious problem for Singapore because the island contains no significant rivers or lakes to collect freshwater and so must rely on reservoirs and storm-water collection ponds to provide freshwater for its 5.4 million inhabitants. Approximately 40 percent of Singapore's water, however, is supplied by Malaysia. Historically, the provision of water by Malaysia to Singapore was guaranteed by two bilateral agreements, although one of these expired in 2011 and was not renewed. The second is set to expire in 2061, providing a sense of urgency for Singapore to become self-sufficient in its water needs.

Singapore has initiated a series of conservation measures. These include public education and publicity programs advocating the rational use of water. Significantly, many of these water-oriented initiatives are coupled with broader environmental campaigns. Thus, public awareness campaigns are directed not simply at the conservation of water, but also toward issues of pollution, hygiene, infectious diseases, and sanitation.

The installation of water-saving devices, such as flow regulators and low capacity flushing systems, is also mandatory. This latter measure is especially significant since household water-use—and particularly the flushing of toilets—accounts for more than 50 percent of water consumption. Within the industrial sector, the recycling of water and the substitution of non-potable water (e.g., sea water, rain water) for potable water has also been encouraged. Other measures, including water consumption taxes, have been implemented.

Aside from the conservation of water, Singapore has pursued other strategies to provide more water, such as the construction of desalinization plants. In total, these plants may provide 40 percent of Singapore's water needs. However, energy requirements are enormous for these plants, making desalinated water exceptionally expensive. Additional reservoirs may also help ease Singapore's thirst, although the development of these projects is hindered by land shortages. Lastly, Singapore has looked beyond its immediate geographic surroundings to obtain sources of water. In 1991, for example, Singapore signed a memorandum of understanding with Indonesia to draw water from Bintan Island in the Riau Archipelago and from the Kampar River in Sumatra. This solution, though, simply replaces Singapore's dependency on Malaysia with one on Indonesia.

One thing is certain: a glass of drinkable water in Singapore will cost more in the future than it does today.

its area because it occupies only a small island. In response, the Singaporean government has utilized land reclamation schemes and has also been working to expand its economic growth beyond its own political boundaries into neighboring Malaysia and Indonesia. In this manner, the government hopes to exploit the comparative advantages of Singapore and neighboring countries. A second, and perhaps more immediate, obstacle confronting Singapore is that of water (Box 10.3).

Kuala Lumpur: Twin Towers and Cyberspace

The skyline of Kuala Lumpur is one of the most recognizable sights in all of Southeast Asia. While Paris has its Eiffel Tower and Shanghai its futuristic TV tower, Kuala Lumpur has the 88-story Petronas Twin Towers. Standing like a giant double-barreled beehive, the Petronas Towers dominate the capital city of Malaysia (Figure 10.15). They are Kuala Lumpur's signature landscape and are symbolic of the lofty goals set by the Malaysian government.

Kuala Lumpur is a relatively young city, having been founded only in 1857 by Chinese tin miners at the swampy confluence of the Klang and Gombak rivers (Figure 10.16). In fact, the name "Kuala Lumpur" translates as "muddy confluence." The settlement grew rapidly, however, and by 1880 had become the capital of the state of Selangor on the Malay Peninsula.

Despite its growing political importance, Kuala Lumpur throughout much of the early twentieth century was still overshadowed in population by other cities along the Malay Peninsula, including Georgetown to the north and Singapore to the south. Although it was designated capital of the Federated States of Malaysia in 1963, Kuala Lumpur trailed both Singapore and Georgetown as preeminent commercial centers on the peninsula. In 1972, however, Kuala Lumpur gained city status and was declared a Federal Territory (similar to Washington's District of Columbia).

Kuala Lumpur experienced tremendous population growth throughout the late twentieth century. Currently, the population is approximately 2 million. However, unlike Bangkok and Manila, Kuala Lumpur has made a more concerted effort to manage urban growth. Planned satellite cities, for example, were designed to the urban congestion of the capital. In the 1950s, the satellite city of Petaling Jaya was established; it is now home to more than 500,000 residents and a major industrial center. Nearby, the satellite town of Shah Alam, initially planned to be half residential and half industrial, was built in the 1970s. Although shantytowns are visible in parts of Kuala Lumpur, as a whole the city exhibits a sedate orderliness more reminiscent of Singapore than of other Southeast Asian cities.

The economy of Kuala Lumpur is an exceptionally diverse mix of manufacturing and service activities. Many of these industries are clustered within the Klang Valley conurbation, an urbanized corridor stretching from Kuala Lumpur westward through Petaling Jaya and Shah Alam to the port city of Klang. Also, indicative of the information economies emerging in Southeast Asia, Kuala Lumpur is a key anchor in Malaysia's "Multimedia Super Corridor" (MSC). Planned to be a setting for multimedia and information-technology companies, the MSC is seen as the catalyst in propelling Malaysia's economy into the global information age. In related developments, two new cities have been constructed to the south of Kuala Lumpur: Putrajaya, the "new" administrative capital of Malaysia, and Cyberjaya,

Figure 10.15 When Kuala Lumpur's Petronas Towers opened in 1999, they became the world's tallest, a title they held until 2004. *Source:* Photo by Jared Boone.

billed as an "intelligent city," complete with a state-of-the-art, integrated infrastructure that attracts multimedia and information-technology companies.

Putrajaya is especially notable as a planned city. Ambitious in scope, Putrajaya has become a national symbol, reflecting the country's determination to become a regional if not global power. Consisting of scores of monumental office buildings, modern shopping malls, convention centers, and private colleges, Putrajaya has become the face of twenty-first-century Malaysia.

Jakarta: Megacity of Indonesia

Most visitors to Indonesia arrive first in Jakarta, a sprawling metropolis situated on the north coast of Java. At two-and-a-half times the size of Singapore, Jakarta is the largest city in population and land area in Southeast Asia. From its origins as a small port

Figure 10.16 A mosque, Jamek Bandaraya, backed by the downtown skyline, now occupies the original site of Kuala Lumpur, a "muddy confluence" of two streams seen in this picture. *Source:* Photo by D. J. Zeigler.

town called Sunda Kelapa, the Special Capital Region of Jakarta (Daerah Khusus Ibukota, or DKI Jakarta) has experienced phenomenal growth over the past five decades. From fewer than 2 million in 1950, Jakarta's population had swollen to more than 13 million by 2000. As with many primate cities in Southeast Asia, urban growth has sprawled into the hinterland. In recognition of this sprawl, in the mid-1970s officials began to refer to the entire region as Jabotabek, an acronym derived from the combination of *Ja*karta and the adjacent districts of *Bo*gor, *Ta*ngerang, and *Bek*asi. When the entire Jabotebek region is considered, the population of the Jakarta metropolitan area includes a mind-boggling 20 million people. Similar to other major cities in the region, Jakarta's population is also impacted by seasonal and daily commuting. Hundreds of thousands of workers, the majority of whom live in new residential communities in the Jabotabek region, commute daily to Jakarta.

Jakarta is Indonesia's largest and most important metropolitan area. It is the national capital and the principal administrative and commercial center of the archipelago. The city also plays a vital role in Indonesia's international and domestic trade and receives a disproportionate share of foreign direct investment. This investment, focused primarily on manufacturing but also on the construction and service sectors, operates as a multiplier effect for Jakarta's economy. It also accounts for a rapid rise in the middle class, with a corresponding impact on the urban landscape. In some respects, though, Jakarta has undergone a period of deindustrialization similar to that of other cities in the world, such as London. Thus, although the city remains an important manufacturing center, economic growth has been accounted for largely by increases in both tertiary and quaternary sectors (especially financial services, communications, and transportation).

Economic and social changes have also contributed to changing land-use patterns. The central core of Jakarta has experienced significant changes over the past decade, such as a conversion from residential to higher-intensity commercial and office land use, as well as the emergence of luxury high-rise apartments. In the Jabotabek region, the development of new towns (e.g., Lippo City, Cikarang New Town, and Pondok Gede New Town) and corresponding large-scale residential subdivisions has also transformed previously agricultural land into urban spaces. Indeed, upwards of eighty thousand new housing units are added each year to the Jabotabek region. Other changes include the emergence of larger industrial estates as well as leisure-related land uses (e.g., golf courses).

Jakarta also reflects the urban woes characteristic of primate cities, including a lack of adequate public housing, traffic congestion, air and water pollution, sewage disposal, and the provision of health services, education, and utilities. Because of the megacity's sheer size, not to mention the heightened political, social, and economic instability of Indonesia, Jakarta's problems are magnified to dangerous levels.

Manila: Primate City of the Philippines

Unlike walking the regimented, disciplined streets of Singapore and Kuala Lumpur, traveling within Manila is an experience unto itself. Indeed, with the possible exception of Bangkok, no other city in Southeast Asia is as famous—or infamous—as Manila for its traffic. Throughout the day, and frequently well into the night, traffic grinds slowly through the rabbit-den of highways and alleyways that constitute Manila's overburdened road network. Diesel-spewing *jeepneys*, rickety buses, and luxury sports utility vehicles all compete in bumper-car-like fashion. Turn lanes and stoplights are largely ignored. Yet beyond the chaos and congestion that puts the freeways of Los Angeles to shame, Manila exhibits its own charm and appeal. Indeed, much of Manila's charm stems precisely from its outwardly confusing appearance.

Politically, socially, economically, and in terms of total population, Manila far surpasses all other cities in the Philippines (Figure 10.17). Indeed, with the exception of Thailand, no other major country in the region has a higher primacy rate than the Philippines. Currently, Manila is approximately nine times as large as the Philippines' second-largest metropolitan area, Cebu. And similar to Jakarta's Jabotabek, Metro Manila is composed of many different political units. In 1975, Metro Manila was formed through the integration of the four preexisting, politically separate cities of Manila, Quezon City, Kaloocan, and Pasay, plus 13 municipalities. The nature of governance and administration of the metropolitan region has changed over the years since its original inception, from a more centralized Metro Manila Commission (MMC) headed by Imelda Marcos as governor to its current monitoring and coordinating form as the Metro Manila Development Authority (MMDA).

The Metro Manila region, reminiscent of Harris and Ullman's multiple nuclei model, is a polynucleated area with many distinct personalities. Binondo, for example, located next to the Pasig River, was originally a Christian Chinese commercial district during the Spanish colonial period and to this day remains the heart of Manila's Chinatown. Nearby is Tondo, today an impoverished, densely

Figure 10.17 Motorbikes are one way of breaking through traffic jams on Bangkok's overcrowded streets. *Source:* Photo by D. J. Zeigler.

populated district of rental blocks; prior to the arrival of the Spanish, it was a collection of Muslim villages. To the south, abutting Manila Bay, is Ermita. Once a small fishing village, the area developed into a prime tourist destination, packed with bars, nightclubs, strip shows, and massage parlors. In recent years, however, these establishments were closed down in Ermita and many have since relocated to other areas of Manila. As a final example, Makati, originally a small market village, is now Manila's major financial center, occupied by banks and multinational and national corporations. Makati also contains some of Manila's most expensive housing subdivisions, sprawling box-like shopping centers, and five-star hotels.

Within the greater Manila Metropolitan region is Quezon City. Named after Manuel Quezon (president of the Commonwealth of the Philippines from 1934 to 1946), Quezon City was the Philippines' national capital from 1948 to 1976. Now it is home to many important government buildings, medical centers, and universities, including the main campuses of the University of the Philippines and Ateneo de Manila University. Quezon City also consists of upscale, gated residential communities patrolled by armed guards. These estates, home to middle- and upper-class residents, are equipped with luxurious air-conditioned homes, tennis and basketball courts, golf courses, and swimming pools. However, reflective of Manila's complex land usage, as well as the highly polarized nature of Philippine society, just outside of these gated communities are numerous squatter settlements.

Characteristic of large cities in Southeast Asia, Manila exhibits an increasing number of consumer spaces. In recent years, large shopping malls have been built, catering to a rising middle class. For many visitors, these malls are remarkably similar to those found in the United States and Europe. Major department stores anchor the malls, while in between are dozens of specialty stores, food courts, and entertainment. The SM Megamall, for example, in addition to its numerous stores and restaurants, contains an ice skating rink, bowling lanes, a 12-screen cinema, and an arcade room. Meanwhile, multiple high-end condominium projects have been built in many parts of Metro Manila that cater to returning Overseas Filipinos and urban professionals. In many cases, these condominium projects were developed beside malls and shopping complexes. For example, the proposed Entertainment City in Manila, a mixed-use leisure complex of casinos, condominium units, and shopping malls, is planned for development on a previously reclaimed area near Manila Bay.

Similar to both Jakarta and Bangkok, the increased concentration of foreign direct investment into the Philippines has translated into rapid changes in the economy of Manila, as well as in land use. During the 1990s, the greater Metro Manila region and the surrounding provinces experienced remarkable industrial and manufacturing growth. This expansion, however, occurred at the expense of Manila's rural and agrarian hinterland. Metro Manila is, in fact, located toward the center of the Philippines' major rice-producing region, and continued urban sprawl is rapidly encroaching on these agricultural areas.

Urban poverty and landlessness continue to be major problems in Manila. The extent of these problems, however, remains a contested issue. Estimates of the number of poor vary widely, ranging from 1.6 million to more than 4.5 million. What is certain, however, is that landownership in Manila is decidedly uneven, with the majority of its population being landless. High urban land values mean that the majority of residents are unable to obtain legal housing, a situation exacerbated by continued high rates of in-migration. Their recourse is to resort to illegal housing and to settle in urban fringe areas, such as along railroad tracks and in vacant lots. Residents of squatter settlements are subject to deplorable health conditions and pollution problems, stemming from inadequate access to sanitary and plumbing facilities. They often must purchase fresh water from itinerant water vendors. Historically, squatter settlements have been demolished and their residents evicted. More recently, the Philippine government has attempted to provide low-cost housing for its urban poor through joint-venture agreements with private developers. But with relocation sites that are far away from the city and insufficient facilities in many of the housing projects, many settlers end up returning to the metropolis.

Aside from poverty, Manila faces other serious problems. Accessibility to water, for example, looms large. Manila is also confronted with serious air and water pollution problems, as well as an inadequate sewerage system. Indeed, during the rainy season many streets throughout Manila, such as those in the port district and Tondo, become impassable due to flooding.

Bangkok: The Los Angeles of the Tropics

Bangkok, at 34 times the size of Thailand's second-largest city, is the textbook example of urban primacy. While only one-fifth of the country is urbanized, fully two-thirds of this urban population is concentrated in the Bangkok Metropolitan Region (BMR). Currently, the core of Bangkok has a population of about 6 million; when the entire BMR is considered, the region's population is more than 10 million. Furthermore, like the ocean's tides, Bangkok's population ebbs and flows, both daily and seasonally. An estimated 1 million people commute daily into Bangkok, while hundreds of thousands of other workers seasonally circulate throughout the city in search of temporary jobs in the informal sector. This seasonal migration is particularly acute in the hot, dry months of February and March, a slack agricultural period.

The official name of Bangkok is Krung Thep, which translates as "The City of Angels" (the same meaning as Los Angeles). And in many respects, notably traffic, pollution, and urban sprawl, Bangkok might be considered the Los Angeles of Southeast Asia. For that matter, Los Angeles may be considered the Bangkok of the United States of America.

Currently, Bangkok remains poised to become an international communications and financial center, as well as a major transportation hub in Asia. Initially, much of Bangkok's growth was tied to massive amounts of investment brought about by the United States' involvement in the Vietnam War. During the 1960s, in particular, Bangkok served as a major military supply base. In subsequent decades, it continued to attract large sums of foreign investment. Between 1979 and 1990, nearly 70 percent of all foreign investment projects in Thailand were concentrated in the BMR. Economically, Bangkok has capitalized on its reserves of cheap labor, favorable tax incentives, and (until recently) political stability.

Similar to Jakarta and Manila, Bangkok is a multinucleated city. And while the "old city" remains the principal administrative and religious core of Bangkok, considerable expansion has occurred into surrounding districts. Bangkok has also experienced a rapid conversion of land use, with many residential areas in the city being converted to commercial use and former small shop houses being transformed into high-rise office buildings and large shopping complexes.

The over-urbanization of Bangkok has resulted in serious environmental problems. Air and water quality have deteriorated in recent years, while the disposal of solid waste is an ongoing problem. Also, Bangkok is sinking. Due to the overdrawing of well water, the city suffers from land subsidence, as its elevation drops at a rate of about 10 cm (3.9 in) per year. Indeed, some areas have subsided by more than 3 feet (0.91 m) since the 1950s. Global warming should be firmly on the minds of Bangkok's urban planners!

Bangkok is also plagued by severe transportation problems (Figure 10.18). The number of motor vehicles (excluding the ubiquitous motorbikes) increased from 243,000 in 1972 to more than a million in 1990; concurrently, only about 50 miles of primary roads were added. As a result, the average speed on most roads in Bangkok is less than six miles per hour. Numerous proposals and strategies have been advanced to rectify traffic congestion, including increased road capacity measures, improvements in public mass-transit systems, improvements in the traffic control system, and strategies to control the volume of traffic (e.g., staggered employment hours to reduce peak commuting traffic). The government is also encouraging the growth of satellite cities as a means of promoting regional economic growth and to the congestion of Bangkok.

Phnom Penh, Ho Chi Minh City, Hanoi: Socialist Cities in Transition

In the early morning, the dusty streets of Phnom Penh are alive with swarms of noisy motorbikes that surge like schools of fish. Luxury cars and sports utility vehicles compete for limited space with the motorbikes. Plodding along the roadsides, in a vain attempt to escape the mechanized frenzy of Phnom

Figure 10.18 Traditional Manila contrast with modern Manila as the city attempts to accommodate the rapidly expanding population by going up and spilling out onto the city's streets. *Source:* Photo by Arnisson Andre Ortega.

Penh's traffic, are converted tractors with wooden trailers that ferry scores of young women—all dressed in identical green-and-white uniforms—to the foreign-owned assembly plants that ring the periphery of the city. Such is Phnom Penh in the twenty-first century: a frantic, disorderly city that is rebuilding after decades of tumultuous revolutions and genocide. The experience (and landscape) of Phnom Penh, combined with those of Ho Chi Minh City and Hanoi, provide vivid proof that urbanization is intimately associated with broader social movements.

On April 17, 1975 the Khmer Rouge marched through the hot and dusty streets of Phnom Penh, bringing to an end years of armed conflict tied to the broader Indochina Wars. Their arrival marked also the beginning of a four-year period known today as the 'Cambodian genocide'. During the Khmer Rouge-era, upwards of one-quarter of Cambodia's population—approximately 2 million men, women, and children—died from torture, execution, famine, and lack of medical care. Today, Phnom Penh still bears the scars of its genocidal past, although these are slowing being repaired. The streets of Phnom Penh, unpaved and pockmarked with potholes in 2001, now shimmer darkly with fresh albeit cracked asphalt. Where once stood hollowed-out buildings, destroyed by war and neglect, now stand freshly painted apartment buildings and shopping complexes. Phnom Penh is indeed rebuilding, though not without difficulties; and this urban growth reflects a new orientation toward the global economy. For example, along the major road linking Phnom Penh and Cambodia's Pochentong International Airport, multinational corporations have established a visible presence, in the form of assembly plants and factories. Also notable are the myriad satellite cities developed or in the process of development (Box 10.4).

In Vietnam, similar changes are underway. The population of Saigon (now Ho Chi Minh City), like that of Phnom Penh, had also increased dramatically through in-migration and refugee flows. By 1975, Saigon had an estimated 4.5 million people. Following the communist victory, the new government of the Democratic Republic of Vietnam relocated about 1 million people.

The process of deurbanization in these socialist countries was often accompanied by a refashioning of the cities. Initially, Western-style establishments and customs were replaced with a Spartan milieu. Cities, most notably Hanoi, were drab and monotonous, composed of row upon row of uniform, boxlike buildings. Conforming to socialist ideology, the new governments attempted to eliminate the private sector; and shops, restaurants, hotels, and services were generally run by government enterprises or cooperatives. Consequently, cities were typically devoid of the mass advertising and consumer spectacles that are commonplace in capitalist cities. The new governments fostered symbolic changes as well. The renaming of Saigon to honor Ho Chi Minh, who led the fight against the French and established communism in Vietnam, provides the clearest illustration. Another visible difference between socialist and capitalist cities was the traffic. In Hanoi, the streets were practically empty of motor vehicles, save for an occasional Soviet-era limousine and a few battered and decrepit buses. Instead, bicycles thronged the streets, especially during peak hours, when residents cycled to and from work and school.

Following this initial stage, socialist governments entered into a second, bureaucratic stage wherein longer-term strategies of socialist urbanization were implemented. Especially in Vietnam, the socialist government developed spatial strategies to ameliorate the problems of large cities, including the provision of adequate food, employment, and housing. Policies were enacted to restrict population mobility, thereby affording relief to the infrastructure of large urban areas, such as Ho Chi Minh City and Da Nang.

Although economic reforms were first introduced in Vietnam in 1979, it was not until 1986, with the initiation of *doi moi* (renovation), the slogan for the government's new development strategy, that substantial economic improvement occurred and with it, rapid changes to the urban landscape. *Doi moi* entails the gradual introduction of capitalist elements, including private ownership, foreign investment, and market competition. Vietnam remains politically committed to socialism, but economically the country is exhibiting a shift toward capitalism and a greater level of integration into the global economy.

Geographically, economic reforms initially focused on the southern region of Vietnam, and especially Ho Chi Minh City, because of that city's much longer tradition with free-market economics and linkages with the outside world. Approximately 80 percent of all foreign investment flowing into Vietnam was directed toward the south. Investments in tourism, assembly, and manufacturing were concentrated in the larger urban areas. Ho Chi Minh City (still called Saigon by many

Box 10.4 Satellite Cities in Southeast Asia

Located on the peri-urban fringe of Ho Chi Minh City, Phu My Hung ("Saigon South") is a newly developed satellite city covering over 3,300 hectares of prime real estate. It is also a highly circumscribed and contradictory space. When fully built, it will include exclusive residential areas, university centers, high-tech zones, and innumerable parks, golf courses, and sports centers. However, surveillance is pervasive, with watchful closed-circuit television monitors, patrolling armed guards, and countless security gates. Visitors are even restricted, if not prohibited, in their ability to take photographs of the new city.

Phu My Hung is one of many new satellite cities being planned or developed throughout Southeast Asia. Ranging in size from a few hundred to several thousand hectares, these modern monuments to the power of money are a both a legacy of past practice and a harbinger of things to come. On the one hand, the development of satellite cities is a continuation of the growth of "new towns" that sprouted up throughout the region in the aftermath of World War II. In Singapore and Malaysia, for example, many new towns were constructed in response to increased population pressures resulting from rapid industrialization and high-levels of rural-to-urban migration. On the other hand, these satellite cities reflect a very different process, symptomatic of deeper neoliberal changes underway throughout Southeast Asia. Notably, the satellite cities of the twenty-first century are being developed mainly by private companies, many of which are located throughout East and Southeast Asia, on a for-profit basis. Phu My Hung, for example, is financed by Taiwanese developers; Camko City, proximate to Phnom Penh, is being constructed by a South Korean firm; and Grand Phnom Penh International City is being developed by an Indonesia Company.

Satellite cities are by their very nature exclusive developments. Marketed to a growing upper-middle-class as well as wealthy expatriates, these cities boast all the amenities of urban living with (supposedly) none of the associated problems. Camko City, for example, is being constructed at an estimated cost of US$2 billion. Set to be completed in 2018, Camko City will host private villas, townhouses, and several high-rise condominiums, all serviced by high-speed information and telecommunication lines, electronic security systems, and sustainable environmental systems. Its commercial area will include a convention center, exhibition center, financing center, and trade center, all served by top-of-the-line hotels.

The governments of Vietnam, Cambodia, and other recipient countries have been hugely supportive of these private initiatives. On the one hand, it is thought (or hoped!) that satellite cities will relieve the congestion of primate cities and, on the other hand, the peri-urban fringe is increasingly viewed as a key site for revenue generation. In this respect, the growth of satellite cities constitutes a twenty-first century landgrab that will irrevocably alter the urban *and rural* landscape of Southeast Asia.

The rapid and pervasive growth of satellite cities throughout Southeast Asia is not unique, in that it mirrors the growth of similar mega-cities throughout the Middle East and other parts of the world. Termed "Dubaisation," after the monumental growth projects of Dubai on the Persian Gulf, the establishment of satellite cities threatens the urban and cultural heritage of Southeast Asia. Simply put, these satellite cities are in many respects "non-places". They are distinguished primarily by their non-distinction. Often modeled on plans evocative of American suburbs, there is a remarkable un-remarkability to these developments. Western-based chains, including McDonald's, Starbucks, Pizza Hut, and Kentucky Fried Chicken intermingle with imposing, glass-enclosed shopping centers and massive, cookie-cutter homes and condominiums.

The establishment of satellite cities has also come with a significant human and environmental cost. Countless hundreds of thousands of families throughout Southeast Asia have been dispossessed, often forcibly, of their lands and livelihoods. Within existent urban areas, less profitable apartment buildings are being demolished to make way for upscale condominiums and exclusive villas. In the peri-urban areas, prime agricultural lands likewise are being lost, paved over to make way for high-rise apartment and concrete highways—some of which are not even open to the general public!

residents) soon returned to its prewar capitalist character, with luxury hotels—Hyatt, Ramada, and Hilton—competing side by side with government-run hotels.

Hanoi, once the sedate, regimented, and subdued political capital of Vietnam, has itself undergone significant economic transformations. By the twenty-first century, Hanoi was increasingly showing the effects of globalization and now exhibits much of the color and dynamism of its rival to the south. Hanoi continues to showcase the symbols of Vietnamese nationalism, such as the Ho Chi Minh Mausoleum and the Ho Chi Minh Museum. But it also has become a bustling metropolis, an urban forest of hotels, restaurants, bars, nightclubs, and discotheques. Tourism, with thousands of global visitors interested in seeing the heart of the Democratic Republic of Vietnam, is leading the change.

URBAN CHALLENGES

Cities in Southeast Asia are not immune to serious problems. Challenges run the gamut from health issues to environmental concerns. Arguably, though, the inability of urban residents to obtain adequate employment and housing looms among the most serious issues faced by Southeast Asia's cities. For example, in the early 2000s in Manila, approximately 40 percent of the urban population was estimated to be living in squatter settlements, with another 45 percent of the population living in slum conditions. In Bangkok, 23 percent of the population was estimated to be living in slum and squatter settlements; and in both Kuala Lumpur and Jakarta, squatters constituted approximately 25 percent of the population. It should be noted, however, that most estimates on slum residents are measured at the national level and that there are, not surprisingly, conflicting numbers of squatter/slum residents.

The rise of squatter settlements is explained by factors other than population increase. Escalating land prices, compounded by real estate speculation, for example, exacerbate the problem of housing. So, too, does the creation of artificial land scarcity. In Metro Manila, for example, large tracts of land, even within the central business district of Makati, remain vacant. And lastly, the demolition of low-cost housing units, replaced by more affluent condominiums and gated communities, results in the rise of squatter settlements.

Sadly, many governments continue to view eviction and demolition as the most effective means of confronting squatter settlements. In the Philippines, for example, more than 100,000 people were evicted from Manila each year between 1986 and 1992. Not surprisingly, a policy of relocating squatters to sites 20–50 miles (32–80 km) outside the city and placing them in high-density residential apartments proved ineffective. Only Singapore has achieved substantial results in the provision of public housing. Other cities, especially Manila, Bangkok, and Phnom Penh, trail woefully behind. While many of these governments have agencies charged with developing public housing, most lack the required economic resources and political resolve to be effective.

Many Southeast Asian governments also are unable to provide adequate services, such as clean water, sewerage, and other utilities. Only about 7 percent of Burma's urban population, for example, has access to piped water. In Jakarta, only a quarter of the population has solid-waste collection; in the remainder of the city, it is collected by scavengers. One effective strategy has been Indonesia's Kampung Improvement Program (KIP). This program is a far-reaching initiative that concentrates primarily on the improvement of infrastructure and public facilities. Specific projects include footpaths, secondary roads, drainage ditches, schools, communal bathing and shower facilities, and health clinics. Since its inception, the KIP has been expanded to more than two hundred cities throughout Indonesia and has benefited more than 3.5 million people.

Both air and water pollution pose serious health hazards to residents and visitors alike in Southeast Asian cities. Jakarta, for example, exceeds the health standards for ambient levels of airborne particulate matter on more than 170 days of the year. Moreover, topographic features may augment pollution problems. The surrounding hills of the Klang Valley around Kuala Lumpur, and the mountains ringing Manila, for example, confine pollutants and thus exacerbate air quality problems.

Water pollution, likewise, remains a major obstacle to the quality of life in Southeast Asian cities (Box 10.5). Many rivers, including the Pasig in Manila, the Chao Phraya in Bangkok, and the Ciliwung in Jakarta, are considered biological hazards. The canals and waterways in Bangkok, especially, are highly polluted from a combination of industrial and household discharge. Only 2 percent of the city's population is connected to Bangkok's limited sewerage system. Consequently, most solid waste is discharged into waterways. Compounding the problem is the fact that more than 15 percent of the garbage disposed of daily is left uncollected.

An additional problem is traffic congestion. The traffic problems of Bangkok and Manila were discussed previously. In Jakarta, likewise, private car ownership has outpaced road construction. Similar problems of congestion and pollution are being felt in Ho Chi Minh City, which is currently home to more than 2.5 million motorbikes, and, increasingly, in Phnom Penh as well. Some governments, including those of Malaysia and Indonesia, have utilized toll roads to reduce traffic congestion. Other efforts concentrate on the development of mass-transit systems, such as the construction of light-rail transit systems in both Manila and Kuala Lumpur. These projects, however, are extremely expensive and many have been temporarily halted. To this day, half-completed overpasses and bridges

Box 10.5 Water Security and Urban Wastewater

Water security is emerging as a key risk factor throughout Southeast Asia. On the one hand, water security is intimately connected to broader economic and geopolitical relations. Ongoing development projects, including the construction of mega-dams along the Mekong River and its tributaries, threaten the livelihood of millions of farmers who depend on the rivers for fishing and irrigation needs. On the other hand, water security is tied to a host of public health issues, namely access to clean, nonpolluted water for drinking, cooking, and cleaning. Here, water security overlaps with myriad other issues related to urban infrastructure, including the collection of solid waste, inadequate flood-control systems, and the treatment of wastewater. In Phnom Penh, for example, studies indicate that upwards of 20 percent of all locally grown vegetables are irrigated with untreated wastewater. Consequently, many of the city's 1.3 million residents are susceptible to untold diseases, such as cholera, typhoid, and shigellosis.

Throughout Southeast Asia, upwards of 40 million men, women, and children are at risk of serious health issues resultant from drinking untreated water, as well as cooking and cleaning in untreated water. People are also at risk of disease through the consumption of fish caught in polluted waterways and of fruits and vegetables grown in wastewater-fed farmlands. Indeed, the threat of urban wastewater is part-and-parcel of a "water-food-energy" nexus, whereby water availability is a serious constraint both on the growing of food—via agriculture and aquaculture—and the preparation of food. For many farmers, such as those living around Phnom Penh's Boeung Cheung Ek Lake, untreated wastewater is a preferred source of irrigation, in that the water is laden with nutrients that serve as a natural fertilizer. However, this same source of energy poses significant health problems. This is but one of the many challenges confronting Southeast Asia's urban and peri-urban areas.

Governments and other stakeholders throughout the region recognize the complexity of solving urban wastewater problems. However, the scale and scope of the bureaucratic response varies greatly, ranging from those governments that exhibit effective and compressive monitoring of the quality and quantity of water resources to other governments that are woefully deficient. Singapore provides the most wide-ranging water-management plan, whereas in Indonesia there is no consolidated water policy for urban areas.

The provision of clean, safe water is an expensive prospect. Water treatment plants are expensive to build and to maintain. So are necessary infrastructural improvements, for example, those designed to control urban flooding and thus minimize contamination of water supplies. There are, however, success stories. In the mid-1990s Metro Manila's water-management was in disarray. Decades of underinvestment had led to poor water and wastewater services that contributed, not surprisingly, to numerous health problems throughout the city. In recognition of this problem, in 1995 the National Water Crisis Act was passed, ushering a new era of government-private partnership that enabled revenues to flow to needed infrastructural development. In turn, this led to improvements both in access to water for millions of residents and—equally important—greater efficiencies in wastewater treatment. Although not a panacea, the efforts of numerous stakeholders throughout Metro Manila provide hope that solutions are possible.

in Bangkok, and incomplete rail systems in Manila, stand as silent reminders of continued underdevelopment.

Amid all these problems, there are efforts by governments, along with local businesses, to promote sustainable urban development. For instance, several environmental laws have been passed in different countries seeking the reduction of pollution or to respond to issues of climate change. In some cities, city planning programs pertaining to urban renewal and the promotion of sustainable urban life have recently been put in place. In Singapore, for example, the Ministry of National Development has actively directed the planning and implementation of policies and infrastructure projects that aim to create a sustainable city of knowledge, culture, and excellence. In other cities, new development projects are advertised as being constructed to satisfy global environmental standards. The real challenge, however, lies in the enforcement of these programs and policies so that they may truly contribute to a sustainable urban life. In fact, many residents are actually displaced in the name of urban renewal projects. Despite the emergence of new urban developments that are intended to promote "urban sustainability," many inhabitants of these cities remain mired in poverty.

Rampant poverty contributes to other serious problems, including political unrest, violence, and terrorist activity. In recent years, Southeast Asian cities have witnessed class-based political tensions and demonstrations. In Bangkok, "red shirt" protesters loyal to the government clashed with "yellow shirt" demonstrators composed of people from the urban middle class. Meanwhile, in Manila, ongoing unrest, reflecting the country's socio-economic inequalities and class-based political tensions, continues to pose serious threats.

Potential violence in cities of Southeast Asia comes in other forms. According to some scholars and regional experts, the main "terrorist" threat to urban life in many Southeast Asian cities is linked to radical Islamist groups, such as Jemaah Islamiah (JI), a group that has links to Al-Qaeda and Abu-Sayyaf. Over the past decade, JI has been linked to bomb attacks in many major Southeast Asian cities. In Manila, the Abu-Sayyaf has also been linked to several bombing incidents. In response to such terrorist threats, many cities in Southeast Asia have established security programs and policies. In Singapore, a large-scale preparedness exercise, named Exercise Northstar V, was conducted. The exercise involved simulated terrorist bomb attacks in multiple locations and included the participation of thousands of government personnel and civilians. In Manila, malls, MRT, and other establishments are guarded by armed security personnel who control the entrance of people.

AN EYE TO THE FUTURE

Like Gregor Samsa in Franz Kafka's novella *The Metamorphosis*, the cities of Southeast Asia have awoken from unsettled dreams to find themselves changed into something potentially monstrous: agglomerations of skyscrapers and street vendors, palatial residential neighborhoods and impoverished squatter settlements, overburdened utilities, and underdeveloped transit systems.

What does the future hold for the cities of Southeast Asia? Three themes come to mind. First, continued population pressures and environmental degradation will most likely accelerate rural-to-urban migration, thereby exacerbating over-urbanization problems. Consequently, the cities of Southeast Asia will continue to expand geographically. How this development occurs, however, and how governments respond or manage this growth, will greatly affect the livability of these cities. Will growth continue unabated, in an unplanned, haphazard manner, or will decentralization strategies and growth-diversion measures effect desired changes? In economically poor countries, and those saddled with massive foreign debts, fiscal capacity, management, and political motivation may hinder these attempts.

A second theme is that these cities will continue to be incorporated into the global economy. This holds true especially for the socialist cities of Phnom Penh, Ho Chi Minh City, and Hanoi. Consequently, manifestations of globalization processes at the local scale will become more apparent. For example, the mushrooming of McDonald's, Starbucks, and Kentucky Fried Chicken franchises will continue (Figure 10.19). But apart from these superficial changes lie deeper, structural transformations resulting from the infusion of foreign capital. Just as political revolutions had impacts on urban areas in the socialist countries, social changes, such as the emergence of a new urban middle class, are likely to stem from and reflect back on urban transformations.

Figure 10.19 If Ronald McDonald wants to sell fast food in Bangkok, he must adapt to Thai culture. Globalization is not a one-way street. *Source:* Photo by D. J. Zeigler.

Southeast Asia, because of its strategic location and long-standing ties to the global economy, is destined to grow ever more important in world affairs. The cities of Southeast Asia will continue to transform, and be transformed by, broader global changes.

SUGGESTED READINGS

Berner, E. 1997. *Defending a Place in the City: Localities and the Struggle for Urban Land in Metro Manila*. Quezon City, Philippines: Ateneo de Manila University Press. An examination of the complex issues of land rights and squatters in overburdened Manila.

Bishop, R., J. Phillips, and W. W. Yeo. 2003. *Postcolonial Urbanism: Southeast Asian Cities and Global Processes*. New York: Routledge. A collection of essays that explores topics such as sexuality, architecture, cinema, and terrorism within the context of global urbanism.

Dale, O. J. 1999. *Urban Planning in Singapore: The Transformation of a City*. Oxford: Oxford University Press. A study of the process of urban planning in Singapore from its early growth on the banks of the Singapore River to the present.

DeKoninck, R., J. Drolet, and M. Girard. 2008. *Singapore: An Atlas of Perpetual Territorial Transformation*. Singapore: National University. A historical geography of Singapore laid out on a series of maps.

Ginsburg, N., B. Koppel, and T. G. McGee, eds. 1991. *The Extended Metropolis: Settlement Transition in Asia*. Honolulu: University of Hawaii Press. A variety of authors look at various aspects of some of the key cities in Asia.

Logan, W. S. 2000. *Hanoi: Biography of a City*. Seattle: University of Washington Press. An exploration of Hanoi's built environment and how the shape of the city reflects changing political, cultural, and economic conditions.

McGee, T. G. 1967. *The Southeast Asian City: A Social Geography of the Primate Cities of Southeast Asia*. New York: Praeger. An urban geography classic.

Ostojic, D. R., et al. 2013. *Energizing Green Cities in Southeast Asia*. Washington, DC: The World Bank. Covers urbanization and sustainable energy use, with case studies of Cebu City, Philippines, Surabaya, Indonesia, and Da Nang, Vietnam.

Sahakian, M. 2014. *Keeping Cool in Southeast Asia: Energy Consumption and Urban Air-Conditioning*. New York: Palgrave Macmillan. An investigation of energy used for air-conditioning against the backdrop of climate change.

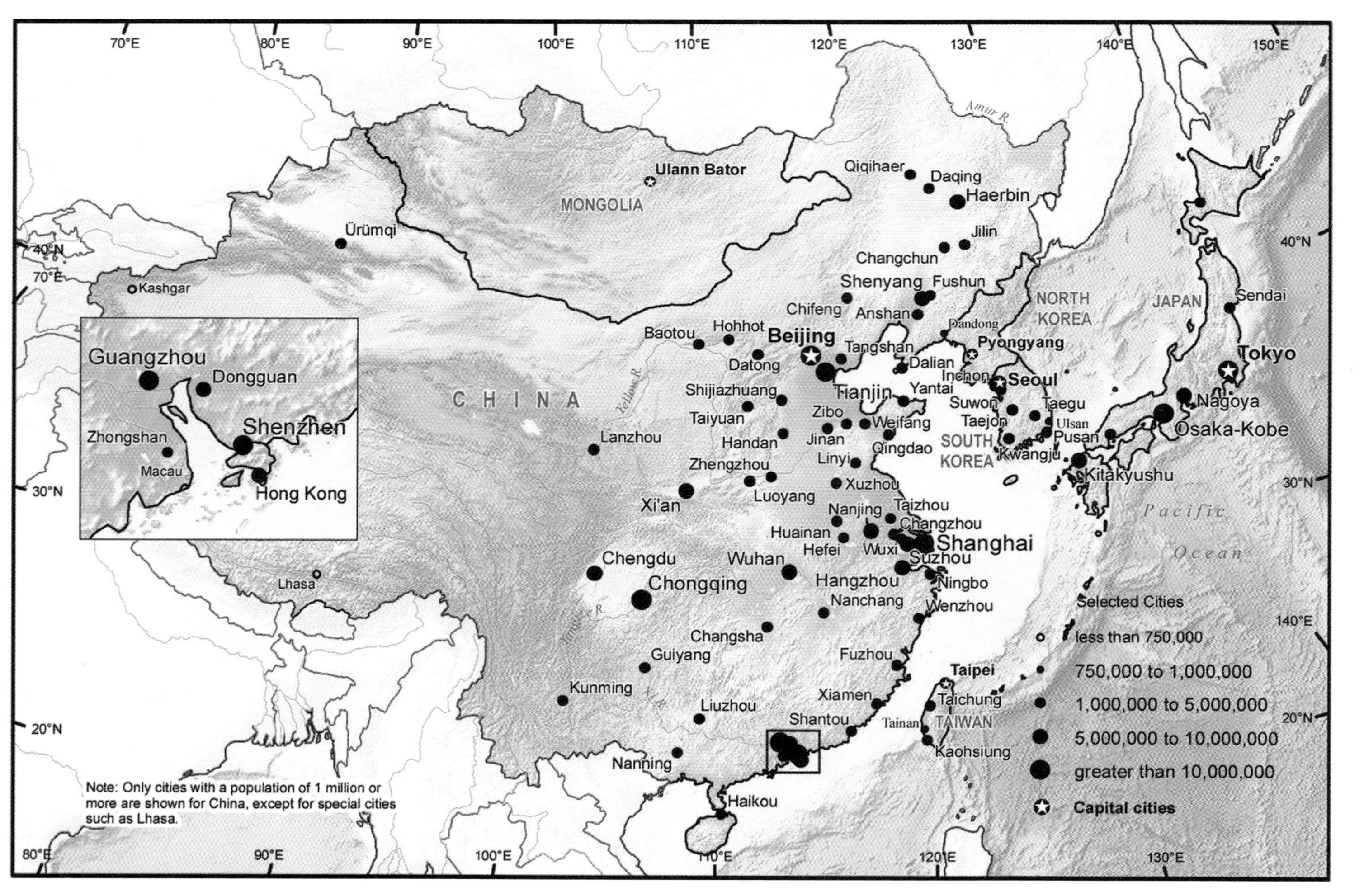

Figure 11.1 Major Urban Agglomerations of East Asia. *Source:* United Nations, Department of Economic and Social Affairs, Population Division (2014), World Urbanization Prospects: 2014 Revision, http://esa.un.org/unpd/wup/.

11

Cities of East Asia

KAM WING CHAN AND ALANA BOLAND

KEY URBAN FACTS

Total Population	1.6 billion
Percent Urban Population	59%
Total Urban Population	960 million
Most Urbanized Country	Japan (93%)
Least Urbanized Countries	China (54%)
Number of Megacities	8
Number of Cities of More Than 1 Million	123 (China has 103)
Three Largest Cities (*Metacities*)	*Tokyo* (38 m), *Shanghai* (23 m), *Osaka* (20 m)
Number of World Cities (Highest Ranking)	6 (Tokyo, Hong Kong, Shanghai, Beijing, Seoul, Taipei)
Global City	Tokyo

KEY CHAPTER THEMES

1. China is one of the original centers of urban development in history and has some of the oldest continuously occupied cities in the world.
2. Colonialism had a less important role in urban development in East Asia compared with world regions, even though many large Chinese cities were treaty ports under colonialism. Hong Kong and Macau were entirely creations of colonialism, which formally ended on the eve of the twenty-first century.
3. Japan, South Korea, Hong Kong/Macau, and Taiwan are already highly urbanized and deeply involved in the global economy, a status reflected in cities that are already in a "postindustrial" phase, with predominantly service-based economies and major high-tech sectors.
4. Since the late 1970s, China has been rapidly industrializing and urbanizing and is now the "world's factory" and a major player in the global economy.

5. China is unique in size, being the country with the largest total population, the largest urban population, and the greatest number of million-plus cities in the world. The population is divided into rural and urban citizens with different types of social welfare and opportunities.
6. Some of the world's most important world cities are in East Asia, especially Tokyo and Hong Kong. Cities such as Shanghai, Beijing, Seoul, and Taipei are in the second tier. The larger cities especially reflect the wealth of the region.
7. North Korea is the lone holdout in East Asia in clinging to a rigid, isolationist, orthodox socialist system, somewhat different from China and Mongolia, which have many market-oriented policies in the last three decades.
8. Urban development was heavily influenced by the Cold War, which lingers in the Korean Peninsula and in the ongoing tensions between Taiwan and mainland China. International trade has become another major driver of development of large cities in the last two decades.
9. Most major cities of the region show evidence of the concentric zone and multi-nucleic models of urban land use.
10. Most cities of East Asia have experienced the usual urban problems: environmental pollution, income polarization, and migration. Recent attention to environmental concerns in the region has reduced pollution, especially of air and waterways.

East Asia exudes power and success. Emerging in the past half century to rival the old power centers of the world in North America and Europe, East Asia's cities have been the command centers for the prodigious economic advances across the region. The region houses two of the world's largest economies, China and Japan, and is also the world's major exporter. Nowhere is this more evident than in East Asia's great cities, such as Tokyo, Beijing, Shanghai, Seoul, Hong Kong, and Taipei (Figure 11.1). They are among the largest cities in the world; indeed, Tokyo has been recognized as the world's largest metropolis for the last three decades, and Shanghai is now the world's largest cargo port. Compared with often-struggling urban agglomerations found in other world regions, especially in developing countries, East Asia's cities have been relatively more successful in coping with rapid growth and large size. Wealth does make a difference.

The region remains fairly sharply split between mainland China and North Korea, both of which are under a one-party system, and the rest, which is not. This dichotomy is reflected in many ways, including the character of the cities and their policies and processes, both past and present, which have shaped them. In the recent two decades, the region has witnessed a rapid rise in China's economic power and in South Korea's technological prowess, while Japan has suffered a series of fiscal and financial problems and was devastated in many ways by the 2011 earthquake-tsunami.

THE EVOLUTION OF CITIES

The Traditional or Preindustrial City

East Asia, especially China, is one of the original centers of urbanism in world history. Many cities here trace their origins back two millennia or more. One can see interesting parallels with the earliest cities in other cultural realms, with their focus on ceremonial and administrative centers planned in highly formal style to symbolize the beliefs and traditions of the cultures involved.

In its idealized form, the traditional city reflected the ancient Chinese conception of the universe and the role of the emperor as intermediary between heaven and earth. This idealized conception was most apparent in national capitals, but many elements (grid layout, highly formalized design, a surrounding wall with strategically placed gates, etc.) could be seen in lesser cities at lower administrative levels. The Tang Dynasty (618–906 CE) capital of Chang'an (present-day Xi'an) was one of the best expressions of the classic Chinese capital city. Inevitably, the demands of modern urban development have necessitated, in the eyes of planners, at least, the destruction of most city walls, thereby removing a colorful legacy of the past. The sites of old walls commonly become the routes of new, broad boulevards. One of the few cities whose original wall has been retained almost entirely is Xi'an, because of its historic role.

Of all the historic, traditional cities, none is more famous than Beijing, the present national capital of China. Although a city had existed on the site for centuries, Beijing became significant when it was rebuilt in 1260 by Kublai Khan as his winter capital. It was this Beijing that Marco Polo saw. The city was destroyed with the fall of the Mongols and reestablished by the Ming dynasty in 1368. The Ming capital was composed of four parts: the Imperial Palace (or Forbidden City), the imperial city, the inner city, and the outer city, like a set of nested boxes. The former Forbidden City can still be partially seen within the walls of what is today called the Palace Museum.

The Chinese City as Model: Japan and Korea

Chang'an was the Chinese national capital at a time when Japan was a newly emerging civilization adopting and adapting many features of China, including city planning. As a result, the Japanese capital cities of the period were modeled after Chang'an. Indeed, the city as a distinct form first appeared in Japan at this time, beginning with the completion of Keijokyo (now called Nara) in 710. Although Nara today is a small prefectural capital, it once represented the grandeur of the Nara period (710–784). Keiankyo (modern-day Kyoto) has survived as the best example of early Japanese city planning. Serving as national capital from 794 to 1868, when the capital was formally shifted to Edo (now Tokyo), Kyoto still exhibits the original rectangular form, grid pattern, and other features copied from Chang'an. However, modern urban/industrial growth has greatly increased the city's size and obscured much of the original form. Moreover, Chinese city morphology, with its rigid symmetry and formalized symbolism, was alien to the Japanese culture. Even the shortage of level land in Japan tended to work against the full expression of the Chinese city model.

Korea also experienced imported Chinese city planning concepts. The Chinese city model was most evident in the national capital of Seoul, which became the premier city of Korea in 1394. The city has never really lost its dominance since. Early maps of Seoul reveal the imprint of Chinese city forms. Those forms were not completely achieved, however, in part because of the rugged landscape around Seoul, which is located in a basin north of the lower Han River. Succeeding centuries of development and rebuilding, especially in the twentieth century during the Japanese occupation (1910–1945) and after the Korean War

(1950–1953), obliterated most of the original form and architecture of the historic city. A modern commercial and industrial city has arisen on the ashes of the old city, reinforcing the popular name for Seoul, the "Phoenix City," after the mythological bird that symbolizes immortality.

Colonial Cities

The colonial impact on East Asia was notable though relatively less intrusive than what occurred in Southeast and South Asia.

First Footholds: The Portuguese and the Dutch

The Portuguese and the Dutch were the first European colonists to arrive in East Asia; the Portuguese were much more important in their impact in this region, the Dutch largely confining themselves to Southeast Asia. Seeking trade and the opportunity to spread Christianity, the Portuguese penetrated part of southern Japan via the port of Nagasaki in the sixteenth century. Their greatest influence was indirect, with the introduction of firearms and military technology into Japan. This led to the development of stronger private armies among the *daimyo* (feudal rulers) of Japan, which in turn led to the building of large castles in the center of each *daimyo's* domain. These castles, modeled after fortresses in medieval Europe, were commonly located on strategic high points, surrounded by the *daimyo's* retainers and the commercial town. These centers eventually served as the nuclei for many of the cities of modern Japan (see Figure 11.4).

The Portuguese also tried to penetrate China. Reaching Guangzhou (Canton) in 1517, they attempted to establish themselves there for trade purposes, but were forced by the Chinese authorities to accept the small peninsula of Macau, near the mouth of the Pearl River, south of Guangzhou. Chinese authorities walled off the peninsula and rent was paid for the territory until the Portuguese declared it independent from China in 1849. With only 12 square miles (30 km^2) of land, Macau remained the only Portuguese toehold in East Asia, especially after the eclipse of their operations in southern Japan in the seventeenth century. Macau was important as a trading center and haven for refugees. Establishment of Hong Kong in the nineteenth century on the opposite side of the Pearl River estuary signaled the beginning of Macau's slow decline, from which it has never fully recovered (Figure 11.3).

In the post-1950 era, Macau survived largely on tourism and gambling (a downscale Asian version of Las Vegas, gangsters and all). In the 1990s, Macau attempted some modest industrialization, as it integrated economically with the Zhuhai Special Economic Zone across the border. Since reversion to the People's Republic of China (PRC) in 1999, emphasis has been on gambling and tourism, with additional investments made by Nevada gambling interests and by the construction of a number of new, gaudy casinos around the reconstructed harbor front that pull in large numbers of gamblers, especially *nouveau riche* from mainland China. Right next to the emerging casino quarter lies the historic heart of old Macau whose colonial-era architecture is now restored and is a pedestrian-only tourist destination.

The Treaty Ports of China

It was the other Western colonial powers, arriving in the eighteenth and nineteenth centuries, which had the greatest impact on urban growth in modern China. The most important were the British and Americans; but the French, Germans, Belgians, Russians, and others were also involved, as were the Japanese, who joined the action toward the close of the nineteenth century.

It all began officially with the Treaty of Nanjing in 1842, which ceded to Britain the island of Hong Kong and the right to reside in five ports—Guangzhou, Xiamen, Fuzhou, Ningbo, and Shanghai. Further refinements of this treaty in succeeding years gave to the other powers the same rights as the British. A second set of wars and treaties (1856–1860) led to the opening of additional ports. By 1911, approximately 90 cities of China—along the entire coast, up the Changjiang (Yangtze) River valley, in North China, and in Manchuria—with a third of a million foreign residents, were opened as treaty or open ports (Figure 11.2).

The treaty ports introduced a new order into traditional Chinese society. The Westerners were there to make money, but they also had the right of extraterritoriality, which meant that they were not subject to Chinese laws. Gradually taxation, police forces, and other features of the municipal government, including infrastructure, were developed by the colonial countries controlling the treaty ports. China's sovereignty thus was supplanted in the concession areas of treaty ports, as foreigners for modest rents paid to the Chinese government leased these areas in perpetuity.

The most important treaty port was Shanghai ("On the Sea"), which had existed as a small settlement for two millennia. By the eighteenth century, the city was a medium-sized county seat with a population of about 200,000 and built in traditional city style, with a wall. The deposition of silt by the Changjiang River over the centuries, however, made Shanghai no longer a port directly fronting the sea. The town was now located about 15 miles (24 km) up the Huangpu River, a minor tributary of the Changjiang.

Shanghai profited from its natural locational advantage near the mouth of the Changjiang delta for handling the trade of the largest and most populous river basin in China. During the twentieth century, Shanghai's manufacturing competed successfully with other manufacturing centers emerging in China, despite the absence of local supplies of raw materials, because of the ease and cheapness of water transport. This was achieved in spite of a relatively poor site—an area of deep silt deposits, a high water table, poor natural drainage, an insufficient water supply, poor foundations for modern buildings of any great height, and a harbor on a narrow river that required dredging to accommodate oceangoing ships. Shanghai became one of the best examples worldwide of how a superb relative location can trump a poor physical site to create a great city.

The Japanese Impact

The Japanese also greatly influenced the urban landscape in their two other colonies in East Asia. During their rule of Taiwan (1895–1945) and Korea (1910–1945), the Japanese introduced Western-style urban planning practices, filtered through Japanese eyes, that they later brought to Manchuria's cities. Taipei was made the colonial capital of Taiwan and transformed from an obscure Chinese provincial capital into a modern city. The city wall was razed, roads and infrastructure were improved, and many colonial government buildings were constructed. The most prominent was the former governor's palace, with its tall, red-brick tower, which still stands in the heart of old

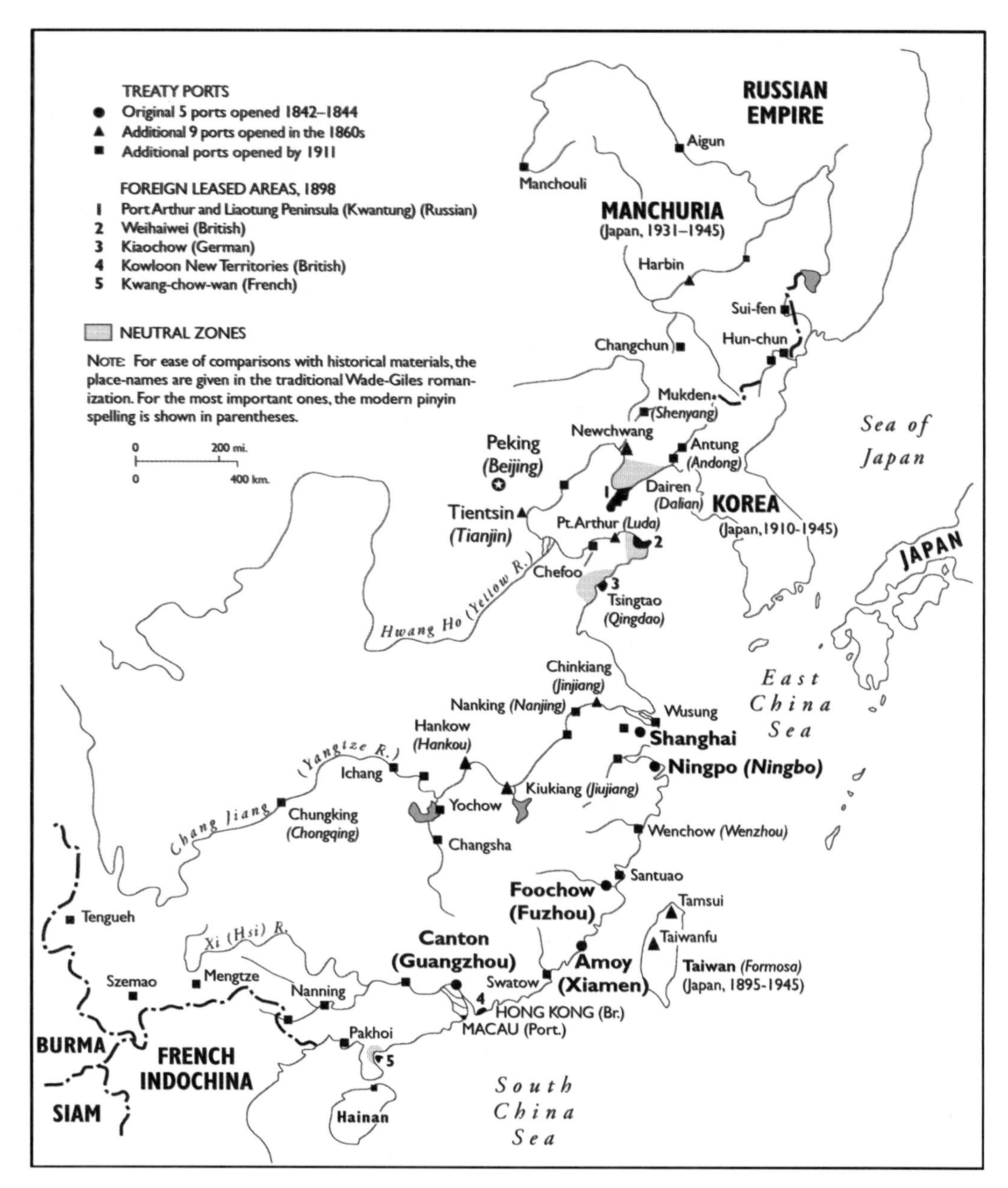

Figure 11.2 Foreign Penetration of China in the Nineteenth and Early Twentieth Centuries. *Source:* Adapted from J. Fairbank et al., *East Asian Tradition and Transformation* (Boston: Houghton Mifflin, 1973), 577.

Taipei, and is now used as the Presidential Office and executive branch headquarters for Taiwan's government. Like Taipei, Seoul was also transformed to serve the needs of the Japanese colonial rule of the Korean Peninsula. In Seoul's case, however, this meant deliberately tearing down traditional palaces and other structures to be replaced by Japanese colonial buildings as part of a brutal effort to stamp out Korean resistance to Japanese rule.

Hong Kong

Hong Kong ("Fragrant Harbor") differed from other treaty ports in that there was little pretense of Chinese sovereignty there (though the Chinese government insisted after 1949 that Hong Kong was part of China). Hong Kong was ceded to Britain at the same time Shanghai was opened up in the early 1840s. Hong Kong became second only to Shanghai as the most important entrepôt on the China coast during the following century of colonialism.

The importance of Hong Kong was not difficult to discover. In 1842, the city began with the acquisition of Hong Kong Island (Figure 11.3), a sparsely populated rocky island some 70 miles (113 km) downstream from Guangzhou. The Kowloon peninsula across the harbor was obtained in a separate treaty in 1858. Then, in 1898, the New Territories—an expanse of islands and land on the large peninsula north of Kowloon—were leased from China for 99 years (hence, reversion to China took place in 1997), creating a total area of about 400 square miles (1040 sq km) for the entire colony. The site factor that so strongly favored its growth was one of the world's great natural harbors (Victoria Harbor), between Hong Kong Island and Kowloon. Indeed, the advantages of the harbor

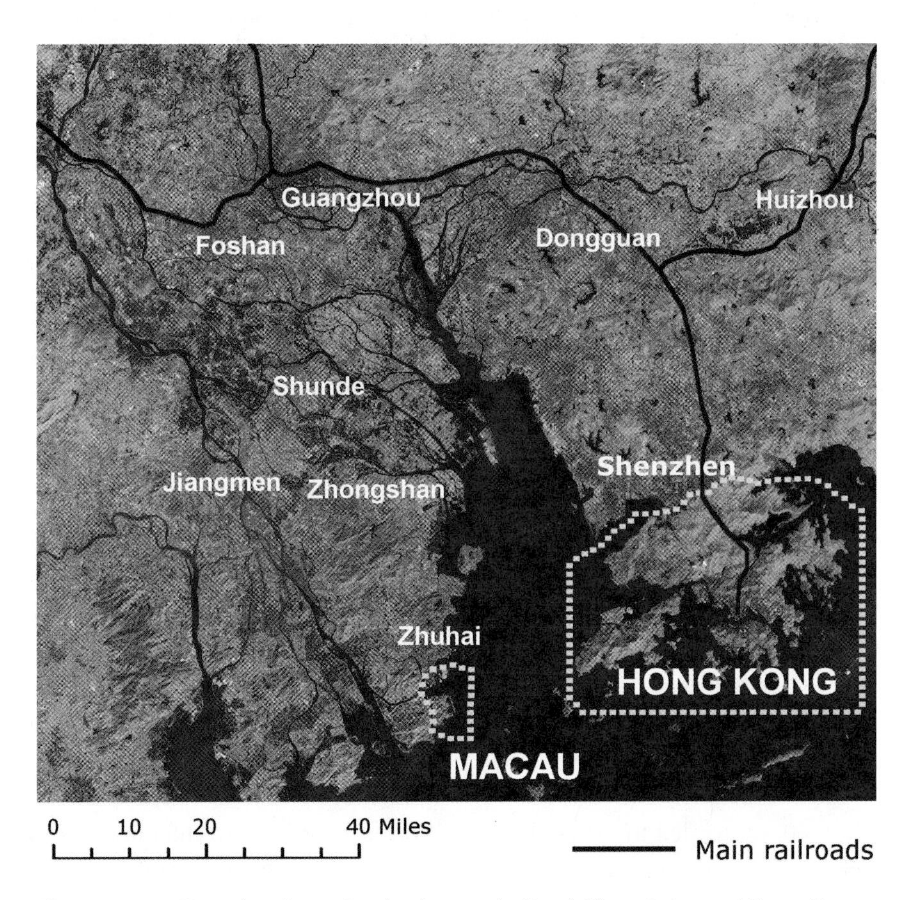

Figure 11.3 Map showing urbanized areas in Pearl River Delta and Hong Kong. Pink represents urban areas. *Source:* Based on 2015 Landsat data.

outweighed the disadvantages—limited level land for urban expansion, inadequate water supply, and insufficient adjacent farmland to feed the population. The city's location at the mouth of southern China's major drainage basin gave Hong Kong a large hinterland, which greatly expanded when the north-south railway from Beijing was pushed to Guangzhou in the 1920s. Thus, for about a century, Shanghai and Hong Kong, two great colonial creations, largely dominated China's foreign trade and links with the outside world.

Japan: The Asian Exception

Following the classic capitals of Nara and Kyoto around the eighth century, other cities followed in Japan, principally the centers of feudal clans. Most of these were transitory, but a sizable number have survived today.

Japan is referred to as the "Asian exception" because it had only a minor colonial experience internally. Indeed, Japan was itself a major colonial power in Asia. Hence, the urban history of Japan involved an evolution almost directly from the premodern, or traditional, city to the modern commercial/industrial city. Japan did have treaty ports and extraterritoriality imposed on it by the Treaty of 1858 with the United States, which led to foreigners residing in Japan as they did in China. However, this colonial phase was shortlived. Japan was able to change its system and reestablish its territorial integrity by emulating the West. Extraterritoriality came formally to an end in 1899, as Japan emerged an equal partner among the Western imperial powers. Gradual political unification during the Tokugawa period (1603–1868) led to the establishment of a permanent urban network in Japan. The castle town served as the chief catalyst for urban growth. One of the most important of these new castle towns to emerge at this time was Osaka. In 1583, a grand castle was built that served as the nucleus for the city to come. Various policies stimulated the growth of Osaka and other cities, including prohibitions on foreign trade after the mid-1630s, the destruction of minor feudal castles, and prohibitions on building more than one castle in each province. These policies had the effect of consolidating settlements and encouraging civilians to migrate to more important castle communities.

The new castle towns, such as Osaka, were ideally located (Figure 11.4). Because of their economic and administrative functions, they generally were located on level land near important landscape features that were advantageous for future urban growth. Thus, Osaka emerged as the principal business, financial, and manufacturing center in Tokugawa Japan. The Japanese cities of that period were connected by a highway network that stimulated trade and city growth. The most famous of these early roads was the Tokaido Highway, running from Osaka to Nagoya, which emerged as another major commercial and textile manufacturing center to the most important city of this period and after—Edo (Tokyo).

Among the major cities of Asia, Tokyo was a relative latecomer. It was founded in the fifteenth century, when a minor feudal lord built a rudimentary castle on a bluff near the sea, about where the Imperial Palace stands today. The site was good for a major city—it had a natural harbor, hills that could easily be fortified, and room on the Kanto Plain for expansion. Tokyo really began a century later, when Ieyasu, the Tokugawa ruler at that time, made Edo his capital. Part of Tokyo still bears the imprint of the grand design that Ieyasu and his descendants laid out. They planned the

Figure 11.4 The Osaka castle in the center of Osaka city played a major role in the unification of Japan during the sixteenth century.
Source: Photo by Kam Wing Chan.

Imperial enclosure, a vast area of palaces, parks, and moats in the heart of the city. Much of central Tokyo's land was reclaimed from the bay, a method of urban expansion that was to typify Japanese city-building from then on, reflecting the shortage of level land and the need for good port facilities. By the early seventeenth century, Edo had a population of 150,000 surrounding the most magnificent castle in Japan. By the eighteenth century, the population was well over 1 million, making Edo one of the world's largest cities.

Edo's growth was based initially on its role as a political center, tied to the other cities by an expansive road network. An early dichotomy was established between Osaka, as the business center, and Tokyo. With the restoration of Emperor Meiji in 1868, Japan's modern era began. The emperor's court was moved from Kyoto to Edo, which was renamed Tokyo ("Eastern Capital") to signify its role as national political capital. This transfer of political functions, plus the great industrialization and modernization program that was undertaken from the 1870s, gave Tokyo the boost that began its growth that continues into the twenty-first century.

INTERNAL STRUCTURE OF EAST ASIAN CITIES

It is not easy to generalize about the internal structure of cities in East Asia. This is partly because of the basic division between socialist and non-socialist urban systems that has characterized the region. It also is because of the lack of fit of Western urban models even for non-socialist cities of the region. In most of East Asia, and increasingly also in China after 1979, the forces that produced and continue to shape cities are similar to those in the West, but with a major difference: China and North Korea remain under a one-party authoritarian system and a highly state-controlled economy. These forces include: (1) rapid urban industrialization combined with increasing urban-rural inequalities, leading to high rates of rural-to-urban migration, rates which have now decreased in more developed economies (Japan, South Korea, Taiwan), but that are still high in China and Mongolia; (2) private ownership of property and dominance of private investment decisions affecting land use; (3) high-density development and heavy reliance on public transport systems despite rising car ownership rates; and (4) significant stratification of socioeconomic classes, especially between locals and migrants. These and other factors impact urban growth, how space is used in cities, and the types and severity of problems.

DISTINCTIVE CITIES

With the exception of British Hong Kong and Portuguese Macau, the colonial era in East Asia ended with the defeat of Japan in 1945. The emergence of communist governments in the late 1940s in China and North Korea, joining the already communist government of Mongolia (established in the 1920s), split the region into two distinctly different camps: the socialist cities of China, North Korea, and Mongolia, versus the market-economy cities of Japan, South Korea, Taiwan, Hong Kong, and Macau. This affected Cold-War era alignments. In the late 1970s, China entered the post-Mao or Reform Era, in which market forces began to play a more significant role in the economy and urban development. Only North Korea remained wedded to rigid, orthodox "socialism."

One can also classify the major cities of the region on the basis of function and size. From this perspective, several cities are distinctive: megalopolises or super-conurbations (Tokyo); recently decolonized cities (Hong Kong); primate cities (Seoul); and socialist cities (Beijing and Shanghai) and regional centers (Taipei) undergoing rapid transformation (Figure 11.5).

Figure 11.5 With Taipei 101, Taiwan's capital reaches for new skylines, in stark contrast to twentieth-century socialist-era development. *Source:* Photo by D. J. Zeigler.

Tokyo and the Tokaido Megalopolis: Unipolar Concentration

Japan illustrates especially well the phenomenon of super-conurbations or megalopolises. A distinctive feature of Japan's urban pattern is the concentration of its major cities into a relatively small portion of an already small country. Despite over a century of industrialization, Japan was not more than 50 percent urbanized until after World War II. By 2010, Japan's rate of urbanization had reached over 90 percent, making it the most urbanized country in the region. As the urban population grew dramatically, so did the number and size of Japan's cities. Small towns and villages (those with fewer than 10,000 people) declined sharply in numbers and population, while medium and large cities grew rapidly, all due to Japan's phenomenal economic postwar growth.

Almost all major cities are found in the core region, which consists of a narrow band beginning with the urban node of the tri-cities of Fukuoka, Kitakyushu, and Shimonoseki at the western end of the Great Inland Sea. Within this core is an inner core, containing more than 50 percent of Japan's total population of 128 million, known as the Tokaido Megalopolis, a 300-mile long regional urban belt connecting the Tokyo and Osaka metropolitan areas. Rapid growth from the late 1950s to the early 1970s occurred through the migration of millions of youth from rural areas. Since then, migration and growth have increased toward Tokyo at the expense of the rest of the country, including Tokyo's longstanding rival, Osaka, a phenomenon dubbed unipolar concentration (i.e., urban primacy). Tokyo continues to expand, draining people and capital investment from other regions, many of which are stagnating. The Osaka region has not seen much industrial growth to replace the smokestack industries, such as steel and shipbuilding, and Osaka businesses continue to relocate to Tokyo. This encourages out-migration and depresses personal consumption. Nagoya has fared better than Osaka by maintaining employment and central city vitality. However, for young people especially, the economic and cultural pull of Tokyo shows little sign of dissipating.

With over 36 million people, Tokyo's nearly 30 percent share of Japan's total population is concentrated on 4 percent of the nation's land area. Tokyo has a disproportionate share of workers, factories, headquarters of major corporations, financial services, institutions of higher education, industrial production, exports, and college students. Tokyo is the center for major governmental functions and all 47 prefectural governments have branch offices in Tokyo to liaise with the national government, further bolstering its national dominance.

During the late twentieth century, Tokyo's city core hollowed out due to suburbanization and deindustrialization (Figure 11.6). There was an increase in office and commercial buildings for business, government, and retail, while residential areas lost numbers, with many elders living alone while children started their families in the suburbs (Box 11.1). A reversal of this migration pattern began in the early 2000s after land prices bottomed out during Japan's recession. People were again able to afford living in Tokyo's central city districts, which were being reshaped by a surge in condominium developments targeting singles, working couples, and wealthier retirees. Government policies promoted conversion of former industrial lands and waterfront development to accommodate housing, research institutions, business and commercial use, and light industry. These changes in Tokyo's urban

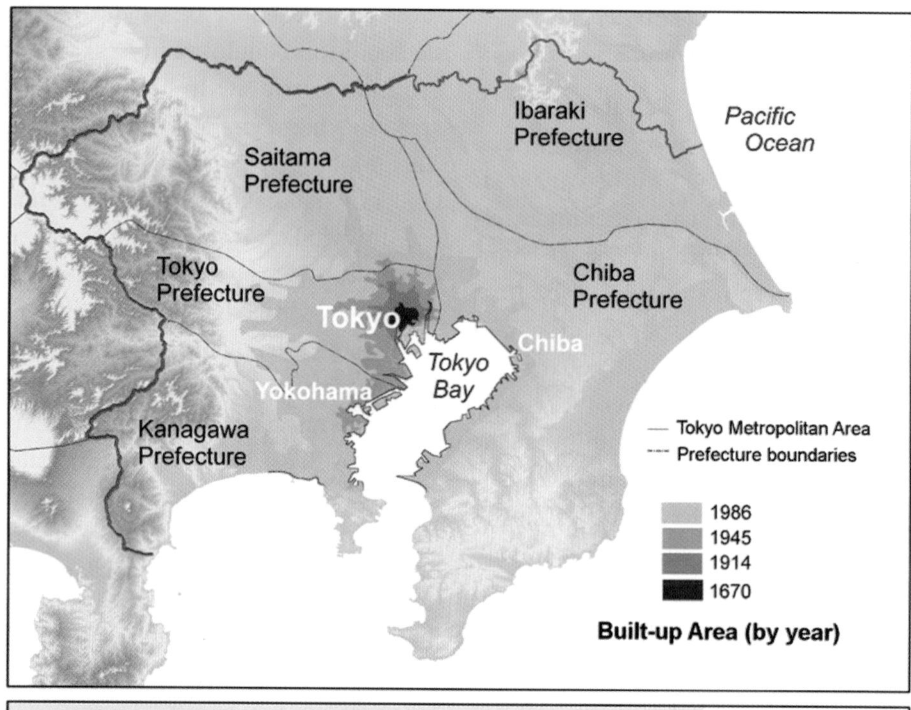

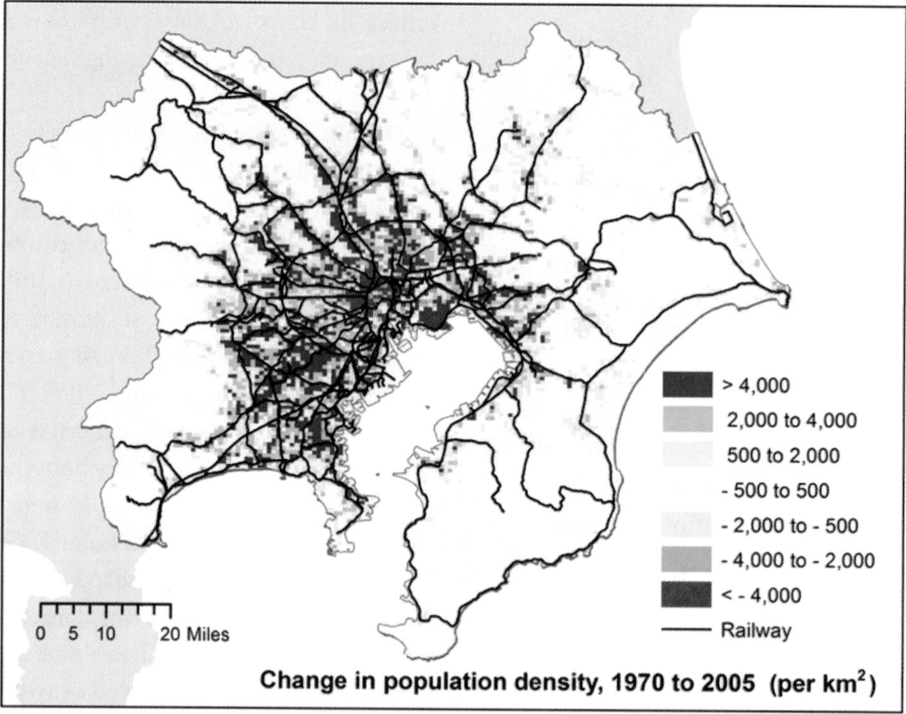

Figure 11.6 Tokyo Metropolitan Area and change in population density, 1970–2005. *Source:* Adapted from H. Bagan and Y. Yamagata, "Landsat Analysis of Urban Growth: How Tokyo Became the World's Largest Megacity During the Last 40 Years," *Remote Sensing of Environment* 127 (2012): 218.

Box 11.1 Japan's Aging Cities

Japan's aging cities are facing a looming crisis. The most serious challenges ahead are not, however, associated with aging buildings and infrastructure, but rather the country's aging population. Japan faces an unprecedented drop in population related to gains in longevity and declining fertility, both occurring without large-scale in-migration to help counteract these trends, like what is happening in other countries facing a similar demographic shift. Projections suggest that close to 30 percent of Japan's population will be 65 and over by 2030. In a country where 90 percent of the population is urbanized, this has profound implications for cities.

The migration out of the central districts of cities like Tokyo that occurred in the 1970s and 1980s was driven primarily by younger couples, many who left their parents to live in the old family home. This made good sense. Older urban neighborhoods offered small-scale streets, local shopping, access to public transit, stronger social ties, and familiarity of place. But this also caused the urban cores to be populated disproportionately with older residents, many living alone in small houses or low-rise apartments. Such a concentration creates problems for governments in terms of social services and health-care provision, but also creates challenges for older inner-city redevelopment. Conflicts often emerge around planning of high-rise condominium construction. Neighbors come together to protest the diminished sunlight for houses surrounding the proposed high-rises. They worry also about changes to the social mix of neighborhoods and loss of intimacy that has developed between people who have lived together for decades. In some cases, these protests against high-rise projects lead also to calls on governments to improve livability in the older areas of the city. Sadly, however, these efforts are sometimes hindered by the very composition of the participants; in what can be quite protracted struggles against large property developers, some elderly activists are unable to sustain their fight to protect the community. And in a sad irony, while local governments have begun to pay greater attention to quality of life issues, in so far that this shift is linked to a global city agenda, the people that Tokyo and other cities are trying retain and attract are not necessarily the elderly.

If one considers metropolitan areas as a whole, including the extended suburbs, the problems of Japan's rapidly aging population become more complex. While inner suburbs have seen some increase in density, especially along commuter rail lines, the outer suburbs have experienced an exodus of younger people wishing to live closer to the city center for economic and social reasons. This is creating a high concentration of elderly living alone in certain suburban areas, mirroring the aging process in the inner-city neighborhoods. However, unlike in the more dynamic and densely populated city center, local governments in the suburbs generally have fewer financial resources—and anticipate fewer resources as their populations decline. It will be difficult to meet the needs the elderly, who are also at greater risk of social isolation given their more dispersed residential patterns. These conditions have many now wondering about the social and economic sustainability of the "graying" suburbs outside of Tokyo and other large cities of Japan.

landscapes mirror recent trends elsewhere: the city is following the path of a global city modeled on vertical, compact, and multifunctional urban features.

Tokyo's expansive transport system has played a decisive role in development of the metropolitan area. Commuter rail lines serving the suburbs were initially privately run and the rail companies owned and developed much of the surrounding land for housing and commercial use. The lines did not run through the city core. Terminal stations were located along a 20-mile loop line that encircled the city's historical central districts and remain the connection points between suburban commuter trains and the subway, bus and other rail cutting through the city center. Many of these stations have developed into major commercial districts with shopping, entertainment, and restaurants, surrounded by large building complexes for office and residential use. Shinjuku is the busiest of these stations, with over 3 million passengers per day. The upscale shopping district Ginza is another famous commercial district that developed near the Yamanote Loop. Such nodes of high-density development give Tokyo elements of a multi-nucleic model intermixed with the radial and concentric ring patterns of commuter rail and freeways. In 2007, 65 percent of trips in the Tokyo metropolitan area were by mass transit, making transit access an important factor in many residents' decision about desirable and affordable places to live.

While residents of Tokyo enjoy access to one of the world's most extensive public transit systems, they struggle with high-density problems, congestion, pollution, and sprawl. Away from busy transit and shopping areas, Tokyo's neighborhoods can feel surprisingly small-scale as low-rise buildings still dominate the landscape (Figure 11.7). Lining major roads are mid-rise office and residential buildings about 10–12 stories tall. Away from the main streets, 2–5 story buildings and grids of narrow streets and alleys cut through densely built-up neighborhoods of small houses, low multiunit residential buildings, stores, and small factories. Pedestrians and bicycles are more common than cars. This helps reduce noise, which is important given that houses are right next to the narrow streets. Railways stations along commuter lines often serve as neighborhood hubs with supermarkets and shopping streets around them. The compact nature of these neighborhoods gives them social vitality and their form is one commonly found throughout Japan.

Beijing: The New "Forbidden City"?

Beijing, the great "Northern Capital" for centuries, was a horizontal, compact city of magnificent architecture and artistic treasures of China's past grandeur when the "New China" began in 1949, although the magnificence of the old city had suffered greatly from general neglect during the century of foreign intrusion and civil wars since the 1840s, and from the "revolutionary reconstruction" and "modernization" in the last 60 years. Centered on the former Forbidden City (Imperial Palace, of course, off limits to the commoner), Beijing was renowned for its sophisticated culture and refined society, a status linked to the city's function as the political center of a vast nation. Illustrating the city's influence, the Beijing dialect (Mandarin) became the national spoken language (*putonghua*) after the collapse of the last dynasty in 1911. However, despite its political and cultural influence, there was little industry and a relatively small population.

In 1949, the city was chosen as the national capital of the new Communist government

Figure 11.7 One of Tokyo's busy narrow side streets, with commercial and residential land use in close proximity. Streets of this size and mix are quite common still even in the busy core of Tokyo and other large Japanese cities. *Source:* Photo by Andre Sorensen.

(Nanjing was the national capital during the Republican era, from the late 1920s to 1949). Since communist takeover, the city has undergone several waves of demolition, construction, and expansion. Today, the administrative area of Beijing covers a large territory, 6,500 square miles (16,800 sq km), encompassing an urbanized core (high-density built-up area), surrounded by numerous scattered towns and large stretches of rural area and with a total population of 20 million. But this *shi* (municipality or city) is a large administrative region and not a "metropolitan area" as it is often mistakenly conceived. Delineation of the approximate commuting zones (the suburbs) and urbanized area would suggest a "metropolitan Beijing" of about 3,700 sq km (~22 percent of the administrative area), and a population of about 17 million. Natural population growth, net migration to metro Beijing, and suburbanization in the last four decades have pushed the metro boundary outward (Figure 11.8). Prior to that, however, migration control to Beijing was among the strictest in the country. Only the well educated and those needed by the central government could move to Beijing; for the rest, it remained a "forbidden city."

In the 1950s, Beijing was transformed into a major industrial center. In the pre-1980 command-economy era, other urban functions such as commerce and services were greatly curtailed. Beijing became even more like Moscow than Moscow was in the former Soviet Union.

The changes to Beijing's traditional urban landscape were enormous in the Maoist era

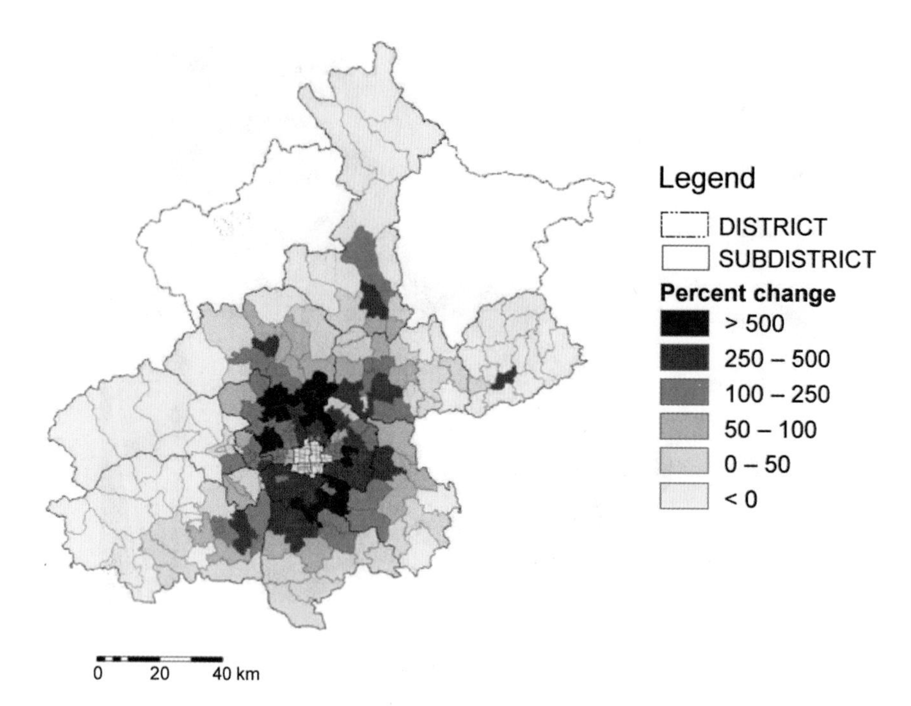

Figure 11.8 Beijing metropolitan area has been expanding outward, fueled by in-migration and local residents moving from the city center to the suburbs. The map shows population growth rates by subdistrict unit in the urbanized part of Beijing based on census data for 1982 and 2010. *Source:* Prepared by Kam Wing Chan, Richard L. Forstall and Guilan Weng.

(1949–1976), but continued to be significant in the last two decades. In the pursuit of "destroying the old and building the new" during the Maoist revolutionary era, many parts of the old city and the city walls were knocked down to make way for a new socialist capital city. These changes shattered the original form of Beijing and forever altered its architectural character. While many parts of the old city contained overcrowded courtyard houses left from the prerevolution era (Figure 11.9), the newly built section of the city under Mao was one with arrow-straight, wide boulevards and huge Stalinesque state buildings, punctuated by seemingly endless rows of unadorned low-rise monotonous apartment blocks for the masses, with emphasis on uniformity, minimal frills, and lowest possible construction costs. The city center lacked a human scale and was deliberately designed to emphasize the power of the party-state. A huge area in front of the *Tiananmen* (Gate of Heavenly Peace) was cleared to create the largest open square of any city in the world. Tiananmen Square became the staging ground for vast spectacles, parades, and rallies organized by the government. Mao and other party leaders would orchestrate the scene from on top of the gate like a latter-day imperial court. After Mao died in 1976, his body

Figure 11.9 Pockets of traditional courtyard houses remain in hutongs, or alleys, in the inner city of Beijing. Many of them have been torn down to make room for high-rise apartments and offices. Some "saved" are converted into shops in main hutongs. *Source:* Photo by Kam Wing Chan.

was embalmed and displayed inside a mausoleum on the south end of Tiananmen Square along the north-south axis running through the Palace Museum. The parallel with the display of Lenin's body in Red Square in Moscow was intentional, as was the attempt to link Mao with the imperial tradition and the role of Beijing as the center of China. Though the square was designed and used mostly by those in power, it also became the staging ground for watershed mass protests organized by students, intellectuals, and workers, from the May Fourth Movement in 1919 to the failed Pro-Democracy Movement in 1989. Elsewhere, the charm of many traditional middle-class courtyard houses in *hutongs*, or narrow alleys, in the old city was almost lost in the need to subdivide housing space for multiple families, but often without necessary updates and maintenance.

China's large cities in the Maoist era were manufacturing centers and administrative nodes of an economic planning system that focused on national, regional, and local self-reliance (Figure 11.10). Most cities tried to build relatively comprehensive industrial structures, resulting in less division of labor and exchanges than in a market economy. The huge surrounding rural areas (often confusingly called "suburban counties") grew food, mainly vegetables, for the cities. Some satellite towns accommodated industrial spillover. Without a land market, many self-contained neighborhoods, based on large enterprises, dominated the landscape of large cities, which expanded in concentric zones. Beijing was no exception.

New policies after Mao were meant to address economic weaknesses and transform Chinese cities via market reforms. Those

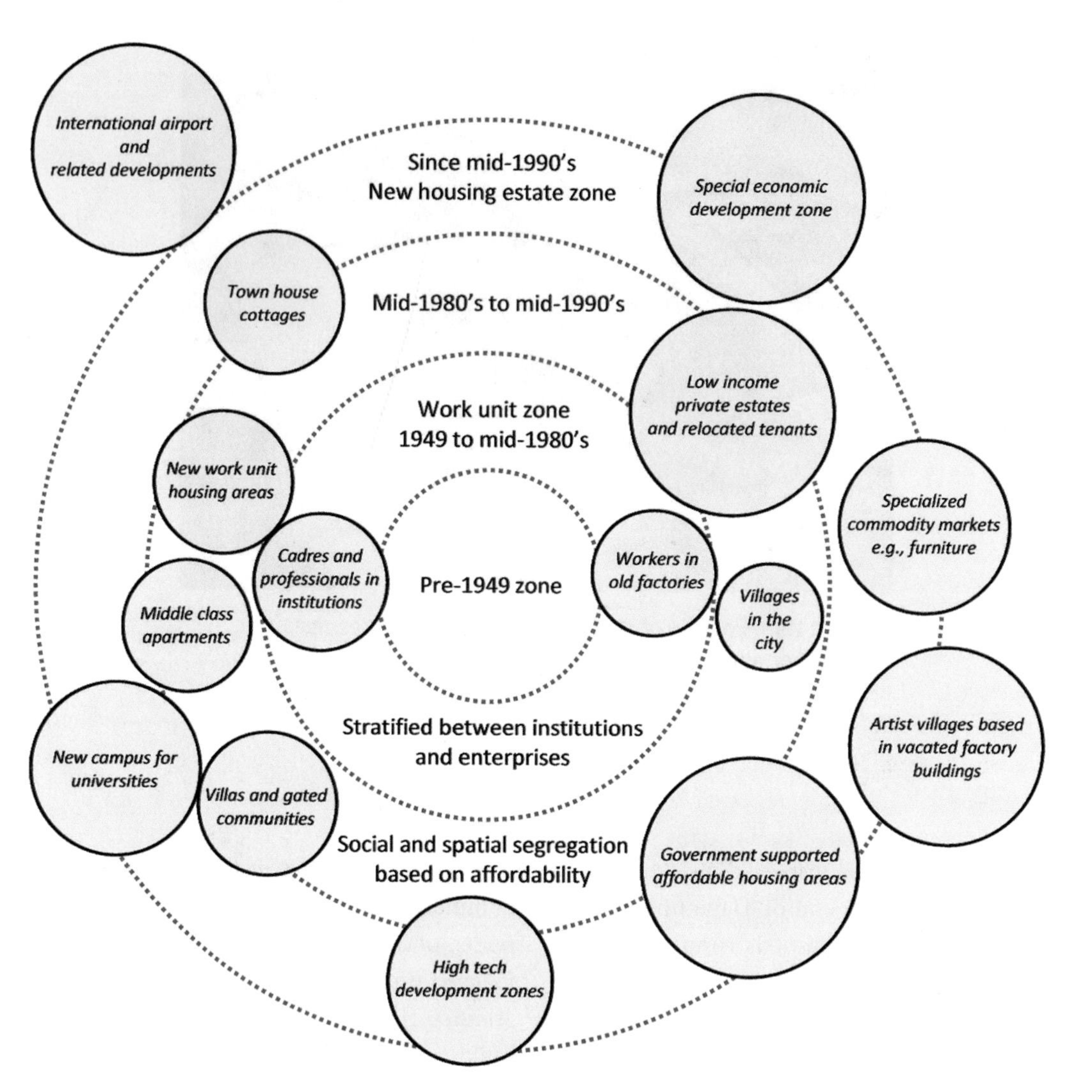

Figure 11.10 Model of the City in the PRC. *Source:* Adapted from Ya Ping Wang, *Urban Poverty, Housing and Social Change in China* (New York: Routledge, 2004).

reforms have improved living standards, especially in coastal cities, and brought an urban consumption boom. In Beijing, this resulted in the proliferation of new stores and restaurants, including mammoth malls. Beijing now also has major commercial/financial districts, such as Xidan, a busy shopping area with ultra-modern architecture and expensive shops, and Wangfujing, an old retail strip that is fully pedestrianized. In the northwestern part of the city is China's "Silicon Valley,"—Zhongguancun. Tech firms have set up main offices there to gain proximity to top universities nearby.

Beijing has pushed outward and has a sizable daily commuting zone consisting of high-rise apartments, luxury detached houses, and often dilapidated "migrant villages" as far as 40 km from the city center (Figure 11.10). Urban expansion parallels a noticeable increase in income disparities and social differentiation.

Figure 11.11 Millions of migrants eke out their living on the urban fringes of Beijing; some live in run-down village houses like this one. The photo was taken after a major rainstorm in summer 2012 in Chengzhongcun. *Source:* Photo by Wilfred Chan.

In the northern outskirts of Beijing, expensive, detached, Western-style bungalow houses have appeared, catering to expatriates and the new rich. At the same time, with relaxed migration controls since the mid-1980s, Beijing now has a large migrant population of over 7 million. These mostly rural migrants fill low-level jobs shunned by the locals. However, these migrants are not given legal residency status (*hukou*) and are often denied access to many urban services (Box 11.2). In addition, young college graduates from other cities that do not have Beijing *hukou* make their homes here. Several migrant communities have sprung up in Beijing's suburbs, such as "Zhejiang Village" and "Xinjiang Village." These communities are named after the provinces from which most of their residents come, creating regionally based urban enclaves. Living conditions in these migrant villages provide a stark contrast with those of wealthier neighborhoods (Figure 11.11).

In the last two decades, the government has implemented many programs to "beautify" and modernize the city. They range from relocating steel plants from the city and implementing strict measures to limit the use of automobiles to bring down air pollution levels, demolishing old *hutong* houses, and displacing hundreds of thousands of the city's poor for what critics call "image projects" of building numerous very expensive ultramodern architectural works. The 2008 Olympic Games provided the greatest stimulus for much beautification effort and for improvements in urban infrastructure—newer expressways and subway lines, a new airport terminal—and construction of world-class sports stadiums.

Under China's national urbanization blueprint promulgated in 2014, the government plans to channel most migrants to small and medium cities. Stringent measures are in place to deter migration to major metropolises. The measures include harder terms for school enrollment for migrant children (Box 11.3). As a result, many of them have lost their school places in big cities. The situation was

Box 11.2 "Cities with Invisible Walls:" the *Hukou* System in China

After the Communist Revolution in 1949, China opted for the Stalinist growth strategy of rapid industrialization-based extraction of agriculture. This industrialization strategy led China to create, in effect, a dual structure: on the one hand, the urban class, whose members worked in the priority and protected industrial sector and who had access to basic social welfare and full citizenship; and on the other hand, the peasants, who were tied to the land to produce an agricultural surplus for industrialization and who had to fend for themselves. This, in turn, required strong mechanisms to prevent peasants from leaving the countryside. In 1958, a comprehensive *hukou* (household registration) system was formalized to control population mobility and exclude peasants from social welfare. Each person has a *hukou* (registration status), classified as "rural" or "urban," and tied to the locale he or she stayed. The decree required that all internal migration be subject to approval by the relevant local government, but approval was rarely granted. While old city walls in China had largely been demolished by the late 1950s, the power of this newly erected migration barrier functioned as invisible but effective city walls.

Since the late 1970s, development of markets and the demand for cheap labor for sweatshop productions for the global market have led to easing of some migratory controls. Rural-*hukou* holders are now allowed to work in cities in low-end jobs shunned by urban residents, but they are still not eligible for basic urban social services and education programs. By the mid-1990s, rural-*hukou* migrant labor had become the backbone of China's export industry and the service sector. In 2014, the size of "rural migrant labor" rose to about 170 million. This two-tier system of citizenship and the unequal treatment of the migrant population have seriously divided China and created many problems. Since 2014, the government has launched another round of *hukou* reform; whether it is real this time remains to be seen.

so grave in Beijing that hundreds of migrant children and parents staged a three-month-long protest against expulsion in front of the local education commission office in summer 2014.

Shanghai: "New York" of China?

Shanghai is considered by many to be China's most vibrant city. This is because of its unique colonial heritage and because it is the center of change and new frontiers in social and economic behavior. Shanghai is the largest and perhaps the most cosmopolitan city in China and has one of the highest standards of living. Shanghai city itself is part of the Shanghai administrative region, comprising an urbanized core, suburbs, and outlying rural areas. It covers a total area of 2,400 square miles (6,300 sq km) with a population of 23 million, including about 10 million illegal migrants. With the rapid development of a national and regional intercity high-speed rail system, Shanghai is now the hub of a larger economic region comprising several metropolises such as Hangzhou and Nanjing (Figure 11.12).

Box 11.3 "Orphans" of China's Urbanization?

On the night of June 9, 2015, four children of the same family were found dead in their home in Guizhou. They were left in the poor village, without any proper care. They committed suicide together by drinking pesticide. The oldest boy was 13 years old, while his youngest sister was only 5.

In China, a new generation of children is growing up in the countryside with only one or no parents around during most of the year. Hence, they are called "left-behind children." There are more than 63 million of them in the country; half are age 6–14. They are left behind because their parents have gone to work in the city, often hundreds of miles away, as part of China's gargantuan army of rural migrant workers, estimated at about 170 million in 2014. While they work in the city, their children often cannot go with them because of various reasons. A 2014 survey estimates that among those children, 10 million did not see their parents for one year or more, and 2.6 million never got even a phone call from their parents within a 12-month period. Many of these children develop psychological problems and some fall victim to bullying, physical or sexual abuse, or even serious accidents.

Under China's current *hukou* policy, migrant workers and their accompanying children are considered only temporary residents in the city. And with very low incomes, they face many obstacles in obtaining education in the cities. Though the central government since 2001 requires local governments to provide education for migrant children in grades 1–9 in places where their parents work, local governments only implement this measure half-heartedly. In many cities, claiming that they lack funding, local governments have erected direct and indirect barriers to deter migrant children from getting a public education.

Under China's 2014 new urbanization blueprint, big cities are asked to limit their population size. As a result, admission to schools has been much harder for migrant children in big cities, forcing thousands to lose their school places. They either go back home or drop out altogether. The situation was so grave in Beijing that hundreds of migrant children and parents staged a three-month-long protest in front of the local education commission office in 2014. The difficulties in getting an education for their kids in the city have forced many migrant parents to leave their kids in the countryside, despite the undesirability and unknown risks of prolonged family separation. Furthermore, public high schools (grades 10–12) are totally off limit to migrant children under China's current policy. With a curriculum for the high school admission exam at home different from the one in the city, migrant school children wanting to continue high school often have to return to home villages years before grade 10 to prepare for the exam. In many instances, that means they are parted from their parents long term in their critical formative teenage years.

The deaths of the four kids in Guizhou have shocked many and drew public attention to their plight. While parents have direct responsibility for protecting their kids, arguably there is a more potent force—the *hukou* system that treats rural migrants differently and the related public school enrollment policy that discriminate against migrant children—that has directly and indirectly contributed to the plight of left-behind children. Some critics have said that these children are orphans of China's rapid urbanization under its peculiar system and discriminatory policy.

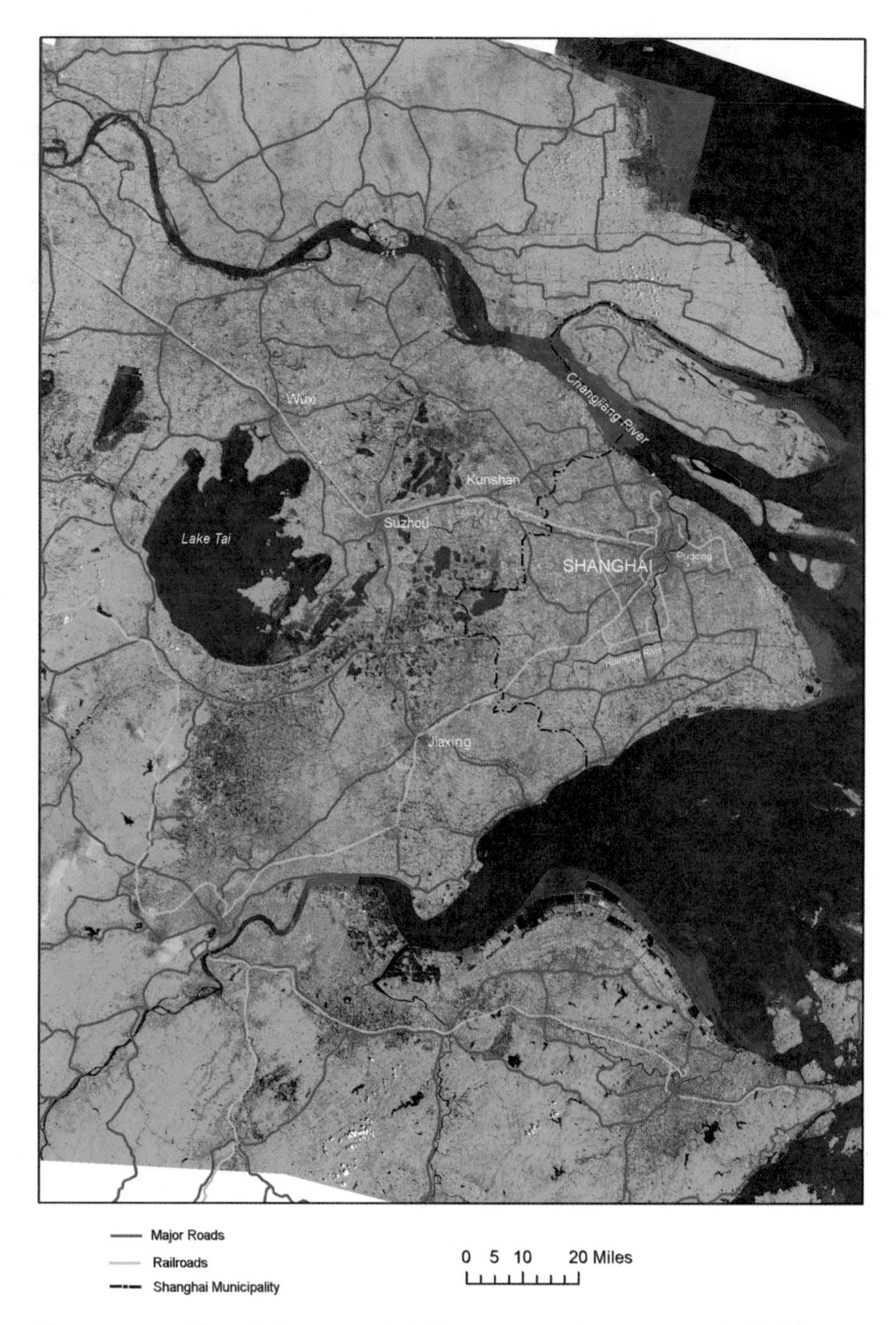

Figure 11.12 Shanghai's economic influence extends to a network of cities and smaller towns beyond its boundaries. In this satellite image, pink highlights areas of concentrated commercial and residential use. *Source:* Based on 2015 Landsat data.

Shanghai came the closest to a true "producer" city in Mao's era. In that era, government revenues relied heavily on remitted revenues from state-owned enterprises (SOEs). Shanghai, being the prime center of SOEs, was a major cash cow for the central government and was heavily protected by the central government. China's Stalinist-type economic growth strategy prioritized industry over agriculture, and that strategy greatly benefited Shanghai, which maintained its lead economic position throughout Mao's years. As with many cities in that time, investment poured into industry but little in "nonproductive" facilities such as housing and infrastructure. The downtown area, particularly around the Bund, or riverfront district, where the major Western colonial settlers built trading houses, banks, consulates, and hotels, had the look of a 1930s Hollywood movie set. In 1934, in the center of the city, the 22-story Park Hotel was built. It was then the tallest building in Asia and remained the city's tallest for almost another half century until 1983, when high-rises were again constructed. It was the relative neglect of many cities, including Shanghai, which contributed to the impression that the Maoist policy was "antiurban," although the reality was quite the opposite.

With the reopening of China in the late 1970s, under the open policy, foreign investors returned to China, this time at the invitation of the Chinese government. Shanghai received a major impetus for development in 1990 in the aftermath of the 1989 Tiananmen crackdown, as the government struggled to regain foreign investors' confidence. China decided in 1990 to open up Pudong ("East of the Pu," i.e., the Huangpu River, which bisects Shanghai), an essentially farming region on the east side of the old city core (Figure 11.13). The aggressively promoted World Expo 2010 was held largely in Pudong with a record number of 73 million visitors.

Figure 11.13 Since the early 1990s, Shanghai's new CBD has arisen across the river in Pudong, centered on the futuristic TV tower around surrounded by ultramodern skyscrapers. Pudong CBD is China's financial district. *Source:* Photo by Kam Wing Chan.

The Pudong development project was one of China's most ambitious undertakings. It included massive investment in infrastructure (including a new airport and a 30-km (18.6 mi) Maglev rail line) and a package of preferential policies, similar to those in China's special economic zones (SEZs), to woo foreign capital. These measures included lower taxes, lease rights on land, and retention of revenues. In Pudong, emphasis was given to high-tech industries and financial services rather than simply export processing. Among the foreign investors, Taiwanese businessmen have several thousand companies in the greater Shanghai region (including nearby cities like Kunshan and Suzhou), with an estimated one-quarter million Taiwanese residing and working nearby. A "Little Taipei" has emerged in the Zhangjiang High-Tech Park in Pudong.

The Shanghai Stock Exchange, opened in 1990, is China's largest stock market. The skyline of Pudong is intentionally futuristic, with flickering neon-lit glass and steel skyscrapers, including a TV observation tower that has become an icon for Pudong and the New China (Figure 11.16). Shanghai's stock exchange is China's largest by market capitalization and ranks the third in the world by market capitalization. It is quite a contrast to the neoclassical Bund on the other side of the river. Shops and architecture in some sections have a very cosmopolitan feel and again there is a sizable expatriate community. However, behind the glistening architecture, there also lives a very large population of migrant poor, often struggling to make a living in this metropolis.

Many problems plague the city and the region: serious interjurisdictional rivalries among local governments, the faraway location of the new international airport in Pudong, overheated real estate development, serious traffic congestion, and severe air and water pollution. Perhaps most important, Shanghai still lacks a well-established legal system that can truly protect citizens' rights and rein in officials from abuses of their powers. These are criticisms that could be directed at all of China today.

Hong Kong: Business Not as Usual

At the stroke of midnight on June 30, 1997, Hong Kong was officially handed over to China and became the Hong Kong Special Administrative Region (SAR). This event marked the end of the colonial era in Asia and the rise of China's power. Hong Kong was one of the last two colonial enclaves left in all of Asia by the late twentieth century. The other colony, Macau, was returned to China by Portugal in 1999, and became the Macau SAR. Hence, as China entered the new century, its humiliating experience with foreign colonialism ended after almost 160 years.

The 1997 handover committed China to guarantee Hong Kong 50 years of complete autonomy in its internal affairs and capitalist system, under a model known as "one country, two systems." The latter refers to the "socialist" system in the PRC and the "laissez faire" capitalist system in Hong Kong. With the exception of defense and foreign relations, the city was to be "ruled by Hong Kong people." In the Basic Law (the SAR's constitution) promulgated in 1990, China even consented that the SAR would choose its own chief executive based on universal suffrage. Nevertheless, worries over what would really happen after 1997 under China's rule triggered an exodus of about half a million Hong Kongers, mostly the wealthy and the professionals, to Canada (especially Vancouver and Toronto), Australia, and the United States.

Earlier, while still under the British, many fled to Hong Kong from China after the communist takeover in 1949. Hong Kong's population soared from half a million in 1946 to more than 2 million by 1950. The flows, both legal and illegal, continue. Squatter settlements appeared in the 1950s and the economy was in a shambles. The British, in collaboration with Chinese entrepreneurs, including many industrialists who had fled Shanghai and other parts of China, turned Hong Kong's economy around by developing products, "Made in Hong Kong," for export. It was a spectacularly successful transformation, with investment pouring in from Japan, the United States, Europe, and the overseas Chinese. Cheap, hardworking labor was available. Site limitations were overcome by massive landfill projects, and fresh water and food were purchased from adjacent Guangdong Province.

During the Cold War period of the 1950s and 1960s, Hong Kong commanded a unique geopolitical position. One paradox of Hong Kong was that China continued to permit this arch symbol of unrepentant Western capitalism and colonialism to exist and thrive on Chinese territory. The Chinese did this partly because Hong Kong made lots of money for them, too—several billion dollars annually in foreign exchange earned from the PRC's exports to Hong Kong and banking and commerce investments. Moreover, a struggling, isolationist, socialist China saw a practical advantage in keeping the door open a crack to the outside world, and also in not being responsible for solving Hong Kong's staggering problems. Banker's Row in Central District came to symbolize the financial powerhouse that Hong Kong had become, with the Bank of China, the Hongkong Shanghai Banking Corporation (HSBC), and the Chartered Bank of Great Britain lined up side by side. The first two are regarded today as among the most important architectural structures of the twentieth century and symbols of Hong Kong's emergence as a world city.

As one of the top tourist meccas in the world, Hong Kong is a stunning sight. The skyline is spectacular, especially at night, with its glittering, ultramodern high-rise buildings packed side by side along the shoreline and up the hillsides (Figure 11.14). Hong Kong's economy is heavily dependent on the property market. Rents are among the highest in the world, which burdens the middle and working classes. The built environment is crafted to fit every possible urban activity. Some of the urban designs are quite ingenious. After the international airport moved to Chek Lap Kok on Lantau Island in 1997, building height limitations in Kowloon ended. The Kowloon side is now taking on a Manhattan-like profile. There seems no limit to the construction boom and demand for new buildings and other structures in this dynamic city.

Less eye-catching to the average visitor, but themselves impressive social accomplishments, are Hong Kong's social housing and new town programs. They were begun in the 1950s to cope with the large influx of Chinese refugees. The programs gradually expanded into some of the world's largest. Today, about half of Hong Kong's population of 7 million lives in social housing. Indeed, because of much lower rents and prices, social housing has been a major mechanism for decentralizing the population outside of the main urban area. Reclamation has been a main strategy for creating new land for the city. Many large new towns, such as Shatin and Tuen Mun, were built almost totally on land reclaimed from the sea.

Hong Kong's export industry gradually declined with China's opening in the late

Figure 11.14 This view of Hong Kong Island, taken from Kowloon across the harbor, dramatically conveys the modernity and wealth of today's Hong Kong. The Central Plaza building towers over the wave-like profile of the Convention Center, where the ceremony of the handover to China took place in 1997. *Source:* Photo by Kam Wing Chan.

1970s. Hong Kong took advantage of the cheap land and labor in the Pearl River Delta and has steadily outsourced its manufacturing to the delta (while company headquarters remain in Hong Kong). Over 100,000 Hong Kong-invested enterprises operate in the delta region and employ several million workers. In economic terms, the delta and Hong Kong are now a highly integrated region, one of the world's major exports centers, with Hong Kong serving as the "shop front" and the delta as the "factory" (Figure 11.3). Tens of thousands of people cross the land border between Hong Kong and Shenzhen daily for work, school, or shopping.

Hong Kong has been the banking and investment center for the China trade, as well as regional headquarters for many international corporations since the late 1970s. It has played a crucial role as the intermediary between China and the world, including serving as middleman for Taiwan's huge economic dealings with the PRC in the 1990s. Tourism remains vital, with many coming from the mainland and skyrocketing from about 4 million in 2001 to more than 41 million in 2013! While tourists shopping in Hong Kong generate sizable revenues for the city, they also bring significant transportation congestion.

The post-handover period has witnessed the city's rapid transition to Asia's premier financial and service center. Simultaneously, a series of mishaps and policy blunders (such as the outbreak of Severe Acute Respiratory Syndrome, SARS, in 2003) have undermined confidence in the new government. Rising economic competition, including from other Chinese cities (especially Shanghai, Shenzhen, and Guangzhou) also put pressure on the city as it struggles to find its role as China rises. More alarmingly, income gaps between rich and poor have risen to a very high level—the highest in the developed world, according to several recent studies. Observers have linked mass protests—some quite disruptive—in the last few years to this widening wealth gap and to the government's pro-business stance with little attention to the welfare of the people.

In addition, there is a more, and probably the most, critical issue for Hong Kong: maintaining relations with mainland China, and protecting its autonomy. Many people in Hong Kong have long expressed concerns over this issue and many have actually voted with their feet and left, damaging the ability to protect the city's cherished political and press freedoms. These rights are not enjoyed by mainland compatriots and are often frowned on by PRC leaders. The difficulties of keeping that rather tenuous balance have surfaced in almost perennial mass protests on July 1 ("The SAR day"), and civil disobediences, culminating in the Occupy Central mass protest in 2014. At the height of the protest, hundreds of thousands of people took part; it lasted for more than two months and paralyzed the central city (Figure 11.15). The popular protest was against PRC's imposition of a screening mechanism of candidates in the proposed election system of the SAR's chief executive based on universal suffrage in 2017. The proposed system did not get enough votes to pass in the SAR legislature in 2015, and Hong Kong's future political system remains uncertain. Whatever may happen, it is clear that the SAR's future is irrevocably and closely tied to that of China. Business in Hong Kong cannot go on as usual, as some had once thought.

Figure 11.15 Also called the "Umbrella Movement," the Occupy Central protest in 2014 was the largest civil disobedience movement since 1967. The protest was against the proposed "universal suffrage" system, which critics consider as not genuine. *Source:* Photo by Wilfred Chan.

Taipei: In Search of an Identity?

Although regional centers are found throughout East Asia, a particularly good example is Taipei, which has been emerging from its provincial cocoon and acquiring some of the aura of a world-class city, following in the footsteps of Hong Kong. There is ambiguity about how to classify Taipei: after 1950, it became the "temporary" capital of the Republic of China (ROC) government-in-exile and as such experienced phenomenal and unexpected growth. If the communists had succeeded in capturing Taiwan in 1950, as they had hoped, Taipei would be a vastly different place today, probably something akin to present-day Xiamen across the Taiwan Strait. Instead, Taipei skyrocketed from a modest Japanese colonial capital city of a quarter million in 1945 to the present metropolis of more than 6 million that completely fills the Taipei basin and spills northeast to the port of Keelung, northwest to the coastal town of Tanshui (now a high-rise suburban satellite), and southwest toward Taoyuan and the international airport. Functionally, the city changed from being a colonial administrative and commercial center to becoming the control center for one of the most dynamic economies in the postwar world.

When the ROC's government retreated from mainland China to Taiwan in 1950, the provincial capital moved to a new town built expressly for this purpose in central Taiwan, not far from Taichung (Figure 11.16). Taipei was theoretically concerned with "national" affairs and hence had all national government offices recreated there (transplanted with administrators and legislators from Nanjing). The provincial capital dealt with agriculture and similar island (local) affairs. This artificial dichotomy, designed to preserve the fiction that the ROC government was the legal government of all of China, held until the early 1990s, when the government finally publicly admitted that it had no jurisdiction over the mainland. Over the decades, Taipei became a large bureaucratic center due to construction of national capital-level buildings in the city. Huge tracts of land formerly occupied by the Japanese were taken over by the government after 1945 and the single-party authoritarian political system under the Kuomintang (KMT) allowed the government to develop the city largely free of open public debate. After President Chiang Kai-shek died in 1975, a huge tract of military land in central Taipei was transformed into a gigantic memorial to Chiang, one of the largest public structures in Taiwan. As Taiwan's political system was democratized over the last two decades, this type of memorial and other commemorative fixtures of the KMT rule under Chiang have been attacked, especially 2000–2008 when the island was governed by a pro-Taiwan independence party, the Democratic Progressive Party (DPP), which remains a strong opposition party.

Taipei's metropolitan area, with an estimated population of about 7 million is thrice the size of Kaohsiung metropolitan area, the second largest in the island and the island's main heavy industrial center. Taipei today remains the center of international trade and investment and includes a large expatriate community. Culture, entertainment, and tourism are all focused on Taipei. Because of its colonial heritage, the city's culture is distinctly Japanese. Manufacturing in Taipei is now concentrated in a number of satellite cities, to the west and south. The old port of Keelung, once the link with Japan, serves as the port outlet for the north. As with Seoul, most of the city's population increase over five

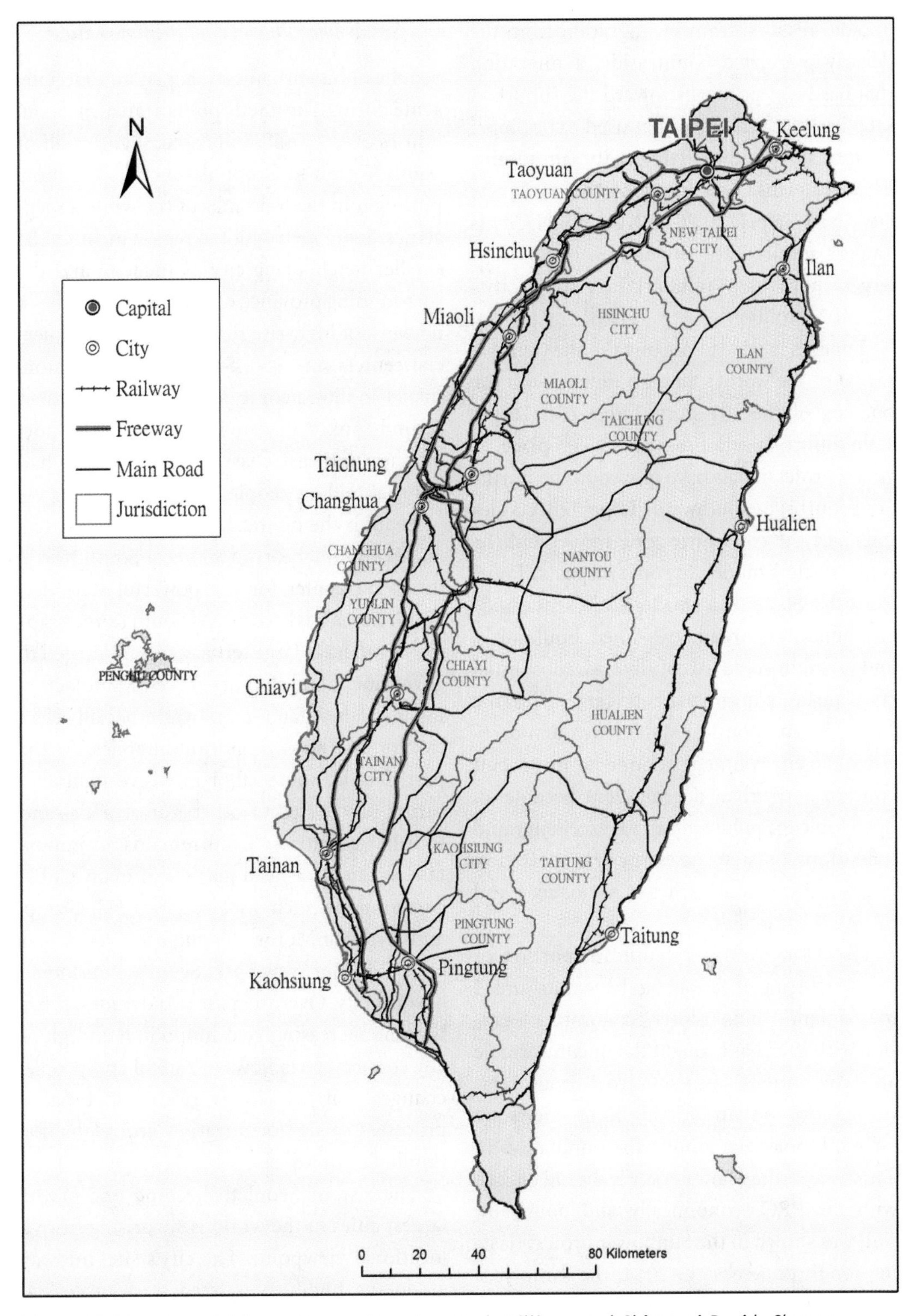

Figure 11.16 Map of Taiwan. *Source:* Based on Jack Williams and Ch'ang-yi David, *Change, Tawian's Environmental Struggle: Toward a Green Silicon Island* (New York: Routledge, 2008).

decades resulted from in-migration from the densely populated countryside, a migration that has been primarily toward the suburban satellite cities recently. What used to be Taipei County surrounding Taipei City (in administrative terms) was incorporated as a new city, "New Taipei," in 2010. New Taipei City is Taipei's suburbs, the population of which is larger (more than 4 million) than Taipei City's (about 2.7 million).

Taipei in 2004 also became the site (temporarily) of the world's tallest building, with the opening of the 101-story *Taipei 101.* Large-scale suburbanization has also taken place, as young professionals have moved to the northern suburbs, or southward. Taipei reflects elements of the concentric zone model and the multi-nucleic model. In some respects, Taipei looks like Seoul on a smaller scale, with modern buildings, broad, tree-lined boulevards, and a high standard of living. Substantial clean-up and improvements came with the 1990s. As the political system was democratized, the environment became an important concern and urban development became an open topic for public input. An excellent rapid mass-transit system eases the transportation crush, composed of hordes of motorbikes and private automobiles.

Like Hong Kong, a significant portion of manufacturing in Taiwan has been outsourced to mainland China. Closer economic integration with PRC has brought the uncomfortable and contentious issue of cross-Strait relations. More so than in Hong Kong, views are far more polarized among the population in Taiwan regarding how closely it should engage with the PRC economically and politically. This was shown in the Sunflower protest (lasting for three weeks) in 2014, the same year Hong Kong had its largest civil unrest since 1967.

Seoul: The "Phoenix" of Primate Cities

Seoul exhibits urban primacy in an especially acute form. The metropolitan area of Seoul houses over 25 million people, slightly half of South Korea's total population of 50 million, putting it in the top ranks of the world's megacities. Seoul metropolitan region includes the smaller neighboring city of Incheon and the surrounding province of Gyonggi, which has a network of high-rise residential and commercial centers. In 1950, Seoul had barely more than 1 million people, just slightly more than second-ranked Pusan (Busan), the main port on the southeast coast. In 2015, Pusan had only 3.4 million people.

Seoul is the political, cultural, educational, and economic heart of modern South Korea, the nerve center for the powerful state that South Korea has become. As South Korea's capital, Seoul has a large tertiary sector devoted to government and military forces. Manufacturing is another major employer, especially electronics, machinery, and automobiles. In the past decade, some labor-intensive manufacturing has shifted to other Korean cities and to other countries. Seoul remains an important location for headquarters of many global corporations such as Samsung, LG Group, and Hyundai Motors. Though not yet on par with Tokyo or New York, Seoul is considered a world city. Over the past two decades, it has become increasingly cosmopolitan and globally connected via flows of capital and people, changes that are closely related to broader processes of democratization and globalization in South Korea.

The rise of Seoul to become one of the largest cities in the world is surprising from a locational viewpoint. The city's site, midway along the highly populated western coastal plain of the Korean Peninsula was a logical place for the national capital of a unified Korea. However, since the division of the peninsula in the late 1940s and the bitter stalemate between North and South Korea since 1953, Seoul's location just 20 miles (32 km) from the demilitarized zone (DMZ) that divides North and South, makes the city highly vulnerable. The city was nearly leveled during the Korean War, when the North occupied the city twice. In the 1970s, a greenbelt was established that encircled the city approximately 9 miles (15 km) from the core and limited its spatial expansion. This was done to contain urban sprawl as well as to protect Seoul from North Korean artillery attacks. Urban planning policies of the 1980s directed developments southward, across the Han River, which was a strategy similarly informed by national defense concerns. Despite concerns about safety, Seoul has experienced an increasing and seemingly unstoppable influx of people and economic activity over the past five decades.

Post-1960s, growth is primarily the result of massive rural-to-urban migration, encouraged by Korea's transformation into an urban-industrial society as South Korea embarked on an export-oriented industrialization strategy concentrated around Seoul. The urban landscape includes a dizzying mix of high-rise apartment blocks, mid-level residential and commercial buildings, with pockets of older 1–2 story buildings; especially southward, the urban landscape now seems "centerless," as dense developments and high-rise buildings create a multi-nucleic pattern throughout the metropolitan area.

Expansion away from the old city (north of the Han River) was driven by a series of policy interventions intended to decentralize the economic functions by establishing new residential and industrial development areas. The first wave of policy-led expansion occurred in the 1970s and produced suburbanization with massive new subdivisions dominated by high-rise apartment complexes, dense retail, new corporate headquarters, and the relocation of many public facilities belonging to national and city governments. A second wave of development in the 1990s responded to demands for more affordable housing. Five large-scale new towns were established 12–15 miles from the city center. Their locations were dictated by the strict greenbelt policy that created a barrier to continued expansion of Seoul's existing suburban areas. Approximately 20 percent of the population, or about 2 million people, moved out from Seoul's central areas between 1992 and 1999. With little industry of their own, these new towns functioned at first much like bedroom communities, linked by highways, and later by transit lines, to the central areas of Seoul. The combination of these two waves of expansion led to a "hollowing out" of the older parts of Seoul, north of the Han River, laying the foundation for an urban renewal program that began in the late 1990s. Older parts of the city have since transformed through a series of large-scale redevelopment projects intended to improve housing quality with construction of new high-rise apartments that allow for more open space within the city center.

Changes in the built form and social landscape of Seoul have occurred in parallel with economic changes. While its early "take-off" occurred through heavy industrialization based on the availability of cheap labor, Seoul today is better described as a postindustrial metropolis whose economic development is centered on financial and corporate services, real estate, and in recent years, high-tech and creative industries. With this shift, city leaders have sought to rebrand Seoul as a leading-edge high-tech and sustainable city. The green belt, which was criticized in the past for failing to curb urban growth and leading to suburban sprawl, is now seen as an important green space winding through a dense metropolitan area. The satellite towns, once largely dependent on Seoul, have become more independent commercial centers, relieving commuter traffic congestion. With improved mass-transit, auto-dependent commuting from suburban areas has decreased. Initially built to meet the needs of the 1988 Seoul Olympics, the impressive transit system is now more extensive, combining subways, trains, and buses throughout the metropolitan area. Mirroring these improvements in transportation networks, Seoul has transformed itself into one of most wired cities in the world. With Seoul's promotion of innovation and creative industries, its economic and cultural dynamism will likely be linked closely to future digital development.

URBAN PROBLEMS AND THEIR SOLUTIONS

The relatively clear-cut dichotomy between the socialist path of China, North Korea, and Mongolia, and the non-socialist path of the rest of East Asia that characterized the region through the 1970s is no longer valid. China has abandoned orthodox socialism though one-party rule remains. North Korea occasionally hints that it might also do so, but then slips back into Stalinist suspicion of the outside world (Box 11.4). Mongolia, like Russia, abandoned not only a socialist system but also single-party rule, and now struggles to join the capitalist world. The colonial era is now over in the region. As a result of these changes, urban problems and solutions take on new guises and, except for North Korea, look somewhat similar across the region.

The Chinese Way

Before 1979, China pursued a Stalinist-type industrialization program, suppressing personal consumption (the "nonproductive" side of cities) and squeezing agriculture to help finance rapid industrial growth. To maintain the huge imbalance between city and the countryside, strict controls over migration to the city through the *hukou* (household registration) system was maintained. Urban residents had some basic welfare and guaranteed jobs, but their lives were closely monitored. Such an approach kept Chinese people, even in the city, at the bottom rank of the living standards among East Asia's countries and has resulted in mass poverty in the countryside. When Mao died in 1976, the system began to change.

In the late 1970s, China's leaders began to make significant policy changes to key economic policies of the Maoist era yet did not abandon the one-party system and authoritarian rule. China became open to foreigners for investment, trade, tourism, technical assistance, and other economic contacts as the policy of self-reliance was set aside. Rapid integration with the global economy profoundly impacted cities and urban development in the coastal region. China first established export-processing zones, such as Shenzhen, with concessionary tax policies to attract foreign investment. By the mid-1990s, practically the entire coastal region contained thousands of "open zones" vying for foreign investments.

Another major change is de-collectivization of agriculture and the return to private smallholdings (under the Household Responsibility System) in the early 1980s. This shift

Box 11.4 Isolation: Peripheral Cities

Isolation can be a huge handicap for cities, but isolation is a relative concept, in that it can be caused by both natural and man-made factors. Four cities in East Asia—Pyongyang, Ulan Bator, Urumqi, Lhasa—play important roles in their respective regions, yet are really isolated, that is, they are peripheral geographically and in terms of their linkages with the rest of the world.

Pyongyang ("Flat Land") is perhaps the biggest anomaly of the four. The government of North Korea rules this austere, reclusive nation of about 25 million from the capital city of Pyongyang. At an estimated 3.5 million in the metro region, Pyongyang is 3–4 times larger, in classic primacy fashion, than the next two largest cities, Chongjin and Hamhung. This is hardly a surprise, given the centrally planned, Stalinist system that hangs on, long after the Soviet Union, Maoist China, and communist Mongolia saw the light. Leveled to the ground during the Korean War (1950–1953), Pyongyang was totally rebuilt in the true socialist city model, with broad boulevards and massive government buildings, a superficially modern showcase of socialist dogma, but a city that gets terrible reviews from the a limited number of foreigners who have managed to visit. Pyongyang is little more than a grandiose monument to the whims of North Korea's autocratic rulers. The city may be geographically sited in the heart of East Asia, but it might as well be in the middle of Siberia.

By contrast, Mongolia's capital city of Ulan Bator (Ulaanbataar, "Red Hero") with its 1.3 million people is the center of a country now doing everything possible to integrate with the outside world. The main problems are Mongolia's tiny population (2.9 million), sprawling land area, and geographical isolation. Ulan Bator is also a primate city. As Mongolia sheds its socialist past and democratizes, the country is rapidly urbanizing and trying to find alternatives to the processing of animal products for its small economy. Tourism is growing, but industry is never likely to be significant here. It will be difficult to overcome the country's geographical limitations, and hence Ulan Bator will likely remain largely a minor regional center.

Urumqi ("Beautiful Pasture") is also a regional capital, for the Xinjiang Autonomous Region in China. An ancient city, Urumqi has become a booming metropolis of about 3 million, with a largely Han Chinese population, as the center of China's administration and development of Xinjiang. As such, Urumqi in recent decades has increasingly taken on the character and physical appearance of a Chinese city, very similar to those found throughout the eastern, more populous part of the country. Although geographically the most isolated of our four peripheral cities, Urumqi is actually very much in touch with the outside world, largely because of China's prodigious economic growth in recent decades. The city is the focal point of large-scale tourism, industrialization, and development of the region's oil and other resources. Urumqi is also the center of efforts by the Chinese government to contain separatist tendencies among Xinjiang's largely Muslim population (especially among the Uigur). Hence, the city's geopolitical importance may well exceed its economic role.

Lhasa ("Place of the Gods,") is the capital city of Tibet and similar in many ways to Urumqi, although much smaller with under 300,000 in the urban area (Figure 11.17). If not for Chinese rule, Lhasa would be even more geographically isolated as one of the world's highest cities (nearly 12,000 feet elevation). Also an ancient city and center of Tibet's unique Buddhist culture under the Dalai Lama (in exile in India), Lhasa was thrust into the modern world with China's takeover in the 1950s and became the focal point of China's efforts to contain Tibetan separatism, drawing much international attention in the process. Like Urumqi, Lhasa is rapidly become essentially a Chinese city, with the Han Chinese population steadily increasing, and Chinese urban forms displacing much that was traditional and Tibetan. Tibet remains one of China's poorest regions, and Lhasa's economy is largely dependent on tourism and services, subsidized by the Beijing government in its determination to ensure that peripheral regions (and their key cities) like Xinjiang and Tibet remain firmly within the PRC. One powerful demonstration of this effort was the opening in 2006 of the first railway linking Tibet with the rest of China (via Qinghai province to the north). Tibetan nationalists view the railway as one more tentacle of Beijing's grip. Beijing, in turn, sees the railway as an essential tool to further bring Tibet into the modern world and irrevocably into the PRC.

In sum, these four cities, in their historical as well as recent development, illustrate that isolation can be imposed by nature or by government, but overcoming isolation is no easy task.

Figure 11.17 The Potala Palace dominates Lhasa, the capital of Tibet. This city used to be the home of Tibet's traditional ruler, the Dalai Lama. *Source:* Photo by Ondřej Žváček.

helped raise labor productivity and brought a better quality of life to hundreds of millions of peasants. As many laborers were no longer needed on the farm, the government was forced to relax internal migration restrictions. The migrant workers, estimated at 170 million in 2014, provide plentiful low-cost labor to make China the "world's factory." Migrants fill many industrial and service jobs shunned by urban workers, but under the *hukou* policy these migrants do not have the same citizens' rights and social benefits as ordinary urban residents. This two-tier urban citizenship system and unequal treatment of the migrant population has become a major urban concern. For example, Shenzhen had been a small village at the China-Hong Kong border and within three decades became a major export-processing center with more than 10 million inhabitants, many of who are young migrants without the local *hukou* status.

The negative side of China's new policy, a combination of market approach and one-party rule, has generated new imbalances between rural and urban areas, provinces and different regions, and socioeconomic classes. To some urbanites, life has unquestionably improved and appears increasingly similar to that of the rest of East Asia. Many big cities now offer a great variety and quality of goods and services, including many luxury ones, a huge contrast from the Maoist years. In fact, some sections of large cities today look like Hong Kong or Taipei. To many others, urban life has also become a hectic struggle to make ends meet, especially with escalating housing prices. Economic and social polarization is definitely rising. An expanding class of urban poor consists of migrants and laid-off older SOE workers. Unemployment exists in a rapidly aging population: China simply has too many people to provide employment for everyone (Figure 11.18). Surplus labor in the countryside, especially in older age groups, remains serious. Moreover, impacting virtually everyone, rich or poor, is the critical state of the environment. In the last decade, cities of China have gained a reputation for serious environmental problems, commonly cited as the world's most polluted cities. This has not gone unnoticed, as governments at all levels have invested in improving urban environmental conditions.

Other Paths in East Asia

As with big cities around the world, the industrial cities of East Asia are experiencing profound problems of overcrowding, pollution, traffic congestion, crime, and shortages of affordable housing and other amenities. This has been a region of impressive economic growth and advancement in recent decades and many urban residents now have standards of living among the highest in the world (except for housing). Retail stores of all types provide consumer goods for affluent residents. At night, cities glitter with eye-popping displays of neon lights, nowhere more dazzling than in Japan. Behind all this, there has been increasing concern about widening wealth inequality. The poor are the elderly, the immigrants, or rural migrants. This is especially serious in China, where internal migrants constitute about 230 million in 2014. Denied access to basic social services, many of them barely eke out a living on urban fringes.

Expensive land is a major constraint to urban development. Thus, the major cities are increasingly following in the footsteps of most other large cities with high-rise syndrome. Growing competition among regional cities exists to build the tallest skyscraper, as if having the tallest building conveys status

Figure 11.18 Migrant workers shine shoes on a street in Wuhan, the largest city in central China. "Rural migrant workers," numbered about 170 million in 2014, are everywhere in China's major cities, doing all kinds of work. The huge army of cheap migrant labor is crucial to China's success in being the "world's factory." *Source:* Photo by Kam Wing Chan.

and superiority. Shanghai, Hong Kong, Taipei, Seoul, and others compete. Even Japan's cities, long characterized by relatively low skylines because of earthquake hazards, have succumbed to high-rise construction, such as in the cluster of the 50-plus-story buildings centered on the city government complex in Tokyo's Shinjuku District, or the new high-rise profile in the port of Yokohama. These cities make maximum use of underground space, with enormous, complex underground malls interconnected by subway systems.

Suburban movement outward from the central city is the only other alternative. New communities have sprung up, including bedroom towns where people can obtain better housing with cleaner air and less noise for less money, even though doing so often means longer commutes to work. Fortunately, most cities have developed relatively good public transport systems. Nonetheless, automobile culture is spreading rapidly, with the private automobile purchased as much for status as for convenience. Automobile culture first took hold in Japan in the 1960s, but other countries have followed and China has now replaced the United States as the world's largest market for automobiles since 2010.

Closing the Gap: Decentralization in Japan

In Japan, Tokyo's dominance as the primate city is indisputable. The concentration of population and power has concerned planners and politicians for decades. Efforts to decentralize Japan's urban system have approached the problem at different levels. One approach to reducing the dominance of Tokyo is to direct industrial investment to other regions. This national strategy was aggressively pursued in the 1970s, stimulating industrial and urban growth outside of the Tokyo area. Labor-intensive work moved to the regions, while research and high-skilled parts of production remained near Tokyo. Unfortunately, it is lower-skilled work that was most vulnerable to wage pressures in the global economy. Tokyo was relatively better positioned to withstand, if not flourish in, these conditions, reinforcing the capital region's economic centrality. Other efforts to discourage over-concentration of economic, political, and cultural functions within Tokyo include policies to disperse these functions within the metropolitan region itself, supported by expansion of infrastructure to connect new high-growth areas. The Bay Aqua Line, an expressway across the middle of Tokyo Bay, was designed to enhance development along the eastern and southern shore. Research and educational institutions disperse growth into less developed areas in the suburbs. While promotion of this multinodal structure for the metropolitan region may address issues such as traffic congestion in the urban core, the new high-growth areas also expand and reinforce Tokyo's centrality.

Tokyo's dominance also reflects strategic choices over time. Despite successive National Development Plans calling for more balanced regional development, momentum for this stalled as Japan struggled during the long economic recession. Governments and businesses return to ideas of agglomeration economies that continue to benefit Tokyo. The Tokyo Municipal Government has consistently opposed repeated calls to relocate the capital to another region of Japan; in the past two decades, Tokyo city leaders pursued urban policies to attract inward investment, increase economic competitiveness, and enhance its status as a global city. Tokyo's waterfront redevelopment and new zoning laws that permit intensification through high-rise construction are two examples of how global aspirations have influenced governments and business leaders and reconfigured city spaces. These factors have enhanced Tokyo's profile as the strongest magnet for people and investment in Japan's urban system.

Seoul: The Problems of Primacy

Somewhat like its larger cousin, Tokyo, Seoul has suffered from problems of urban primacy such as traffic congestion and housing shortages. Dispersed development over the past few decades was meant to address these and related problems by channeling expansion into master-planned new towns outside of central Seoul in the 1990s, though the areas surrounding these new towns also saw haphazard and unplanned development due to land speculation. Gradually, these new towns have developed their own commercial, business, and educational facilities, meaning that residents are making fewer trips into the central areas. This, in combination with the ongoing expansion of mass-transit networks outside of the city's core and the establishment of dedicated bus lanes, has seen improvements in the traffic conditions throughout the Seoul metropolitan region. Interestingly, while the planned new towns are dense enough to support mass transit, this is not the case with the smaller, more sporadic suburban developments that have sprung up around the new towns. It is these elements of Seoul's suburban landscapes that are a source of many of the development-related problems that plague the city.

New-town developments have influenced the city's internal structure and have helped Seoul become the central node in the national economy. Seoul's ongoing spatial expansion made it a magnet for investment, leading to uneven regional development with a concentration of large corporations, state institutions, and people throughout the country. To counter this concentration, political leaders have recently pushed forward on a controversial plan to establish Sejong, a brand new city 75 miles south of Seoul that will be the home to many national government agencies and ministries. In moving many of central administrative functions to Sejong, over 10,000 civil servants and their families will also relocate, leading the way for the development of a high-tech cluster and city of 500,000 by 2030. By 2015, Sejong had over 100,000 residents, but questions remain about efficiencies of this decentralization strategy, as some important government functions will still be based in Seoul, the political capital.

While the city and national governments have sought to counterbalance the centralizing tendency of Seoul's development since the early 2000s, the city government has also been trying to avoid decline in the urban core. However, efforts to revitalize older urban areas through densification (by replacing two- to five-story buildings with high-rise apartments) and providing more outdoor open spaces have displaced many lower-income residents from their old neighborhoods. Some areas have seen massive displacement of lower-income tenants and small family-run businesses, unable to purchase or rent newly constructed units. Disputes about the gentrification process are centered on how compensations packages are awarded. A 2009 protest against forced evictions in the Yongsan area of Seoul turned violent, leaving six people dead. As with other cities undergoing rapid and dramatic redevelopment, it is not clear whether Seoul will retain the social mix that had traditionally defined life in its smaller, more traditional neighborhoods as these areas are replaced by the high-rise apartments and broad boulevards that have become the trademark of the new East Asian city.

Taipei: Toward Balanced Regional Development

Taipei has made dramatic progress toward solving some of its urban problems. Completion of the Mass Rapid Transit system and stepped-up enforcement of traffic rules have brought order to one of Asia's worst traffic nightmares. Air pollution has been drastically cut through various programs. Housing is still expensive, but the city is cleaner and decidedly a better place. Although many people have moved to the suburbs, a large residential population still lives within the central city. The doughnut model does not fit Taipei, but the multiple nuclei model is applicable.

To address overcrowding in Taipei, the national government embarked on island-wide regional planning in the early 1970s, resulting in a development plan that divided the island into four planning regions, each focused around a key city. The Northern Region, centered on Taipei, has about 40 percent of the island's total population. Through multiple policies about rural industrialization, massive infrastructure investment, and programs to enhance the quality of life and the economic base of other cities and towns, Taiwan has managed to slow the growth of Taipei and diffuse urbanization. It built a high-speed rail in 2006 connecting two "rivalry" cities: Taipei and Kaohsiung. Historically, these two cities have been controlled by the two main competing political parties which have their power bases in the North and the South, respectively. The current central government policy under the KMT has facilitated significant outsourcing of Taiwan's industry to mainland China, which has accelerated the economic structuring in the island. It will have a noticeable impact on Taiwan's urban and economic structures and even political future.

The Greening of East Asian Cities

Cities of East Asia have faced many environmental challenges as they experienced different development paths. Growing populations, rapid industrialization, and the recent general shift toward high consumption lifestyles have strained air, water, and land resources. Cities in this region respond to these challenges variably, influenced by local conditions and national policy priorities, while reflecting broader global trends toward sustainability. Driving the increased attention to quality-of-life concerns of residents is a growing awareness of the health and social costs of environmental degradation (Box 11.5). Greening of cities is linked to broader transformations in urban economies, dominated now by service and high-tech industries. A similar urban development pattern has occurred worldwide, but for East Asia it is the speed of change that is most remarkable. And perhaps, nowhere is this truer than in China's cities.

In various listings of the world's most polluted cities, China is well represented. The public is so familiar with these rankings that when a list of the world's most polluted cities released by the World Health Organization in 2015 did *not* list a Chinese city, this prompted many bloggers to ask about China. The surprise is not unfounded and all levels of the government recognize the seriousness of the environment degradation in China's cities. Absence of Chinese cities in the said list is due in part to the worsening conditions elsewhere, most notably the heavily polluted cities of South Asia, but cities in China have also seen some improvements in the last two years after the "war against pollution" was launched in 2014. Economic slowdown has also helped to reduce pollution. One common strategy has been to close or move polluting factories currently located in urban areas and, in northern cities, to replace coal-burning boilers with natural gas. Other programs, often with international funding, involve water improvement initiatives such as construction of new wastewater treatment plants and clean-up of urban waterways through dredging and improved controls on industrial and agricultural activities. Cities have increased space for parks and local greenery as another high-profile strategy for improving the environmental quality of cities, while new subway construction has increased public transit capacity in many major cities. Shanghai has been the most ambitious city, building one of

Box 11.5 A Stream Returns to the City of Seoul

Hae-Un Rii, Dongguk University

Flowing through the center of Seoul for centuries was a stream originally known as Gaecheon (meaning "open stream") and now known as Cheonggyecheon. In terms of Pungsu (the Korean equivalent of fengsui in Chinese), it played an important role as a natural waterway, flowing from west to east in the middle of Hanseong (or Hanyang), the Joseon dynasty's name for their capital city. The Joseon emperors kept control of the stream, used it for sewage disposal as early as the fourteenth century, and planted trees along both of its banks.

Cheonggyecheon began to disappear underground as Seoul developed during the four decades of Japanese rule in Korea. Shortly after the Korean War, as the city's population was rapidly expanding, the process of paving over Cheonggyecheon continued. By 1977, there was no longer a visible stream of water in center city. The bridges that crossed it were gone, asphalt covered the surface, elevated roadways soared overhead, people who lived near the stream were moved out, and commercial uses took over nearby land. Modernization had triumphed over what had become an eyesore, an offense to the nose, and something of an open sewer. However, a bit of nature had also disappeared from the landscape and so had a piece of Korean history.

In concert with Seoul's drive to become one of the world's greenest cities, reopening Cheonggyecheon for 5.84 km and turning its banks into parkland started in 2003 and finished

Figure 11.19 Cheonggyecheon Stream Restoration project in downtown Seoul during the Lantern Festival. *Source:* Photo by D.J. Zeigler.

in 2005. But it resurfaced as a natural stream no longer. Its flow now consists of purified water, and the sewage system is located under the stream. Plus, there is a big problem of flooding during the rainy season. Consequently, once the rains start, the city government prohibits people from walking along the stream. One of the threats during the rains is a possibility that the door from the emergency sewage system might be open up resulting in the sewage water getting into Cheonggyecheon.

Although it was controversial at first, returning Cheonggyecheon to the citizens of Seoul and making it available to visitors from around the world is now regarded as a success. Many people now walk along Cheonggyecheon for exercise during all the four seasons; they enjoy the cool water and surrounding natural environment during the summer; and many cultural exhibitions take advantage of the open space. One of the most popular festivities is the Lantern Festival, which combines the traditions of Korean history and aesthetics with elements of popular culture from around the world (Figure 11.19).

Just imagine that you are walking along Cheonggyecheon: where flowers are growing along its banks, where you can sit in shady areas and watch ducks and fish, and where, from time to time, you can participate in a festival, enjoy a fashion show, or just have a rest with nature. For a megacity like Seoul, Cheonggyecheon is a kind of dream in the middle of downtown. In fact, foreign governments visit this site to learn how to manage an open-space corridor like this one, which has reemerged as an urban resource.

the world's largest subway networks during a 15-year period of rapid expansion beginning in the mid-1990s.

Clean-up efforts have paid off in many cities, particularly at the street level, where residents experience direct impacts of environmental quality every day. However, in the switch from "productive to "consumptive" cities, some problems are more intractable. Most notable in this regard have been the incredible rise in automobile use and the seemingly endless proliferation of solid waste on urban outskirts. There are also problems with agricultural land being pulled out of food production to construct new housing developments, golf courses, and large-scale factories. Other urban problems have origins beyond the city boundaries, such as the dust storms that plague Beijing, cities of northern China, and the Korean Peninsula. These notorious dust storms have on occasion also swept through to more southern cities, including Shanghai. However, while perhaps most visible, dust from construction and sand storms is less worrisome than small-sized pollutants emitted by private automobiles and factories operating within or near urban areas. Levels of air pollution in recent years have risen to alarming levels in winter, especially in northern China. Beijing's pollution in 2014 was so severe that headlines around the world referred to it as an "airpocalypse," with images of commuters wearing protective facemasks. City governments scramble to respond, introducing strict limits on new car registrations and plans to move pollution-causing industries away from urban centers.

Similar trends characterize the mixed environmental records in other cities of East Asia.

Hong Kong, Tokyo, Taipei, and Seoul have all undergone transformations that include a move toward less polluting forms of economic activity matched by increasing investment in and demand for cleaner air and water, and more environmental amenities such as public green spaces. In Seoul, the restoration of Cheonggye Stream in 2003 was part of a strategy to rebrand the city by adopting a sustainability urban management paradigm. This project fit well with urban revitalization efforts that sought to bring new life to what had become the somewhat barren streetscapes of the city's older core. After just over two years, 3.6 miles (5.8 km) of highway cutting through the city center were replaced by a restored stream, flanked by a linear park and walkway for pedestrians and cyclists, with car traffic limited to roads on the sides of the greenway (Figure 11.19). To compensate for the loss of the freeway, the city expanded bus lanes and improved links to Seoul's existing subway system. The restoration of Cheonggye Stream has earned praise for its direct environmental effects and the indirect benefits associated with increased flows of people back into the city core, which had lost much of its street-level life due to earlier waves of suburbanization. Large-scale greening projects can also accelerate gentrification. Urban renewal through the Cheonggye Stream Restoration Project increased property values and rents, driving out small industrial enterprises that were located in this less desirable area. Many of these enterprises, along with other low-cost businesses and housing, are being replaced through "green gentrification" by more affluent uses, including offices and new commercial businesses such as cafes, hotels, restaurants, and retail.

While restoration of Seoul's urban stream represents an ambitious multipurpose urban redevelopment scheme with local environmental benefits, East Asia is becoming known for even more ambitious urban environmental initiatives based on the creation of eco-cities, some of which are to be built from scratch. These master-planned cities include the New Songdo City (near Seoul), Dongtan (near Shanghai), and Tianjin Eco-City (near Beijing). While better known as the new administrative capital for South Korea, Sejong shares a similar futuristic image as a city that showcases sustainable infrastructure and environmental amenities. In most cases, eco-city plans call for integrating residential, commercial, and industrial developments, incorporating cutting-edge green technologies and emphasizing high-tech research and development. These eco-city projects are meant to be large and spectacular, with residential population targets in the hundreds of thousands. Such large numbers are consistent with the high-rise and high-density development pattern that is characteristic of regional urbanization. Among these planned eco-cities, Dongtan was the most high-profile project, though little has been done to move the project forward; many question if it will ever be completed. Other projects have begun construction, working their way through planning and consultations, often in collaboration with international design and engineering consortiums. Proponents of these projects argue that even if they are slow to be realized, they have positive impacts on urban vision in other cities and help drive development of green urban technologies, such as low-carbon and resource-efficient heating. Some critics worry that these eco-utopian communities will not achieve their ambitious goals because they are more about improving the image of government officials and the design firms than promoting the projects. Others worry that even if built as planned, they will be accessible only to highly educated and wealthier residents, providing limited opportunity for the poor to enjoy the benefits of East Asia's sustainable urban futures.

PROSPECTS FOR THE FUTURE

Urban residents in many poorer cities around the world must look with some envy at the more prosperous cities of East Asia. To citizens of the region, however, and especially to urban planners, the overall problem of most cities of East Asia is how far short the cities still fall from expectations. For example, leading Japanese observers note that Japan's foremost urban centers lack anything resembling the character and depth of their European counterparts. Instead, they seem to be forever under construction. They also see that there is huge potential demand for urban redevelopment, yet Japanese city planning shows little vision regarding a living environment. These may be excessively harsh criticisms from idealistic planners. Much the same could be said of the rest of the region. Continuous demolition and construction that leaves little history and character are the prices of rapid growth and economic success. But it is fair to say that cities of East Asia also respect ingenious designs to create hospitable urban habitats in the relatively unfavorable environments of high population pressure and scarce land.

So where do these countries and cities go from here? East Asia will likely continue to be a region of continuing relatively high economic growth with most countries pursuing a pro-business strategy. Urbanization will continue in China, but it will require some new thinking to integrate the migrant population. There are vast amounts of capital for urban development and for expanding information technology, but marshaling capital and talents to improve the quality of urban life and promote more equitable growth will remain major challenges. Undoubtedly, state-of-the-art technology, such as the bullet trains first pioneered in Japan, are now spreading in China and other parts of the region, but many question whether they will benefit the masses or just the affluent.

An analysis of China's current mammoth high-speed train project, almost totally government funded, shows that the project has benefited the rich and the middle class (i.e., about the top one-third of the population), who can afford the far higher fares. The project, rife with corruption from the beginning, has negatively impacted the lower-income groups because many cheaper "slow trains" have been taken off the rails. The enormous hardship that many low-income migrants must endure in the annual *chunyun*—the "spring movement" home and back during the Chinese New Year break—in the "bullet-train age" clearly demonstrates the regressive nature of some top-down "modern" projects.

Increasingly louder voices from the grassroots population in many cities of East Asia, especially in Hong Kong and Taiwan, cannot be always ignored as in the past. The cities of East Asia may be destined to play leading roles in world affairs in the twenty-first century, belonging as they do to one of the three power centers of the global economy. The top player of all may well be China—if things are done right—with its great cities superseding those of Japan, which dominated the region in the twentieth century.

SUGGESTED READINGS

Bruno, M., S. Carena, and M. Kim. 2013. *Borrowed City: Private Use of Public Space in Seoul.* Seoul,

Korea: Damdi Publishers. Fascinating visual catalog and discussion of everyday street life in Seoul, documenting the small ways that people appropriate public space and challenge urban order.

Hein, C., and P. P. Schulz. 2006. *Cities, Autonomy and Decentralization in Japan.* New York: Routledge. Overview and case studies focusing on decentralization of Japan's urban system and governance within cities.

Miller, T. 2012. *China's Urban Billion.* London: Zed Books. An up-to-date, well-written account of China's urbanization and its various complications, with many interesting snippets from the field.

Naughton, B. 2007. *The Chinese Economy: Transitions and Growth.* Cambridge, MA: MIT Press. A comprehensive introduction to the Chinese economy and its various mechanism and government policies.

Solinger, D. 1999. *Contesting Citizenship in Urban China.* Berkeley: University of California Press. An insightful study of problems faced by millions of migrant workers in Chinese cities at the turn of the century. Many points are still relevant today.

Sorensen, A. 2002. *The Making of Urban Japan: Cities and Planning from Edo to the Twenty-First Century.* London, New York: Routledge. An examination of Japan's urban development from earliest times to the present.

Wang, Y. P. 2004. *Urban Poverty, Housing and Social Change in China.* New York: Routledge. A comprehensive treatment of urban social issues in China in the reform era, with focus on urban poverty and housing.

Wu, W., and P. Gaubatz. 2013. *The Chinese City.* New York: Routledge. A systematic overview of urban development in China, including two chapters on the traditional period.

Yusuf, S., and K. Nabeshime. 2006. *Post-Industrial East Asia Cities.* Stanford: Stanford University Press and World Bank. A book on technologies and innovations in several major cities in East Asia as they move away from manufacturing.

Zhang, L. 2010. *In Search of Paradise: Middle-Class Living in a Chinese Metropolis.* Ithaca, NY: Cornell University Press. An ethnography of the different ways in which China's new middle class is affected by the experience of home ownership.

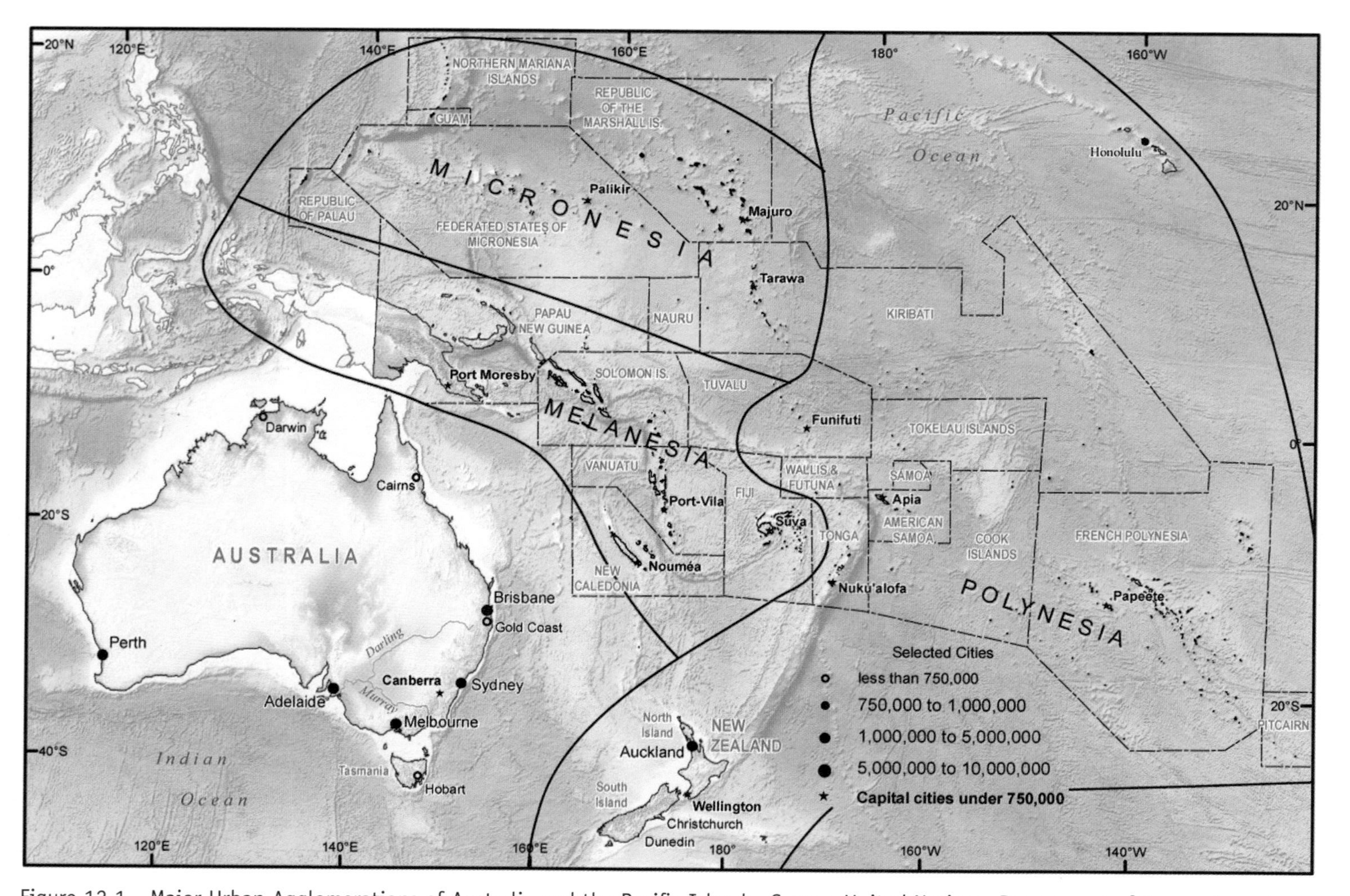

Figure 12.1 Major Urban Agglomerations of Australia and the Pacific Islands. *Source:* United Nations, Department of Economic and Social Affairs, Population Division (2014), World Urbanization Prospects: 2014 Revision, http://esa.un.org/unpd/wup/.

12

Cities of Australia and the Pacific Islands

ROBYN DOWLING AND PAULINE MCGUIRK

KEY URBAN FACTS

Total Population	39 million
Percent Urban Population	71%
Total Urban Population	28 million
Most Urbanized Countries	Australia (89%)
	Northern Mariana Islands (89%)
	New Zealand (86%)
Least Urbanized Countries	Papua New Guinea (13%)
	Solomon Islands (21%)
	Samoa (19%)
Number of Megacities	None
Number of Cities of More Than 1 Million	6 cities
Three Largest Cities	Sydney (5 m), Melbourne (4 m), Brisbane (2 m)
World Cities	4 (Sydney, Melbourne, Auckland, Brisbane)

KEY CHAPTER THEMES

1. Cities in this region may be divided into two regional groups—those of Australia and Aotearoa/New Zealand and those of the Pacific Islands—each with distinct characteristics.
2. Australia and Aotearoa/New Zealand exhibit many of the urban characteristics of other developed countries, such as the United States and Canada.
3. The urban character of Pacific Island cities is similar to that of less developed countries though they are smaller and have considerably lower rates of population growth.
4. All countries in this region are dominated by primate cities, but in the case of Australia, primate cities are the capitals of states in the federal union.
5. Many of the cities in the region were established as colonial or national capitals, and urban patterns and character are tied to this political influence.

6. Sydney is by far the most globally linked city and the key economic center in this vast realm, though the global economic, cultural, and social connections of all cities have increased dramatically.
7. In Australia, a popularly documented "sea change" phenomenon is drawing people away from the big cities toward small but growing coastal towns.
8. Suburbanization and gentrification remain key residential forces in Australia and Aotearoa/New Zealand, and globalization is a central driver of urban economies.
9. A multicultural population is increasingly the norm in most cities in the region, especially in Australia and Aotearoa/New Zealand.
10. Awareness of the environmental impacts of urbanization is rising, and environmental vulnerability, especially to the direct and indirect consequences of climate change, is a key issue confronting the future of cities in the Pacific Islands.

The Pacific region is a constellation of islands of varying sizes (Figure 12.1). Australia (the island continent) and Aotearoa/New Zealand (now carrying both Maori and Pakeha, or settler, names) dominate the region geographically and economically. However, many smaller islands are found in those vast realms of the Pacific Ocean known as Melanesia, Micronesia, and Polynesia. Socially, politically, economically, and biophysically, this is a diverse region with diverse cities.

In this part of the world, it is easiest to understand cities as forming two main groups: those of Australia and Aotearoa/New

Table 12.1 Australia and Aotearoa/New Zealand: Changes in Distribution of National Population

Nation/Cities	*% National Population 1981*	*% National Population 2013*
	Australia	
Sydney	21.8	20.9
Melbourne	18.6	19.1
Brisbane	7.2	9.8
Perth	6.2	8.7
Adelaide	6.3	5.7
Hobart	1.1	1.0
Darwin	0.4	0.6
Canberra	1.6	1.7
	Aotearoa/New Zealand	
Auckland	26.1	33.4
Christchurch	10.1	8.0
Wellington	10.8	4.5
Dunedin	3.6	2.8

Sources: New Zealand, *Census of Population and Housing 2013*; Australian, *Bureau of Statistics Estimated Resident Population 2013.*

Zealand, and those of the Pacific Islands. The former includes cities with characteristics of more developed countries: industrialized, with a generally high level of affluence, and connected to global flows of people, money, information, and services. There are two key urban characteristics shared by both these nations. First, they are urban. Currently, over 89 percent of Australia's and 86 percent of Aotearoa/New Zealand's population live in urban areas. Second, they are, and long have been, nations of urban primacy: their urban pattern is dominated by a small number of large cities. Approximately one-fourth of all Aotearoa/New Zealanders live in just one city—Auckland—and Australia's two largest cities—Melbourne and Sydney—are home to more than 40 percent of the nation's population (Table 12.1).

The islands within Micronesia, Polynesia, and Melanesia have starkly different urban characteristics. They have highly nonurban

Table 12.2 Population of Pacific Island Cities

ISLAND NATION/City	*Population (2015)*	*% of Country's Population*
FIJI	909,389	
Suva	77,366	8.5
Nadi	42,284	4.6
Lambasa	24,187	2.6
KIRIBATI	105,711	
Tawara	40,311	38.1
Betio Village	12,509	11.8
Bikenibeu Village	6,170	5.8
MARSHALL ISLANDS	72,191	
Majuro	25,400	35.2
VANUATU	236,486	
Vila	35,901	15.2
Luganville	13,397	5.7
Norsup	2,998	1.3
TONGA	106,501	
Nukualofa	22,100	20.7
Neiafu	4,320	4.1
Havelu	3,417	3.2
SOLOMON ISLANDS	622,469	
Honiara	56,298	9.0
Gizo	6,154	1.0
Auki	4,336	0.7
SAMOA	197,773	
Apia	40,407	20.4
Vaitele	5,631	2.8
Faleasiu	2,592	1.3
PAPUA NEW GUINEA	6,672,429	
Port Moresby	283,733	4.3
Lae	76,255	1.1
Arawa	40,266	0.6

Source: Country Watch 2015 *Country Profiles*, http:// www.countrywatch.com/

populations. Although reliable statistics are difficult to obtain, it is estimated that 35 percent of the population lives in urban areas, with a projected increase to over 50 percent by 2025. There are 35 towns and cities with a population greater than five thousand. Two-thirds of the southwest Pacific realm's urban dwellers are to be found in Papua New Guinea (PNG) and Fiji, the most populous nations in the region (Table 12.2). The region's largest cities—Port Moresby (PNG), Nouméa (New Caledonia, still a French possession), and Suva (Fiji)—are tiny by world standards. Negligible population growth is occurring in these cities, where economic opportunities remain limited. In Pacific Island nations, prestige and status are still very much tied to the land and the rural, rather than to cities and the urban.

HISTORICAL FOUNDATIONS OF URBANISM

Australia, Aotearoa/New Zealand, and the Pacific Islands have indigenous peoples with long histories of settlement, up to 40,000 years in the case of Australian Aboriginals. Cities in this part of the world are, however, very young. Urban settlement began with the advent of numerous colonizers in the eighteenth and nineteenth centuries. Australia became a penal colony of the British in 1788, with the arrival of convicts to Sydney and Port Arthur (near Hobart, Tasmania) and later to Brisbane (Figure 12.2). The continued arrival of convicts to these coastal towns and the establishment of additional settlements like Melbourne and Adelaide, for purposes of colonial administration, commerce, and

Figure 12.2 One of The Travelers on Melbourne's Sandridge Bridge represents the convict era in Australian history. The former railroad bridge is now a pedestrian crossing and sculpture garden. *Source:* Photo by D. J. Zeigler.

trade cemented metropolitan primacy. The political independence of each of the British colonies (later to become Australia's six states) also meant that the capital cities operated independently of each other throughout the nineteenth century, providing services to their rural hinterlands, acting as ports for the import and export of commodities to and from Europe, and functioning as centers of colonial administration. Indeed, competition between the capitals further worked to bolster primacy. With each capital focused on ensuring continued economic growth, backed by political force within their respective territories, the establishment of alternative, prosperous, and comparable urban centers was made more difficult.

Two major events of the mid- to late-nineteenth century further enhanced the size, functions, and importance of Australia's six colonial capitals. Railroads focused on the capitals, facilitating more efficient connections between the cities and their hinterlands. Industrialization similarly occurred within (rather than beyond) these coastal centers of colonial administration, though there were to be later exceptions like Wollongong and Newcastle in New South Wales, and Whyalla in South Australia. By 1900, Australia had a total population of a little less than 4 million; Sydney and Melbourne each had populations of approximately half a million; Adelaide, Brisbane, and Perth more than 100,000 each; and Hobart remained small at 35,000 people. Colonialism had hence been responsible for this uniquely Australian urban primacy and settlement pattern in at least two ways. First, the sites of European settlements (either convict or free), with their coastal locations and trading functions, formed the foundations of the colony and its growth (Figure 12.3). Second, the functions of colonial administration, and

Figure 12.3 Adelaide is the state capital and primate city of South Australia. It was founded as a planned capital city for a new British colony in the 1830s. *Source:* Photo by D. J. Zeigler.

competition among the capitals fueled the growth of existing rather than new urban centers.

The first half of the twentieth century saw urban Australia grow in the spatial pattern established by British colonialism. A manufacturing boom that began in the 1920s reinforced the primacy of each state capital. This era also saw the beginning of the systemic suburbanization of Australian cities. The establishment of middle-class suburbs in attractive surroundings away from the central city was facilitated by the development of public-transport lines radiating out from the city center, as well as the activities of land developers and house builders. With the absence of inner-city slums on the scale of those in Britain, the social differentiation of Australian cities took on characteristics of the sector model related to transport links and features of the natural landscape.

The turn of the twentieth century did see one challenge to the existing state capitals: the planning of the new city of Canberra. The federation of Australia's colonial territories in 1901 was designed to both create and unite a nation. The six colonial capitals became capitals of states in the newly formed Commonwealth of Australia, and a new national capital—Canberra—was established between the two cities that dominated the national urban hierarchy—Melbourne and Sydney. The decision to locate the national government in the newly formed Australian Capital Territory (ACT) between the country's two largest cities was a compromise. The Australian Parliament

Figure 12.4 Canberra's distinctive but controversial Parliament House is difficult to appreciate from the outside because much of the structure is underground. The inside is breathtaking, filled with beautiful art and materials native to Australia. *Source:* Photo by D. J. Zeigler.

did not formally relocate from Melbourne to Canberra until 1927, and even today, the ACT remains comparatively small, with fewer than 400,000 inhabitants (Figure 12.4). Its dominating characteristic is the prominent role played by formal urban planning. A master plan developed by an American, Walter Burley Griffin, guided its development as a "garden city" built around a large lake, with a central focus on a "parliamentary triangle," and satellite suburbs with town centers of their own. Canberra's expansion was slow—only 16,000 people lived there in 1947—and its early economy was reliant on public service and diplomatic functions. Today, its economy is supplemented by a large student population that attends the relatively large number of public and private institutions of higher learning, including the Australian National University.

In Aotearoa/New Zealand, European settlement and modern urbanization began in 1840 with the signing of the Treaty of Waitangi between the British and the native Maoris. Unlike the convict bases of Australia's settlements, free setters in Aotearoa/New Zealand were encouraged to migrate and invest, with the resultant economy largely dependent on pastoral activities like grazing sheep and cattle. Unlike Australia, urban primacy was not a nineteenth-century phenomenon here, due to the originally more dispersed settlement pattern and more diverse reasons for urban settlement. For example, early towns like Wellington and Christchurch were established by trading and/or religious interests; Auckland's natural harbor made it an ideal port (Figure 12.5); and gold rushes underpinned the growth of Dunedin. Thus, by 1911, Auckland had a population of 100,000, Christchurch 80,000, Wellington 70,000, and Dunedin 65,000. Over half of the non-Maori population lived in urban areas. In contrast,

Figure 12.5 Built on an isthmus and connected to a rich hinterland, Auckland now hosts many activities found in major world cities, including the famous Sky Tower that dominates the skyline. *Source:* Photo by D. J. Zeigler.

throughout the nineteenth century and the first half of the twentieth century, Maori settlement was predominantly rural.

Like Australia and Aotearoa/New Zealand, Oceania has had a long-established indigenous population. Similarly, it was the colonial context that underpinned the urban system of the region. Oceania was one of the last regions of the world to be colonized, with British, French, American, and Dutch powers establishing presences in countries like Fiji, Samoa, Tonga, and Vanuatu at various times across the nineteenth century. Towns first developed as trading ports, usually close to existing villages, good harbors, and viable anchorages. These towns grew slowly, and some, like Levuka in Fiji, declined over time because of relative inaccessibility. They were never large: in 1911, Suva had a population of only 6,000 people, about 5 percent of Fiji's population.

The first half of the twentieth century saw a diversification of urban functions and sporadic urban growth. Although widespread industrialization did not occur, the processing of agricultural commodities like sugar, and the extraction of resources through mining, diversified the economic base and saw the growth of cities in Fiji and New Guinea, where the mining towns were nearly as large as the colonial capital of Port Moresby. In Micronesia, intense Japanese colonialism saw cities like Koror, on the island of Palau, grow substantially; other administrative capitals grew slowly. By the middle of the twentieth century, urbanization remained limited.

CONTEMPORARY URBAN PATTERNS AND PROCESSES

The contemporary urban systems of Australia, Aotearoa/New Zealand, and the Pacific Islands are based upon the patterns established in previous decades. Economic, social, and political influences across the region have consolidated urban primacy. Urbanization processes, the overall urban pattern of the region, and the characteristics of cities within it, are far from uniform. For cities of the Pacific Islands, tourism, political independence, instabilities, migration, and environmental hazards play significant roles. In Australia and Aotearoa/New Zealand, in contrast, industrialization followed by deindustrialization, globalization, international immigration, urban governance, and rural/urban population dynamics are the primary influences.

The Pacific Islands

The historical pattern of urban primacy in a largely nonurban region remains a hallmark of the Pacific's urban geography (Table 12.2). By 1960, only Suva (Fiji) and Noumea (French Caledonia) had populations greater than 25,000, and even today the size of the cities remains small. Political independence from colonial powers began in the 1970s. Only a few territories, such as New Caledonia, remain in colonial hands. Independence had a number of significant impacts on the region's urban system. Colonial administration was no longer the primary purpose of the largest cities in the region, but processes associated with independence cemented the primacy of these towns. In some, like Port Moresby, PNG, independence fostered urban growth because of new investment in urban housing and services (Figure 12.6). Across the region, accelerated urban growth followed independence because of, for example, the removal of negative perceptions of urban living, or the establishment of some countries as tax havens (e.g., Port Vila, Vanuatu). Independence also

Figure 12.6 The Papua New Guinea High Commission, with its distinctive Pacific aesthetic, is located in Australia's national capital, Canberra. Members of the Commonwealth of Nations exchange High Commissioners instead of Ambassadors. *Source:* Photo by D. J. Zeigler.

required bureaucracies in national capitals, and encouraged education and urban living in general.

Land and land tenure systems are a defining characteristic of Pacific cities. In Melanesia, Polynesia, and Micronesia, customary land tenures pose significant challenges for urban growth, housing, and infrastructure provision as well as the quality of urban life. In Port Moresby, for example, traditional owners hold one-third of the city's total area, and land is seen as a communal resource. However, customary land tenure places limits on the land available to house urban residents and is associated with higher housing costs. It also provides a disincentive to invest in land development and urban infrastructure. A number of possible solutions to the limitations that customary land tenure places on capitalist urban growth have been proposed. These include proposals to lease customary allotments, or the ability to use land to generate income through means other than compensation. Such proposals have been severely hindered by the limited capacity of urban governance across the islands.

Connected to issues of land tenure are the general housing characteristics of the urban Pacific. Palatial houses exist, but they are often built by expatriates and in gated communities. Formal housing of the type commonly found in Australian and Aotearoa/New Zealand cities exists as well. Far more common, however, are informal settlements. The great demand for housing, in the context of substantial urban poverty and limited employment opportunities, means that informal housing is common. Public housing is available, though waiting lists are extremely lengthy.

Finally, the present and future of the cities of the island Pacific cannot be understood without reference to environmental contexts and threats. Urban settlement has involved degradation of islands' fragile coastal environments. The waste and water requirements of growing urban populations threaten to overwhelm already stressed ecosystems. Urban water is typically sourced from freshwater lenses, and if these are over pumped, saltwater contamination can occur and render the water unsuitable for human use. Because of the geology of the islands, waste disposal also affects the environment. Other forms of water supply contamination can occur (e.g., by chemicals, sewerage), which in turn affects human health. The most important environmental issue for these cities in the twenty-first century is climate change, especially global warming. The low-lying islands, and their cities, are at risk of inundation because of sea-level rise. Climate change is also believed to involve increased storm activity, accelerated coastal erosion, saltwater intrusion into reserves of fresh water, and increased landward reach of storm waves. Each of these events has the potential to dismantle city infrastructure and threaten urban livelihoods. Environmental hazards are further exacerbated by social vulnerabilities, especially limited institutional capacities for urban planning. In 2010 at the UN Framework Convention on Climate Change meeting in Cancun, the Deputy Prime Minister of Tuvalu classed climate change as a "life or death survival issue," threatening the very existence of this Pacific Island nation. The highest point on Tuvalu's capital island, Funafuti, is less than 14 feet (4.3 m) above sea level.

The global economic context is crucial to urban economies in the Pacific. Many nations, like Fiji, have turned to tourism for economic survival, with urban consequences. Global commodities and mining, as well as the presence of wealthy expatriates, underpin the urban hierarchy of PNG. And finally, international migration, in particular emigration, can relieve some of the social, economic, and environmental pressures in cities. In Tonga especially, migration to Aotearoa/New Zealand, Australia, and the United States operates as an urban "safety valve," allowing Tongans to realize economic opportunities overseas rather than in overcrowded and economically limited urban areas. This safety valve has also become part of new, informal, urban economic activities.

In sum, cities of the island Pacific are places of vulnerability and opportunity. In a largely nonurban context, in which effective urban planning and coordination are nonexistent at worst and problematic at best, urban living is still sought as a chance for a better quality of life. Though officially derided, life in informal settlements remains attractive.

Australia

The dominance of state capital cities remains the defining characteristic of Australia's urban system. The primary drivers of urban development in the twentieth century—industrialization, migration and, latterly, globalization—have only reinforced the importance of state capitals and fueled their population growth. Between 1947 and 1971, the population of Australia's five largest cities doubled, and growth has continued since then. Historically, Sydney (capital of New South

Wales) and Melbourne (capital of Victoria) have been the island continent's largest and most economically dominant cities. Australia's manufacturing growth after World War II was centered in Melbourne, which, until recently, housed the majority of Australian corporate headquarters (Figure 12.7). Other state capitals served their rural and resource-based hinterlands, with smaller and less diversified economic bases. In the immediate postwar period, Adelaide was somewhat of an exception, as the center of Australia's car industry.

Aboriginal Australians are much less likely to be urbanized than the broader Australian population. They are also more likely to live in small towns rather than large cities. Indeed, a little over 1 percent of Sydney's total population, and 1.7 percent of Perth's population are indigenous. Indigenous movement to capital cities is often temporary, and linked to kinship and friendship ties with rural areas. Aboriginal people have long dwelled on the fringes of cities, often in substandard housing. Places of residence within the city are related to the provision of public housing and also localities with strong identification for indigenous Australians. One of these places is "The Block," in Sydney's inner-city Redfern, where housing and other cultural services are concentrated.

The past 25 years have seen some shifts in the distribution of economic and population growth across Australia's large cities. Two factors underpinned these slight alterations in the urban system. The first was the influx of people into Australian cities through international migration. For the past 20 years, more than 100,000 people annually have migrated to Australia from around the world, most of these to the capital cities, particularly Sydney, Brisbane, and Perth. Cities that have not received substantial numbers of migrants, like Adelaide and Hobart, have declined in relative

Figure 12.7 Melbourne's traditional image is being shattered today by skyscrapers like Eureka Tower (world's tallest residential building when built) and Deborah Halpern's Angel, a sculpture with roots in the aboriginal aesthetic of Australia. *Source:* Photo by D. J. Zeigler.

terms. The second factor was globalization, or more specifically changing urban functions as the Australian economy became increasingly tied to, and driven by, global flows of commodities and money, and increasingly reliant on globally networked business services. Globalization has seen Sydney rise in prosperity and prominence to become Australia's only world city. The headquarters of Australian-based businesses, and the regional offices of multinationals, are now more likely to be in Sydney than in Melbourne. The relative growth of Brisbane and its surrounding region during the same period can be attributed to internal migration (principally from Sydney), the rise of a tourist-based economy, growing economic ties between Brisbane and the Asia-Pacific region, and Queensland government incentives for business to relocate to Australia's sunbelt. Connections to Antarctic tourism and scientific activities are emphasized in the southern-most capital of Hobart (Box 12.1).

Australia's state capitals are highly suburbanized and geographically expansive by international standards (Box 12.2). Historically, the predominant housing preference is for a detached house, producing sprawling suburban conurbations (Figure 12.8) like that between Brisbane and the Gold Coast, 37 miles (60 km) away. The continued proliferation of suburban housing is currently under some threat. The high energy demands of suburban life—use of the private car, heating, cooling, and the water-use demands of large houses—are increasingly questioned. Limited availability of land and the high costs of servicing the social and physical infrastructure needs of new suburbs have led to policies of urban consolidation across the nation, with an emphasis on sustainable building practices (Box 12.3). Mixed-use residential and commercial developments on old industrial land are increasing, and in some years the construction of new apartments outstrips that of detached houses. Equally important is a cultural and economic reevaluation of living in Australia's inner cities. Australian inner cities are vibrant, cosmopolitan spaces, with a wealth of retail, social, and recreational opportunities; and they are highly accessible by public transport (Figure 12.10).

The internal structure of Australian cities has changed over the past three decades. Based on an analysis of social and economic characteristics, metropolitan localities may be divided into seven types of places (Figure 12.11): three advantaged and four disadvantaged. In *new economy* localities are found people employed in new global industries and many educated professionals. *Gentrifying* localities are found across Australia's inner cities, and are home to those with ties to the global economy but with a sizable proportion of low-income residents as well. *Middle-class suburbia* houses many educated professionals, though with a low density of connections to the global economy. *Working-class battler* communities have trades people, often homeowners, while *battling family communities* have above average levels of single-parent and nonfamily households. In *old-economy* localities, primarily suburban and especially in Adelaide, the decline of manufacturing has seen concentrations of unemployment. Finally, *peri-urban* localities on the fringe of the capitals attract low-income people seeking cheaper housing or homes for retirement.

While state capitals have, on average, been growing, small towns in rural and regional Australia have exhibited divergent patterns. Many rural towns, traditionally operating as service centers for surrounding farms, have experienced population declines. Decreasing

Box 12.1 Hobart as a Gateway to Antarctica

Hobart is Australia's southernmost capital city, located at approximately 43 degrees south of the equator on the island of Tasmania. While the majority of Australian and New Zealand capital cities have strong connections to the Pacific and its islands, Hobart's location and history provide foundations for strong links to the Antarctic frontier, and the designation "gateway to Antarctica." Historical ties, research connections, and tourism underpin this designation.

Building upon a long history as a sealing and whaling port in the first half of the nineteenth century, Hobart became a key staging point for Antarctic explorations. French and British expeditions of the 1830s were pioneering, though there was a lull until a flourishing of scientific and exploratory visits in the 1890s. Hobart was involved in most of the storied Antarctic explorations of the early twentieth century, including those of Roald Amundsen (1910–1912) and the first Australian expedition by Sir Douglas Mawson (1911–1914). Hobart was used to gather supplies before ships departed, and also as a site from which to announce the success (or otherwise) of voyages upon their return.

Hobart remains a hub for Antarctic scientific exploration today. A number of key research bodies concerned with Antarctica and its surrounding oceans are either based in or networked through Hobart, such as the International Antarctic Institute and the Australian Antarctic Division. The latter is responsible for overseeing Australia's engagement with the Antarctic territories, both scientific and more broadly.

Tourism is an increasingly critical element of Hobart's economic fortunes. While the majority of visitors to Antarctica leave from South America, a small number depart via ship or plane from Hobart, typically destined for East Antarctica. These journeys take between 7 and 14 days by ship and 4.5 hours by plane. For those unable to afford the time or expense of such journeys, Hobart also offers visitors recreations of the Antarctic expeditions of the twentieth century. Hobart's Constitution Dock houses the Mawson's Huts Replica museum, a series of buildings that recreate the physical sensations of the huts lived in by Douglas Mawson and his team during their expedition of 1911–1914. Built from the same materials, and with the use of digital audio to recreate a windy Antarctic landscape, the huts enable visitors to experience what life was like for Mawson and his team of 18 men. This theme of replicating Antarctic experiences is found at a number of other sites in Hobart, such as a sub-Antarctic plant house at the botanical gardens, and a walking tour of the significant sites and moments in Antarctic exploration.

farm incomes, the closure of many public and commercial services such as banks, and limited employment and education opportunities for young people have encouraged migration out of these towns and into larger regional centers or, more commonly, capital cities. A counter trend of growth in Australia's coastal towns is also evident. The twenty-first

Figure 12.8 Sydney is known as a city of suburbs and single-family homes such as this one. *Source:* Photo by Robyn Dowling.

century boom in resource prices has meant that towns in coastal Australia have grown rapidly, instigating severe housing shortages and consequent escalations in-house prices. The "sea change" phenomenon, in which city dwellers swap a hectic city lifestyle, transport congestion, and high housing costs for a slower pace of life and cheaper housing in coastal towns is also important. Initially confined to older people, principally retirees and those nearing retirement, sea changes are now undertaken by young professionals able to run businesses outside the major cities, as well as less affluent families seeking cheaper home ownership. Towns like Byron Bay, Coffs Harbour, and Port Macquarie in New South Wales, Barwon Heads in Victoria, and Denmark in West Australia are commonly identified sea-change locations. "Tree change" is a more recent but similar phenomenon in which urban dwellers move to greener locations like rural Tasmania, inland New South Wales (e.g., Orange, Mudgee) or Victoria (e.g., Daylesford).

Aotearoa/New Zealand

After World War II, the growth trajectories of the cities in Aotearoa/New Zealand largely paralleled those of Australia. The four largest cities of Auckland, Wellington, Christchurch, and Dunedin continued to grow, as did the primacy of Auckland (Table 12.1). A number of processes underpinned this pattern. Market reforms since the 1980s have strengthened global economic, cultural, and social ties, which in turn have transformed large cities. Second, immigrants, initially from the Pacific Islands but also more recently from China and India, have flowed into the large cities, especially Auckland and Christchurch. The third factor is the internal shift in economic activity. While a general process of deindustrialization in Aotearoa/New Zealand occurred in the late twentieth century, employment losses in manufacturing were more severe in Wellington, Christchurch, and Dunedin; and some manufacturing relocated to Auckland. Finally, entrepreneurial urban governance processes were deployed to make cities more attractive

Box 12.2 The Geography of Everyday Life in Suburban Sydney

Figure 12.9 New roles for women, and new problems, have emerged in Australian cities over the past three decades. *Source:* Courtesy of Robyn Dowling.

Australia is a suburban nation. Despite increasing urban consolidation and gentrification, more than 72 percent of Sydney's population live in detached housing, and 33 percent in areas more than 9 miles (15 km) from the city center. Suburban Sydney, unlike its North American counterparts, is heterogeneous. Sydney's greatest concentration of migrants is found in its suburbs, and hence we see pockets of affluence and poverty neighboring each other. What is everyday life like in this differentiated world city?

Suburban Sydney residents live in houses of varying age and design. New houses are more likely to be large—27 percent of houses have four or more bedrooms, a double garage, formal and informal living areas, separate rooms for each child, perhaps a games/media room and a backyard that may just be able to accommodate a cricket pitch. Family members—both adults and children—typically know their immediate neighborhood and participate in local sporting and recreational activities. The family shops locally, sometimes at a small corner shop or on a main street, but more likely at a supermarket in a large shopping mall. Here, not only can they pick up their weekly provisions, but they can also eat a meal and see a movie.

Daily travel patterns are increasingly complex spatially and socially. One adult (more likely male) will commute to the CBD for his job in the finance or business sector or to another suburb for manufacturing employment. The woman is likely to work in this or a nearby suburb, most likely in retailing or a similar service-sector job in banking, hospitality, or education. The limited availability of public transport in certain parts of suburban Sydney, and the generally poor servicing of cross-suburban travel, mean that these journeys to work are most likely to be undertaken by car. For mothers of young children, the importance of the car is even more pronounced, as she drops children at school/childcare on her way to work, and takes them to social and sporting activities on the way home (Figure 12.9). For these suburbanites, the time and cost of car travel is becoming an increasing burden, though with no relief in sight.

Box 12.3 Green Buildings

The challenges of reducing consumption of finite resources—especially water and fossil fuels—in cities of Australia and Aotearoa/New Zealand are great. National scale policies that encourage the reduction of demands for energy and water and/or promote the use of renewable sources of energy are sparse. At the building scale, the picture is more positive. Encouraged by local government policies and building regulation, new housing developments in the inner cities are innovatively embracing low-energy infrastructures. The Central Park development in inner Sydney is a salient example. Occupying the 5.8-hectare site of a former brewery, Central Park consists of 8 residential, commercial, and heritage precincts with an eventual expected occupation of 4,000 residents. Building a neighborhood that was sustainable across multiple dimensions was a goal of the development. The use of nonrenewable sources of energy is achieved with an onsite tri-generation facility that provides cooling, heat, and power to site. Water supply is harvested from rainwater collected in tanks at various parts of the site, while wastewater from commercial, residential, and garden uses is collected and recycled for use in cooling systems, toilets, and landscape irrigation. Finally, the buildings have both green walls and green roofs, planted with native vegetation and watered with recycled water, with the aim of providing not only more visually appealing facades but also natural means of cooling. Importantly, this neighborhood was reliant on financial and regulatory support from various state agencies. This includes subsidies and grants from the New South Wales Government, as well as the sustainability measures implemented by the City of Sydney.

Source: Central Park Sydney. 2015. *Central Park Sydney*. [ONLINE] Available at: http://www.centralpark-sydney.com

Figure 12.10 The advantage of high population density and compact urban form is that you can walk or bike to Old Victoria Market in Melbourne for the freshest of fruits, and vegetables. *Source:* Photo by D. J. Zeigler.

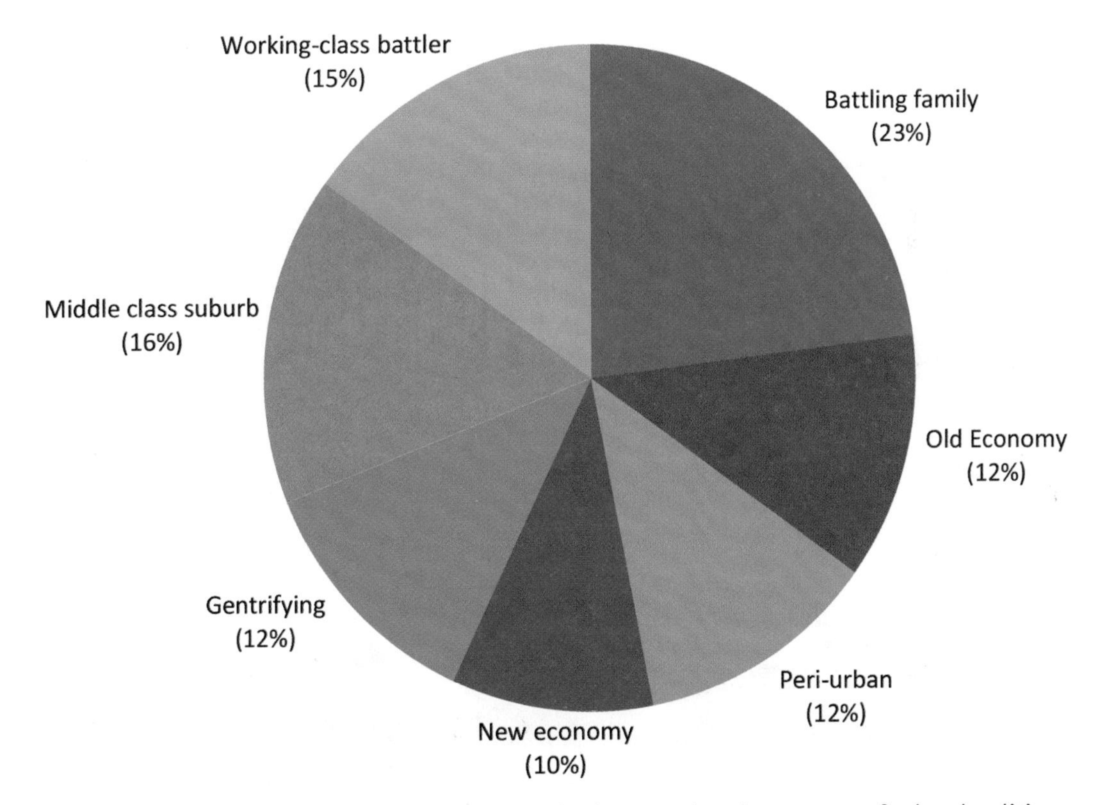

Figure 12.11 Changes over the past three decades have produced new types of urban localities in Australia. *Source:* Compiled by authors from statistics in Scott Baum, Kevin O'Connor and Robert Stimson, *Faultlines Exposed* (Melbourne: Monash University ePress, 2005).

and to stem population decline. In Wellington, New Zealand's capital, the waterfront was redeveloped using both public and private sector investment. The aim was for the city to become an international conference venue, and the government also located the new Te Papa National Museum there.

Aotearoa/New Zealand cities are low density, though suburban living is no longer the only residential option as high- and medium-rise apartments are becoming more common. The proportion of Maoris living in urban Aotearoa/New Zealand is now almost on par with that of the non-Maori population, because of the loss of Maori land and consequent rural-to-urban migration. Maoris face significant disadvantages in the cities, with high rates of unemployment and lower levels of home ownership and education. Increasing ethnic diversity is also an important urban characteristic.

DISTINCTIVE CITIES

Sydney: Australia's World City

With a population currently of about 4.5 million, and projected to reach 5.7 million by 2031, Sydney is the most populous and most prosperous city in Australia. The city is home to some of Australia's most widely recognized iconic landmarks: the Harbour Bridge (Figure 12.12), the Opera House (Figure 12.13), and Bondi Beach. It is an international finance market; it attracts a growing concentration of corporate headquarters; and

Figure 12.12 Completed in 1932, the Sydney Harbour Bridge opened up the city's North Shore. Tourists, tethered by lifelines, have been climbing the arch since 1998. *Source:* Photo by D. J. Zeigler.

Figure 12.13 Now a UNESCO World Heritage site, the Sydney Opera House has become a symbol of the island continent. *Source:* Photo by D. J. Zeigler.

it is Oceania's highest value-generating economy and dominant world city. Equally, it demonstrates some of the defining characteristics of contemporary Australian urban life: suburbia, urban-based prosperity arising from an advanced service economy, multiculturalism, and environmental threat (Box 12.4).

Sydney entered the twentieth century as the primate city and highest order service center in the state of New South Wales. By 1911, just 123 years after first European settlement, it had a population of 652,000 and was already a city of suburbs. Sydney's post–World War II "long boom" brought unprecedented economic and population growth and set in motion the formative settlement patterns that have shaped the contemporary city. Between 1947 and 1971, population expanded by 65 percent to reach 2.8 million; it grew to almost 5 million by 2015. The vast majority of growth has been accommodated in expansive suburban developments, including large-scale public-housing estates built mainly across the city's western suburbs. Despite planned expansions of public-transport networks, the rate of urban expansion and rising levels of car ownership meant that the city quickly assumed the car-oriented form of autosuburbia, connected by networks of freeways rather than public-transport corridors. Speculative developers' and housing consumers' preferences for low-density, detached dwellings meant that the city assumed a sprawled metropolitan form, poorly served by the existing rail network radiating from the Central Business District (CBD). Twenty years of urban

Box 12.4 Multiculturalism and Local Government in Australia

Cities in Australia have long been immigrant cities. After World War II, labor migration to Australia was dominated by people from the United Kingdom, Ireland, and southern Europe. The 1980s and 1990s saw a shift to the countries of southeast Asia, and more recently toward refugees from Africa and the Middle East. Hence, cities like Sydney are characterized by considerable cultural diversity. It is largely within urban neighborhoods that "everyday multiculturalism," the ordinary living of cultural diversity, occurs. Sometimes, this engenders conflict, as seen in the following excerpt from an article by a religious affairs reporter for the *Sydney Morning Herald*.

A Muslim centre built in the heart of Sydney's Bible Belt is facing fresh opposition—over its plans to host midnight prayers. But plans to extend the Annangrove prayer centre's hours and permit it to open late at night on three holy days have attracted four objections—well short of the thousands of complaints that almost blocked its construction four years ago. . . . The trustees [of the Imam Hasan Centre] want permission to open the doors until midnight three times a year, an increase in capacity from 120 to 150 people and a 45-minute extension in operating hours to permit cleaning and the occasional committee hearing. "Can you tell me any church that has any time restriction or limit on numbers?" said Abbas Aly, one of the centre's trustees. . . . If you ring up our neighbours they'll tell you they hardly notice us here. It's hardly used midweek and most of our programs are on a Saturday."

The [Baulkham Hills] council originally refused to approve the centre when more than 900 local residents claimed its existence threatened the ambience and character of the semi-rural suburb, in Sydney's north-west. Mr Aly appealed to the Land and Environment Court, which reversed the decision on the grounds that the local objections were not based on facts.

Once construction started, the site was vandalised, sprayed with racist graffiti and smeared with animal offal. Pigs' heads were impaled on wooden stakes. Mr Aly said tensions between local people and the centre had long since dissipated, except for the occasional persistent critic, especially as it had become clear that the centre looked more like a community centre than a mosque. "We've had quite a positive response to our latest development application from neighbours, compared to the 8500 complaints to our construction. We get quite a number of people who have come in to apologise. I asked them did they see the plans, they said, 'No, we just believed what we were told', and I take my hat off to them for coming in and making their peace." The Mayor of Baulkham Hills, Tony Hay, said four complaints had been lodged against the variation in consent orders, mainly expressing concern that creeping changes were undermining the intent of the original Land and Environment Court proceedings. No decision had been taken yet. . . .

Source: Linda Morris, "Midnight prayers raise objections," *Sydney Morning Herald*, September 10, 2007.

consolidation policy has contained the extent of sprawl, but strong population growth (50,000 per year since the late 1990s) has meant that fringe expansion has continued. Sydney's employment, retailing, and services have been decentralizing since at least the 1970s. The development of regional centers of commercial activity, such as Ryde, North Sydney, Parramatta, Penrith, and Liverpool has given the city an increasingly polycentric form. Indeed, the current metropolitan planning strategy labels Sydney as a "city of cities."

Despite Sydney's predominantly low-rise suburban form, the city center is characterized by high-rise office towers, global tourist landscapes and, lately, residential towers tightly grouped on the edges of one of the world's most spectacular natural harbors (Figure 12.14). Since the late 1960s significant waves of international property investment—in commercial office and hotel developments—have transformed the CBD's built environment, as has the transformation of Sydney's economic base to one dominated by increasingly globally connected financial and other advanced services. Sydney has become one of the most significant financial centers in the Asia-Pacific realm, making up 40 percent of Australia's telecommunications market. Employment in the global city sectors of finance, insurance, property, and business services is concentrated in and around the city center where many of the estimated 600 multinational companies who run their Asia-Pacific operations from Sydney are clustered, along with the headquarters of approximately 200 of Australasia's top companies. The economy of the city center now generates 30 percent of the value of metropolitan

Figure 12.14 Sydney's skyline, typical of a world city, dominates the capacious harbor. Can you identify Sydney Tower? *Source:* Photo by D. J. Zeigler.

Sydney's economic output and contains 28 percent of all metropolitan employment, with high concentrations in the highly paid professional and managerial occupations.

Concentrated in Sydney's city center are high-paid, advanced-services workers, as increasingly globalized connections have driven long-standing processes of gentrification, the recent resurgence of high-rise luxury residential dwellings, and the multiplication of globalized consumer spaces. Inner suburbs of nineteenth-century housing have been revitalized. New upmarket residential locales have been built in high-density, previously used land on the edges of the CBD, and in a host of high-rise high-density towers throughout the CBD. These developments have meant that the resident population of Sydney's inner city has increased by 40 percent since 1996. The development of a range of globalized consumer spaces, catering both to global tourists and to inner-city residents, has also transformed the city center. In the 1980s the New South Wales government redeveloped Darling Harbour container terminal as an international conference center, festival shopping, and entertainment precinct. In the 1990s, special legislation was passed to enable redevelopment of heritage wharves at Walsh Bay as an exclusive residential, commercial office, and restaurant precinct. Currently, the redevelopment of the Green Square precinct, located halfway between the airport and the CBD, is transforming the residential and commercial space of this former industrial precinct.

Sydney's world-city status is also reflected in the fact that about 40 percent of all migrants to Australia settle there, thus deepening and diversifying the long-established multicultural nature of the urban area's population. Eight out of every ten residents of Sydney were either born overseas or are the children of immigrants. The United Kingdom, China, and Aotearoa/New Zealand are the dominant source countries, though there are also substantial numbers of residents born in Vietnam, Lebanon, India, Philippines, Italy, Korea, and Greece. Historically, particular migrant groups—especially those of non-English-speaking backgrounds—have tended to settle initially in particular Sydney suburbs: Greeks in Marrickville and Italians in Leichardt in the 1950s and 1960s, Vietnamese in Cabramatta in the 1970s and 1980s, and Lebanese in Auburn in the 1990s. However, recent research has shown that Sydney's settlement is characterized more by multiethnic suburbs, such as Auburn, rather than ethnic minority concentrations, and by intermixing of different ethnic minority groups both with each other and with the host society rather than by ethnic segregation. Over time, spatial and social assimilation of migrants into a predominantly multicultural city has been the dominant pathway.

Whether growing evidence of social polarization in Sydney will produce more entrenched socio-spatial segregation along lines of class and ethnicity is a concern both to Sydney's planners and citizens. In a trend common to many global cities, Sydney's median dwelling price doubled between 2004 and 2014, with the consequence that housing stress (i.e., paying more than 30 percent of household income on housing costs) affects approximately 200,000 households across the city. As the median house price has crept up, lower-income groups, including recent migrants, have been increasingly confined either to rental housing or to less accessible suburbs removed from employment opportunities and services. It remains to be seen whether Sydney's social divides, traditionally nowhere near as pronounced as in U.S. cities, are set to become increasingly stark.

Nonetheless, Sydney remains renowned for its quality of life. It habitually enjoys a top-ranking position in international benchmarking exercises assessing physical and cultural lifestyle assets. However, the city's beautiful natural environment, open spaces, and national parks belie the environmental challenges generated by Sydney's car-dependent nature and population pressure, especially regarding air quality and water supply. Car ownership is ubiquitous and 71 percent of work trips are taken by private motor vehicle. Consequently, air quality suffers due to photochemical smog-producing ozone at levels that, while improving, still regularly exceed the 4-hour standard for ozone concentration on 21 days a year. In addition, despite falling rates of water use per capita, Sydney's population growth is challenging the adequacy of the city's water supply. As of the early twenty-first century, Sydney's water consumption was at 106 percent of the amount that can be sustainably drawn from the drainage basin. Continuing urban development poses a significant threat to Sydney's water quality.

Perth: Isolated Millionaire

With a population of 1.7 million, Perth may be the world's most isolated large city. Located on Australia's west coast, Perth was established in 1829 along the banks of the Swan River and laid out according to a grid pattern commonly associated with colonial planning. As the colonial capital of Western Australia until 1901 (when the states were united as a Commonwealth), Perth grew slowly for its first one hundred years. Throughout its history, Perth served both a rural and mining hinterland, with much of Australia's key mineral resources located in Western Australia—gold and bauxite, for example. It is mining and other global connections that have shaped the city over the past 50 years. The mining boom of the 1960s and 1970s, coupled with immigration (primarily from the United Kingdom but also from parts of southeast Asia), instigated an acceleration of the city's economic and population growth. The location of offices of mining companies and associated services saw tall buildings emerge on the city skyline (Figure 12.15). In Australia, the 1980s were a decade characterized by an entrepreneurial spirit embraced by both government and business. A consumption and leisure-based economy emerged, aided by the city's hosting of the 1987 America's Cup Challenge. For the past three decades, the Perth economy has continued to thrive on its economic base of mining and tourism, boosted by substantial immigration.

Now capital of the state of Western Australia, Perth today is far removed from its colonial beginnings. It has a modern skyscraper-dominated skyline; and the entrepreneurial governance of the late twentieth century involved substantial redevelopment of older parts of the city as tourist and leisure spaces. The redevelopment of the Old Swan Brewery site in inner Perth is one example of these processes. The Government of Western Australia's development corporation chose to redevelop the site that was once home to the factory making Perth's famous beer. The Old Swan Brewery complex now hosts a myriad of leisure activities including theaters, dining, and office space, as well as car parking. Across Australian cities such redevelopment plans are invariably contested. Conflict over the Swan Brewery redevelopment project is representative of indigenous struggles to claim space within urban Australia. In this particular case,

Figure 12.15 Kings Park in Perth offers a view of the skyline that serves the commercial interests of Western Australia and the Indian Ocean rim. *Source:* Photo by Stanley Brunn.

Aboriginal protesters drew attention to the symbolic significance that the site held for them; and they wanted the brewery buildings demolished and the land returned to parkland. Their point was made in a variety of ways, including an 11-month period in which they camped on the site. The protests were unsuccessful, with the government authority going ahead with the redevelopment and incorporating elements of Aboriginal culture into the design. At another level, the protest was successful for the ways in which it brought an Aboriginal presence into the urban world.

Perth, like other Australian cities, is a sprawling city. Population growth has spawned metropolitan growth, initially to the east and more recently southward toward the municipality of Mandurah. Population growth also instigated increased demand for water in an environment of minimal rainfall. In 2006, a desalination plant for the city was opened to bolster the city's water supply. For much of the twentieth century, it was presumed that the private car would adequately cater to the transportation needs of this growing population. More recently, however, the need for better public transportation has been recognized. A new, profitable and well-patronized suburban railway line to East Perth was opened. Perth is also home to a wide variety of other sustainable transport initiatives. Foremost here are "TravelSmart" programs, run by employers, schools, universities, or workplaces. These programs encourage individuals to consider non-car travel, and sometimes provide incentives to do so. Like Auckland's "walking school buses," they have been successful in reducing private automobile travel in Perth, and in raising awareness of the city's precarious environmental future.

Gold Coast: Tourism Urbanization

Australian sociologist Patrick Mullin has used the term "tourism urbanization" to describe a scenario of tourism-sustained urban growth, where (1) urban development is based primarily on tourist consumption of goods and services for pleasure, and (2) urban form is shaped by the city's function as a leisure space. The Gold Coast on Australia's Queensland coast can be understood in these terms.

In Australia's Gold Coast—25 miles (40 km) south of Queensland's capital city, Brisbane—white settlement began in the 1840s with timber getting and agricultural development. By the 1870s, wealthy Brisbane residents were already discovering the area as a leisure destination, known simply as the South Coast. The development of a rail connection from Brisbane in the 1930s saw the area's appeal broaden and some minor beach resorts emerge. But it was not until the boom of the 1950s that the area took on the name "Gold Coast" and began its development as Australia's highest intensity, high-rise tourist destination. Through many cycles of boom and bust, intense real estate investment in tourist accommodation, retail, restaurants, and entertainment ventures along this 35-mile (56 km) strip of spectacular surfing beaches, transformed the Gold Coast into the most intensely developed coastal tourist strip in Australia and a key international tourist destination.

By the 1980s, the area—especially Surfers Paradise at the heart of the Gold Coast—had gained a dubious reputation as a place of relaxed social norms, brashly opulent neon-lit landscapes, and get-rich-quick real estate deals. Nonetheless, the region matured as a tourist destination. Large-scale foreign direct investment in real estate, especially from Japanese interests in the 1980s and more recently from the Middle Eastern, brought significant diversification to the array of tourist products and consumption landscapes in Surfers Paradise and its hinterland. The area developed a series of integrated tourist resorts such as the Marina Mirage and the golf-themed Sanctuary Cove; large-scale retail malls such as Pacific Fair, Conrad Jupiters casino; multiple golf courses; and multiple theme parks, including Movieworld, Sea World, Dream World, and Wet'n'Wild Waterworld.

The Gold Coast (incorporated as a city since 1959) has had a rapidly expanding resident population, which now stands at about half a million. Its more than 13,000 accommodation rooms in hotels and serviced apartments accommodate an additional 3.5 million domestic visitors and 800,000 international visitors annually, primarily from Asian countries and Aotearoa/New Zealand. But more than a consumption-driven tourist economy today underlies the Gold Coast. It is also one of the most rapidly developing cities in Australia, characterized by sustained rapid population growth rates of around 2 percent annually. Its growth is largely migrant driven as lifestyle attractions have drawn in-migrants from across Australia, many of whom have found housing in low-density canal-estates built behind the high-rise coastal strip. More recently, as the Gold Coast has expanded, more conventional forms of suburbia have developed including a major new-town development in Robina to the southwest. As this has occurred, the initial dominance of retirees among in-migrants—prompting one author to label the city "God's waiting room"—has subsided such that the largest in-migrant group now ranges between 20 and 29 years old. The city's population is expected to reach nearly 789,000 by 2031; but the Gold Coast is also blending into the extended urban region of southeast Queensland (SEQ), a conurbation that stretches 150 miles (240 km) from Noosa southward through Brisbane and the Gold Coast to Tweed in northern New South Wales. SEQ's population is approaching 3 million, representing more than two-thirds of Queensland's population. The population of SEQ is projected to reach 4.4 million by 2031.

As the Gold Coast blends into this urban region, its economy is diversifying. Tourism-related industries have tended to support lower-skilled occupations and low-paid and/or casual employment, prone to seasonal fluctuation. Now, the state-supported Pacific Innovation Corridor initiative aims to promote the region's hi-tech, biotech, computing, and multimedia industries that will integrate the region into a globalized knowledge economy and improve rail and road connections to Brisbane's larger economy. Nonetheless, Gold Coast is still one of the lowest income cities in Australia and has higher levels of socioeconomic disadvantage than other Australian cities, in part a product of its occupational structure. The tourism-dominated economy is reflected in lower-skilled occupations, low rates of higher education, high rates of low-paid casual employment, and high rates of unemployment. As the conurbation expands, challenges emerge: managing disadvantage, enabling economic diversification, building roads and transit systems, developing sustainable communities, and balancing environmental protection against development.

Auckland: Economic Hub of Aotearoa/New Zealand

While not the nation's capital, Auckland has dominated Aotearoa/New Zealand's urban system since overtaking Dunedin and Christchurch as the country's largest city in the late-nineteenth century. Like Sydney, it developed on an aesthetically and economically advantageous harbor, and is similarly renowned for its natural beauty. Historically, it too served a rich agricultural and forested hinterland. The deregulation of the national economy in the 1980s paved the way for the transformation of Auckland. It is Aotearoa/New Zealand's largest, most prosperous and economically active city. By the late twentieth century, it hosted more than a third of the nation's employment in manufacturing, transport, communication, and business services. It increasingly occupies a strategic position in the national economy, through its operation as a place in which and through which the global economy operates. It is the location of multinationals, international financial transactions, global property investments, and a hub for international tourists. Global rather than local connections are also important in explaining a number of other facets of urban life in Aotearoa/New Zealand.

The 1980s saw the transformation of the Auckland residential and commercial landscapes. High-rise residential towers (like the famous Sky Tower, tallest building in the Southern Hemisphere) were built around the city's CBD, often financed in foreign currencies, designed by architects outside Aotearoa/New Zealand, and managed by global property conglomerates. High-rise residential living has become increasingly popular. The building of medium-density housing has added to the city's density. Sometimes modeled on "new urbanist" ideas imported

Box 12.5 Gentrification and Ponsonby Road, Auckland

Figure 12.16 Ponsonby Road is now a focal point of chic eateries and boutique shopping in Auckland. *Source:* Photo by D. J. Zeigler.

Whether the claim that gentrification is now a global phenomenon is valid or not, this urbanization process has certainly reshaped the inner suburbs of many of Australia's and New Zealand/Aotearoa's cities. The process has witnessed middle-class renovation and resettlement of formerly working-class housing in inner-city neighborhoods in all the major metropolitan centers, as well as in regional cities such as Newcastle and Wollongong in New South Wales. Gentrification is not merely a residential phenomenon but one involving refashioning local shopping streets, leisure and recreation facilities, and neighborhood services as residents' aesthetics, ethos, and consumption patterns combine to mold local streetscapes. These impacts are evident on King Street in Sydney's Newtown, Brunswick Street in Melbourne's Fitzroy's, Boundary Road in Brisbane's West End, Darby Street in Newcastle's Cooks Hill, and Ponsonby Road in Auckland's Ponsonby.

The suburb of Ponsonby is located less than a mile west of Auckland's CBD (Figure 12.16). After World War II, many of Ponsonby's more prosperous residents relocated to the expanding outer suburbs and were replaced by lower-income Pacific Island and Maori migrants. However, waves of gentrification commenced in the 1970s, as diverse groups of young, well-educated, Pakeha (white New Zealanders of European descent) were attracted to the area by its cheap property, low rents, and social and ethnic diversity. Ironically, that diversity can be threatened by the very process of gentrification. In Ponsonby's case, gentrification overlapped with an Auckland-wide housing boom and property price inflation in the 1990s; the result has been significant displacement of lower-income, less-educated inhabitants, driven out by rising rents and spiraling house prices. Ponsonby's population has, proportionately, become distinctly "whiter" and higher income. Nonetheless, despite price inflation, the area has maintained a relatively young population and a significant proportion of rental housing.

Certainly, diversity is characteristic of the dramatic transformation of the consumption spaces and public culture of Ponsonby Road. Gentrification has combined with changes to licensing laws to see the birth of a thriving agglomeration of over 90 cafes, restaurants, and bars, interspersed with specialty stores, greengrocers, butchers, and newsagents. Mark Latham's (2003) research has shown how the plethora of cafes and bars—often flamboyantly and expensively styled and open to the street—depart from the more traditional, enclosed spaces of pubs and the culture of hard-drinking masculinity that they accommodate more readily than other forms of sociality. As a result, gentrification has seen Ponsonby Road develop a range of more ambiguous spaces for consumption and sociability that are welcoming to women, gay-friendly, and less confined to traditional norms of gendered identity. In this particular site of gentrification, working-class displacement and middle-class colonization have been accompanied by the development of a diverse public culture that, while definitely accessible (most easily to those with disposable income), is open to diverse expressions of identity and diverse ways of inhabiting the city.

Source: A. Latham, Urbanity, lifestyle and making sense of the new urban cultural economy, *Urban Studies* 40 (2003): 1699–1724.

directly from the United States, these new suburbs modify the conventional suburban way of life with smaller houses, a gridded street pattern, and sometimes a communal open space. Though not gated communities in the strictest sense, the role of these new suburbs in fostering social exclusion is an ongoing issue. In fact, the same issue often arises as inner-city neighborhoods undergo gentrification (Box 12.5).

Lifestyle television programs and home-focused magazines are hugely popular and foster expenditure on household items and renovation projects in Auckland and its suburbs. Suburban backyards may be getting smaller, but they still serve the important purposes of providing a place for children to play, domestic vegetable cultivation, and the fulfillment of aesthetic and economic aspirations. Some new groups of migrants do aspire to and do fulfill these suburban ideals, like residence in a detached house. Migration has also transformed suburban landscapes. Suburbs like Sandringham, with new places of worship and retail landscapes, have been the destination of many migrants from Asia.

The sustainability of a large, dynamic city like Auckland is attracting ever more scholarly and policy attention. Contradictions between reliance on the private motor vehicle and a strong environmental consciousness have seen the widespread adoption of "walking school buses." Rather than children being driven individually to school, they congregate at locations along a designated route and walk to school with other children, all under parental supervision. The average walk takes about 30 minutes. About 200 active walking school buses now operate throughout Auckland and its suburbs, but they are more likely to serve middle-class neighborhoods. They have been credited with removing cars from the road, reducing air pollution, reducing obesity, and enhancing community. Official urban policies of sustainability have already influenced the building of medium-density housing and housing with a small ecological footprint. A more widespread implementation of urban sustainability in Auckland has also recently been discussed.

Figure 12.17 Located on Auckland's North Shore, Devonport's landscape has been almost completely transformed by suburbanization. Nevertheless, a few visual reminders of the original inhabitants remain, including this Maori warrior. *Source:* Photo by D. J. Zeigler.

Port Moresby and Suva: Island Capitals

Port Moresby and Suva are the largest cities, and political capitals, of their respective nations of Papua New Guinea and Fiji. They have parallel histories, urban patterns, and contemporary influences. While their current political instabilities may be unique, their other characteristics are broadly representative of cities in the island Pacific.

Neither PNG nor Fiji has a prosperous economy. They have weak manufacturing sectors, are reliant on an agricultural enterprises beset with inefficiencies and at the mercy of low globalization, and are plagued by political instability. Hence, both Suva and Port Moresby have fragile economic bases. While population has been steadily growing in both cities, employment opportunities have not. Consequently, unemployment is high, with one estimate putting unemployment in Port Moresby at around 60 percent. These fragile economic circumstances underpin the most salient characteristics of Pacific Island cities: a large informal sector, including informal settlements, plus political problems and unrest (Figure 12.17).

Informal, squatter-like settlements are common in these cities. Port Moresby has at least 84 agglomerations of substandard, poorly serviced housing, in which urban poverty is concentrated; Suva has just a little less. Basic urban infrastructure—water, sewerage, electricity, and garbage collection—is either completely lacking or minimally provided in such settlements. Problems are exacerbated by a lack of formal employment opportunities. Urban poverty is rising, exemplified by the increasing number of street children. Informal employment, particularly prostitution, has arisen to counter the lack of formal-sector employment opportunities.

Policy responses to urban poverty and marginalization in Suva and Port Moresby have been small and problematic. The under funding of basic infrastructure has contributed to the problem. There is widespread opposition to the urban poor and street prostitution. The government's response to prostitution, the prevalence of street children, and informal settlement has been largely negative. In PNG, problem settlements have been bulldozed rather than adequately resourced. More generally, these cities have been sites of social and political unrest, which has had implications for the internal structures of these cities. In Port Moresby, for example, security concerns have seen European and other expatriates withdraw further into barricaded residential estates on the hillsides of the city.

TRENDS AND CHALLENGES

Many of the cities in Australia and the Pacific are cradled by fragile ecosystems and are extremely vulnerable to the multifaceted impacts of climate change (Figure 12.18). Australia's largest cities are further challenged by the fact that they are all located in areas where climate change is inducing significant declines in rainfall levels. All of the capital cities (except Hobart) have desalination plants in operation or nearing completion, to convert seawater to drinking water, though most are currently finding it difficult to cover costs. Desalination may be one solution to water supply, but its voluminous energy demands impose other impacts on the environment. In addition, the geographic expansion of urbanized areas involves the loss of productive land, loss of biodiversity, and increased energy use. The imperatives in all cities have thus become reduced energy consumption and emissions reduction alongside increased use of renewable energy.

Urban governance provides many challenges across the region. The challenge is the establishment of effective urban governments able to meet environmental and security challenges and fashion positive outcomes (Figure 12.19). Governance processes that contribute to social cohesion are also key. In Australia and Aotearoa/New Zealand, urban governance is now characterized by a variant of neoliberalism in which market processes and solutions underpin policy. Waterfront redevelopments in many cities are classic

Figure 12.18 In Newcastle, NSW, this ClimateCam billboard broadcasts figures on the city's electricity consumption. These are updated hourly as a way of raising awareness about the city's contribution to resource use, GHG emissions and climate change. *Source:* Photo by Kathy Mee.

Figure 12.19 One of the challenges of urban governance in Australia is maintaining safe streets. Signs like this one in Sydney have been increasing rapidly as people everywhere become more security conscious. *Source:* Photo by D. J. Zeigler.

outcomes of neoliberal policies. The extent to which such governance is equitable remains questionable, and ways to produce more "just" cities within such a framework are still being sought. Equitable outcomes for indigenous peoples of these cities are especially important.

Finally, the provision of adequate, appropriate, and affordable housing is a pressing issue for all cities in the region. In Sydney and Melbourne particularly, where house-price escalation has been intense, affordability has now reached historic lows. Mortgage stress—where households are paying more than 30 percent of gross household income on housing—has risen, most particularly in the suburbs. The impacts of the affordability crisis include displacing younger people and lower-paid workers from high-cost urban areas, labor shortages, and growing debt burdens on households with mortgages.

SUGGESTED READINGS

Connell, J., and J. P. Lea. 2002. *Urbanisation in the Island Pacific*. 3rd ed. London: Routledge. An overview of urbanization in 11 independent island states.

Connell, J., and P. McManus. 2011. *Rural Revival? Place Marketing, Tree Change and Regional Migration in Australia.* Surrey: Ashgate. Examines urban-rural migration using numerous case studies across Australia, to understand how urban-to-rural migration can be achieved and offer approaches for wider applications.

Forster, C. 2004. *Australian Cities: Continuity and Change*. South Melbourne: Oxford University Press. Explores the urban experience across the Pacific Islands, including the role of cities in national development and as centers of globalization.

Jacobs, J. M. 1996. *Edge of Empire: Postcolonialism and the City*. London and New York: Routledge. An analysis of how the connections built by globalization and the postcolonial world shape and reshape the composition of cities.

Le Heron, R., and E. Pawson. 1996. *Changing Places: New Zealand in the Nineties*. Auckland: Longman Paul. Provides an overview of the transforming geography of New Zealand including its cities and regions in the context of globalization.

Major Cities Unit. 2013. *State of Australian Cities 2013*. Infrastructure Australia. Canberra: Australian Government. Provides a detailed empirical snapshot of demographic, economic, social, environmental, and governance dynamics of Australia's major cities.

McGillick, P. 2005. *Sydney, Australia: The Making of a Global City*. Singapore: Periplus. A photographically illustrated history of Sydney's built environment.

McManus, P. 2005. *Vortex Cities to Sustainable Cities: Australia's Urban Challenge*. Sydney: University of New South Wales Press. Examines the histories and planning decisions that have contributed to the unsustainability of Australian cities.

Newton, P., ed. 2008. *Transitions: Pathways toward Sustainable Urban Development in Australia*. Canberra: CSIRO Publishing. Examines demographic and development trends and their implications for resource demands and sustainable urban development.

Weller, R., and J. Bolleter. 2012. *Made in Australia: The Future of Australian Cities*. Perth: University of Western Australia Publishing. Uses the concept of visionary cities within urban megaregions to project the social and sustainable future of Australia's urban population.

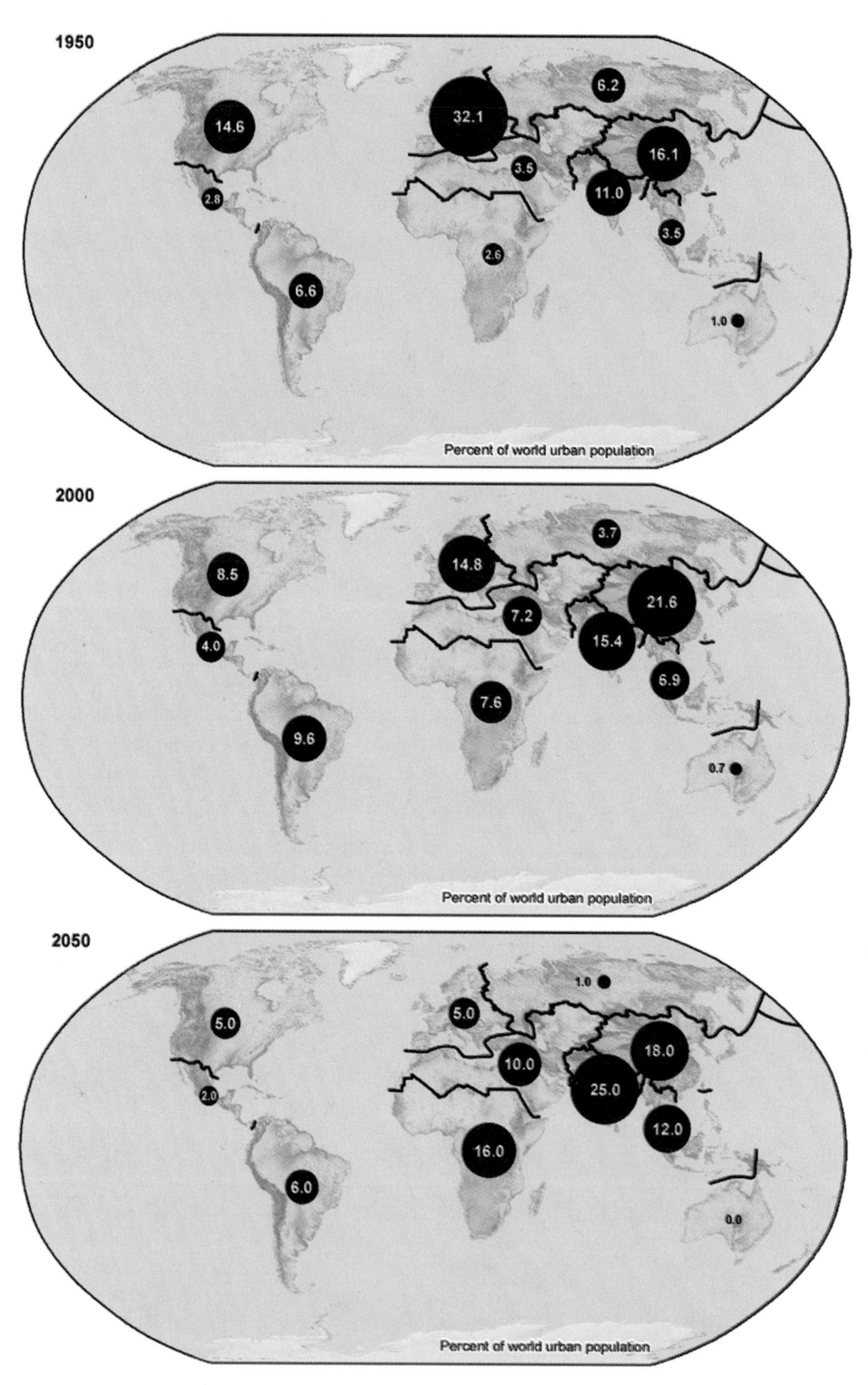

Figure 13.1 Urban Populations: 1950, 2000, and 2050. *Source:* Data from United Nations, *World Urbanization Prospects*, 2001 Revision (New York: United Nations Population Division, 2002), www.unpopulation.org. Projections for 2050 by authors.

13

Cities of the Future

BRIAN EDWARD JOHNSON AND BENJAMIN SHULTZ

KEY URBAN FACTS

Largest in Population (in millions: 2010–2025 projections)	Tokyo (37.1), Delhi (28.6), Mumbai (25.8), São Paulo (21.6), Dhaka (20.9), Mexico City (20.7)
Urban Areas Adding Most Residents (in millions: 2010–2025 projections)	Delhi (6.4), Dhaka (6.3), Kinshasa (6.3), Mumbai (5.8), Karachi (5.6), Lagos (5.2)
Fastest Annual Growth Rates (%) (2010–2025 projections)	Ouagadougou (8.5), Lilongwe (7.1), Blantyre-Limbe (7.1), Yamoussoukro (6.9), Niamey (6.7), Kampala (6.6)
Slowest Annual Growth Rates (%) (2010–2025 projections)	Dnipropetrovsk (−0.25), Saratov (−0.20), Donetsk (−0.17), Zaporizhzhya (−0.15), Havana (−0.11), Volgograd (−0.09)
Sharpest Declines (in thousands: 2010–2025 projections)	Dnipropetrovsk (37), Havana (35), Saratov (24), Donetsk (24), St. Petersburg (18), Zaporizhzhya (17)

KEY CHAPTER THEMES

1. General growth in overall world urban populations will continue, with rapid growth in the less developed countries and slow growth in more developed countries.
2. Urban growth in developing countries will present many challenges, including provision of basic infrastructure and human services.
3. Young people and seniors will fuel the growth of city centers in developed countries, while larger numbers of minorities and immigrants will move to the suburbs.
4. The world's fastest growing cities are in countries with weak environmental standards, putting the health of the world's poorest urban residents at risk and putting greater strain on the global environment.

5. Cities in coastal areas are likely to face rising sea levels and more frequent and strong storms due to climate change, forcing many cities to upgrade infrastructure and update land-use plans.
6. As new industrial cities continue to develop, they will likely go through a period of deindustrialization as they transition to a cleaner service economy, just as the revitalized former industrial centers of the developed world have done.
7. Contrary to predictions that the world will become "flatter," cities that are communication and transportation hubs will thrive, as will cities with concentrations of highly skilled workers.
8. Some cities located near newly discovered natural resources will thrive, as will cities in Sub-Saharan Africa and other regions, as they are able to attract manufacturing jobs due to East Asia's transition to a service economy.
9. Innovative approaches to urban governance, including extensive use of geographical information systems (GIS), will continue to be important as urban governments plan for future change.
10. Security provision and location-based mass surveillance will increase in urban areas, especially in the wake of terrorist attacks in major cities.

Populous and diverse cities now exist throughout the world. In fact, since 2007, for the first time in history, the majority of the human population is living in cities (Figure 13.1). This chapter describes cities of the future with predicted trends in worldwide urban growth, distributions, compositions, economies, and landscapes. In addition, challenges posed by urban growth and change are explained, including providing for the residents of booming cities in the developing world, or global south, and managing urban revitalization and suburban diversification in the developed world, or global north. How cities will attempt to be more sustainable and adapt to climate change is also discussed, along with deindustrialization, urban spaces as both communications and transportation crossroads, and resource-based cities of the near future. Then, the chapter addresses changing urban governance, which includes interurban collaboration and geographic technologies. The chapter concludes with a look at cities that are predicted to have the highest quality of life (Box 13.1).

URBAN GROWTH IN THE GLOBAL SOUTH

Going forward, the percentage of humans living in cities worldwide is predicted to continue increasing. By "cities" or "urban areas," researchers commonly mean cities as well as their suburban and exurban areas. Since becoming 50 percent urban in 2007, the human population has increased to 54 percent urban in 2015. This proportion puts the worldwide urban population at almost 4 billion people. By 2050, the United Nations predicts that the global urban population will have risen to 66.4 percent or over 6.3 billion people (Figure 13.2).

Box 13.1 Engineering Earth Futures

Stanley D. Brunn, University of Kentucky

The planet is replete with mega-engineering projects of varying sizes, financial costs, and environmental impacts. While we generally would associate these projects with extensive transportation schemes, dams, airports, river diversion, and irrigation projects, they also can include theme parks and leisure spaces (golf courses and sports arenas), new capital cities, and towering skyscrapers. But engineering the earth can and does include projects of a social nature as well. Are not genetically modified foods, Google Earth, GIS, the Internet, and Facebook also ways in which we choose to engineer the earth? And what about social engineering projects such as gated communities, international "cookie cutter" suburbs, resettlement projects for "alien" newcomers and socially divisive projects designed by governments that separate groups based on skin color, religion, and ethnicity? Examples of these projects exist on all continents and in their social, environmental, and political impacts will only increase in importance in the coming decades.

Three issues are paramount in looking at the future of megaprojects. First, is "bigger" necessarily "better"? Contemporary society, whether in the developed or developing world, seems almost addicted to projects that are huge. China's Three Gorges Dam, Dubai's Burj Khalifa, the Trans Amazon highway, plus gigantic nuclear power plants, offshore oil rigs, postmodern skyscraper skylines, and outlandishly designed new capital cities, all come to mind. There is no shortage of mega-dreams and mega-schemes to try and resolve an immediate energy or transportation problem or to appease the superego of government leaders. We seem to think the size "fix" will solve the problem or appease potential users.

Second, what are the impacts of mega-projects? For each project, there are social as well as environmental externalities. What are the short- and long-term consequences of relying on nuclear power or altering the course of a river or planning residential areas in environmentally sensitive wetland zones or on erosion-prone slopes? Or, what about a society whose youth and wealthy are addicted to the disembodied worlds of Facebook, iPods, cell phones, and the internet? Or, a world where everything and everyone is a coded into GIS databases? Do we really want a world *where* everything can be mapped and *where* everyone is mapped 24/7? With our fixation on "technology coming to the rescue," we often forget the environmental and human impacts of financial and engineering solutions to problems. Social, environmental, and civil engineering often seem to operate in parallel universes. Hazards to future generations may originate in the social and physical engineering projects of today.

Third, what about our future? Futures, like pasts and presents, will be engineered. The questions are *who* will be masterminding the planning and *what* do we expect from it? Postmodernists like to remind us that we have the freedom and flexibility to design the futures we want. Perhaps that is true; but, perhaps this view is also elitist, self-serving, materialistic, and unresourceful. Present and future worlds will be comprised of many living in survival conditions in rich and poor worlds whose daily lives are and will be full of uncertainty, frustration, desperation, and without a moral compass. Where these conditions exist, how can one "engineer hope"? Perhaps the human condition will be improved by placing greater focus on economic and social micro-engineered projects that are local, sustainable, and community based rather than highly visible and heavily financed projects by the international banking community. What is certain is that global warming, biological species decline, financial meltdowns, social underachieving, geopolitical impotence, surface conformities, and social restlessness are futures with us now. It remains for the youth and elders, leaders and followers to decide what kind of engagement, empowerment, or engineering of the planet's resources and population we wish to pass on the coming generations.

More than one hundred mega-projects are discussed in the multivolume *Engineering Earth: The Impacts of Megaengineering Projects*, which was edited by S. D. Brunn and published in 2011 by Springer (Dordrecht, Netherlands).

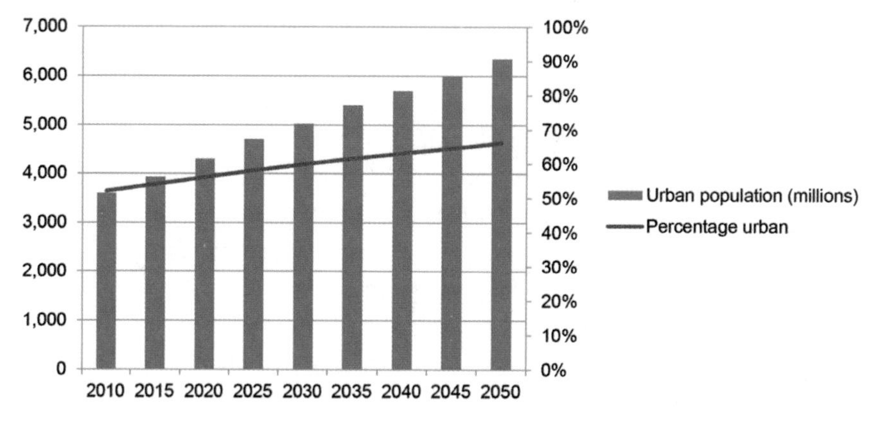

Figure 13.2 Global Urban Population: 2010–2050. *Source:* Population Division of the Department of Economic and Social Affairs of the United Nations Secretariat, 2010.

Not all urban areas are predicted to grow at the same pace, however. Urban areas in the less developed world, also known as the global south, are predicted to grow more rapidly than those in more highly developed countries, or global north (Tables 13.1 and 13.2). For example, Tokyo is currently the world's most populous urban area, with 38.0 million people. It is predicted that Tokyo will still top the list in 2030, having shrunk slightly to 37.2 million people. In contrast, Delhi, the second most populous urban area at present, is predicted to grow by 40.3 percent, from 25.7 million in 2015 to 36.1 million in 2030.

Populous cities in the more highly developed countries are predicted to grow slowly or plateau in terms of the number of inhabitants, while cities in less developed countries are predicted to grow substantially. Cities in very highly developed countries, such as Osaka, New York, Los Angeles, Paris, and London, will thus drop down the ranking list of the world's

Table 13.1 World's Most Populous Cities in 2015

		World's Most Populous Cities in 2015	
Rank	*City*	*Population*	*Development Level (2012)*
1	Tokyo	38.0 million	Very high human development
2	Delhi	25.7 million	Medium human development
3	Shanghai	23.7 million	Medium human development
4	São Paulo	21.1 million	High human development
5	Mumbai	21.0 million	Medium human development
6	Mexico City	21.0 million	High human development
7	Beijing	20.4 million	Medium human development
8	Osaka	20.2 million	Very high human development
9	Cairo	18.8 million	Medium human development
10	New York	18.6 million	Very high human development

Source: Population Division of the Department of Economic and Social Affairs of the United Nations Secretariat, 2015; United Nations Development Program, 2014.

Table 13.2 World's Most Populous Cities in 2030

		World's Most Populous Cities in 2030		
Rank	*City*	*Population*	*Development Level (2012)*	*% change from 2015*
1	Tokyo	37.2 million	Very high human development	−2.1
2	Delhi	36.1 million	Medium human development	40.3
3	Shanghai	30.8 million	Medium development	29.5
4	Mumbai	27.8 million	Medium human development	32.1
5	Beijing	27.7 million	Medium human development	35.9
6	Dhaka	27.4 million	Low human development	55.5
7	Karachi	24.8 million	Low human development	49.5
8	Cairo	24.5 million	Medium human development	30.5
9	Lagos	24.2 million	Low human development	84.7
10	Mexico City	23.9 million	High human development	13.6

Source: Population Division of the Department of Economic and Social Affairs of the United Nations Secretariat, 2015; United Nations Development Program, 2014.

most populous, as will other cities in Europe and North America; while cities in East Asia, South Asia, and Sub-Saharan Africa are predicted to climb the ranks. The very populous cities of the world are predicted to be increasingly located in East Asia and South Asia.

Causes of Urban Growth in the Global South

Why are urban areas burgeoning in the global south? In these countries, rural-to-urban migration is currently occurring and is predicted to persist into the next several decades. People are moving from the countryside to cities, swelling the populations of urban areas. These moves are caused, in large part, by broader trends in population and economic geography.

The demographic transition model and the related migration-transition model help to explain how population trends are linked to urbanization. They suggest that families in less developed rural areas desire large families, in part, because children provide additional help working the land, potentially increasing a family's earnings. As farmers in developing countries become able to purchase labor-saving tractors and other farm machinery, some of their children become redundant with respect to labor needs, and commonly move to cities to seek work. As industrialization creates a demand for labor in the cities, unemployed or underemployed in-migrants from rural areas fill that demand.

Agricultural mechanization and urban industrialization are driving the growth of cities in many developing countries. The changing geographies of labor markets are causing a surge in city populations today, and they are predicted to persist into the near future (Figure 13.3). In developed countries, this agricultural mechanization/urban industrialization trend played out in previous eras when cities in today's developed countries grew rapidly from rural-to-urban migration. Today, however, the cities of developed countries do not receive many rural newcomers because farmers in these countries have utilized labor-saving machinery for decades and have adjusted their fertility downward. For example, the United States became a majority-urban population in 1920 at a time when the trend of agricultural mechanization was sweeping across the country's farms and ranches, and when industrialization was transforming urban economies by creating jobs in the secondary sector of the economy.

Figure 13.3 At close of business on Fridays in Portland, Oregon, placards are out to remind commuters to enjoy their weekend. It's good for their health. *Source:* Photo by D. J. Zeigler.

In addition, public health and medicine has improved in many developing countries in recent decades. Families who, for generations, had many children to offset predicted high infant and child mortality rates are now finding that through improvements in sanitation, maternal health care, vaccinations, and overall medical care, more children are surviving to adulthood. Higher survival rates commonly result in more grown children living in a rural area than can be supported there, so those people commonly move to cities. Again, this trend occurred in developed countries long ago, and their fertility rates in both rural and urban areas have been low for generations.

Changing cultural attitudes also contribute to urbanization in the developing world. As culture globalizes and many traditional cultures begin to adopt Western values, achieving social status becomes more a matter of individual accomplishment rather than being linked to having a large family. Differences in income, education, and professional titles, for example, establish social hierarchy. Access to higher education and better paying positions in the tertiary sector are overwhelmingly concentrated in cities, attracting thousands of aspiring young professionals each year.

Challenges Posed by Urban Growth in the Global South

The high growth rates of cities in developing countries present challenges. With millions moving to rapidly growing urban areas of the developing world each year, city services, infrastructure, and housing provision have not kept up with the population increase, resulting in large informal settlements characterized by poor housing conditions in many areas of the developing world. Many factors contribute to urban poverty, including tenuous employment in the informal sector, lack of effective urban networking to find jobs, low wages of multiple-member households, low rates of literacy, and lack of job skills. The urban poor are also those who have to carry water to their residences (equivalent to eight suitcases a day); they experience irregular and unexpected losses of water and electricity (if such services are available at all), lack readily available and safe fuel sources to heat food or homes, and use various outdoor facilities for toilets. They live in constant fear of eviction, loss of income and employment, injuries from work, and random violence brought on by criminal gangs or police.

For example, a rapidly growing city like Dhaka in Bangladesh must find a way to provide infrastructure and services for the half million new residents arriving, on an average, each year. This includes adding more water mains, sewers, schools, hospitals, train lines, buses, roads, energy supplies, dwellings, and jobs. Newcomers having a difficulty finding housing sometimes end up living in informal settlements, which go by different names in different parts of the world: shanty towns, *bustees*, *barrios*, *bidonvilles*, *favelas*, and *villas miserias*. If residents have no legal right to the land, which is almost always the case, they are called squatter settlements, and if the built environment is deteriorated and unhealthy, they may also be referred to by the now-pejorative word, slums. In such settlements, utility service is typically not provided, land tenure is uncertain, and housing is self-help. Governments have financial and administrative difficulties in formalizing such settlements and providing city services. Accommodating over 500,000 new residents every year would be a tremendous challenge even for cities in wealthy countries, let alone many of the cities in less wealthy regions that will need to address these issues in the coming decades.

URBAN CHANGE IN THE GLOBAL NORTH

Metropolitan areas in the developed countries are predicted to grow slowly: there are fewer people to draw from rural areas and populations are at or below replacement level fertility. What growth does occur will be due largely to international immigration. Within metropolitan areas, however, population rearrangement is occurring, with the growth in city-center populations and diversification of the suburbs being evident in present and future trends.

Central cities have been attracting population recently, which is a notable change because such cities lost population, largely to their suburbs, for several generations. Depopulation of core cities and moves to the suburbs and exurbs was especially evident in North America, but occurred in Western Europe as well. It is notable that Greater London's population has been rebounding since the 1990s and is now, at 8.5 million, larger than at any previous point in its history including its former high point, which was hit right before World War II. London has clearly bucked the decentralization trend. So have other large European cities, but much less vigorously.

Inner-city revitalization in the United States and Canada has occurred in part because the largest age cohorts in many developed countries are Baby Boomers (born 1946–1964) and Millennials (born 1982–2002). These generations have larger proportions of those seeking residences in or near city centers compared to other more suburban-inclined generations. Boomers are seeking smaller residences to better fit empty-nest households and minimize home upkeep obligations and expenses as they age. In the United States, Boomers no longer have children at home, so there is little impetus to remain in or move to what they perceive to be higher-quality suburban school districts. Many Boomers have come to desire the shorter commutes provided by living close to workplaces in the core cities and shorter travel time to city-center cultural attractions and dining (Figure 13.4).

Near the other end of the age spectrum, North American Millennials are showing a preference for core cities or urban neighborhoods, rather than the overwhelmingly preferred suburban areas of earlier generations. Having come of age during the Great Recession (beginning in 2008), Millennials want to live close to large downtown labor markets in order to increase job search success and to facilitate the frequent job switching that they predict will mark their careers. Childlessness has increased, particularly among younger generations, and those who do have children have fewer and at later ages. Millennials are living in diverse households made up of roommates, cohabitating couples, or singles. Thanks to Millennials, divorcees, widows, and widowers, the fastest growing household type in many developing countries is single people living alone. At the same time, young generations such as Millennials struggle to save to purchase houses amid economic and job market difficulties. They are buying homes later and in fewer numbers compared to previous generations. Population and economic trends are funneling this generation to core cities with their large supply of smaller dwellings and rental residences. Also, it has become more common, especially in Western Europe, for Millennials to reside with their parents well into adulthood, further slowing

Figure 13.4 Repurposing old buildings to serve as apartments and condominiums in the heart of downtown is bringing life back to central cities. Every CBD has signs like this, but this one happens to be in Cincinnati, Ohio. *Source:* Photo by D. J. Zeigler.

the demand for new housing in suburban and exurban areas.

Suburbs in Europe, North America, and Australia continue to grow, just more slowly than in the second half of the twentieth century. Many suburbs in North America and Europe were overbuilt leading up to the Great Recession, and the glut of homes has caused suburban housing prices to stagnate or decline. In Florida, Nevada, and Arizona, approximately 15 percent of all residences were vacant during the Great Recession, casualties of overbuilding and the housing crash. In Ireland, 300,000 new homes, mostly located in suburban and exurban locations, stand unoccupied and often unfinished, contributing to a 50 percent decline in overall property prices since the peak in 2007. These houses, like those in North America, were built when mortgages were easily had, buyers (incorrectly) predicted that prices would continue to rise, and growth in wages did not keep up with increases in housing prices.

In the developed countries of North America, Europe, and Oceania, international immigrants increasingly are moving to suburbs. European suburbs have long been home to immigrant neighborhoods, including North Africans in the suburbs of Paris and Turks in the suburbs of Berlin. Increasingly, newcomers to the United States and Canada are moving to enclaves in the suburbs rather than to ethnic neighborhoods in core cities. The U.S. Census Bureau has also reported on the sustained movement of African Americans to the suburbs. The large cities of New York, Los Angeles, Chicago, and Detroit, among others, have all seen a decline in the black population. Both push and pull factors seem to be at work in their moves: Some are responding to gentrification of inner-city neighborhoods, and others are responding to their own

Figure 13.5 2015 commemorated the 50th anniversary of the "March on Washington for Jobs and Freedom" led by Martin Luther King, Jr. Here at his memorial on the National Mall, a new generation looks up to Dr. King. *Source:* Photo by D. J. Zeigler.

suburban dreams (Figure 13.5). Suburbs are clearly becoming more diverse, ethnically and economically.

URBAN SUSTAINABILITY AT CENTER STAGE

The rapid growth of cities has many costs and benefits in terms of sustainability (Figure 13.6). On the one hand, urbanization is often taken as a sign of economic development, as the world's economic engine is powered by cities. Concentrating a larger number of people into a smaller area also makes the delivery of important services (e.g., health care) more efficient. On the other hand, the United Nations estimates that cities account for more than three-fourths of global energy consumption and greenhouse gas emissions. The fact that 67 percent of the world's population is expected to live in cities by 2050 means that understanding how to make urbanization sustainable will be one of the greatest challenges in the twenty-first century.

The concentration of industrial activity in and around cities is one of the primary reasons why cities account for such a high percentage of the world's greenhouse gas emissions. In the developed world, there has been a concerted push toward "green" technologies and cleaner forms of energy to decrease pollution and reduce emissions. While these are necessary and important measures intended to create a more sustainable future, new industry is increasingly attracted to metropolitan areas that offer lax environmental standards, especially in China and India.

Pollution Problems and Urban Futures

Powered by its vast deposits of coal and other minerals supplying factories in its metropolitan areas, China experienced an average economic growth rate of over 10 percent each year from the 1980s into the first decade of the 2000s, lifting hundreds of millions of people out of poverty. The environmental cost of this growth, however, has been severe. The World Bank estimates that only one percent of the country's over 600 million urban residents breathe air that is considered safe by European Union standards. Research by the World Health Organization indicates that dirty air and water already account for hundreds of thousands of premature and preventable deaths each year.

Figure 13.6 In Seoul, Korea, open space is green space. Although it's one of the world's megacities, Seoul has made living with nature a priority of life and governance. *Source:* Photo by D. J. Zeigler.

India, with over a billion people of its own, has also experienced severe environmental degradation as it continues to urbanize and industrialize at a startling pace. Delhi has the highest level of urban air pollution of any of the world's megacities. Nine other Indian cities are among the top 20 in the world with the highest levels of airborne particulate matter. Like China, dirty energy sources and heavy industry fuel India's rapid ascent. Unlike China, however, India's relatively high fertility rate means that the population will continue to grow, thereby placing even more pressure on the already overcrowded cities in the future.

The United Nations warns that the persistence of current urbanization trends depends upon a city's ability to provide adequate services as well as viable employment opportunities. In South and East Asia, where urbanization has been accompanied by economic growth, cities will be increasingly able to meet those basic needs in the near future, becoming wealthier and more environmentally sustainable. Urbanization in Sub-Saharan Africa, by contrast, has been mostly in the form of slums growing around cities that already lack sufficient resources. Currently, Sub-Saharan Africa is the least urbanized region in the world. With many cities growing between 5 and 10 percent per year, the region is expected to be more than 50 percent urban by 2030.

According to research by the World Bank, more than half of Africa's urban residents do not have access to proper sanitation, and that figure exceeds 80 percent in a few countries like Sierra Leone and Niger. The same study found that 20 percent of urban Africans do not have access to clean drinking water, and more than three out of four residents in western African cities live in substandard housing. All of these factors combine to severely challenge the sustainability of Sub-Saharan Africa's urbanization. Inadequate disposal of urban waste and trash in LDC cities around the world affects the spread of diseases as well as the overall quality of urban life. Hardly a year goes by without the reported outbreak of disease or the contamination of an urban water supply. While European, North American, Australian, and Japanese cities implement programs designed to protect the quality of the living, working, and leisure environments, for cities in less developed countries, health issues are often considered of secondary importance vis-à-vis economic development goals. Polluted air from automobiles or from industries is a fact of daily life just as much as polluted drinking water and piles of garbage. Ecological problems abound including the outbreak of diseases, insufficient health vaccinations, and sporadic public safety warnings.

The UN's Sustainable Cities Programme operates in scores of cities, mostly capitals. It aims at developing local capacities for environmental planning and management as poorly managed urbanization creates serious environmental and social problems. At the municipal level, development issues generally include water resources and water supply management, environmental health risks, and solid and liquid waste management/on-site sanitation, air pollution and urban transport, drainage and flooding, industrial risks, informal-sector activities, and land-use management in the context of open-space/urban agriculture, tourism and coastal area resource management, and mining.

Climate Change and Urban Futures

While rapid urbanization brings challenges of its own, cities of the future will be faced with yet another unprecedented event: the effects of climate change. Many of the world's largest and fastest growing cities are in coastal areas. This fact makes them particularly vulnerable to climate change for two reasons. First, weather events like strong hurricanes and typhoons (once considered extraordinary) are expected to become more frequent and more intense in the near future. Major storms like Hurricane Katrina on the Gulf Coast and post-tropical cyclone Sandy in the Northeastern United States are likely to become the norm rather than the exception. Second, sea levels are expected to rise anywhere from one to six feet by the end of the century (Box 13.2). When one considers that a rise in sea level of just eight inches has the potential to displace millions of people currently living in low-lying coastal areas, it becomes apparent that cities must take a proactive stance to protect residents and infrastructure. A report by the Worldwatch Institute noted that the growing numbers of environmental disasters are resulting in more loss of life each decade. In the past 25 years, 98 percent of the people injured by natural disasters lived in the 112 countries classified as low or middle income by the World Bank. These countries accounted for 90 percent of lives lost to natural disasters. The same is true for the melting of sea ice, which results from global warming. Compounding the problem, less than 3 percent of LDC residents have insurance compared to about 30 percent in the rich world.

The problems that climate change presents to cities are compounded when they expand geographically to meet the demand for new housing, services, transportation, and industries. New construction gets pushed into low-lying areas and other places that are not suitable for building because they are more vulnerable to sea-level rise and flooding caused by storms. This is especially problematic in the

Box 13.2 Living With Water

Michael Allen, Old Dominion University

As a historic, coastal community, Norfolk, Virginia, has a long-standing relationship with water. Situated at the mouth of the Chesapeake Bay near the Atlantic Ocean, the city is bisected by the Elizabeth River and penetrated by various "creeks." European merchants settled in the region 400 years ago. Since then, Norfolk's 144 miles (232 km) of coastline have encouraged trade and commerce. Tourism and seafood industries thrive. The city is the home to the world's largest naval base. The Port of Virginia provides thousands of jobs, millions in revenue, and services to 45 countries worldwide. And, the nearly 250,000 Norfolkians have learned to live with water.

Figure 13.7 Even short rainstorms bring flooding to Norfolk's streets and underpasses. The problem promises to worsen as sea levels rise and much of Norfolk subsides. *Source:* Photo by Michael Allen.

However, living in an urban, coastal community also includes challenges. Nor'easters and tropical cyclones, most notably Hurricane Isabel in 2003, often impact the region. And, whether the result of storm surge, sea-level rise, or simple downpours, flooding plagues Norfolk's residents (Figure 13.7). According to the National Climate Assessment, global sea level has increased about 8 inches (200 mm) since the late 1800s. However, in Norfolk, sea level has increased a total of 1.51 feet (0.47 m) just since 1927. And, for Norfolk, it's a one-two punch. While the sea is rising due to greenhouse gases and melting ice, the land is also sinking due to prior land-use policy. Throughout its history, many creeks, wetlands, and marshlands have been filled with debris, and with time, this land naturally sinks. Consequently, the federal government recognizes Norfolk as the second most vulnerable city in America to sea-level rise, behind only New Orleans.

Despite overwhelming scientific evidence, the discourse surrounding climate change is often politicized in the United States, and even denied. Yet, in Norfolk, residents share stories of climate-change reality and narrate complexities that involve morals and values, economics, and the role of government. Who is responsible for environmental change? What should the government do? Through the National Flood Insurance Program, FEMA subsidizes rates for high-risk, flood-prone regions. Following natural disasters, money is filtered to these regions for redevelopment. Some argue these subsidies encourage coastal development in high-risk zones. Norfolk has invested millions of dollars to protect property from rising seas. In 2012 alone, $7 million was used to raise roadways and homes, yet the ability to control nature has proven to be almost impossible: flooding still occurs. Forums, such as the 2015 Dutch Dialogues, suggest that the city (and region) is beginning to think more creatively in terms of mitigation and adaptation.

With an estimated $300 million price tag, shared investment in climate mitigation and adaptation requires all levels of government, the military, and various nongovernment entities to partner and collaborate. It's a shared responsibility. Norfolk's mayor, in fact, recently suggested that there may come a time in the near future when Norfolk will have to create retreat zones—areas that will be fully inundated by water and unsuitable for development. Significant political willpower to address climate change is needed. Named as a top-100 resilient city by The Rockefeller Foundation, Norfolk is situated to serve as a model for designing cities of the future, for teaching us how to live with the water while supporting economic prosperity, and for addressing social injustice while building more resilient communities.

developing world, where many people moving to megacities are poor and set up housing that is both substandard and in hazard-prone landscapes.

Infrastructure to Mitigate Climate Change

Cities in the near future will be forced to step up their investments in technology and infrastructure that protect the built environment from flooding and storms. The Organization for Economic Cooperation and Development (OECD) estimates that in 50 years, the value of the infrastructure assets that are at risk from storm damage (buildings, transportation, and utilities) in New York City, Miami, and Guangzhou will be worth around ten trillion dollars combined. Tokyo and Nagoya in Japan and Amsterdam and Rotterdam in the Netherlands are already taking proactive measures to protect their infrastructure assets. Comparing a map of the world's most populous and fastest growing cities with one showing shallow continental shelves reveals that many prosperous U.S. cities from Boston to Miami to Houston will be affected by a two-meter rise in sea level. Cities in northern and northwest Europe are also vulnerable, as are cities at the mouth of the Ganges, on the Malay Peninsula, in insular and mainland Southeast Asia, eastern China, and Japan.

Ho Chi Minh City, for example, is expected to grow from its current population of 8–20 million by 2050. To accommodate this growth, the city will look to expand into surrounding agricultural and forested areas, thereby removing vegetation that acts to mitigate natural flooding. Rising temperatures in the adjacent South China Sea are expected to bring more frequent, heavier rains, bigger storm surges, and more tropical storms. Other cities in the region have already begun feeling the pressure. New industrial developments around Bangkok have expanded into low-lying wetlands and former rice paddies, where flooding is a regular feature of the landscape. In 2009, heavy rains from a tropical storm inundated 80 percent of Manila, and even stronger storms are expected in the near future.

At the other end of the climate-change spectrum, droughts are also expected to become more frequent and more severe. As urban areas in the American southwest like Los Angeles, Las Vegas, and Phoenix add land area and residents, they will put intense pressure on the already strained water supply. California is currently in the midst of a prolonged mega-drought that will likely go unabated for the foreseeable future, amid high population growth and agricultural output. Growing cities, therefore, will have to find new and innovative solutions to meet the water demands of the population.

Large-scale desalination of seawater is one proposed solution that is already being developed. For coastal cities, this is an expensive option that could meet some of the demand. Cities will also have to increase their investments in water treatment plants that recycle wastewater. The main adaptation, however, will be to rethink land-use planning and urban and suburban lifestyles in regions that are starved for water. Rather than having sprawling cities with wide roads that create an urban heat island, cities will need to have more compact pedestrian zones. Instead of planting trees in drought-prone areas to provide shade, buildings will need to be constructed closer together in a manner that provides human-made protection from the sun. Narrower alleys between buildings also create breezes that can lower surface temperatures. Urban residents will have to replace their grassy lawns with vegetation that is more suited to a dry, desert climate. They will also have to shift from single-family housing to multistory buildings to make water use more efficient.

Deindustrialization and Urban Futures

Cities in the developed world underwent the same sort of growing pains during their development phase in the nineteenth and early twentieth centuries as cities in the developing world are facing today. Improper sanitation and water pollution led to cholera outbreaks in cities like London, Paris, and New York. Smog rose to lethal levels in London and elsewhere in the United Kingdom in the nineteenth century. London's nickname, in fact, was "The Smoke." In the developed world, many countries passed legislation to rein in pollution and improve the environment only as recently as the 1970s.

Just as the developed world implemented tougher environmental regulations, the developing world was opened up to economic globalization, resulting in drastic change in urban economic functions. This change has been marked by deindustrialization in former industrial powerhouses like Pittsburgh, Milwaukee, and Cleveland in the United States;

and Glasgow, Liverpool, and Manchester in the United Kingdom. Over the past three decades, these and other former industrial centers have undergone a marked shift away from industry and toward services. This transition resulted in the loss of hundreds of thousands of manufacturing jobs, depriving people of a living wage. At the same time, deindustrialization made way for a much cleaner, more environmentally sustainable urban economy based on knowledge and information. It also allowed for urban revitalization by repurposing millions of square feet of former warehouse and factory space.

Pittsburgh, for example, was once the heart of the U.S. steel industry. Skies darkened by industrial smoke and buildings blackened by soot were characteristic images of the city. When the last steel mill closed in the mid-1980s, Pittsburgh transformed itself into a center of engineering and medical research and integrated itself into the modern knowledge economy. Abandoned buildings from its industrial heyday became valuable real estate in the form of loft apartments, boutiques, and coffee shops. More young professionals are moving to the urban core rather than to low-density, auto-dependent suburbs as their predecessors did, thus reducing the necessity for long commutes. As a result of these changes, the air is now remarkably clean and clear, especially when one considers that just a few decades ago smog levels in Pittsburgh were no different than those observed in today's Chinese cities. Similar trends are occurring in cities throughout the developed world, which is a positive sign as far as the sustainability of urban living is concerned (Figure 13.8).

New industrial cities in China and India are likely to undergo the same type of transition in the near future. In these and other developing countries, the urban middle class

Figure 13.8 Is this carbon-neutral office building in Melbourne, Australia, the future of sustainable urban architecture? The colorful panels on the outside are components of the sun-shade system. What you can't see are the night cooling windows, the green roof, the vacuum toilets, and the anaerobic digester. *Source:* Photo by D. J. Zeigler.

is growing as wages and education levels rise. These trends open the way for a transition to a cleaner, more sustainable future. At the same time, a rise in the middle-class population in developing countries means that more people are living in larger residences, buying automobiles, demanding reliable city utilities, and requesting that governments build roads and highways to support increased automobile travel. It will be a challenge for such cities to dedicate funds and space to the preferences of their growing metropolitan middle classes.

Urban Gardening and Urban Futures

In addition to repurposing old industrial buildings, urban gardening is another attempt to make cities of the future more sustainable. Many older neighborhoods in more established North American cities feature numerous vacant lots where a building or parking lot once stood. Programs in these cities allow local residents to use the vacant plots for community gardens, which can bring numerous benefits. First, urban gardening helps reduce the heat island effect that is common in cities constructed almost entirely of asphalt and steel. Second, gardens are a low-cost way to beautify the urban landscape, especially in deindustrializing cities where the surrounding buildings and infrastructure have fallen in disrepair. Third, by giving local residents a chance to grow some of their own food, urban gardens cut down on the number of food-miles produce needs to travel before it reaches urban consumers. Fourth, people come into closer contact with their neighbors and their neighborhoods through the act of tending to a community garden. Finally, urban gardens offer a serene place to relax in the city. They can even attract tourists and create jobs.

While the benefits of urban gardening may apply to all types of cities, they are especially relevant to the older and deindustrializing cities of North America and Western Europe. Urban population shrinkage has already transformed many Rust Belt cities. Managing this shrinkage with creative and environmentally sustainable solutions has been and continues to be a primary concern as deindustrialization continues for the foreseeable future. It is even possible that horizontal surface plots that currently characterize urban gardening may transition into multistory, vertical urban farms capable of producing far more volume and variety of crops than currently possible. In addition to supplying food, these vertical gardens could act as a natural temperature control and add color to the urban skyline.

THE GEOGRAPHY OF CONNECTIVITY AND TALENT

In the developed world, one of the most important developments of the twenty-first century has been the rise of the knowledge economy. While the twentieth-century economy traded on tangible inputs and manufacturing, the knowledge economy trades on information. Since advanced information and communications technology (ICT) decouples access to information from place, this economic shift has important implications for the future of cities the world over (Box 13.3).

When ICT was first becoming mainstream and transforming communication, many technology writers predicted the demise of cities because decentralized communication seemed to remove the need for clustering by industries and people. Telecommuting was expected to transform the modern workplace and give workers the opportunity to

Box 13.3 Human Geographies of the Twenty-first Century

Stanley D. Brunn, University of Kentucky

Economic, social, and political futures in any region are affected by local cultures and events as well as regional and global actions and institutions.

1. *Continued urbanization at the global scale.* Urban agglomerations will grow bigger and urban institutions will increasingly dominate even rural areas. Interactions and associations are likely to be increasingly between cities near and distant rather than between cities and rural trade areas.
2. *Urban connectedness, anomie, and placelessness.* Faster transportation and communication technologies may lead people to devalue their sense of place, thus increasing anomie, alienation and social instability.
3. *Meshings of the local and global in daily life.* Scale meshings will be evident in transactions and interactions—where one works, with whom one works, and the destinations of goods and services. While some urban residents will interact predominantly at local levels, others are will operate at transnational scales.
4. *Asianization and Africanization of Europeanized worlds.* An ongoing contemporary cultural global process is the impress of Asian and African diasporas on traditionally Europeanized worlds. These new cultures will add new layers of food, music, entertainment, and intellectual diversity to city life.
5. *Increased regional and global awareness.* The diffusion of ICTs and instant global reporting of crises will increasingly transcend political boundaries and raise the awareness of planetary concerns. One example is the growing emphasis on the basic human rights of women, children, elderly, disabled, and cultural minorities.
6. *Competitive K-economies.* "K" is for knowledge. It symbolizes the transition from handware to brainware and the importance of images and symbols in product consumption. These brain economies will be of increased significance in globally competitive and creative cities.
7. *Contested legal structures.* Increased volumes and densities of transborder urban networks and circulations will raise questions about the effectiveness of traditional governments in daily life: individual versus group rights, temporary verses permanent residents, those with and without property, and possibly with and without statehood.
8. *Redefining norms and abnorms.* Urban cultural and political clashes (subtle and violent) are likely to continue to emerge among groups: those calling for tolerance and diversity in work, living, lifestyle, and social spaces versus those seeking to retain traditional norms based on religious and rural values or outdated modes of authority.
9. *Limits on technological breakthroughs.* Prepare for more constraints on the adoption and dissemination of new technologies because of adverse social impacts on a culture, high fixed costs, and government security. Will technology "gaps" between haves and have-nots begin to narrow or will technological solutions be too expensive?

10. *Veneers of homogeneity amidst diversity*. The "McDonaldization of the World," or the creation of a Western globalized consumer world dominated by Western food, music, fashion, and entertainment, will reflect a certain visible sameness. But beneath these landscapes and icons of Westernization will be the rich historical and cultural mosaics of enduring regional cultures.

finally leave the city in favor of the cheaper, pastoral countryside. Although ICT certainly adds geographic flexibility to the traditional workplace, its impact on the future of cities is still unclear. On the one hand, major cities are places with superior capacity to receive, process, and transmit the information that fuels the knowledge economy (Figure 13.9). In addition, being near competitors in the same field makes it possible to keep up with the state of the art. As Richard Florida has pointed out, industry insiders stay abreast of the latest rumors, gossip, advancements, and innovations in their respective fields by being in constant contact with their competitors, whereas those far from the primary cluster of activity may find it difficult to access the same type of information in a timely manner.

On the other hand, because of its ability to decentralize information, the internet makes it possible for a wider range of places to participate in the knowledge economy. Business interactions can now take place remotely and instantaneously because commodities and information are no longer bound to physical space. Therefore, it is possible that smaller cities in the global periphery can take advantage of the decentralized nature of the internet and insert themselves into the modern knowledge economy. The intersection of the cultural industries with advanced ICT is particularly illustrative of this point.

Historically, innovations in music, fashion, art, and in other cultural industries have come from major centers of global cultural production like New York, Los Angeles, Paris, Milan, and London. These world cities provide both a large and sophisticated local marketplace and a critical mass of creative individuals with whom to share new ideas and collaborate on innovative projects. In the digital age, the internet decentralizes the marketplace for cultural products in addition to providing a platform for mass sharing and collaboration. As a result, new producers have entered the cultural industries en masse, regardless of their location. If a new music group lives in a remote location with few performance venues, they can easily distribute their music to a global audience online. Likewise, fashion designers can create new items, advertise them online, sell them, and ship them to consumers around the world.

Elite cultural industries in large production centers, such as fashion in Milan and film in Los Angeles, will likely continue to dominate the cultural economy into the future. However, the internet creates new opportunities for a wider range of production centers in smaller cities to compete in niche markets, even if one or a few large centers remain dominant. The ability to remotely access and participate in the cultural economy bodes well for newly urbanizing places, especially those in the developing world. At the same time, it is important to note

Figure 13.9 The Shard, completed in 2012, is the latest addition to London's collection of skyscrapers and the tallest building in the European Union. Globalization has bid a whole new generation of skyscrapers into construction. *Source:* Photo by D. J. Zeigler.

that the internet is far from being a spaceless, placeless phenomenon. Advances in ICT have shrunk the global time-space continuum, but the economic landscape is still highly uneven. The best access to high-speed internet and other components of the digital infrastructure are overwhelmingly concentrated in urban areas, leaving many small towns and rural areas on the other side of the digital divide. In addition, service provision favors wealthy places as companies compete for consumers who are able to afford the highest possible prices.

Cities as Virtual Crossroads

How can we expect ICT to impact the size, shape, and function of cities in the coming decades? How much will decentralized information chip away at the economic dominance and informational privileges that currently characterize cities? The answer to those questions is not so straightforward. The modern world of business and communication is no longer a "space of places" but rather, in Manuel Castell's words, a "space of flows." In other words, even though the economy is already highly globalized, it will become even more interconnected in the near future. Cities that are best able to accommodate flows of information and commodities will thrive, while those lacking this ability will struggle. The most successful cities will establish more intensive connections with each other, and some peripheral places that were previously less touched by globalization will also come into the network.

Urban form has always been shaped, in part, by communication and transportation systems and structures. Canals, tramlines, subways, and expressways have all significantly altered the time it takes to travel between any two connected places. The time that it takes to move people, goods, and information within or between cities affords places distinct advantages and disadvantages.

In the past, cities thrived or atrophied based on their relative positionality within global flows of trade. Many large urban agglomerations still owe their prosperity to transportation facilities, including important roads, canals, and airports. While positionality will be no less important for cities of the future, prosperity will increasingly rely on an alternative structure. Telecommunications networks, as well as the social, economic, and political opportunities for people in cities to access such networks, are currently very unevenly distributed across the globe. Broadband and fiber optic cables are heavily concentrated in (and link) East Asia, Europe, and North America. Heavily wired cities include Seoul, London, and Boston; and they are much more likely to experience change brought about by time-space compression than cities lacking high network connectivity, such as Pyongyang or Kinshasa.

Many cities have been, and continue to be, restructured as wired and wireless cities in order to compete for a vast array of high-technology digital transfers that link businesses and households. San Francisco, Seattle, San Diego, San Jose, Los Angeles, New York, Washington, DC, Chicago, Boston, and Miami are among the top U.S. cities in internet penetration. The internet is being utilized not only for luring capital and businesses, but also for security purposes. For example, web cameras monitor London's streets and its financial district. Internet penetration in some LDCs is also significant and is a deliberate strategy to attract economic investment from the MDCs. Delhi, Mumbai, and Bangalore are all wired cities in India's globalized economy.

Examples of countries making heavy investments in national and urban digital economies are Singapore, Estonia, Slovenia, countries of the Persian Gulf (as part of their post-petroleum economic planning), Hong Kong, Ireland, Jamaica, and Trinidad. Digital cities invest in computer hardware and software design and support large and small companies performing various digital tasks, as well as the many nongovernmental organizations and governmental institutions. These include libraries, courts, hospitals, and employment and environmental centers.

Certain cities are more tightly connected to global networks than others. However, physical connections do not ensure that a majority of a city's inhabitants will benefit from globalized networks. Large groups in a highly connected city, as measured by bandwidth, can remain disconnected. Unfamiliarity with world languages that dominate the internet, especially English, but also Chinese and Spanish, can render network connections without meaning and value. Similarly, various cultural and political restrictions can exclude people, based on gender, ethnicity, or religion, from regular internet use. For instance, in certain cities it is highly inappropriate for women to spend time in male-dominated internet cafes. Finally, economic barriers may be the most powerful exclusionary force. Both the means (a computer and a mobile device) and the cost of access (hourly or monthly fees) can be prohibitively expensive for many residents even in the most connected cities. Questions of inclusion and exclusion are becoming ever more important for cities, because unequal access to distant information and communications will likely cause other increasing urban inequalities.

Ultimately, highly connected cities will continue to be command-and-control points for the global economy, housing headquarters of far-reaching corporations, particularly in the financial, insurance, and real estate fields; as well as government and nongovernmental organization offices. Cities will thrive because they are well-connected hubs in the global flows of information and travel. Those cities that have the satellite dishes, fiber optic cables, and internet routing switches will prosper, as will cities that provide wireless internet and mobile device networks that enable large proportions of the population to be connected and become economically productive (Figure 13.10). Cities that have successfully transitioned to idea-based economies, specializing in research and innovation, will thrive because these cities are where ideas are born and content is created.

Currently, London and New York are the most outstanding examples of such highly connected world cities, but with high population and economic growth, cities such as Shanghai and Dubai may increase in global economic status. These and other command-and-control cities will be increasingly connected to each other, rather than to their local hinterlands. Cosmopolitan migrants, business people, and government leaders will commonly move between them, over great distances, rather than moving between world cities and their outlying regions. Larger airports, longer-range commercial jets, and high-quality communications networks enable such distant movement of people as well as ideas to occur quickly.

Figure 13.10 Wireless networks, cell phones, and matrix barcodes bring urban landscapes to life, tell the stories of times past, and signal advances in technology that mark world cities. London is so wired, you can even talk to the long-gone goats. *Source:* Photo by D. J. Zeigler.

Cities as Nodes of Globalization

Globalization is, and will likely continue to be, one of the most frequently repeated themes of the twenty-first century. The idea that cities are becoming economically, culturally, and politically more globally connected is grounds for hope, desire, fear, and despair. Globalization can lead to shared advances in science and technology, economic growth, and the exchange of philosophic, political, and artistic ideas between people and places that were once isolated from each other. However, increased connections can also result in economic and political exploitation and the destruction of cultural bonds.

Globalization is not something that spreads across the surface of the globe like fog rolling over a chain of hills. Rather it is evident only in specific times and places. In other words, if we take the example of any city experiencing transformations because of global connections, that city is not necessarily spatially proximate to the sources of those transformations or to other cities experiencing similar transformations.

A widely repeated idea is that the absence of globalization is responsible for global inequities. In other words, cities that remain disconnected from global processes and flows of trade are unlikely to have high living standards. Such ideas are usually based on economic theories that rely on "the logic of the marketplace." By allowing the global market to regulate society instead of being regulated by society, it is argued that market forces will enrich people in poor cities by effectively governing and creating wealth for all participants. The counterpoint of this argument is the idea that ever-increasing globalization and integration of cities will only exacerbate global inequalities. Within the globalized economy, capital and jobs can now be moved rapidly from place to place, but the actual populations remain rooted in their home cities. The effects of globalizing processes have led some scholars to refer to "a race to the bottom," in which people have to accept increasingly lower wages and benefits in order to perform the same jobs. These concerns have fueled a massive global movement of protest and highlight the inequalities caused and intensified by globalization, including antiglobalization demonstrations at biannual World Trade Organization conferences wherever they occur.

One of the most important ways that globalization can be measured is through the movement of people and cargo by air. However, the most globalized airports are not always located in cities that most people would associate with a high degree of global connectivity. Of the world's three largest cargo airports in 2010, one comes as a surprise. Memphis, Tennessee, is outranked only by Hong Kong as the world's second-largest cargo-handling airport, due to FedEx's primary hub being located there. Following Memphis are Shanghai and Seoul, and, then, surprisingly, Anchorage, Alaska, which serves as the crossroads between North America and Asia. In Europe, the leading cargo-handling airports are Frankfurt and Paris; Dubai leads in the Middle East.

When passenger flows between cities are considered, we get a more familiar picture of cities networked into the global economy. Europe and particularly North America dominate the rankings. While some highly ranked cities have high connectivity because they serve as major airline hubs, others including London and New York are highly connected due to their central position in global economic and tourist flows. Among the world leaders for city airport systems by passenger traffic are London, New York, Tokyo, and Atlanta.

Cities Beyond the Networked Core

Large cities are home to the longest life expectancies in developed countries because of the presence of public health infrastructure, research hospitals, safe occupations, and the near absence of heavily polluting industry. These characteristics attract people from across the lifespan, but they may be especially appealing to seniors, who comprise an ever-larger share of total city populations. In fact, the graying generation, by virtue of its disposable income and consumer-buying habits, may be drawn to such cities and serve as a key to growth and economic development. The idea that cities are the healthiest places to live is rather remarkable when compared to previous centuries during which cities were unhealthy places due to manufacturing-generated pollution and diseases caused by inadequate sanitation, ventilation, and hygiene.

In addition to cities that specialize in health care, we can also expect some smaller cities with low global profiles to boom because they are situated near newly exploitable natural resources. Examples include towns that have grown into cities near the burgeoning tar sands of central Canada, like Fort McMurray, Alberta, and around natural gas fracking in the United States, such as Williston, North Dakota. Increasing affluence around the world means more people are consuming energy, which is driving energy prices higher, causing resource exploitation in remote regions that have previously been cost-prohibitive. The long-term viability of cities that are dependent upon natural resources is unclear, however. "Boom and bust," rather than long-term prosperity, are typical of natural-resource-based economies. Cities in the petro-states of the Persian Gulf, especially Dubai in the United Arab Emirates and Doha, Qatar, are now testing the proposition that economies built on petro-dollars can transition to economies based on high-end services. At the present time, the price of crude oil is dropping and all cities with an economic base in the petroleum sector may be in for a coming bust.

Other cities outside the world's core network will thrive because they are located near inexpensive labor, taking over labor-intensive manufacturing tasks for the global market in a process called offshoring. As African cities, for instance, procure reliable information-technology infrastructure such as fiber optic cables, and as electricity provision and shipping connections improve, labor-intensive manufacturers will likely relocate labor-intensive operations in this inexpensive region. It is probable that cities in Africa will industrialize and grow in the coming years, similar to the recent experience of Asian cities.

Will this movement of manufacturing jobs to Africa come at the expense of Asian cities that are currently growing? When economic growth tapers in today's developing countries, will cities in those countries experience real estate busts, especially in China? Or will developing countries transition from export-driven economies to domestic-demand economies? People in developing countries are starting to earn enough money to travel, and transportation services are growing, as are casinos, theme parks, golf courses, and other tourist activities.

In the near future, some cities will decline, losing population and influence because they have not successfully navigated a changing globalized economy. Cities based on labor-intensive manufacturing located in developed countries will decline because rich countries do not have low-cost labor. Instead, cities in developed countries that transition to capital-intensive economic activities that use high-skilled labor will thrive. Examples of cities that have not negotiated the deindustrialization transition well include Detroit in the United States and Newcastle-upon-Tyne in the United Kingdom. Both saw their population peak in 1950 and decline for decades after that. Now, the rate of decline has leveled off for the former, and population has begun to increase for the latter.

GOVERNANCE, GIS USE, AND SECURITY PROVISION

Governmental Cooperation

New and modified forms of urban governance will likely emerge in coming decades to deal with interurban economic connectivity as well as larger and changing city populations. City and metropolitan-region governments have increased in importance and are predicted to do so into the future. Port authorities that operate airports and seaports, development authorities that promote economic growth, and transit and toll road authorities that build and operate transportation systems are commonly metropolitan-area based, rather than being housed at the provincial or federal level.

In metropolitan areas that cross an international border, such as El Paso/Juarez and San Diego/Tijuana, city governments are cooperating on some of the issues noted here. World cities are increasingly cooperating with other large city governments across the globe because of shared challenges and potential solutions. For example, New York City looks to London for traffic management ideas such as congestion tolling instead of calling on New York City's state government or the U.S. federal government. Increasingly, cities are finding that they have more in common with other cities and are aligning government cooperation accordingly. Cities are collaborating with other cities across international borders rather than with their provincial hinterlands, mirroring economic realignments that have been occurring for the past several decades.

As countries throughout the world become more urban, politics is also likely to change. Generally, regions that have a higher proportion of their populations living in urban areas are more politically liberal, progressive, and open-minded compared to regions with more nonurban residents. Urban residents, who live in compact, diverse, and highly socially connected places tend to be more tolerant of diversity and change. Urban populations have driven movements to legalize same-sex marriage, marijuana use, casino gambling, and assisted suicide. These urban voters are very different politically than their rural counterparts or the countries in which they are located; instead, they share more in common with other cities.

Geographic Information Systems

Digital technologies, especially geographical information systems are used to plan, build, and manage the urban environment. Since the early 1970s, GIS has become routine in urban planning tasks in North America, Western Europe, Japan, and Australia. These systems are also becoming adopted in the cities of some less developed countries, particularly India, China, South Africa, Senegal, Ghana, Brazil, and Mexico. GIS is the science and technology that integrates vast databases with georeferenced data (exact latitudinal/longitudinal coordinates) to prepare maps, satellite images, and aerial photographs. These databases allow researchers and planners to perform a range of statistical and spatial analyses including modeling, visualization, and simulation; to design, plan, and manage urban environments; and to propose and evaluate future scenarios. GIS can easily be linked to the Global Positioning System (GPS) to ensure highly accurate spatial analysis. Also, mobile and hand-held GPS units integrated into GIS programs can be taken to remote areas to conduct real-time research.

Usage of GIS has become popular with citizen-based grassroots organizations because GIS enables them to be informed—and powerful—participants in community planning efforts. Key GIS applications executed with technocrats working in tandem with grassroots groups include natural resource management and conservation efforts, community-based planning and neighborhood revitalization in urban areas (Figure 13.11), and activism organized at local, national, global, and multiscalar levels. In such efforts, local, qualitative knowledge is integrated with quantifiable public data sets, and mental maps and sketch maps are integrated with official digital maps, aerial photographs, and satellite images. These practices have been used in a number of projects among indigenous societies in the non-Western world.

In the context of neighborhood revitalization in the Western world, community organizers use public data sets through GIS to inform and legitimate local knowledge to obtain action and formulate strategies, monitor and predict neighborhood, prepare for organizational tasks, fund recruitment efforts, enhance service delivery tasks, and explore how spatial relations shape urban policy. GIS

Figure 13.11 What would you build here? Let your voice be heard. Here, people along 14th Street in Washington, DC, are being challenged to create the neighborhood they want by voting on ideas that they themselves come up with. *Source:* Photo by D. J. Zeigler.

analysis has enabled communities to accomplish such tasks as:

- fighting blighted housing conditions by tracking down absentee landlords through property records;
- providing police assistance by identifying crime hot spots across time and space;
- analyzing land-use data to track vacant and boarded up houses;
- keeping track of school-age populations to optimize assignments to schools and plan new school bus routes;
- combining mortgage-lending data with demographic data to address discriminatory investments; and
- generating sophisticated, multiscalar maps to measure a community's changing well-being.

Community GIS activities have been aided by the emergence of internet mapping sites such as Google Earth, in which high-resolution satellite images of places across the world can be accessed in minutes for free by anyone. The power of Google mapping is evident in the case of southern Brazil's Surui people, whose chief approached Google for high-resolution satellite imagery to monitor illegal loggers and miners on the tribe's 600,000 acre reserve. While high-tech mapping has already been used to track illegal activity and record knowledge, this is the first time an Amazon tribe will share their own vision of their territory with the rest of the world via Google Earth. In essence, they have a technology that allows them to stand up against globalizing forces that are biased in favor of urban economies. In the western world, various government agencies provide easy access to public databases through Internet GIS sites. One site, COMPASS, enables citizens of Milwaukee to access, view, query, and map detailed property data, health data, crime data, community asset data, and demographic data at no cost. Such Internet GIS sites are particularly useful for resource-poor organizations that have difficulties in creating and maintaining in-house GIS.

Surveillance of Public Space

At the neighborhood scale, street-level governance in cities will continue to evolve. In terms of law enforcement, community policing has been credited with driving street crime down to historically low levels, particularly in North American central cities. Some of this decline is likely due to gentrification, as residents with higher incomes and levels of education comprise a greater proportion of downtown and inner-city neighborhoods compared to past decades. These populations are also demanding better city services and have flexed their political muscle to secure enhanced police, public transit, and sanitation services. New York has utilized a controversial stop-and-frisk policing policy, but this method has been challenged in court for allegations of racial profiling.

City governments have recently responded, and continue to respond to, the threat of terrorism. The attacks on September 11, 2001 in New York; March 11, 2004 in Madrid; July 7, 2005 in London; November 26, 2008 in Mumbai; April 15, 2013 in Boston; and November 13, 2015 in Paris, are some instances in which international and domestic terrorists have targeted civilian urban spaces such as office buildings, hotels, shops, streets, and transit systems. City police departments have responded by adding antiterrorism divisions to supplement more traditional patrol, detective, transportation, and organized crime units.

Almost all cities, with London and New York being among the first, have increased surveillance via networks of police cameras mounted throughout the central city. In addition to providing a record of street activity, these cameras can also track license plate numbers and employ facial recognition software. At a macroscale, unmanned drones have the ability to capture high-resolution photos and videos from a distance, unbeknownst to passersby on the street. Terabytes of surveillance can be sent, received, and analyzed in a matter of seconds by technicians in remote locations. At the microscale, police offers are beginning to wear body cameras, and that trend is likely to extend to other professions as well. Cities of the future will have an even greater number of cameras, in a greater variety of locations, and operating at a greater range of scales. Urban residents will either adapt to this new reality or stand up against becoming a surveillance society.

Cities in the developed world justify mass surveillance by maintaining that it gives law enforcement officials far greater ability to make cities safer. Critics counter that mass surveillance programs create new realities that are also undesirable. For example, as biometric recognition software increases in sophistication, it becomes possible to track an individual's every movement. Civil liberties activists have concerns about the possibility to abuse such an extensive and powerful tool. Urban governments, especially in democratic countries, must find an appropriate balance between managing security concerns and protecting an individual's right to privacy.

Human security issues are emerging as important topics for those in the social and policy sciences. They are also important for those studying urban worlds as security concerns affect the daily lives of individuals and influence where they reside, work, and play. Human security concerns include dealing with terrorism in its many dimensions. Transnational terrorism and biological and environmental terrorism are considered high-priority threats for urban governments. Public officials must insure that dangerous individuals, goods, and substances do not cross their borders or enter their country's airports and ports. They are ever vigilant for contaminated products, whether foods, live plants and animals, or other materials, that might endanger their food supplies, livestock, public water systems, and the unique natural and built environments that attract tourists. The security industry will remain one of the urban growth industries for much of this century, as evidenced already in airports, train stations, harbors, and border crossings.

CONCLUSIONS

The cities of the future will be more populous and more numerous than those that exist today. In fact, one prediction has the world's metropolises, merging into megalopolises, and then into one giant ecumenopolis (Figure 13.12). Until that time comes, the most populous cities will increasingly be located in today's developing countries. This urban growth in the developing world is being driven by dropping mortality amid high fertility, industrialization, and the accompanying urban in-migration as redundant labor leaves rural areas to fill newly created jobs in cities. The same urban population transition has already played out in the developed world, where city populations have plateaued. In the developed world, low birth rates mean that

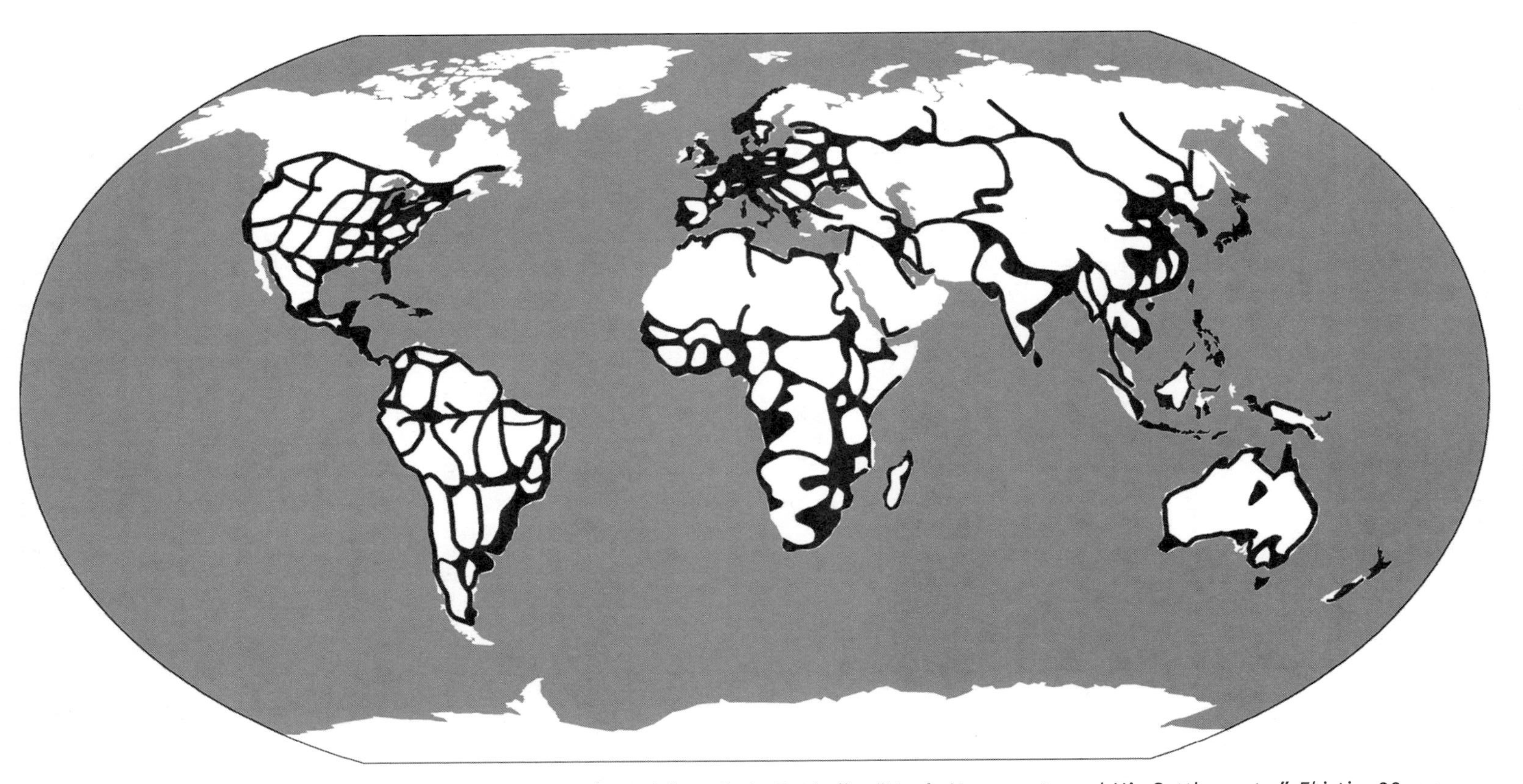

Figure 13.12 Ecumenopolis: The Global City. *Source:* Adapted from C. A. Doxiadis, "Man's Movements and His Settlements," *Ekistics* 29, no. 174 (1970): 318.

growth will come primarily from immigration, with most immigrants now showing a preference for suburban areas. Urban revitalization is occurring in the developed world as smaller, older, and wealthier households relocate closer to city centers, along with young adult households. Amid these changes, cities in less developed countries will grapple with providing adequate infrastructure, services, and housing to their residents. And, urban areas in more developed countries will accommodate urban revitalization, diversification of their suburbs, and the postindustrial transition to services.

Sustainability and climate change will be near the forefront of challenges facing cities the world over in the near future. Urban living is more energy efficient, but it concentrates pollution. The urban poor, especially in less developed countries, commonly reside in low-cost and environmentally degraded neighborhoods often characterized by informal housing. These residents often lack access to education and high-quality jobs and often experience health challenges associated with poor environmental quality. Cities will have to develop ways to become more sustainable as well as provide opportunity for their residents. This will happen against the backdrop of climate change, which will force cities to adapt to withstand more fierce and frequent storms, rising sea levels, and increased drought.

Demographically and economically growing cities will follow one of several paths. In the more developed world, the most successful cities will continue to navigate the transition from deindustrialization to a high-tech, creative, service economy based on the command and control of the increasingly globalized economy. These cities will continue to augment their positions as important nodes in the global information technology and transportation networks. Other cities will thrive by offering cheap labor, proximity to a natural resource, or a particular specialty like advanced medical treatment. Cities with such nondiversified economies will need to prepare for a possible future when their singular competitive advantage wanes, such as when wages rise or drilling wells become played out.

Cities are increasingly cooperating with each other, even across international borders, in activities from preventing terrorism and crime to operating ports. They are becoming more linked to each other in a governmental sense and less connected to their provincial or national governments. GIS is commonly used to foster more responsive and efficient governance. Such technologies are employed not only by city officials, but also by community groups and organizations, and the private sector.

Urban Living at Its Best

What cities will successfully navigate the geographic, demographic, economic, environmental, and political challenges of the coming decades? Cities that provide high-paying, service-sector jobs at the same time that they tend to the welfare of the underprivileged; cities that guarantee a healthy environment backed up by a flourishing health-care industry; cities that invest in infrastructure to facilitate transportation and communication; cities that assure housing markets respond to both private and public sector demands; cities that look after public safety at the same time that they affirm personal freedoms; and cities that realize the role of education, recreation, culture, and the arts in competing with other cities in a globalized economy.

Many organizations are interested in quality-of-life measures, which can be used to rank cities and to promote an individual city's ability to attract investment, conferences, and sporting events, as well as specific groups such as artists, scientists, wealthy retirees, and tourists (Figure 13.13). The Mercer Quality of Life Index provides one widely respected ranking of cities that offers the highest quality of living in the world (Table 13.3). The Mercer methodology includes measures of personal health and safety, the economy and physical environment, transportation and communications, public services, and the overall political climate. Not surprisingly, the top cities were in high-income European countries, plus similar cities in Canada, Australia, and New Zealand. Vienna topped the list. The highest U.S. cities were Honolulu, which ranked 31st, and San Francisco, which ranked 32nd. The cities with the lowest rankings were Baghdad followed by the African cities of Brazzaville, Bangui, and Khartoum.

In 2010, Mercer also developed an Eco-City Ranking based on water availability, water potability, waste removal, sewage, air pollution, and traffic congestion (Table 13.3). The top five "eco-cities" were Calgary, Honolulu, Ottawa, Helsinki, and Wellington. Nordic cities ranked especially high; those in Eastern Europe ranked somewhat below their counterparts in Western Europe. United States and Canadian cities ranked high. In the Asia-Pacific region, Adelaide, Kobe, Perth, and Auckland ranked the highest, and Dhaka the lowest. In the Middle East and North Africa, Cape Town, Muscat, and Johannesburg ranked high; Antananarivo and Baghdad were at the bottom.

Will some cities in today's developing world ascend these rankings? Will others fall? As the world becomes more urban and more interconnected, how will cities, their landscapes,

Figure 13.13 The creative class responds to culture and the arts. Without them, cities decline. That's why the Chrysler Museum in Norfolk, Virginia, just invested $24 million in an upgrade and brought to town Florentijn Hofman's Rubber Duck, at least for a short visit. *Source:* Photo by D. J. Zeigler.

Table 13.3 Quality of Living and Eco-City Rankings

The World's Top Twenty Cities, by Rank			
Mercer Quality of Living Survey 2015		*Mercer Eco-City Ranking 2010*	
1	Vienna, Austria	1	Calgary, Canada
2	Zurich, Switzerland	2	Honolulu, USA
3	Auckland, New Zealand	3	Ottawa, Canada
4	Munich, Germany	3	Helsinki, Finland
5	Vancouver, Canada	5	Wellington, New Zealand
6	Düsseldorf, Germany	6	Minneapolis, USA
7	Frankfurt, Germany	7	Adelaide, Australia
8	Geneva, Switzerland	8	Copenhagen, Denmark
9	Copenhagen, Denmark	9	Kobe, Japan
10	Sydney, Australia	9	Oslo, Norway
11	Amsterdam, Netherlands	9	Stockholm, Sweden
12	Wellington, New Zealand	12	Perth, Australia
13	Bern, Switzerland	13	Montreal, Canada
14	Berlin, Germany	13	Vancouver, Canada
15	Toronto, Canada	13	Nuremberg, Germany
16	Ottawa, Canada	13	Auckland, New Zealand
16	Melbourne, Australia	13	Bern, Switzerland
16	Hamburg, Germany	13	Pittsburgh, USA
19	Luxembourg, Luxembourg	19	Zurich, Switzerland
19	Stockholm, Sweden	19	Aberdeen, U.K.

Source: Mercer Quality of Living Ranking 2015; Mercer Eco-City Ranking 2010.

Box 13.4 Seeing Cities on the Soles of Your Feet

Donald J. Zeigler, Old Dominion University

The cities of the world await your visit. Start learning urban geography the way it should be learned: on the soles of your feet. Be attentive to details, but also use the wide-angle lens that your education in geography has provided. Look for patterns and processes on the landscape and document them for the future. When geographers are traveling, they are doing research. Here's how:

Take pictures. Photograph people and landscapes—ordinary and extraordinary. You can't cover everything, but you can find a few topics of personal interest and follow them from city to city: skylines, waterfronts, signs, buskers, open space, monuments, maps on the landscape, etc. Meander off the "high streets" and slow down. What story is the city trying to tell you? Help tell that story in pictures. To do that, you will have to become a master of field notes as well.

Keep a journal. Document your observations in words. When you stop for coffee or lunch, take out your journal, activate Notes on your iPhone, or open your tablet computer. Before you forget them, record a few observations and impressions. Then, at the end of the day, synopsize what you have learned. It will take discipline but your essays will bring back the experience like nothing else you can do. And remember the first axiom of learned travel: writing it down helps you think it out.

Open your ears. Attune your ears to new languages, new words, new music, and new cacophony. Each city has an auditory signature of its own. Listen to how the local place names are pronounced. Pick up some slang. Find out what music is saturating the air waves. Talk to people you meet, give ear to their accents, and pay attention to *how* they tell their stories, not just *what* they have to say. Record what you can.

Educate your taste buds. Eating is a learning activity, and the cities you visit will be anxious to educate you. Learn about the dishes that define a nation, the regional cuisines, and favorite desserts. Patronize the microbreweries; search out the farmers' markets; take as much time as the locals do to finish their meals; and become a locavore. And always ask questions about what you are eating. Use words and pictures as gustatory *aides memoires*, field documents that will activate your taste buds' memory.

Acquire some mementos. But be selective. Look for post cards, buy a local newspaper, find a local language dictionary or children's book, save a few coins or bills, and buy a few stamps at the post office (the visit alone will be worth the experience). Go to the supermarket for a box of cereal: eat the cereal and save the box. Go to the newsstands and buy a copy of *National Geographic* in a foreign language. Go to the tourist kiosks for a souvenir, and make sure it was actually made in the country you are visiting.

and their populations change? The best way to find out is to go see for yourself (Box 13.4).

SUGGESTED READINGS

Brunn, S. D. 2011. "World Cities: Present and Future." In *Geography for the 21st Century*, ed. J. P. Stoltman, pp. 301–14. Thousand Oaks, CA: Sage Publications, Surveys major population trends and problems facing the world's cities.

Castells, M. 2009. *The Rise of the Network Society: The Information Age: Economy, Society and Culture*, Vol. I, 2nd ed. Oxford: Wiley-Blackwell. Examines the "network society" and changes that gave rise to the "new economy."

Davis, M. 2006. *Planet of Slums*. London: Verso. Explains how informal urban settlements develop and describes the poor living conditions experienced by their residents.

Ehrenhalt, A. 2013. *The Great Inversion and the Future of the American City*. New York: Vintage Books. Discusses the movement of people to urban centers in the developed world and how suburbs are trying to stay appealing by becoming denser.

Florida, R. 2014. *The Rise of the Creative Class, Revisited*. New York: Basic Books. Describes the shift to a flexible, creative, innovation-based and collaborative economy.

Nicholls, R. J., et al. 2008. "Ranking Port Cities with High Exposure and Vulnerability to Climate

Extremes: Exposure Estimates," *OECD Environment Working Papers*, No. 1. OECD Publishing. Examines the extent to which 136 port cities around the world will be affected by climate change.

United Nations Development Programme (UNDP). 2014. *Human Development Report 2013: The Rise of the South*. The UN Human Development Index quantifies the status of health, education, economies, and gender equity across the globe.

Urban Land Institute. 2002. ULI on the Future: Cities Post 9/11. Washington, DC: Urban Land Institute. Examines the challenges that cities face in dealing with terrorism in the post-9/11 world.

World Bank. 2013. *Global Monitoring Report 2013: Rural-Urban Dynamics and the Millennium Development Goals*. Washington, DC: World Bank. An annual report that examines how urbanization helps countries achieve the UN's Millennium Development Goals.

Worldwatch Institute. 2007. *State of the World 2007: Our Urban Future*. Washington, DC: Worldwatch Institute. Reports on challenges facing urban humankind, accompanied by valuable tables, footnotes, and references.

Appendix

World's 100 Largest Urban Agglomerations in 2015

including population in 1950, 1975, 2000, and 2030 (projected)

Urban Agglomeration	Country	Annual Population of Urban Agglomerations				
		1950	1975	2000	2015	2030
Tokyo	Japan	11 275	26 615	34 450	38 001	37 190
Delhi	India	1 369	4 426	15 732	25 703	36 060
Shanghai	China	4 301	5 627	13 959	23 741	30 751
São Paulo	Brazil	2 334	9 614	17 014	21 066	23 444
Mumbai (Bombay)	India	2 857	7 082	16 367	21 043	27 797
Mexico City	Mexico	3 365	10 734	18 457	20 999	23 865
Beijing	China	1 671	4 828	10 162	20 384	27 706
Osaka	Japan	7 005	16 298	18 660	20 238	19 976
Cairo	Egypt	2 494	6 450	13 626	18 772	24 502
New York-Newark	United States of America	12 338	15 880	17 813	18 593	19 885
Dhaka	Bangladesh	336	2 221	10 285	17 598	27 374
Karachi	Pakistan	1 055	3 989	10 032	16 618	24 838
Buenos Aires	Argentina	5 098	8 745	12 407	15 180	16 956
Kolkata (Calcutta)	India	4 513	7 888	13 058	14 865	19 092
Istanbul	Turkey	967	3 600	8 744	14 164	16 694
Chongqing	China	1 567	2 545	7 863	13 332	17 380
Lagos	Nigeria	325	1 890	7 281	13 123	24 239
Manila	Philippines	1 544	4 999	9 962	12 946	16 756
Rio de Janeiro	Brazil	3 026	7 733	11 307	12 902	14 174
Guangzhou, Guangdong	China	1 049	1 698	7 330	12 458	17 574
Los Angeles-Long Beach-Santa Ana	United States of America	4 046	8 926	11 798	12 310	13 257
Moscow	Russian Federation	5 356	7 623	10 005	12 166	12 200
Kinshasa	Democratic Republic of the Con	202	1 482	6 140	11 587	19 996
Tianjin	China	2 467	3 527	6 670	11 210	14 655
Paris	France	6 283	8 558	9 737	10 843	11 803
Shenzhen	China	3	36	6 550	10 749	12 673
Jakarta	Indonesia	1 452	4 813	8 390	10 323	13 812
London	United Kingdom	8 361	7 546	8 613	10 313	11 467
Bangalore	India	746	2 111	5 567	10 087	14 762
Lima	Peru	1 066	3 696	7 294	9 897	12 221
Chennai (Madras)	India	1 491	3 609	6 353	9 890	13 921
Seoul	Republic of Korea	1 021	6 808	9 878	9 774	9 960
Bogotá	Colombia	630	3 040	6 356	9 765	11 966
Nagoya	Japan	2 237	7 318	8 740	9 406	9 304
Johannesburg	South Africa	1 653	2 975	5 605	9 399	11 573
Bangkok	Thailand	1 360	3 842	6 360	9 270	11 528
Hyderabad	India	1 096	2 086	5 445	8 944	12 774
Chicago	United States of America	4 999	7 160	8 315	8 745	9 493
Lahore	Pakistan	836	2 399	5 452	8 741	13 033
Tehran	Iran	1 041	4 273	7 128	8 432	9 990
Wuhan	China	1 069	2 265	6 638	7 906	9 442
Chengdu	China	646	1 860	4 222	7 556	10 104
Dongguan	China	92	125	3 631	7 435	8 701
Nanjing, Jiangsu	China	1 037	1 589	4 279	7 369	9 754
Ahmadabad	India	855	2 050	4 427	7 343	10 527

Hong Kong	China, Hong Kong SAR	1 682	3 906	6 835	7 314	7 885
Ho Chi Minh City	Viet Nam	1 213	2 336	4 389	7 298	10 200
Foshan	China	103	429	3 832	7 036	8 353
Kuala Lumpur	Malaysia	262	661	4 176	6 837	9 423
Baghdad	Iraq	579	2 620	5 200	6 643	9 710
Santiago	Chile	1 322	3 138	5 658	6 507	7 122
Hangzhou	China	610	1 083	3 160	6 391	8 822
Riyadh	Saudi Arabia	111	710	3 567	6 370	7 940
Shenyang	China	2 148	3 291	4 562	6 315	7 911
Madrid	Spain	1 700	3 890	5 014	6 199	6 707
Xi'an, Shaanxi	China	575	1 063	3 690	6 044	7 904
Toronto	Canada	1 068	2 770	4 607	5 993	6 957
Miami	United States of America	622	2 590	4 933	5 817	6 554
Pune (Poona)	India	581	1 345	3 655	5 728	8 091
Belo Horizonte	Brazil	412	1 906	4 807	5 716	6 439
Dallas-Fort Worth	United States of America	866	2 234	4 168	5 703	6 683
Surat	India	234	642	2 699	5 650	8 616
Houston	United States of America	709	2 030	3 847	5 638	6 729
Singapore	Singapore	1 016	2 262	3 918	5 619	6 578
Philadelphia	United States of America	3 128	4 467	5 156	5 585	6 158
Kitakyushu-Fukuoka	Japan	1 403	4 609	5 421	5 510	5 355
Luanda	Angola	138	599	2 591	5 506	10 429
Suzhou, Jiangsu	China	457	530	2 112	5 472	8 098
Haerbin	China	727	1 738	3 888	5 457	6 860
Barcelona	Spain	1 809	3 679	4 355	5 258	5 685
Atlanta	United States of America	513	1 386	3 522	5 142	6 140
Khartoum	Sudan	183	886	3 505	5 129	8 158
Dar es Salaam	United Republic of Tanzania	84	572	2 272	5 116	10 760
Saint Petersburg	Russian Federation	2 903	4 325	4 719	4 993	4 955
Washington, D.C.	United States of America	1 298	2 626	3 949	4 955	5 690
Abidjan	Côte d'Ivoire	65	966	3 028	4 860	7 773
Guadalajara	Mexico	403	1 850	3 724	4 843	5 837
Yangon	Myanmar	1 302	2 151	3 553	4 802	6 578
Alexandria	Egypt	1 037	2 241	3 546	4 778	6 313
Ankara	Turkey	281	1 709	3 179	4 750	5 875
Kabul	Afghanistan	171	674	2 401	4 635	8 280
Qingdao	China	751	933	2 940	4 566	5 920
Chittagong	Bangladesh	289	1 017	3 308	4 539	6 719
Monterrey	Mexico	396	1 633	3 405	4 513	5 471
Sydney	Australia	1 690	3 118	4 052	4 505	5 301
Dalian	China	716	1 294	2 833	4 489	5 851
Xiamen	China	193	445	1 416	4 430	6 911
Zhengzhou	China	196	645	2 438	4 387	5 900
Boston	United States of America	2 551	3 233	4 036	4 249	4 671
Melbourne	Australia	1 332	2 732	3 461	4 203	5 071
Brasília	Brazil	36	827	2 932	4 155	4 929
Jiddah	Saudi Arabia	119	594	2 509	4 076	4 988
Phoenix-Mesa	United States of America	221	1 117	2 923	4 063	4 808
Ji'nan, Shandong	China	576	1 150	2 592	4 032	5 234
Montréal	Canada	1 343	2 791	3 429	3 981	4 517
Shantou	China	270	380	2 931	3 949	4 899
Nairobi	Kenya	137	677	2 214	3 915	7 140
Medellín	Colombia	376	1 536	2 724	3 911	4 747
Fortaleza	Brazil	264	1 136	2 875	3 880	4 551
Kunming	China	334	515	2 600	3 780	4 805

Source: United Nations Department of Economic and Social Affairs/Population Division. *World Urbanization Prospects: The 2014 Revision.*

Cover Photo Credits

Clockwise from top left:

New construction in cities around Russia (Vladivostok is pictured) relegates Soviet urban landscapes to the background as new commercial and residential buildings vie for valuable real estate locations. Photo by Sergei Domashenko.

As cities fill up with people, streets become more congested with not only cars, but bicycles and camels as well. Photo by George Pomeroy.

Three young girls find time for fun as they assist their mothers who labor as ambulantes (street-vendors) in the informal economy of Huancayo, a city in the Peruvian central Andes. Photo by Maureen Hays-Mitchell.

Since the early 1990s, Shanghai's new CBD has arisen across the river in Pudong, centered on the futuristic TV tower surrounded by ultramodern skyscrapers. Pudong CBD is China's financial district. Photo by Kam Wing Chan.

The Dome of the Rock (venerated by Muslims) and the Western Wall (venerated by Jews) are symbols of a religiously divided Jerusalem. Photo by D. J. Zeigler.

African cities located in low-elevation coastal zones, such as Monrovia, Liberia, are vulnerable to severe flooding from sea-level rise. Photo by Robert Zeigler.

The traditional markets of Marrakech, Morocco, are some of the most well-known in the world. In Arabic-speaking countries they are known as souks or suqs. Photo by D. J. Zeigler.

Angkor Wat, built between 1113 and 1150 by Suryavarman II, is one of but hundreds of wats spread throughout Cambodia. Because it symbolizes Cambodia's golden age, its image can also be found on the nation's flag. Photo by James Tyner.

The citadel, or cale, of Gaziantep, Turkey, occupies a strategically located hilltop that dominates the fertile agricultural region near the Turkish-Syrian border. Photo by D. J. Zeigler.

Slater's Mill is today an historical landmark in Pawtucket, Rhode Island; it marked the beginning of the factory system in the United States. Photo by D. J. Zeigler.

The Pelourinho historic district, named for the "pillory" formerly used to punish slaves, indicates the strong Afro-Brazilian influence in Salvador da Bahia. Photo by Brian Godfrey.

This view of the 9/11 Memorial shows one of the two reflecting pools which sit within the footprints where the Twin Towers once stood. Photo by John Rennie Short.

The Church of Our Savior on the Spilled Blood, in St. Petersburg, was built on the spot where Emperor Alexander II was assassinated in March 1881. Built from 1883-1907, the Romanov family provided funds for this glamorous cathedral. Photo by Jared Boone.

Paris evolved around an island in the Seine River: Île de la Cité. Today, it is most famous for the cathedral of Notre Dame, whose spire is barely visible here. Photo by D. J. Zeigler.

Now a UNESCO World Heritage site, the Sydney Opera House has become a symbol of the island continent. Photo by D. J. Zeigler.

The Sikhs, neither Hindu nor Muslim, are a major part of India's cultural diversity, seen here in their main gurdwara, the place where they worship. Photo by D. J. Zeigler.

Geographical Index

Index to Subjects

About the Editor and Contributors

Roberto I. Albandoz is an academic adviser at the Pennsylvania State University's World Campus. He holds a bachelor's degree in geography from the University of Puerto Rico, a master's degree in environmental and urban systems from Florida International University, and a PhD in geography from the Pennsylvania State University. His primary interests lie in historical geography, cultural geography, and geography-based travel. He has worked as a member of the United States Coast Guard and as an engineer at the Applied Research Center at Florida International University. He also writes travel articles for *El Nuevo Día* newspaper in Puerto Rico.

Amal K. Ali is associate professor in the Department of Geography and Geosciences at Salisbury University in Maryland. She teaches and conducts research in land-use planning, smart growth, and international development planning.

Lisa Benton-Short is associate professor and Geography Department chair at George Washington University. She teaches courses on cities and globalization, urban planning, and urban sustainability. She has research interests in urban sustainability, environmental issues in cities, parks and public spaces, and monuments and memorials.

Alana Boland is associate professor in the Department of Geography and Planning at the University of Toronto. She teaches courses on China, environment, and development. Her research examines how the changing relationship between the economy and environment has influenced the management of resources and governing of spaces in Chinese cities. Her current work focuses on urban environmental infrastructure in China during the 1950s and 1960s.

Timothy S. Brothers is associate professor emeritus of geography at Indiana University Perdue University Indianapolis (IUPUI). He specializes in human-environment relations in the Caribbean. His book *Caribbean Landscapes* is an interpretation of the most characteristic natural and human landscapes of the region based on satellite imagery, photographs, and essays.

Stanley D. Brunn is emeritus professor of geography at the University of Kentucky. He has taught classes on world cities, future worlds, political and social geography, and the geographies of information and communication. He has traveled in more than seventy countries and taught in seventeen. He has written and edited books and chapters on U.S. elections, Walmart, post-9/11 worlds, ethnicity, geography/technology interfaces, time/space cartographies, religions, and mega-engineering projects. Currently, he is editing a book on world languages, studying disciplinary history questions, and writing poetry.

Kam Wing Chan is professor of geography at the University of Washington. His main research focuses on China's cities, migration, employment, and the household registration system. He has served as a consultant for the World Bank, Asian Development Bank, United Nations, and McKinsey & Co. His recent commentaries and interviews have appeared in the *Wall Street Journal, New York Times, The Economist, South China Morning Post*, BBC, *Caixin*, and *China Daily*.

Ipsita Chatterjee is assistant professor in the Department of Geography, University of North Texas. She is interested in issues relating to globalization, neoliberalism, urban transformation and renewal, conflict and violence, and social movements. Her research has focused on issues of class and ethnic segregation, ghettoization, and other forms of urban exclusions.

Megan L. Dixon is instructor of writing, geography, and environmental studies at The College of Idaho. Her research in urban and cultural geography focuses on high-profile development in St. Petersburg and Shanghai as well as on the impacts of Chinese migration to Russia. Current research involves overlapping Latino/Anglo narratives of place and water use in the intermountain west (central Idaho). Her persistent interest in shifting paradigms of place began with a visit to Kalinin/Tver in 1989, just before the fall of the Berlin Wall.

Seth Dixon is associate professor in the Geography Program at Rhode Island College. He regularly teaches courses in urban geography, cultural geography, GIS, and world regional geography. Past research of his has centered on urban landscapes of identity in Mexico as well as the politicization of heritage. He has served as the coordinator for the Rhode Island Geography Education Alliance and as a blogger for National Geographic Education. His main project is to archive and share digital geography education resources through social media networks to other educators.

Robyn Dowling is an urban cultural geographer at Macquarie University in Sydney, Australia. Her primary research interests are in cultures of everyday urban life, focusing on gender, home and suburbs. She publishes widely on issues such as home ownership, suburban gender identities, and cultures of transport. Her current research explores the contours of privatization and privatism in Sydney's residential life.

Ashok K. Dutt is professor emeritus of geography, planning, and urban studies at the University of Akron in Ohio. His research focuses on religion, language, development, crime, and medical geographies of Indian cities. He has authored, coauthored, edited or co-edited 23 books, and authored or coauthored more than 80 journal articles and 60 book chapters.

Alexis Ellis is a research urban forester with the USDA Forest Service. Her work in this field focuses on the research and development of tools for analysis of urban tree growth and mortality and their associated ecosystem services. She has a background in geospatial analysis and is currently developing tools for the spatial analysis of urban tree populations and their associated benefits.

Irma Escamilla is a doctoral candidate in geography at the National Autonomous University of Mexico, where she serves as an academic technician in the Institute of

Geography. Her research focuses on urban-regional geography, urban labor markets, population geography, gender and geography and historical geography.

Brian J. Godfrey is professor of geography at Vassar College. He studies urban and regional change with a primary focus on the Americas, including North America, Latin America, Brazil, and the Amazon Basin. Favoring the analytical lens of historical geography, his scholarship has examined such topics as global cities and urbanization, neighborhood change and gentrification, ethnic geography and racial segregation, politics of memory and historic preservation, urban political ecology, and public space. His current research explores the heritage-based redevelopment of Brazil's historic cities.

Jessica K. Graybill is associate professor of geography and director of Russian and Eurasian Studies at Colgate University. Ongoing interdisciplinary research on coupled human and natural systems investigates environmental change due to socioeconomic and political transformation, natural resource extraction, and climate change in multiple regions of the Former Soviet Union, especially the Far North and East. With students, she also examines the human and natural ecologies of shrinking cities in upstate New York.

Maureen Hays-Mitchell is professor of geography at Colgate University. Her scholarly interests center broadly on the gendered dimensions of development in the Global South, with a primary focus on Latin America where she conducts grassroots fieldwork. She applies a feminist lens to issues involving the urban informal economy, the role of women's associations in peace-building, the work of truth commissions, and post-conflict reconciliation. Her current research explores memory on the post-conflict landscape of several Latin American countries.

Ishrat Islam is professor and chair of the Department of Urban and Regional Planning at the Bangladesh University of Engineering and Technology (BUET). Her teaching and research interests are spatial planning, wetlands preservation, disaster management, and indigenous architecture.

Brian Edward Johnson is director of the University of Alabama at Birmingham's Office for Study Away, where he also teaches introductory, urban, population, and regional geography courses. His research focuses on urban planning in cities experiencing revitalization and demographic change. He serves as chair of the AAG's Stand Alone Geographers Affinity Group and is a City of Birmingham planning commissioner.

Corey Johnson is associate professor of geography at the University of North Carolina, Greensboro. He teaches courses on Europe, the European Union, and political and urban geography. He has lived and worked in Germany. His current research is on geopolitics of energy and borders in the European Union.

Nathaniel M. Lewis is lecturer in human geography at the University of Southampton. His work focuses on relationships between gender, sexuality, health, and the urban environment. He completed his PhD at Queen's University in Kingston, Ontario, and a postdoctoral fellowship at Dalhousie University in Halifax, Nova Scotia. His recent publications can be found in journals such as *Health*

& Place, *Annals of the Association of American Geographers*, *The Canadian Geographer*, and the *Journal of Homosexuality*.

Linda McCarthy is associate professor of geography and urban studies at the University of Wisconsin–Milwaukee. She is also a certified planner. Her teaching interests include Europe, cities, and globalization. Her research focuses on public policy and economic development. Her recent publications have been on government subsidies to corporations, globalization, and brownfields.

Pauline McGuirk is professor of human geography and director of the Centre for Urban and Regional Studies at the University of Newcastle, Australia. Her research focuses on the politics, development, and governance of metropolitan cities, but she has also published widely on aspects of urban development, city politics, planning, urban identity, and place marketing. She is currently investigating new forms of governance and residential housing in Sydney.

Garth A. Myers is distinguished professor of Urban International Studies at Trinity College in Hartford, Connecticut, and director of the Urban Studies Program. He is the author of four books and co-editor of two others. He has also published more than 60 articles and book chapters on African urban development, and he teaches a variety of courses on African geography and urban studies.

Arnisson Andre Ortega is assistant professor at the University of the Philippines, Diliman. His research interests include urban geography, migration, and critical demography. His current projects focus on spatialities of peri-urban transition, dispossession and gentrification, and social politics of transnational migration.

Francis Owusu is professor and chair of the Department of Community and Regional Planning at Iowa State University. He has conducted research in several African countries and has published extensively on urban livelihood strategies, development policy, and public sector reforms. He teaches geography and planning courses, including the geography of Africa and a course on world cities. He is originally from Ghana.

George Pomeroy is professor of geography-earth science at Shippensburg University of Pennsylvania. He teaches land-use planning, environmental planning, and courses related to both South and East Asia. His research focuses on new-towns planning and on building capacity for local government planning activities.

Zia Salim is assistant professor of geography at California State University, Fullerton. He teaches courses on urban geography, geographic thought, and social and environmental issues in urban areas. His research interests include urban inequality, uneven development, and cultural landscapes. More specifically, his work has examined central city redevelopment, affordable housing, homelessness, gated communities, nonprofit service provision, and public art.

Joseph L. Scarpaci is associate professor of marketing at the Gary E. West College of Business in West Liberty, West Virginia. He is also the executive director of the Center for the Study of Cuban Culture and the Economy. He is the coauthor of *Marketing Without Advertising: Brand Preference and Consumer Choice* *in Cuba*, among other works focused on Cuba and the Caribbean.

Benjamin Shultz is assistant professor at International Balkan University in Skopje, Macedonia, where he teaches courses on American society and politics, nationalism, and cultural industries. His current research interest is in the processes of integration of the Balkan states into Europe. As a side project, he is interested in applying spatial analysis methods to sports statistics.

Thomas Sigler is lecturer in human geography at The University of Queensland in Brisbane, Australia. He holds a bachelor's degree in geography and international relations from the University of Southern California, and a master's degree and PhD in geography from Penn State University. His primary interests lie in economic and urban geography, specifically in the evolution and formation of contemporary cities through historical interurban networks. He has conducted research on the United States, Panama, Hong Kong, and Australia. Prior to finishing his PhD, he worked in the Democratic Republic of the Congo and Honduras for governmental and nonprofit organizations.

Angela Gray Subulwa is associate professor in the Department of Geography and Urban Planning at the University of Wisconsin–Oshkosh. Her research interests include examining the processes of forced displacement and identity as they relate to development, gender, political, and cultural geographies. Most of her research focuses on southern Africa, with a concentration on Zambia. In addition to numerous geography courses, she also teaches in the Women's and Gender Studies Program.

James Tyner is professor of geography at Kent State University, where he teaches East and Southeast Asia, political geography, and other courses on social, population, and urban geography. His current research focuses on political violence in Southeast Asia and international population movements.

Donald J. Zeigler is professor of geography at Old Dominion University in Hampton Roads, Virginia. He teaches courses on the Middle East, urban and cultural geography. He is an avid traveler and photographer, a past president of the National Council for Geographic Education, and the current president of the Southeast Regional Middle East and Islamic Studies Seminar.